THE SAMPLING AND ASSAY OF
THE PRECIOUS METALS.

THE SAMPLING AND ASSAY

OF THE

PRECIOUS METALS:

COMPRISING GOLD, SILVER, PLATINUM,
AND THE PLATINUM GROUP METALS
IN ORES, BULLION, AND PRODUCTS.

BY

ERNEST A. SMITH, Assoc. R.S.M.,

DEPUTY ASSAY MASTER OF THE SHEFFIELD ASSAY OFFICE;
FORMERLY OF THE ROYAL SCHOOL OF MINES, LONDON, ETC.; MEMBER OF THE
INSTITUTION OF MINING AND METALLURGY, OF THE INSTITUTE OF METALS,
AND OF THE SOCIETY·OF CHEMICAL INDUSTRY, LONDON.

With 166 Illustrations.

LONDON:
CHARLES GRIFFIN & COMPANY, LIMITED.
PHILADELPHIA: J. B. LIPPINCOTT COMPANY.
1913.

PREFACE.

"THE term 'precious metals' has usually been applied to gold and silver only, its use dating back to a time when no regard was paid to platinum. As platinum is now in common use and is more valuable, weight for weight, than gold, it may very well rank as a precious metal." Moreover, platinum is not infrequently associated with gold, and the methods of assaying gold and platinum are so closely allied that they cannot be separately treated. Within recent years a considerable amount of research has been conducted in connection with the precious metals, especially platinum and the rare metals which constitute the "platinum group." Much of this research has been of considerable value to assayers in enabling them to work out suitable methods for the accurate sampling and assay of the precious metals.

It may, therefore, not be out of place here to point out that the assayers and chemists of to-day who have gained distinction as the result of researches upon the properties of platinum, iridium, and other metals of the platinum group, owe a great debt of gratitude to practical metallurgists like Mr Edward Matthey, A.R.S.M., and the late Mr George Matthey, F.R.S. (of Messrs Johnson, Matthey & Co., the well-known platinum refiners), who have enabled them to obtain the metals in the desired form and condition for research.

As a recent writer has said, "The physicist (and also the chemist) to-day is able to begin his account of research with *I took* where a Davy, a Wollaston, or a Faraday would have been constrained to write *I prepared*," and the difference is much greater than is generally recognised.

In the preparation of this work the aim has been to provide a full description of the various methods of sampling and of assaying—both by the dry and wet methods,—the precious metals contained in ores, bullion, and metallurgical products, and to produce a book useful alike to the student and to the assayer in practice.

So far as the author is aware, no attempt has been made to cover the same ground as that covered by the present volume. The subject is treated more fully than is usual in text-books on assaying generally or in works dealing with the metallurgy of the precious metals.

Special attention has been devoted to sampling, a subject which, in the author's opinion, has not hitherto received in text-books or in the curriculums of mining schools the attention which its great importance deserves. Sampling operations being common ground, and by no means peculiar to the treatment of the ores of any one particular metal, all reference to them is very frequently omitted both in works on metallurgy and in text-books on assaying.

Short descriptions of the principles of ore and bullion valuation and sale have also been included.   A special chapter has been devoted to the assaying connected with the cyanide process of gold extraction, including the determination of cyanides on account of their importance in the extraction of gold and because their determination has become a part of the ordinary work of an assayer in a cyanide mill.

As text-books on assaying are used chiefly as works of reference, the author has aimed at making each section of the present volume as complete as possible in itself, so as to avoid as far as possible the necessity for frequent reference to other parts of the book.   The adoption of this course has necessarily resulted in a certain amount of repetition, but it is hoped that this will be amply justified by the greater convenience in reference thus secured.

The work is primarily intended for students, having been written chiefly at the request of the author's old students at the Royal School of Mines, London ; hence more detail has been given than would be necessary in a work intended only for the use of experienced assayers.

The author cordially acknowledges the assistance he has received in the preparation of this work from Beringer's *Text-Book of Assaying*, Dr T. K. Rose's *Metallurgy of Gold* and *Precious Metals*, Fulton's *Manual of Fire Assaying*, and to the articles on Assaying by the author's father, the late Richard Smith of the Royal School of Mines, written for Dr Percy's *Metallurgy of Silver*.   He is especially indebted to Mr H. Dean, A.R.S.M., M.Sc., of Durham University, for executing many of the drawings, reading the proofs, and preparing the index, and to Mr George Sims for assistance in reading the proofs.

The majority of the illustrations are original, but the author is indebted to the President and Council of the Chemical, Metallurgical, and Mining Society of South Africa for permission to reproduce fig. 1, to Dr T. K. Rose for fig. 102, and to the Richmond Gas Company, Ltd., Warrington, for the drawing for fig. 16.

It is hoped that the work may be useful to students and to assayers engaged in the precious-metal industries, and that it may form a fitting companion volume to the series of treatises on Metallurgy written by Associates of the Royal School of Mines, and inaugurated by the late Sir W. C. Roberts-Austen.

ERNEST A. SMITH.

SHEFFIELD, 1913.

CONTENTS.

CHAPTER I.

INTRODUCTORY.

CHAPTER II.

THE DESIGN AND EQUIPMENT OF ASSAY OFFICES. ASSAY OFFICE RECORDS.

Design and Equipment.

Assay Office Records.

CHAPTER III.

ASSAY FURNACES.

A. Furnaces employing Solid Fuel.

B. Furnaces employing Gaseous Fuel.

C. Furnaces employing Liquid Fuel.

CHAPTER IV.

FURNACE IMPLEMENTS. APPARATUS USED IN FURNACE OPERATIONS.

I. Furnace Implements.

CHAPTER V.

BALANCES AND WEIGHTS: WEIGHING.

Balances.

CHAPTER VI.

GOLD AND SILVER: PHYSICAL AND CHEMICAL PROPERTIES.

Gold.

CHAPTER VII.

PRECIOUS METAL ORES: VALUATION OF ORES.

CHAPTER VIII.

SAMPLING.

CHAPTER IX.

PREPARATION OF SAMPLES FOR ASSAY.

CHAPTER X.

FLUXES AND OTHER MATERIALS EMPLOYED.

CHAPTER XIV.

THE ASSAY OF GOLD ORES.

CHAPTER XV.

THE ASSAY OF COMPLEX GOLD AND SILVER ORES.

CHAPTER XVI.

SPECIAL METHODS OF ORE ASSAY.

CHAPTER XVII.

BULLION: VALUATION AND SAMPLING.

CHAPTER XVIII.

THE ASSAY OF GOLD BULLION.

CHAPTER XIX.
ASSAY OF SILVER BULLION.

CHAPTER XX.
THE ASSAY OF BASE BULLION (LEAD AND COPPER).
I. Base Lead Bullion.

II. Base Copper Bullion.

CHAPTER XXI.
THE ASSAY OF INDUSTRIAL GOLD AND SILVER ALLOYS.
ASSAY OFFICES FOR HALL-MARKING.
I. Industrial Gold Alloys.

II. Industrial Silver Alloys.

CHAPTER XXII.
ASSAY OF AURIFEROUS AND ARGENTIFEROUS
METALLURGICAL PRODUCTS.
I. Base Metals.

II. Metallurgical Products.

III. Various Bye-Products.

CHAPTER XXIII.

LABORATORY WORK IN A CYANIDE MILL.

CHAPTER XXIV.

PLATINUM AND THE METALS OF THE PLATINUM GROUP.

I. Platinum.

II. Metals of the Platinum Group.

CHAPTER XXV.

THE ASSAY OF PLATINUM.

APPENDIX.

THE SAMPLING AND ASSAY OF THE PRECIOUS METALS.

CHAPTER I.

INTRODUCTORY.

Definition of Assaying.—Assaying was the term originally applied to the art of determining the quantity of metal contained in an ore, in alloys, or in certain other compounds by the dry or "fire" method. At the present time the term assaying has a much wider meaning, being applied to the determination by whatever method the operations are carried out. Assaying is said to be either "dry" or "wet" according as the agency of "fluxes" and fire or of liquid solvents is employed. Formerly, in order to obtain the result with the greatest expedition, the dry method (also called a furnace method) was chosen for producing the chemical reaction; but, as this was frequently done at the expense of accuracy in the result of the assay, the wet method is also used in modern times. It has not, however, entirely displaced the dry method, for the latter is employed in all cases where sufficiently accurate results can be more quickly reached, as in assaying lead, tin, gold, and silver ores, or where suitable wet-assay methods cannot be conveniently substituted for it.

Furnace methods, in combination with wet methods, may frequently be practised with advantage. It must be admitted, however, that except in a few cases such as those given, the tendency of modern assay practice is to discard purely furnace methods in favour of wet methods.

During the last few years rapid strides have been made in the application of quick and reliable chemical methods in the assay of ores, metals, etc., and assaying is also becoming wider in its scope, consequently the line of demarcation between it and chemical analysis has now almost disappeared. Indeed, assaying has become so identified with the latter that the two terms are often used synonymously, and even the determination of an *organic* constituent of a mixture is not infrequently called "an assay."

History of Assaying.[1]—Assaying was the earliest-known branch of chemistry, and in fact that branch which afterwards, through the medium of the various investigations of alchemy, drew attention to theoretical chemistry, and thus founded that science.

The art of assaying is so ancient, and it has so constantly and imperceptibly been added to and improved, that its origin has been almost entirely obscured. A cupellation process for gold is mentioned by Diodorus Siculus as being in use in the second century B.C. Georgius Agricola[2] (1494–1555) was one of the first

---

[1] I am indebted to the Cantor Lectures on "Alloys," by the late Sir W. Roberts-Austen, read before the Society of Arts in 1884, for many of the references to the early history of assaying.

[2] *De re Metallica*, libri xii., Basle, 1556.

writers to deal systematically with the art of assaying minerals.   Much valuable information on early methods of assay is also given by Lazarus Erckern [1] (1574). There appears to be little doubt that the art of assaying originated with methods for testing the purity of the precious metals.

In the early stage of the civilisation of a nation coins of pure metal have usually been adopted, but the pure metal was subsequently mixed with one of less value, probably because the union of two metals by fusion produced a hard and durable alloy ; but it is also very probable that these attempts to alloy the precious metals were frequently of a fraudulent character, and carried out with the object of producing a metal which might be substituted for the pure metal. How long these fraudulent but successful practices were carried on without any adequate means of detection it is impossible to say.

The want of a method for ascertaining the degree of purity of gold and silver, or for determining the amount of precious metals in their alloys, must have been felt as soon as the use of metals for currency was established.   The history of assaying has yet to be written, but in rapidly reviewing the methods of assay which have been practised, it will be well to consider them in an order that is, in the main, chronological, but which enables the physical methods, as distinguished from the chemical, to be dealt with first.

Probably the earliest method of ascertaining approximately the composition of an alloy of gold consisted in rubbing the metal on a hard stone of dark colour known as "Lydian Stone," or, as it is called in more recent times, the "Touchstone," the streak thereby produced being compared with similar streaks derived from alloys of known composition.   In later times the streaks were subsequently tested with acid.   The use of such touchstones comes to us from very early times, and its use has survived for approximate assays until the present day (see Touchstone, p. 280).   According to Riche,[2] the touchstone was well known to the Romans, and perhaps even more anciently.   It was in use in England in the twelfth century and probably much earlier, and Prof. Gowland [3] has stated that this method of testing the quality of gold was used in the first Government Mint established in Japan in the sixteenth century.   The only other physical method, as distinguished from chemical methods, that has been widely adopted, is that of ascertaining the purity of metals by their density, devised by Archimedes, 212 B.C., for ascertaining whether the crown of Hiero was or was not genuine.   It consists in comparing the weight of the metal with that of an equal volume of water, and is still sometimes resorted to when the metal to be examined must be preserved intact.

Pliny states that in his time a method was in use for estimating the amount of silver in an alloy of silver and copper by the degree of discoloration or blackening which attends·the heating of the alloy in air.   This method, long practised in France, and known by the name of *essais à la raclure*, is described by Rochon,[4] who says that it was generally recognised in the Roman mints in the time of Marius Gratidianus, "Triumvir of the Money"; but recent experiments have shown that it ceases to be useful for alloys of silver and copper which contain less than 80 per cent. of silver, as such an alloy becomes quite black when heated in air, and in such alloys as are poorer in silver, variations in tint cannot be detected.[5]

---

[1] L. Erckern, *Allerfurnemisten Mineralischen Eerzt u. Bergwerks Arten*, Frankfort, 1574, translated by Sir John Pettus, 1683 (*The Laws of Art and Nature in Assaying Metals*).
[2] A. Riche, *Monnaie, Médailles et Bijoux*, Paris, 1889, p. 277.
[3] *Journ. Inst. of Metals*, vol. iv. 1910, p. 11.
[4] *Essais sur les Monnaies*, p. 17, 1792.
[5] See Chaudet, *L'Art de l'Essayeur*, p. 77, Paris, 1835 ; also Percy, *Metallurgy of Silver*, p. 157.

In very early times the need must have been felt of some chemical method of isolating the precious metals—of actually separating them from their base associates, so that the gold or silver set free could be weighed, and the amount originally present in the mass deduced by calculation. The crude method of assaying silver already described, which depends on the change of colour produced by oxidation of the baser constituent of an alloy, finds a parallel in the method of cupellation used from very early times, which also depends on the principle that precious metals will resist oxidation, while the metal with which they are usually associated will not. The process of cupellation is very ancient. That the Greeks were familiar with the process appears probable from the fact that lead has been found in ancient ornaments both of gold and silver, discovered by Schliemann at Mycenæ. Pliny also teaches that the Roman metallurgists used lead for the purification of gold and silver. Geber the Arabian, who died A.D. 777, the greatest of the early alchemists, gives, if medieval translations of his works are to be trusted, a sufficiently accurate description of the process to enable it to be conducted at the present day with no other aid than that which he gives. The operation, as described by Geber, would, however, more nearly correspond to a refining operation conducted on a large scale, with a view to the extraction of silver from lead, rather than to the method of assay as practised at the present day on a few grains of metal. In Geber's work "On Furnaces" there is no mention or illustration of the "muffle furnace," so that he seems to have conducted the process in a cupel surrounded by incandescent fuel.

The exact-period at which the process of cupellation was first used for assaying is not accurately known, but the method of conducting assays by the cupel (or "coppels" as given in old treatises on assaying) on what would at the present day be considered a very large sample of metal, seems to have been held to be necessary in the twelfth century, for in certain official trials of coin in the time of Henry II., 1154–89, the "Miles Argentarius" or Assayer of Silver, is instructed to make tests by cupellation.[1] Thus it is shown that the process was officially recognised in this country in the reign of Henry II., and in France the first official mention of it occurs about the year 1314.

The "parting assay," in which gold is parted or separated from an alloy of gold and silver by the action of nitric acid, was certainly known to Geber and the early alchemists, but the first official mention of the use of the parting assay appears to be in a decree of Philippe de Valois, in the year 1343, confirming its use in the French Mint.

"The methods of procedure in the seventeenth century have been briefly described by Savot[2] and by Reynolds,[3] and more fully in the Compleat Chymist. In 1666 Pepys saw the parting assay being practised at the Mint in the Tower of London, and from his description it is clear that the method then employed bears a surprisingly strong resemblance to that of the present day" (T. K. Rose).

With regard to the application of the early methods of assaying to the determination of metals in ores very little appears to have been recorded. It is not, however, unreasonable to suppose that the ancient process of cupellation originally suggested the idea of fusing precious-metal ores with lead or lead compounds to collect and concentrate the gold and silver as a preliminary to cupellation. It would appear that the art of assaying as applied to ores received

[1] Quoted by Lowndes, *Essay for the Amendment of the Silver Coins*, p. 155 (London, 1695), from the Black Book of the Exchequer, written in the time of Henry II. cap. 21, "Officium Milites Argentarii et Fusoris" (see Cantor Lectures, Roberts-Austen, *loc. cit.*).
[2] *Discours sur les Médailles Antiques*, Paris, 1627, p. 72.
[3] *A New Touchstone for Gold and Silver Wares*, London, 1679, p. 362.

attention in Germany at a very early date. The methods used by German assayers of the sixteenth century are fully described by Erckern in the work already referred to.

The method of assay by "scorification" is mentioned, and was preferred to fusion in a crucible, which is also described. When fusion in a crucible was recommended, it was only as a preliminary to scorification.[1] Erckern describes many of the methods and precautions used at the present day, such, for instance, as the treatment of cupriferous precious-metal ores with nitric acid, the use of common salt as a cover to the charge in the crucible, and the use of "proof-centners," the old German system of weights which corresponds to the present-day "assay ton" system. It is also of interest to note that he mentions the forerunner of the buck-board for the fine pulverising of ores, and instances assay offices in which 200 assays of ores were made in a week. These references are sufficient to indicate the advanced state of the knowledge of assaying at that period.

With the advance of metallurgical science and the extraction of the common metals by smelting processes on a large scale the need must have been felt of methods of assaying the ores of these metals. A study of the dry methods of assay, many of which were in use until quite recent years, shows that they may be regarded as smelting processes in miniature, as the older methods of dry assaying frequently follow, on a small scale, the operations of smelting. These methods, as previously stated, are now to a very large extent obsolete in good assay practice as methods of determining with accuracy the value of ores, etc., and have been generally replaced, when possible, by the more accurate and usually more rapid methods of chemical analysis. The dry methods of assaying iron, copper, zinc, nickel, and cobalt, etc., are now only taught as exercises in furnace manipulation or to instruct the student in the principles of metallurgy; occasionally they are used in the works laboratory to test the smelting qualities of ores.

To-day the metallurgical chemist relies very largely on volumetric methods for the analytical determinations required in metallurgical work. Many volumetric assays can be performed in a shorter time than gravimetric determinations for corresponding constituents, which is an important item where much work has to be done. The results yielded are either very accurate or at least sufficiently exact for metallurgical work; and they are in most cases less expensive, but require special apparatus of accurate construction.

Of the few dry methods of assay still in use those employed for the determination of the precious metals are by far the most important. The quantity of gold and of silver contained in ores is too minute to allow of its being accurately determined by wet methods. An impression prevails that the results obtained by furnace methods are less accurate and reliable than those obtained by other methods. Although this is true of some of the dry methods that are now obsolete, the accuracy with which small amounts of gold and silver can be determined by the fire-assay probably exceeds that of any known method of analysis for any other metal.

The assayer is frequently called upon to determine the gold in samples that yield only one-twentieth of an ounce to the ton, or even less.

Qualifications of Assayers.—It may be well here to make brief reference to the qualifications of assayers. Before a student enters a course on assaying he should have gone through a course of practical chemistry in which he has been taught to work with the greatest accuracy, and has acquired that degree of confidence in his work which every operator ought to possess, so that he may be

---

[1] Erckern, *loc. cit.*, book i. chap. x, and book ii. chap. viii. Quoted by Dr Rose, *Metallurgy of Gold*, 5th edition, 1906, p. 484.

prepared to commence training in the special methods necessary for the guidance of metallurgical operations as well as for the valuation of ores and metallurgical products.

It would be impossible to insist too strongly upon the importance of an assayer receiving sound instruction in the specific department of analytical chemistry which relates to assaying. Occasionally a large sum of money may depend upon an assayer possessing a profound knowledge of certain departments of analytical chemistry. Thus, material submitted for assay might contain small quantities of some of the metals of the platinum group, which may not be detected by an assayer possessing only a limited knowledge of chemistry. An assayer of the precious metals may be called upon to determine the presence or absence of iridium and of other metals also in gold, which, for their satisfactory detection, require no ordinary skill, and involve some of the most difficult operations of analytical chemistry. But although a knowledge of chemistry is absolutely necessary, this of itself would be a very insufficient qualification for the man who has to perform assay operations : no one can attain to any degree of eminence in assaying unless he has had some amount of practice in the laboratory. Theoretical knowledge counts for much, but skill counts for more ; and we cannot get skill ready made. It can only come as the result of long and steady practice. Without experience an assayer could not decide whether the requisite degree of heat was produced in any of the furnaces, or whether any of the various operations of assaying had been properly effected. It is well, however, to bear in mind that while scientific knowledge alone will not qualify a man to take charge of an assay laboratory, so neither is empirical knowledge the only qualification desirable in such cases : it is clearly the combination of scientific with practical knowledge that will render the head assayer of a large metallurgical works in the highest degree competent for his responsible position.

A man once brought his son to the Royal School of Mines, London, with the request that he might be taught to "do copper." He did not want his boy to "waste his time learning about oxygen and hydrogen and all that," but he wished him simply to learn to "do copper." [1] Although seldom expressed with such refreshing candour, the desire to "do assaying" without learning more than the minimum amount of chemistry is still very prevalent ; and, unfortunately, assaying is a subject which may be, and frequently is, taught and practised in such a manner as to degrade it to the level of a purely mechanical and often quite unintelligible series of rule-of-thumb operations.

The author fully endorses Prof. J. O. Arnold's [2] statement that students "cannot be too strongly urged to remember the fact that an assayer or metallurgical chemist and an 'analytical machine' who turns out so many estimations per day are two very different personalities. Assayers deficient in a thorough knowledge of at least elementary chemistry, physics, and mathematics, and the principles of qualitative and quantitative analysis, can claim to rank only with skilled artisans : in both assaying and chemical analysis the head and the hands should always work together. Even with students really well grounded in pure chemistry the possession of knowledge and the ability to apply it are two widely different things. After the preliminary training, all that can be recommended, in other respects, to one who desires to perfect himself in assaying, is the most scrupulous cleanliness, order, and precision as regards his assays and implements, and the most unwearying adroitness and attention in performing the various manipulations required. Students should bear in mind that the whole aim of assaying is to get the most accurate result with economy of time and material, and, when these factors are incompatible, to sacrifice everything to the most accurate result."

[1] Quoted by G. S. Newth, in his preface to *Manual of Chemical Analysis.*
[2] Preface to *Steel Works Analysis*, J. O. Arnold, of the University of Sheffield.

Some assayers after a few years' experience hold the view that there is nothing to be learnt in assaying, and that the work is humdrum and monotonous. But, as Mr A. Whitby [1] has pointed out, a man's work in any call of life is what he makes it, and the art of assaying especially calls for the exercise of the highest qualifications of the chemist—irreproachable honesty and perseverance under difficulties within and without the office. It must not be permitted to sink to the rule-of-thumb state, and the interest evoked by indefatigable attempts to attain excellency is unending. There is no real truth in the common assumption that the assayer cannot err, but it is open to him to endeavour to live up to the assumption, which can only be accomplished by an intelligent application and use of all his faculties.

It is generally recognised that assaying has a very important bearing on the mining and metallurgical departments of mineral enterprises, but it must be admitted that the work of the assayer is seldom valued as it should be. The operations of mill and smelter are being more and more directed according to the results obtained in the laboratory, and the metallurgical chemist is now required to make daily a number of determinations that would have appalled his predecessor of even a few years ago. The time allowed for making individual determinations is also being steadily reduced. Under these conditions it is imperative that the man placed in charge of the assay department of a large works or mill should be thoroughly qualified for the important position he holds, and his qualifications and position should be duly recognised.

It is the chief function of the assayer and metallurgical chemist to *interpret* results; it is he alone who can be trusted with the correct interpretation of results and the practical carrying out of his deductions. He should, therefore, be valued chiefly in his advisory capacity, and should be accorded a position and power commensurate with the importance of his functions.[2] If it be worth while to take aid from chemical science at all, it is cheapest and best done by providing a competent analyst and assayer, with the best and most convenient instruments and apparatus that money can procure. It is evident, however, in many cases in metallurgical works, that this view is as yet not acted on. As pointed out by Sir Philip Magnus, "Few manufacturers in this country yet realised the economy and industrial advantage of attaching to their works intelligence departments staffed with scientific experts. He was glad to say, however, even in this respect the outlook was improving."[3]

The importance of the mine-assay department has long been recognised among practical mining men, and in South Africa the demand for good assayers and a fully equipped assay office has lately increased owing to the close attention being paid to the lower-grade ores and to the demands of the reduction works.[4]

---

[1] "Routine Assaying on the Rand," *Proc. Chem. Met. and Min. Soc. S.A*, 1906, p. 272.
[2] Harrison and Wheeler, *Journ. Iron and Steel Inst.*, 1908.
[3] *Nature*, November 26, 1908.
[4] A. M'Arthur Johnston, *Rand Metallurgical Practice*, vol. i., 1912, p. 286.

CHAPTER II.

THE DESIGN AND EQUIPMENT OF ASSAY OFFICES.

IN assay offices where quick and accurate work is required, the general arrangement and convenience of the office itself is of the greatest importance. The assayer has usually to adopt the quickest methods consistent with the necessary degree of accuracy, and such details as space and apparatus have to be carefully considered, and are considerably influenced by the design and equipment of the assay office.

It may be well here to remind the student who has received his training in a well-equipped laboratory supplied with every convenience that his success as an assayer will be largely dependent on his ability to adapt himself to new and less favourable conditions of work such as would exist in a small laboratory unprovided with many of the appliances which he has been in the habit of using, and regards as indispensable.

The general arrangement and equipment of an assay office will necessarily vary somewhat according to the amount and class of work to be done, and also according to the individual preferences of the assayer. The city assayer seldom has much opportunity of planning his office to suit his particular requirements: usually he has to adapt existing buildings as best he can. When, however, no such restrictions exist, it is true economy of time and labour, and also conducive to accurate working, to have the assay-office buildings specially planned to meet the requirements of the work to be done. No design of assay office can be given that is suitable for all requirements, but a few general remarks may be made.

The most important consideration in planning an assay office is so to economise space, by the careful arrangement of the positions of the several rooms and of the appliances and apparatus, that no unnecessary steps will be required in moving from one room to another and from one piece of apparatus to another. If possible, all the rooms should be on one level.

A well-equipped assay office for general assaying should consist of at least six rooms, which are allotted as follows :—

1. Grinding and sampling room.
2. Furnace room.
3. Laboratory for wet work.
4. Weighing or balance room.
5. Store room.
6. General office.

One or more of these rooms may of course be omitted according to circumstances.

In many cases the whole of the assay work has to be performed in two rooms, or even one room. Unless, however, circumstances render it unavoidable, the several operations should be performed as far as possible in separate rooms, so as to avoid the possibility of the contamination of samples from dust resulting from crushing and grinding and from the manipulation of the furnaces. When

7

a large number of samples have to be assayed by furnace methods, a separate room is sometimes kept solely for "fluxing" or preparing assays for fusion, but as a general rule this is performed in the furnace room.

The appliances for the equipment of assay offices are described subsequently in Chapter III., page 17, and Chapter IV., page 36.

The size of the various rooms will depend on the amount and character of the work to be done in them, but the furnace room and laboratory for wet work should be of ample proportions, well lighted and ventilated, and 15 or 16 feet high.   Hot-water pipes or gas fires are best for heating the rooms, because they create no dust.

1. **Grinding and Sampling Room.**—This room should contain all the appliances for crushing, grinding, and sampling.   It should be divided by a screen, and the crushing and grinding machinery kept in one section so that the dust produced by these operations shall not contaminate material that is being sampled.   Motor or other power should be available for driving the machinery. The floor of this room should be of concrete or similar material.   Ample bench room with cupboards should be provided; but shelves on the walls should be dispensed with as far as possible, as they retain considerable amounts of fine ore dust.

The sampling room is frequently absent, but it will amply justify its existence in an assay office, unless the samples are always received in a condition suitable for assay purposes.

2. **Furnace Room.**—The furnace room, as of primary importance in many cases, should be as large as possible, considering the sedentary and unhealthy nature of the work, especially in hot climates.   The ground-floor of a building is the most convenient, so that a good height of chimney or stack may be obtained for the furnaces.   The floor near the furnaces should be of stone or concrete; all other parts may be boarded, or, if a concrete floor extends all over the furnace room, it is advisable to have wooden stages in front of the work-benches so as to avoid damp.   The work-benches should be of thoroughly seasoned non-resinous wood, and stoutly built.   They should be fitted with drawers and cupboards with shelves above for reagents and fluxes.   The tops of the benches require to be made of hard wood such as teak, or American walnut.

The furnaces usually needed, various types of which are described subsequently, are one or more muffle furnaces for cupellation and scorification, etc., and one or more assay-melting furnaces.   When furnaces burning coke or coal are used, it is advantageous to have facilities for firing and for clearing the ash-pit from the back of the furnace, when this can be conveniently arranged, so as to keep as much dust and dirt out of the furnace room as possible.   For the same reason, even when the furnaces are fired from the front, the coke- or coal-bin should be placed if possible outside the furnace room and behind the furnaces, but communicating with the furnace room by means of a trap or shoot through which the fuel is drawn with a shovel as required.   The furnaces should not be placed where they will be subjected to direct sunlight, as this will occasion error in judging of the degree of heat.   As a general rule, the main light in the furnace room, except in hot climates, is best derived from a large top-light.

An iron table on which to stand the moulds when pouring assays is frequently desirable, especially when furnaces with sloping tops are used.   It is made of boiler-plate $\frac{3}{8}$ inch thick, and, say, $2\frac{1}{2}$ feet wide and 4 to 6 feet long.   It is supported on a frame of angle iron or on brick pillars about 30 inches high.

A hot plate, described later, should be provided in the furnace room for drying samples, etc., and is best contained in a cupboard or recess surmounted by a hood so as to be out of the reach of draughts.   A good sink is also very desirable. Provision should be made near the furnaces for drying cupels, crucibles, and scorifiers, etc.

3. **Laboratory for Wet Work.**—A room of about 30 feet by 20 feet, and 15 or 16 feet high, is large enough for much work. A ceiling sheathed with wood is very desirable in the laboratory, as this prevents the possible falling of pieces of white-wash or plaster loosened by the action of acids. The windows should be all on one side, as cross-lights are objectionable.

The working benches are made of well-seasoned wood as described for the furnace room, and should be placed close against the side-light in preference to the middle of the room, where the light is not so good.

A good draught cupboard should be provided to carry off acid fumes, etc. This is lined with slate, white tiles, or white opal glass, and fitted with glazed sashes balanced so as to throw up towards the ceiling.

Heating or "hot" plates are almost indispensable if much wet work has to be done. These should be at least $\frac{3}{8}$ inch thick, otherwise they are very liable to buckle with constant heating. The plates, made any convenient size, are supported on quadrupods or on bricks, and heated by several ring or other burners placed underneath them. It is advisable to lay a sheet of $\frac{1}{4}$-inch asbestos millboard or cloth on the top of the plate. The top of the plate should be at a convenient level from the floor, say $2\frac{1}{2}$ feet.

Sand baths are used by many assayers in preference to hot plates; they can be made of any convenient size to accommodate the number of assays to be made. In a large laboratory sand baths may be 3 feet by 2 feet, and 3 to 4 inches high. Hot plates and sand baths are best contained in a good draught cupboard.

A good earthenware sink is required, and an additional tap connected with a good head of water will be needed, if a filter-pump is to be used.

Ventilation may be effected preferably through louvres on the ridge of the roof or by means of swinging upper windows.

4. **Weighing or Balance Room.**—The balance is the most important part of the assayer's outfit, and in all properly equipped assay offices a balance room is provided. The sensitive balances used in assaying are very liable to be injured by fine dust and acid fume, so that a separate room for them is necessary. This room, if possible, should be quite apart from the laboratory used for wet work. The north aspect, if obtainable, is the best for the windows of the balance room; direct sunlight coming into the room is very inconvenient. The room should be as far away as possible from the crushing and grinding machinery, so that the balances may be placed where they will be subject to the least possible vibration. The bench on which the balances are placed should rest on concrete or brick pillars, say 18 inches square at the bottom, and 12 inches at the top, set in the ground underneath the floor, and projecting a few inches above the floor, which is cut away from the pillars by a space of about half an inch. By this arrangement the balance bench is free from all vibrations of the building. The bench should be placed parallel to the window, so that the balances face the window with the light behind the operator. The top of the balance bench is best made of hard wood, well seasoned: slate is sometimes recommended, but is objectionable as it is very cold to the hands for continuous weighing, and is very liable to cause breakage of the bottles, etc., into which substances are being weighed if these are accidentally knocked over or placed hurriedly on the bench. The walls of the balance room are best painted some light colour, or tiled with cream-coloured tiles.

5. **Store Room.**—This is best attached to the main working rooms, and not in a basement, where the damp is always objectionable. It is desirable to have it divided by a partition, so that apparatus and reagents, etc., can be kept quite apart from samples that require to be stored.

6. General Office.—This should be arranged near the main entrance to the assay office so as to be readily accessible. It should be provided with a writing-

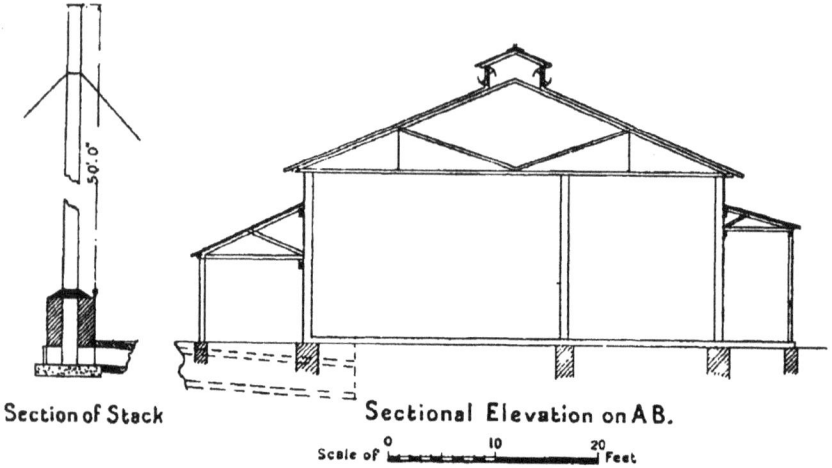

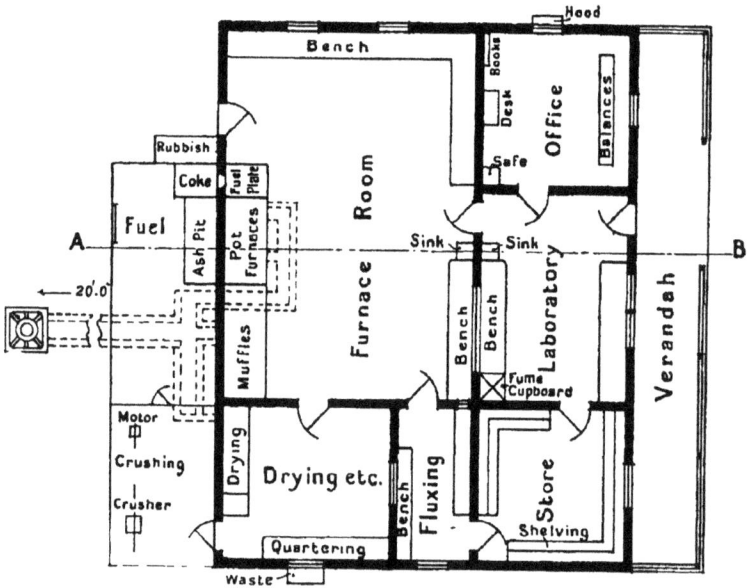

Section of Stack   Sectional Elevation on A B.

Scale of 0   10   20 Feet

Fig. 1.—Plan of Mill assay office (A. Whitby).

table or desk, cabinets for filing correspondence, typewriter, safe, and a book-case to contain standard works of reference, etc. Suitable arrangement should be made in the lobby for accommodating coats and hats.

Mill-Assay Office.—The accompanying plan (fig. 1) illustrates the arrangement suggested by A. Whitby,[1] slightly modified by E. J. Laschinger, for an assay office for a 200-stamp proposition (i.e. a mill with 200 stamps crushing capacity) on the Rand in South Africa, and using the cyanide process. The plan may serve as a guide in the building and fitting up of new laboratories.

Where convenient, the building should have foundations quite distinct from those of the battery, and as far as possible removed from it. It should be in as central a position as can be obtained, and it should not be attached to any reduction plant. With the exception of the balance room, each working room in the proposed structure is under observation from some other. This Mr Whitby considers most important, as a greater number of assays can be controlled with a smaller staff (native). The assay office consists of six rooms contained within the main building, viz. :—

1. Drying and sampling room.
2. Fluxing room.
3. Furnace room.
4. Laboratory for wet work.
5. Balance room and general office.
6. Store.

In the lean-to outside the main building provision is made for the crushing and grinding appliances, as suggested by E. J. Laschinger in the discussion on Mr Whitby's paper: in all other details the plan is in accordance with that submitted by Mr Whitby. The building is of brick, with concrete or granolithic floor throughout, excepting the balance room, which has a wood floor and a wood ceiling. The furnace and sampling rooms are not ceiled, but it is of the utmost importance to have a dust-proof ceiling for the balance room, if not quite so necessary for the wet laboratory and fluxing rooms. The roof trusses are of steel. Laschinger[2] gives the estimated cost of erecting such an assay office on the Rand (in 1906) as £2000. The following details are given.

The crushing appliances should consist of a small, fine crusher with motor or other convenient power. An auxiliary hand machine for crushing should also be supplied.

1. *Drying and Sampling Room.*—This should be provided with an ample drying hearth at one end for drying such samples as slimes, residues, etc. A long bench is also provided for mixing and quartering samples. It communicates with an outside bin as shown, for the reception of waste or rejected ore from sampling.

2. *Fluxing Room.*—This room is used for "fluxing" the samples and preparing them for fusion. Under the window which overlooks the sampling room is a small trap through which the samples, after thorough mixing, are passed to the counter behind the fluxer, who works at the bench on the opposite side of the room.

The fluxing room, in addition to the necessary balance, fluxing bench, and a few drawers, is provided with one or more small benches or shelves to take the day's work only, i.e. samples not yet reported on. After the fluxer has added all his ingredients to the samples, the tins into which they are placed are passed through a small trap into the furnace room, where the contents are mixed and transferred to the crucibles.

3. *Furnace Room.*—The furnace room, being of primary importance, should be given ample proportions. It is provided with wind and muffle furnaces,

[1] *Journ. of the Chem. Met. and Min. Soc. of S. Africa*, March 1906, p. 266. The plan is reproduced by kind permission of the Council of the Chem. Met. and Min. Soc. of S. Africa, and of Mr E. J. Laschinger. See also plans for small and large assay office, by A. M'Arthur Johnston, *Rand Metallurgical Practice*, vol. i., 1912, pp. 287–8.
[2] *Journ. of the Chem. Met. and Min. Soc. of S. Africa*, June 1906, p. 368.

but Mr Whitby makes no definite statement as to the type of furnace most suitable. The furnaces are connected with a stack 50 feet high, which is separate from the building, and connected, as shown, by an underground flue, into which all the flues from the furnaces are led. The stack is high enough to give a strong draught without the aid of fans. The fuel for the furnaces is stored in a shed outside the main building, but it communicates with the furnace room and supplies fuel to the fuel plate as required. The ashpits to the furnaces are cleared from the outside, so as to keep dirt out of the furnace room. A cast-iron plate floor is provided in front of the furnaces. Ample bench accommodation and a sink are provided. Anvil blocks are supported on solid timber to the foundations. The rolls are better placed in the furnace room than in the wet room, where they are affected by the acid fumes, etc.

4. *Laboratory.*—No special comment is made regarding this room. It has good benches with cupboards for apparatus, and is fitted with a sink and fume cupboard with good draught.

5. *Balance Room.*—As previously stated, it is of the utmost importance to have a dust-proof ceiling in this room. The balance counter should be away from the walls and supported on solid pillars resting on the foundations. With these precautions any kind of top may be used. The use of slate tops is deprecated by Whitby. The light for the balances should be ample, and come from behind the operator, but no direct sunlight should ever fall on the cases.

Three balances are necessary : (1) a chemical balance with a minimum sensibility of $\frac{1}{4}$ milligramme used for weighing out bye-product charges, (2) an assay balance, sensible to $\frac{1}{20}$ milligramme for bullions and mine samples, and (3) another sensible to $\frac{1}{100}$ milligramme for surface or reduction work.

The balance room is also used as a general office in which all the assay books, etc., are kept.

6. *Store Room.*—The samples, when reported on, are stored away in numbered bags on shelves allotted to them in the store room. In this room the fluxes are also kept. The various materials used by the fluxer are stored in small bins which communicate with the fluxing counter in the adjoining room.

Custom Assay Offices.—In many of the mining districts of America and Australia there have been erected mills that undertake to crush, sample, and treat parcels of ore for the general public. These mills are known as *Custom Mills*, and the assay offices connected with them are known as *Custom Assay Offices.*

Assay-Office Records.—A careful record should be made of all samples received for assay, and of the results obtained. The record must be clear and neat, so that reference, even after an interval of years, should be certain and easy. One method should be adopted and adhered to. In some assay offices the card-index system and other modern methods of keeping records are adopted. Where a large number of samples have to be dealt with, three books are required, viz. : (1) sample book, (2) laboratory book, (3) assay book. These books will necessarily vary in detail in every office according to the character of the samples to be dealt with, but the following examples will serve to illustrate the method of record.

Sample Book.—The sample book contains particulars of the samples (marks, etc.), which are entered by the office clerk as they arrive. The clerk at the same time puts on each sample the distinguishing number against which it has been entered and by which it is known in its course through the various assay operations. The marks on the samples very frequently consist of the initial letters of the name of the mine or smelting company, or the name of the boat in which the ore was shipped.

Example of Page of Sample Book.

Date.	No.	Description of sample.	Remarks.
1912.			
Nov. 8	95	R. J. T., argentiferous lead ore . . .	For Pb and Ag.
,, 9	96	Silver ore, ex "Sorata"	For Ag and Au.
,, 9	97	Silver ore, ex "Potosi"	For Ag.
,, 9	98	A. M. Co., concentrates	For Au.
,, 10	99	R. J. T., argentiferous lead ore . . .	For Pb and Ag.
,, 10	...	...	Report Ag per { ton of ore and ton of lead.
,, 11	100	T. Min. Co., gold ore, X.	For Au.
,, 11	101	T. Min. Co., gold ore, XX. . . .	For Au.
,, 11	102	S. S. Co., burnt ore, Spain . . .	For Ag, Cu, and S.
,, 12	103	Argentiferous copper ore, ex "Liguria" .	For Ag and Cu. Test for Au.

Laboratory Book.—This is the assayer's note-book, in which he enters clearly the particulars of his work—the results obtained, and details as to how these results were arrived at. The calculations should be done on scrap-paper, and only sufficient detail entered in the laboratory book to enable the results to be recalculated should this become necessary. The results of each assay should be recorded as soon as obtained, and not left to memory only.

Example of Page of Laboratory Book.

95. Argentiferous lead ore.
   Clay-pot fusion, 25 grammes gave 15·3 Pb = 61·2 per cent. Pb.
   0·615 Ag = 80·367 ozs. per ton.

96. Silver ore.
   Pot fusion, 25 grammes gave 0·159 Ag = 207·76 ozs. per ton.
   0·002 Au = 2·613 ,, ,,
   (Moisture, 1·3 per cent.)

97. Silver ore.
   Scorification, 3 grammes gave 0·0352 Ag = 383 29 ozs. per ton.
   (Moisture, 2·2 per cent.)
   Metallics from 100 grammes = 0·0051 gramme = 1·666 ozs. per ton.

100. Gold ore (quartzose).
   50 grammes gave Au + Ag ·0043 gramme.
   Au 0·0027 gramme = 1·764 ozs. per ton.
   Ag 0·0016 ,, = 1·045 ,, ,,

101. Gold ore (pyritic).
   50 grammes gave Au + Ag 0·0018.
   Au 0·0015 gramme = 0·9800 oz. per ton.
   Ag 0·0003 ,, = 0·1960 ,, ,,

The Assay Book.—The assay book is the official book for the regular entry of all assay results, and is a combination of the sample and laboratory books. It corresponds with the report form given below. The assay book should contain sufficient detail to distinguish each sample, but excessive detail should be avoided. When only a small number of assays have to be dealt with, the sample book may be omitted, and the entries made directly in the assay book as the samples arrive.

Example of Page of Assay Book.

Date.	Date reported.	Description of sample.	Reference No.	Gold, ozs. per ton.	Silver, ozs. per ton.	Copper.	Lead.	Other metals.[1]	Remarks.
						Percentage.			
1912. Nov. 8	9	R. J. T., argentiferous lead ore.	95	...	80·367	...	61·2		
,, 9	14	Silver ore, ex "Sorata"	96	2·613	207·76	...	...	...	Moisture, 1·3 %.
,, 9	14	Silver ore, ex "Potosi"	97	...	384·96	...	...	...	{ Moisture, 2·2 %. Metallics, 1·666 ozs.
,, 9	18	A. M. Co., concentrates	98	52·15	10·69	...	...	...	Contained Cu.
,, 10	12	R. J. T., argentiferous lead ore	99	...	105·40	...	50·0	...	{ Ag per ton of Lead, 210·8 ozs.
,, 11	14	T. Min. Co., gold ore, X.	100	1·764	1·045				
,, 11	14	T. Min. Co., gold ore, XX.	101	0·980	0·196				
,, 11	20	S. S. Co., burnt ore, Spain.	102	...	1·630	0·13	...	...	Sulphur, 0·15 %.
,, 12	20	Argentiferous copper ore, ex "Liguria"	103	...	15·190	5·3	...	...	Au trace only.

[1] Separate columns are added for other metals or other constituents such as S, As, FeO, SiO$_2$, etc., according to requirements.

Report Forms or Assay Certificates.—These are printed forms on which the result of the assay is reported to the person from whom the sample was received. They should entail as little writing as possible in making out the report. Report forms differ considerably according to the character of the samples to be reported upon. Almost every assayer has his own particular form for certificate of assay ; but so long as the certificate states clearly the results of his work, any little differences of detail are unimportant.

The two accompanying report forms may be taken as examples of those generally used by assayers :—

Assay Office, London, E.C.

Reference No.............   ......, ...19......

Richard Smith & Son,
Assayers and Consulting Metallurgists, etc.

CERTIFICATE OF ASSAY.

To M.........................

....

We have assayed the sample of ....

.... marked as under, and find the following to be the

result.

Mark of Sample

..........................., ex " ...."

Assay Office, London, E.C.

........................19......

Reference No..........

To M..............................

....

CERTIFICATE OF ASSAY.

I hereby certify that the samples of...
herein described have been assayed, with the following results :—

Sample number and mark.	Description of sample.	Gold.		Silver.		Total value per ton.	Percentage.		Remarks.
		Ozs. per ton, 2240 lbs.	Value at ....s. per oz.	Ozs. per ton, 2240 lbs.	Value at ...d. per oz.		Copper.	Lead.	

Assayer...........................

CHAPTER III.

ASSAY FURNACES.

Two kinds of furnaces are used in assaying, (1) "wind" or crucible furnaces in which the assay is in direct contact with the fuel, and (2) muffle furnaces, in which the assay is contained in a muffle or small fireclay oven heated externally.

There are many designs of assay furnaces, varying chiefly with the kind of fuel to be used. They may be conveniently classified as follows :—

    A. Furnaces employing solid fuel.

    B. Furnaces employing gaseous fuel.

    C. Furnaces employing liquid fuel.

The choice of furnaces is usually dependent on the nature of the fuel that is available, and this varies with different localities. Furnaces may be permanent or portable according to requirements.

The furnaces described below may be taken as types of the different classes.

A. Furnaces employing Solid Fuel.

The furnaces of this class are more generally used than those in which gaseous or liquid fuel is employed. The solid fuels used are coal, coke, wood, and charcoal.

The furnaces are built of good bricks, solidly cemented with clay, and tightly bound with iron hoop and rods. Fire-bricks are used for the interior, which is exposed to the greater heat. The furnaces are connected with a tall chimney to provide a good draught.

Wind or Crucible Furnace.—Fig. 2 is a vertical section through the middle of the fire-grate of a permanent wind furnace ordinarily used for assaying. It consists of the fireplace, 8 inches square by 12 inches deep; and the ashpit, provided with two sliding doors for the regulation of the draught. The doors are made of sheet iron, running in a stout wrought-iron frame, built into the brickwork. By means of the sliding doors the size of the opening may be varied according to requirements, and in this way it is possible to regulate the admission of air with the greatest nicety. The capacity of the ashpit should be at least equal to that of the fireplace. It is a good plan to have a small iron tray in the ashpit for receiving and removing the ashes. The fire bars, made of bar iron, 1 inch square, are 21 inches long, and project a short distance beyond the front of the furnace so as to be easily removable for drawing the fire: their ends rest loosely on iron supports. The short flue from the fireplace communicates with a main flue, connected with a chimney not less than 30 feet high, with which other similar furnaces may be connected as described later. The interior of the furnace is of firebrick; the exterior and the part below the grate are built of ordinary brick of good quality. The whole of the brickwork is kept firmly bound together by means of cast-iron plates and wrought-iron tie rods. The furnace mouth can be closed by means of two fire-bricks, each of which is clamped with a piece of flat bar-iron firmly wedged at one end : these bricks are of two sizes, as shown, and the larger one only need be removed when the crucible is taken out of the furnace. The furnaces are sometimes provided with a sloping top closed with a sliding iron plate, some assayers preferring this arrangement to a flat top. The draught may be regulated not only by the sliding doors, but also by means of a damper in the main flue or by placing a piece of firebrick in the short horizontal flue in the fireplace. The damper may consist of an iron plate sliding in grooves as shown, or a fire-clay tile may be used as shown in fig. 6, page 22. When an iron damper is used, the handle should be hinged so that it drops when the damper is fully open.

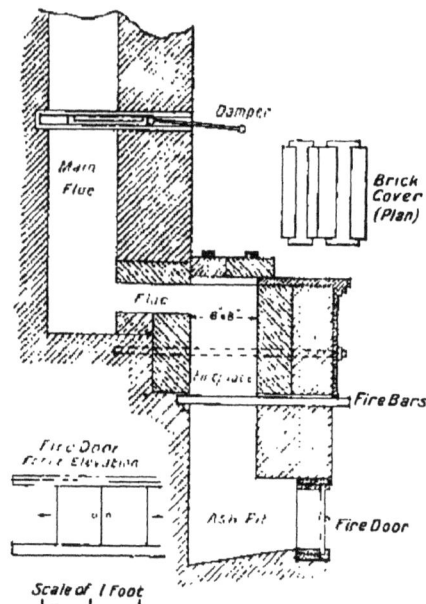

Fig. 2.

With such arrangements perfect control may be obtained over the temperature of the furnace ; it can be kept below a dull red, or increased sufficiently to melt iron or nickel if so desired.

The fuel employed in these furnaces is almost invariably coke. Charcoal may be used, but it is more costly, and does not give such a high temperature as coke.

Free-burning (flaming) coal is sometimes employed in places where good coke is very expensive, and in this case the crucible should always be supported on a piece of fire-brick about 3 or 4 inches high, resting on the fire-grate, so that it is completely surrounded by the flame.

In many cases an old crucible inverted will serve as a convenient support.

Wind furnaces are generally square, but they may be made rectangular to accommodate a larger number of crucibles, the flue being placed at one of the longer sides.

In assay offices where the amount of furnace work to be done is large, the furnaces are frequently made 12 inches square with a flue, say, 4 inches by 5 inches, so that a number of fusions can be made simultaneously in one furnace.

For the general run of instructional work in mining schools, however, the smaller sizes are preferable, as each student should have a furnace to himself, so that he alone is responsible for the temperature employed in his assays.

Beginners should not attempt more than one fusion at a time, so as to learn the exact conditions of the fusion.

Furnace Manipulation.—Wind furnaces require to be properly manipulated to obtain satisfactory results. The degree of temperature possible to attain depends on the depth of the furnace, i.e. distance between the grate and the flue, the height of the chimney, and the quality of the fuel used. "A very essential condition in obtaining the maximum heating effect of a furnace, the importance of which can alone be appreciated by experience, is to choose pieces of fuel of a suitable size. If, on the one hand, a shovelful of coke be taken at random, it generally contains the dust and dirt found in most fuel, which, by filling the interstices, prevents the air from passing as required, and consequently retards the combustion. On the other hand, if a furnace be filled with large pieces, considerable spaces are left between them, so that only a comparatively small surface is exposed to the action of the atmospheric oxygen, and a correspondingly small quantity of fuel is consumed in a given time, so that the maximum heat can never be obtained.

In order to produce the desired result, it is necessary that the pieces shall have a certain uniform size, and experience has proved that pieces about 1 to $1\frac{1}{2}$ inches in diameter produce the best effect.

These may be selected by sifting the coke through two strong wire screens, one of which has meshes about $1\frac{1}{4}$ inches square, and the other about 1 inch square. The coke which passes through the larger one, but will not go through the smaller screen, will be the right size for use in the furnace." (Mitchell.)

The coke used should be of good quality and not too dense; it should be broken into lumps of a uniform size before being brought into the furnace room.

In lighting a furnace a start is made with wood, as this readily ignites and starts the combustion of the coke, which of itself does not ignite readily. A little charcoal, when obtainable, is a great help at the start. Before commencing to work the furnace the fuel should be well packed by stirring, raising the coke and not ramming it, and it should be uniformly heated, not hot below and cold above.

Crucibles must be placed in the middle of the fire, at equal distances from the sides and bottom of the furnace, and must be completely surrounded with the fuel, so that they are uniformly heated. The fuel should be packed sufficiently to give a good solid foundation on which the crucible can be placed.

If the fire has not been well packed, the fuel, in burning away, allows the crucible to fall down, and may cause the loss of the contents. A plan adopted by many assayers is to make a hollow in the fire, place an old crucible into it, and pack round with coke, so that the surface slopes upwards from the mouth of the crucible to the sides of the furnace. When the empty crucible is uniformly heated, it is taken out and the crucible containing the assay substituted. It is rarely advisable to have a very hot fire at first, for the reasons stated on page 153, where fuller details of the fusion of assays are given.

The highest temperature is found at about 2 inches above the grate, which should be taken into consideration in placing the crucibles in the furnace.

The temperature is regulated in the manner indicated above, and, if necessary, fuel is added from time to time; but before this is done, the glowing coke must be poked down to do away with empty spaces.

In commencing a second assay immediately in the same furnace, certain precautions must be taken to ensure success. All ash and clinker must be removed from the grate by means of a poker; the fire must be well stirred to pack the fuel, and fresh fuel added.

A great saving of fuel is effected by working off the assays continuously so that the furnace is not allowed time to cool.

Carr's Crucible Furnace.—Carr's patent crucible furnace [1] (fig. 3) for coke fuel and natural draught is now being adopted in many assay offices and mining schools.

The outer case of the furnace consists of wrought-iron plates securely bolted to the angle framework, and as the internal linings are supplied in a solid form, the furnace can be quickly fitted up and put into operation.

[1] Morgan Crucible Company, London.

The fire-bars are placed lower than the bottom of the fire-brick lining of the furnace, and so give an air passage on each side, in front, and at the back of the bars, allowing a current of air to pass equally on all sides; and from the way in which the air is admitted to the fuel, a perfect and regular combustion of the coke takes place, and a high uniform heat, of the required intensity, completely surrounds the crucible. The air passing through the bars is reduced to a minimum, which prevents the unnecessarily rapid consumption of the fuel bed, so that the crucible remains in its proper position, and is not chilled at the base by the incoming cold air.

The solid lining rests on the bottom plates, and is kept in a central position by a backing of non-conducting material, and the heat is thereby effectively retained within the furnace and prevented from destroying the metal casing, which will last for many years.

Two fires in one frame are shown in the illustration. The furnaces are connected with a chimney shaft, and the draught regulated by means of a damper in the flue in the ordinary way.

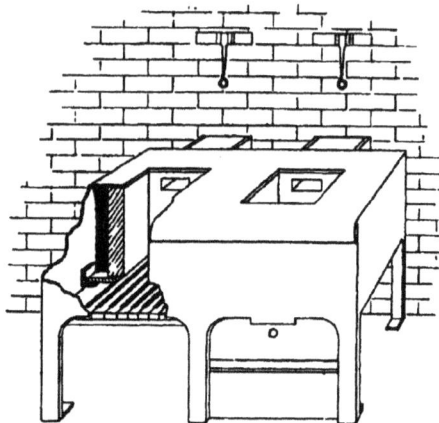

Fig. 3.

Muffle Furnaces.—The principal part of these furnaces is the muffle (fig. 4), which is a ⌂-shaped vessel of fireclay open in front and closed at the back. It serves for the reception of the assays, and is heated from the outside either by direct contact of glowing fuel such as coke or by the flame of flaming fuel such as bituminous coal. These furnaces are absolutely necessary for cupellation and other oxidising processes described subsequently.

Muffles are made in varying sizes and shapes. The best shape for general use is one of nearly rectangular cross-section, with a slightly arched top, as shown in fig. 101. This shape is invariably used when crucible fusions are made in the muffle. The largest muffles ordinarily used in coal-fired furnaces are 19 inches long, 12 to 14·5 inches wide, and 7·75 inches high (outside dimensions).

The muffle is provided with a slit or a number of small holes about ⅜-inch diameter, at the back, in order that by means of the furnace draught a current of air may be kept constantly passing through the muffle. It is necessary that the position, number, and size of the holes in the muffle should be carefully adjusted according to the draught of the furnace and other circumstances.

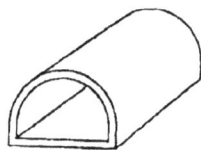

Fig. 4.

Instead of holes the muffle is sometimes provided with a fireclay or graphite tube at the back, through which the air is withdrawn. The tube connects with an iron tube, furnished with a damper for regulating the air-supply through the muffle. There is no other opening in the muffle, so that the draught through it is quite independent of the draught through the furnace.

Muffles provided with tubes as described are illustrated in section in fig. 5 and also in the gas muffles in fig. 18, page 33.

This arrangement has been adopted for the muffles used at the Royal Mint, London.

The mouth of the muffle is closed during working by means of a thin fire-brick standing on edge or some other form of door.

The essential dimensions in a muffle furnace are (1) the area of the fire-grate, (2) depth of the fireplace, i.e. the distance from the fire-bars to the bottom of the muffle, (3) the "fire-space," i.e. the distance between the muffle and the walls of the furnace. These dimensions are dependent on the nature of the fuel used. The solid fuels usually employed for heating muffle furnaces are bituminous coal (i.e. long-flame coal), anthracite, and coke. The fuel preferred by many assayers and usually the one most easily obtained is bituminous coal, giving a fairly long flame. Anthracite and coke, either alone or a mixture of the two, are very frequently used, but as they burn without flame they must surround the muffle.

The stoking of muffle furnaces is done from the front or from the back. With the latter arrangement much dust and dirt are kept out of the furnace room, and the operator, working in front of the furnace, is not exposed to the direct heat; but it also prevents him from giving immediate attention to the regulation of the draught and the firing should the assays require it.

Coal-fired Muffle.—A common form of muffle furnace for coal with a stoke-hole at the back is shown in cross-section in fig. 6A, and in longitudinal section in fig. 6B.

The external dimensions of the muffle are 19 inches long, 12 inches wide, and 8 inches high. In front the muffle rests upon the front wall of the furnace, and at the back or at the bottom it rests on two or more firebrick supports as shown. It should be clayed well round the mouth, so that no air may enter except through the mouth itself.

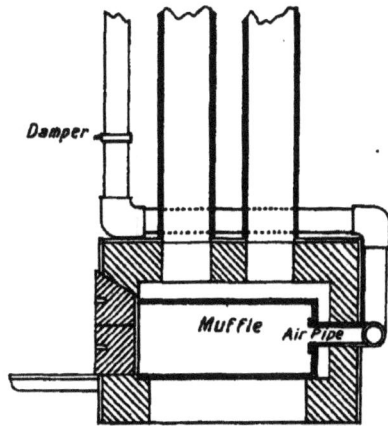

FIG. 5.

The luting material used for this purpose is a mixture consisting of 1 part of raw fire-clay and 3 parts of burnt fireclay, i.e. firebrick or crucibles crushed to pass a sieve of 8 to 10 mesh. Sufficient water is added to make a plastic mass, but the mixture should be used as dry as possible. Raw fireclay has too great a shrinkage when drying to be used alone.

The dimensions of coal-fired muffles vary somewhat with the nature of the coal employed, i.e. whether long or short flame, but the fireplace is always deeper than in the case of muffles heated by coke.

The distance from the fire-grate to the bottom of the muffle varies from about 12 inches to 18 inches: in fig. 6 it is 15 inches. The fire-space round the sides and top of the muffle is 2½ inches. The fire-space above the muffle is slightly arched.

The flue is best placed over one end of the muffle as shown, but it may also be placed over the middle of the muffle. The flue-area in muffle furnaces is usually from ⅓ to ¼ of the grate-area.

The three middle fire-bars should be made of square bar-iron, and should be removable, to facilitate discharge of the fuel when closing down the furnaces.

The ash-pit is closed with two sliding doors, as described for the wind furnace, which are used for adjusting the draught. The draught is also regulated by the damper consisting of a movable fireclay tile inserted in the flue. The walls of the furnace should be about 13 inches thick to prevent loss of heat by radiation. A similar type of furnace is used for heating two muffles placed one above the other, or three muffles, two below placed side by side, and one above.

Coke-fired Muffle.—The method of firing constitutes the chief difference between coke-fired and coal-fired muffles. In the case of the coke the firing is somewhat more difficult, as it has to be done from the top of the furnace, and the fire requires careful attention.

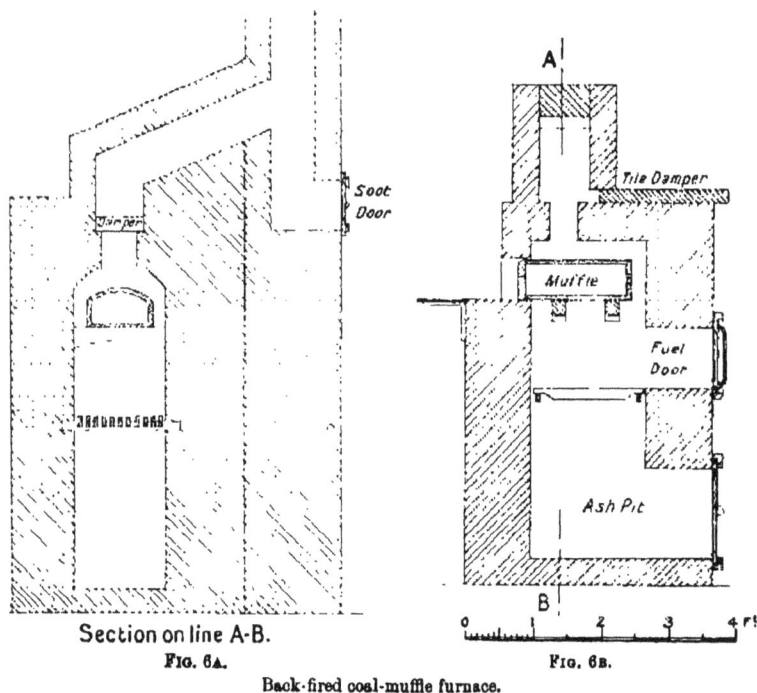

Section on line A-B.

FIG. 6A.   FIG. 6B.

Back-fired coal-muffle furnace.

A coke-fired muffle is shown in front elevation in fig. 7A, in vertical section on line A, B (figure) in fig 7B, and in vertical section through the middle of the furnace and the axis of the muffle in fig. 7C. The muffle is about 12 inches long, 8 inches wide, and 7 inches high; the distance from the bottom of the muffle to the fire-grate is 6 inches, and the fire-space at the sides of the muffle 3 inches.

The fuel is introduced through an opening in the top of the furnace, which is closed with a fireclay slab bound with iron and fitted with a handle. The fuel is carefully packed round the sides and the top of the muffle. The coke should be broken to pieces not larger than 1½ inches in diameter, otherwise the fuel is liable to get wedged at the side of the muffle and tend to break it during stoking. The fire is stoked through the opening in the front wall of the furnace immediately below the muffle door, which is closed with a loosely fitting tile: it is also stoked through the opening in the top. The draught is regulated by the

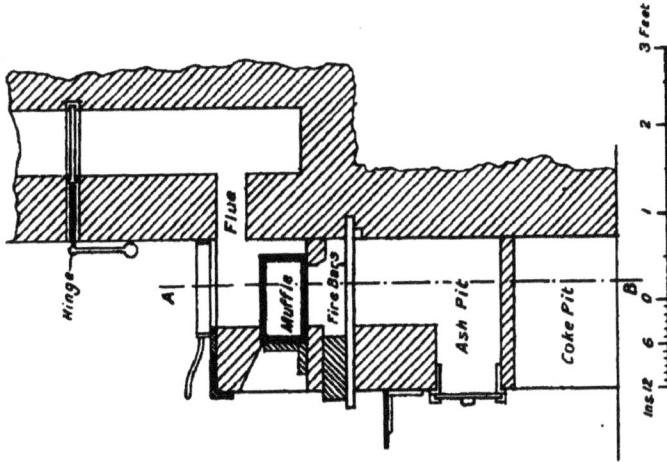

Fig. 7c.—Longitudinal section.

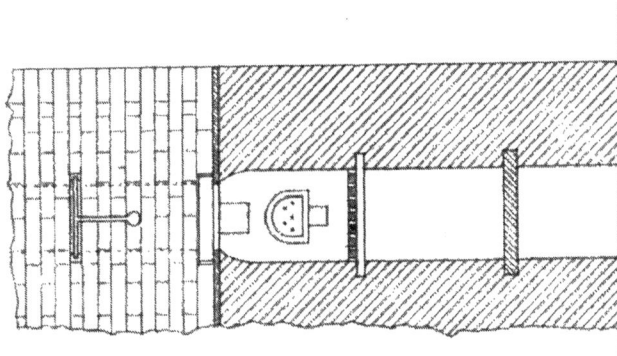

Fig. 7b.—Cross-section.

Coke-muffle furnace.

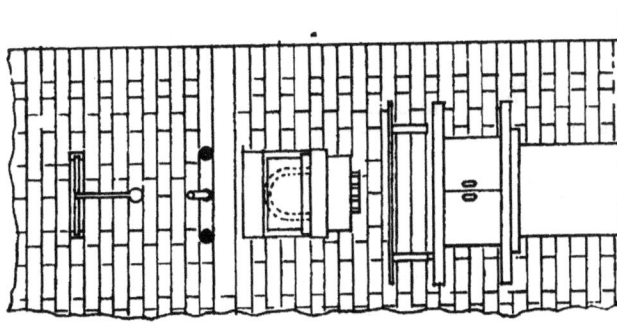

Fig. 7a.—Front elevation.

iron damper in the flue and the sliding doors in the ashpit.   Some of the fire-bars are removable for the reason already given.

Coal- and coke-fired furnaces are in very general use for assaying purposes. When the amount of furnace work to be done is not large, the coke-fired wind furnace is usually preferred for fusions and a coal- or coke-fired muffle furnace for muffle work.   The furnaces may be separate or combined.[1]   When a large number of assays have to be made, the fusions are generally performed in wind furnaces of large dimensions or in a large muffle.   However, according to M'Arthur Johnston,[2] coke-fired fusion and muffle furnaces are gradually being eliminated in mine-assay offices in South Africa, and coal furnaces, chiefly of the reverberatory type, substituted.   The coal used is a long-flaming, bituminous coal.

The advantages of furnaces of the reverberatory type are considerable reduction in cost of fuel, less wear and tear of muffles, and the heating of a large number of crucibles at the same time.

The furnaces can be fitted into the side of the building with the working door opening into the room and outside firing adopted, thus adding considerably to comfort in working.

Four styles of furnaces have been introduced on the Rand, viz.: (1) the Tennant furnace, (2) the Johnston furnace, (3) the Rival furnace, and (4) the Rusden furnace.   They may be briefly described as follows:[3]

Tennant Furnaces.—In the Tennant pot furnace the crucibles are placed on bars immediately above the coal fire, whilst another lot can be introduced above these, also on bars.   The crucibles are thus heated from below, the flame playing round each pot.

In the Tennant muffle furnace the muffles are arranged in a large vertical flue down which the combustion gases pass after being deflected by a bend in the flue.   The muffles (two or more) are placed at different intervals to permit of uniform heating as far as possible, but the one nearest the fire-grate gets by far the strongest heat.   This constitutes one of the drawbacks to the furnace, as it frequently happens that when the lower muffle is ready for cupellation the top muffle is too hot and its use has to be discontinued.   Before entering the main chimney the gases from the downcast pass beneath a drying plate, which can be used for drying wet slime, etc.

Johnston's Furnace.—For pot fusions the furnace consists of a small ordinary flat-bedded reverberatory furnace with fire-grate at one end, the crucibles being arranged on the bed of the furnace.   A similar arrangement is adopted for the muffle furnace, the muffles being placed over the bed so that the flame surrounds them.

Rival Furnace.—This is a combined pot and muffle furnace.   The reverberatory chamber containing the muffles is placed above that holding the crucibles, whilst by-passes allow of the waste gases from the latter being used to initially or supplementarily heat the muffles.

Rusden Furnace.—This furnace differs from the others mainly in that the roof of the crucible reverberatory forms the door for charging and withdrawing the crucibles.   The door consists of two firebrick slabs.   The muffle furnaces are on either side with a by-pass to divert the flame to them when the crucible furnace door is open.

Arrangement of Furnace Flues.—The proper construction of the furnaces is

---

[1] A combined crucible and muffle furnace built of ordinary firebrick is described by G. T. Holloway, *Trans. Inst. Min. and Met.*, 1907, vol. xvi. pp. 341-9.   Drawings and specifications are given.

[2] *Rand Metallurgical Practice*, vol. i., 1912, p. 297.

[3] For fuller details, with sketch of Johnston furnace, see *Rand Met. Prac.*, vol. i. 1912, pp. 297-299.

a matter of first importance, as a slight defect in the arrangement of the flues will considerably affect the draught, and prevent the temperature requisite for assay operations from being attained.

The short flue of an assay furnace is horizontal, and generally connected with a larger main flue which will vary in length according to the distance of the stack from the furnace. The size or cross-section of the flue has a great influence on the working of the furnace, for, if too narrow, the draught will be slow owing to friction of the air current on the walls.

For an active and strong draught the flue must be wide and the chimney large and high. As a general rule the section of the flue is from $\frac{1}{4}$ to $\frac{1}{5}$ the area of the fire-grate. When several furnaces are built side by side so that all

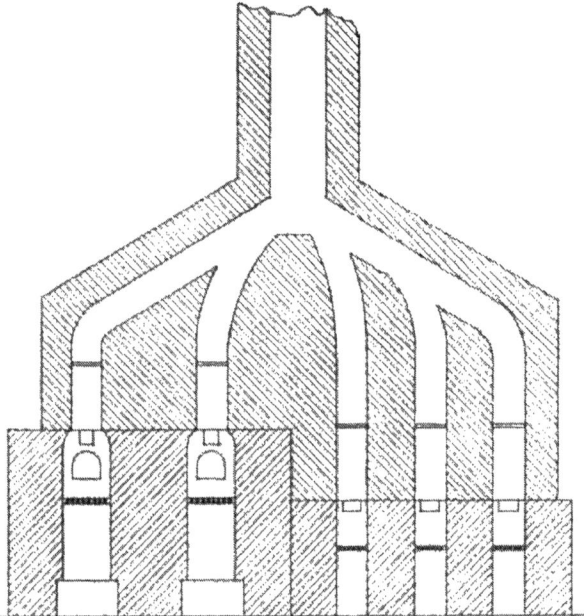

Fig. 8.

the flues open into a common conduit or main flue, special attention must be paid to the arrangement of the flues, otherwise defective draught will result. For example, if the flues from several furnaces are connected with one horizontal conduit, each opening into it at right angles, the stack being at one end of the conduit, the draught will be defective owing to the air currents from the several flues interfering with one another. The furnace near the stack will draw well, while the draught in those farthest from the stack will be so feeble that only a low temperature will be maintained.

For producing an equally high temperature in each of a series of similar furnaces, so that each furnace will act as well as if it were connected with a separate chimney, the arrangement of the flues should be such that the separate currents are diverted into one uniform stream.

The arrangement shown in section in fig. 8 for two muffle furnaces and three wind furnaces placed side by side has been found to be satisfactory.

Another very essential structure connected with a wind furnace is the stack or chimney, which is used for creating a draught and for carrying off the products of combustion. It is on the height and size of the stack that the draught depends, and, in consequence, the degree of heat produced within the furnace.

In general, a draught sufficiently strong to produce a temperature high enough for all assay purposes can be obtained with a chimney from 30 to 40 feet high, but higher chimneys may be sometimes used with advantage.

The height of the stack will depend on the position of the furnace. A height of 30 feet is usually regarded as the minimum that can be allowed in order to ensure a good draught, but, according to T. K. Rose,[1] very satisfactory results can be obtained with a stack only 16 feet high.

Chimneys are generally made square or rectangular, and in the case of small wind furnaces have interiorly the same dimensions as the body of the furnace in which the crucibles are placed.

When the chimney is connected with several furnaces having an arrangement of flues as shown in fig. 8, so as to produce a current in one direction, a single opening will suffice; but if furnaces are arranged on opposite sides of the stack, and the flues open into it at right angles, the chimney shaft must be divided for a certain distance so as to divert the two streams into one vertical direction as shown in the accompanying fig. 9, otherwise they will interfere with each other and check the draught.

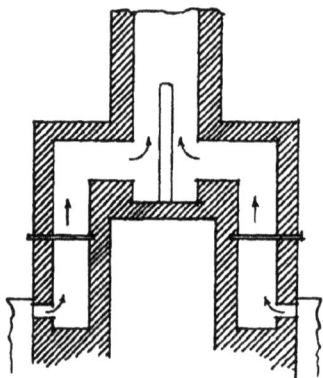

Fig. 9.

To attain the maximum of temperature in wind furnaces the air must be allowed to pass with the greatest possible facility into the furnace; and if the furnace is in a confined space with little access of air, it is necessary to lead air to the furnace from outside the building through an air channel or pipe under the floor with an opening, covered by an iron grating, immediately in front of the furnace. The channel should be furnished with a valve or door at the outer end so that the quantity of air admitted may be regulated as desired.

Portable Furnaces for Solid Fuel.—These furnaces consist of a casing of stout sheet-iron lined with firebrick, or are made entirely of fireclay, strapped with iron bands.

A portable prospector's furnace made by the Morgan Crucible Co. is shown in fig. 10.

It is a crucible and muffle furnace combined. The casing is of stout sheet-iron with removable fittings, lined with strong fireclay tiles which are kept in position by a top plate firmly screwed down. The tiles have rebated edges to ensure tight joints.

The grate consists of nine wrought-iron bars fitted diagonally into a cast-iron frame. The bars can readily be removed to clean the grate, and, being of ordinary size and section, they can be replaced without difficulty.

A small iron plate can be supported on the projecting ends of the fire-bars, to form a convenient shelf on which to place scorifiers, etc.

Cast-iron draught doors, fitted with regulators, are provided for adjusting the air-supply.

[1] *Metallurgy of Gold*, 5th edit., 1906, p. 377.

A chimney collar is fitted to the back of the furnace body so that a chimney of suitable length can be added.

The top of the furnace is closed with a fireclay cover, provided with a sight-hole, through which the contents of the crucibles can be inspected from time to time, thus obviating the necessity of removing the cover. The sight-hole is closed by a fireclay stopper. The muffle (12 inches by 6 inches by 4 inches) is placed in the middle of the furnace and rests on suitable supports back and front. The muffle door is provided with a mica sight-hole through which the contents of the muffle can be watched. The furnace can be used for crucibles without removing the muffle.

The total height of the furnace is 28 inches; depth, back to front, 14 inches; and width, 16 inches: diameter of chimney collar, 5 inches. Total weight of furnace complete, 130 lbs.

Portable furnaces are also made with a circular body of stout sheet-iron lined with three solid fireclay rings 2 inches thick, which cannot fall in pieces when fired, as frequently happens to the linings made of firebricks or tiles cemented together by a clay lute such as are used in square furnaces.

To reline these circular furnaces a mould of wood is placed in the middle and the open space between the surface of the mould and the old lining is filled with a stiff paste of refractory clay, each layer being well rammed down. When the space is filled, the case is withdrawn, and the crust of clay dried very slowly. Any cracks that result from the drying must be filled up.

A good mixture for the lining consists of 1 part of refractory clay and 3 or 4 parts of sifted quartz sand. The old lining should be wetted before the new lining is added.

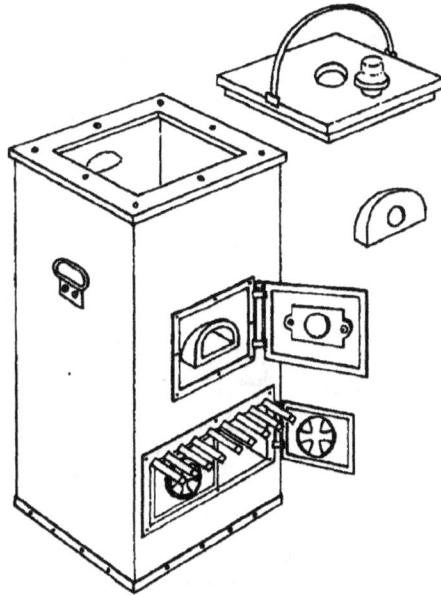

Fig. 10.—Portable assay furnace.

The wind furnace ordinarily used for assay purposes being of simple construction can be built by any bricklayer acting under instructions, and it is advantageous to erect a brick furnace in all cases where the assayer is likely to be more or less permanently settled in any locality.

No mortar is used in its construction, the bricks being cemented with clay, mixed with an equal bulk of sand. The stack must be of brick for a distance of 2 or 3 feet from the furnace, but may be of wrought-iron piping in its upper part.

A usual type of portable coke-muffle furnace is described and illustrated in the *Metallurgy of Gold* by T. K. Rose in this series. It consists of an outer casing of wrought-iron plates about $\frac{1}{4}$ inch thick, united by angle iron. The lining consists of Stourbridge firebricks.

B. Furnaces employing Gaseous Fuel.

Where illuminating gas or other gaseous fuel is available, assay furnaces, heated by gas, are now in general use. They are easily manipulated, and the work can be carried on in a very cleanly manner.

The use of gas instead of solid fuel has many advantages to recommend it, the chief drawback being the increased cost of working where illuminating gas is burned, as gas-furnaces are useless for assay purposes without a good pressure and a large supply-pipe. When these are secured, the gas-fired muffle furnaces will be found very convenient for assay-office use. For crucible assays, however, the ordinary wind furnace burning coke is preferable to a gas-fired crucible furnace, especially where a large number of fusions have to be made.

When coal-gas is used as a source of heat the combustion should be as complete as possible, so as to obtain the maximum temperature and prevent smoke. This is secured by using a burner constructed on the bunsen principle, and surrounding the flame with some refractory and non-conducting material such as fireclay. The gas-furnaces described below are constructed on this principle.

For successful working of gas-fired furnaces the gas-supply tap and pipe must be large and clear, so as to give as great a pressure of gas as possible at the burners. It is best to connect the furnaces throughout with iron piping, but if indiarubber tubing is used it must of necessity be perfectly smooth inside. The tubing made on wire, whether the wire is removed or not, will not feed the burners satisfactorily.

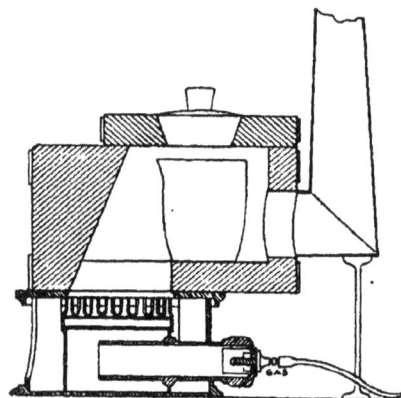

FIG. 11.

Gas-fired Crucible Furnaces.— Fig. 11 represents, in section, a Fletcher draught crucible furnace taking crucibles not exceeding 4 inches by 3·5 inches. It consists of a circular fireclay body bound with sheet-iron hoops, and a burner. The furnace is covered with a lid which can be lifted by a handle or can be pushed sideways sufficiently to enable the crucible to be lifted out. During fusion the contents of the crucible can be inspected by removing the small fireclay plug in the middle of the cover. The flame from the burner surrounds the crucible and then passes up the sheet-iron chimney.

The gas-supply required is 60 cubic feet per hour, equal to ½ inch clear-bore gas pipe and tap. When used for ordinary crucible assays the weight of the ore taken for assay is limited to 25–35 grammes, as this is the largest charge, with the necessary fluxes, that can be accommodated conveniently in the crucibles employed.

Gas-fired Reverberatory Furnace (Fletcher).—This furnace (fig. 12) consists of a rectangular fireclay chamber with a draught burner at one end and a chimney at the other. The floor-space in the furnace is 14 inches long, 6 inches wide, and 7·5 inches high, and will permit of several crucibles being heated at the same time. The furnace can be placed with the opening either at the side or the top, as it works equally well in both positions. The opening is closed with two fireclay doors. The furnace can be used for crucible fusions only, or can be made to take one muffle and two or three crucibles, or two muffles, at the same time. The muffles are 4 by 7 by 4¼ inches, and are placed across the furnace after removing the doors. With a chimney draught these furnaces give a temperature about equal to the fusing-point of silver, but with a blast cast iron can be melted.

The burner is at one end, out of the way of injury in case of accident to a crucible: it should be placed close up to the furnace when in use.

When the draught burner is used the blue cones of flame should be clearly seen on the burner: if they disappear, the gas-supply must be increased, or the slide over the burner air-tube closed until they reappear. In the latter case the furnace works with a smaller gas-supply at a lower temperature, and by closing this slide and reducing the gas-supply any temperature required can be obtained.

Fig. 12.

If the adjustment of gas and air is neglected, the burner grid becomes red hot, and is quickly rendered useless. When properly used, the grids will last for some years.

These furnaces are very convenient for general assay-office work when the amount of furnace work is not large.

Gas-fired Muffle Furnaces.—These furnaces are being very generally adopted for bullion assays, as they are capable of giving very uniform temperatures when proper attention is given to the adjustment of the burners. Fig. 13 illustrates a type of Fletcher gas-muffle furnace used in assaying. It is made in several sizes; the No. 8 size, with a muffle 8¼ inches by 5¼ inches by 14 inches, clear inside working space, is very suitable for general assay work. The furnace is made of stout sheet-iron, lined with refractory material, and standing on an iron frame with four legs. The top of the furnace is removable to permit of a new muffle being fitted when necessary. In some designs of gas-muffle furnaces the muffle is introduced through an opening or door at the back of the furnace instead of through the top. The muffle is heated by means of eleven large burners of the bunsen type, mounted on an iron frame and arranged in a row beneath the muffle.   A single burner is represented on a larger scale in fig. 14.

The flame from the burners passes through a rectangular hole in the bottom of the furnace, heats the muffle on all sides by direct contact, and passes finally into the chimneys at the top.

The furnace is provided with two chimneys 2 feet 6 inches in height, each having a cast-iron foot to enable it to stand steadily, and a damper by means of

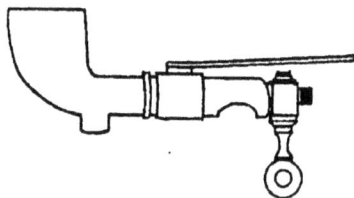

FIG. 14.

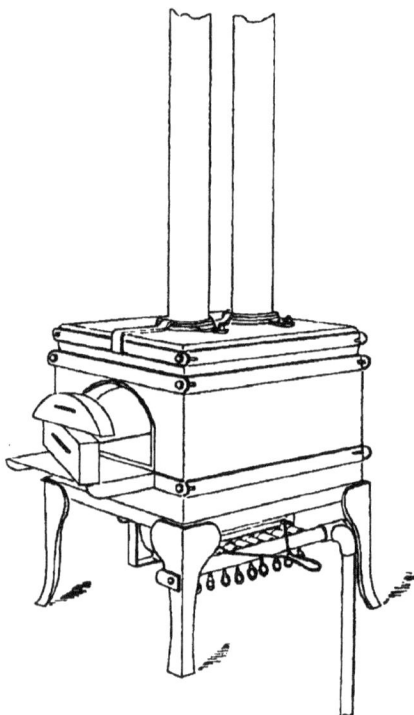

FIG. 13.

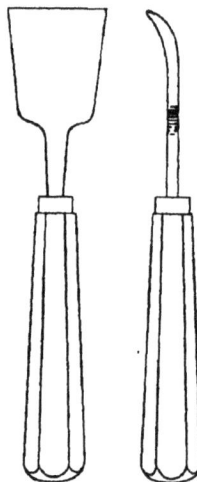

FIG. 15.

which the draught can be regulated.   The muffle door is made in two parts, each provided with a slot for handling by means of the holder shown in fig. 15.

The tang end of an old file fitted to a handle makes a very suitable holder for the purpose.

For exact temperatures a pressure governor is necessary, and it is very advisable to use a quadrant gas tap, with arm and pointer, for exact adjustment, as shown in fig. 18, page 33.   The most uniform temperature is obtained by working with some of the middle burners turned off slightly, and two or three burners at each end left full on : the burners are soon adjusted after a few trials. Each burner is provided with a slide for the regulation of the air-supply, the slides being fastened to a wire handle as shown, so that the air-supply may be adjusted or cut off from all the burners simultaneously.   When lighting the furnace, the air openings are covered with the slides, a lighted taper put to the

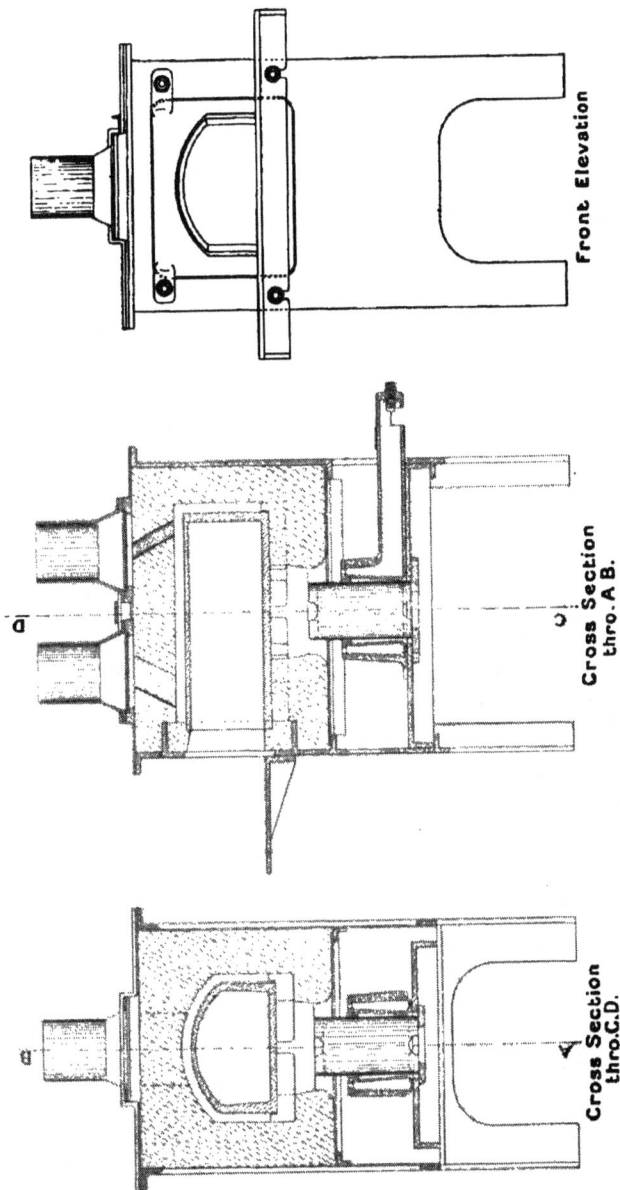

Fig. 16.—Natural draught muffle furnace (Richmond's, Warrington).

Front Elevation

Cross Section thro. A B.

Cross Section thro. C.D.

burners, and the gas turned on slowly.   When the gas is full on the air openings
are uncovered.

It is very important that a light should be put to the burners before turning
on the gas, or an explosion will take place, which may break the thin muffle or
otherwise damage the furnace.   The burners must be kept clean : care should
therefore be taken to cover them when a new muffle is being fixed in the furnace.

When the furnace is working at its full power, the flame must be just visible
in the chimney.   Muffles always require great care in heating and cooling, otherwise they very soon crack and become useless.   When the furnace is cooling
down after turning off the gas, the dampers should be shut to prevent a current
of cold air from cooling the muffle too quickly, and the doors should not be
closed too tightly, otherwise they are very liable to
stick fast owing to the contraction of the muffle.

Another form of gas muffle for assay purposes
is the Richmond natural draught muffle furnace
shown in two cross-sections and in front elevation
in fig. 16.   The furnace consists of a body of non-
conducting material encased in a casing of stout
sheet-iron braced together with flat and angle iron.

The muffle is heated by a large circular special
form of high-power bunsen burner, the burner
mouth being sealed from the air at the point of
entrance to the combustion chamber.   The primary
air is admitted to the burner in limited proportion.
As will be seen from the sectional drawings, the
secondary air is drawn into the furnace in such a
way as to serve two distinct purposes: (1) it is
heated in its passage before mingling with the
bunsen flame, and (2) the points of admission are
so arranged as to materially assist the up-draught
from the burner and intensify the heat.

The burner is so shaped as to emit separate
tongues of elongated flame and extract the utmost
duty from the secondary air.

For ensuring perfect combustion the burner is
fitted with an air regulator, adjusted *in situ.*

Each furnace is fitted with two flue pipes, as
shown, the length of which is adjusted to give
the correct draught.   The casing in the front of the furnace can be removed
by unscrewing the four nuts shown in the front elevation, thus enabling a new
muffle to be fitted when necessary.

The muffle and burner are enclosed so as to prevent loss of heat by radiation :
the operator is also protected from the heat, and the life of the muffle is prolonged.   ·

The muffle is closed by firebrick doors similar to those described above, but not
shown in the drawing.

The furnace is supplied in several sizes suitable for assay purposes.

Gas-furnaces should be worked under a suitable hood connected with a flue
to draw off the products of combustion, etc.   A sheet-iron hood for a two-chimney
muffle furnace is shown in fig. 17.   Instead of a hood the furnace may be
connected up direct with a flue in the wall by means of sheet-iron elbow-joints.
A convenient arrangement for a series of Fletcher gas muffles is illustrated in
fig. 18.

The furnace bench is of iron plates or of fireclay or stone slabs resting on brick
supports, the plates or slabs being placed as shown so as to give an open-air channel

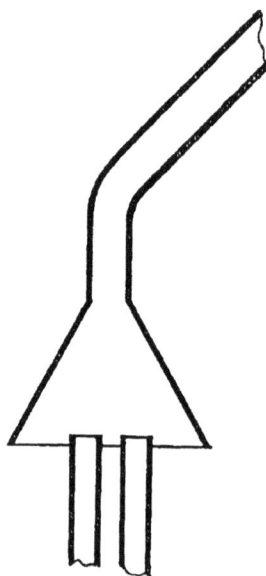

FIG. 17.

under the furnaces, which extends the whole length of the bench and is open at both ends. By this arrangement more space is given below the muffles, and the gas taps are more readily accessible for adjustment: it also allows of the gas-supply pipes being placed under the front part of the bench, where they are out of the way, thus leaving the working part of the bench free.

The hood is of sheet-iron and supported on brackets, the gases and fumes being drawn through three flues in the wall, the flues being connected with a chimney stack. To shield the operator from the heat radiated from the furnaces

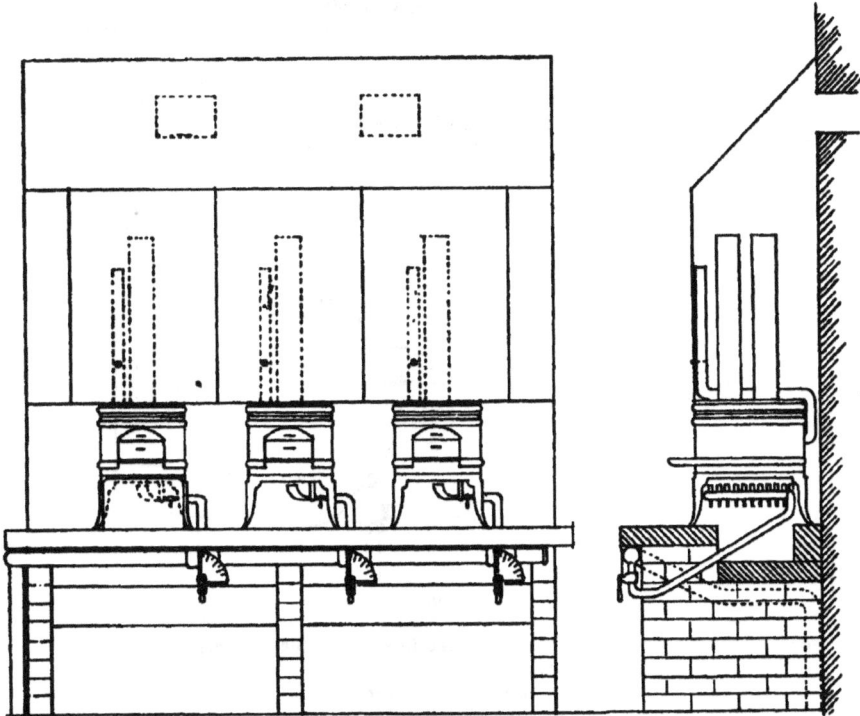

Fig. 18.

sheet-iron screens are suspended from the hood. These are hinged so that they can be thrown up out of the way when a new muffle is being fitted. A small movable sheet-iron screen is also placed in front of each furnace to keep the heat off the working bench: one of these screens resting against the legs of the furnace frame is shown in the end muffle on the left. A quadrant gas tap, with arm and pointer for exact adjustment, is fitted to each furnace and is almost indispensable in cases where the muffle has to be slowly cooled to prevent the "spurting" of silver assays.

When a new muffle has been inserted the best point for heating up and for cooling down the furnace should be carefully ascertained by a few trials and then recorded in some convenient place for reference.

C. Furnaces employing Liquid Fuel.

Furnaces heated by petroleum, gasoline, etc., are used when coal, coke, and gas are alike difficult to obtain.

Gasoline-fired (or hydrocarbon) furnaces are most extensively employed in out-of-the-way districts, as gasoline is easily transported and has great heating power. Furnaces of this type are seldom employed in this country, but in America they are in more or less common use for small assay offices where the amount of furnace work to be done is comparatively small.

The apparatus required for heating a furnace with gasoline consists of a special form of burner, and a steel tank, of 5 or 10 gallons capacity, provided with an air pump to furnish the necessary pressure. A pressure gauge is attached to the tank. The furnaces are similar in construction to those used for gaseous fuel, and are made of fireclay bound with sheet-iron. There are many varieties of furnaces: some are intended for crucible fusions only, others for muffle work

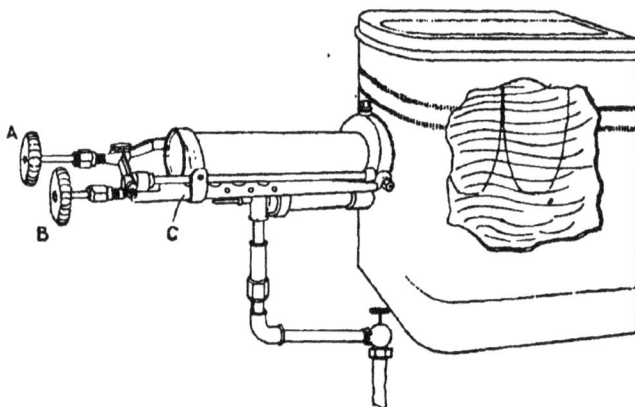

FIG. 19.

or for both muffle and crucible work, the furnace in this case being divided into crucible and muffle compartments.

The Carey burner shown in fig. 19 is in very general use with these furnaces. It is made of brass and copper, and provided with a generating device. The upper valve A controls the main gasoline supply, and the lower valve B controls the generator. "The burner is heated by the generator, so that the gasoline issuing from the main needle-valve is vaporised, and in its passage to the furnace draws in air through the burner tube, the mixture igniting and burning at the mouth of the burner in the hot furnace." [1]

The needle-valve is ordinarily used in burners of this type: it consists of a pointed hard steel needle which works in a circular valve orifice, making an annular opening for the escape of the gas. This annular opening varies in dimensions according to the position of the needle, and may be closed completely by screwing the needle up as far as it will go.

To generate or start the Carey burner a small quantity of gasoline is allowed to flow into the trough C by turning the wheel B, and the gasoline ignited with a match and allowed to burn. The valve B is turned on and off at intervals of a

---

[1] Fulton, *Manual of Fire Assaying.*

few seconds to keep the flame going until a clear blue flame burns around the head of the burner, which blue flame indicates that the burner is sufficiently generated and ready for use. This requires about three minutes. The burner should fit tightly against the fireclay ring or boss in the opening of the furnace, so that all the air from the combustion of the gasoline is drawn in through the burner tube. The gasoline is fed to the burner under a pressure of 10 to 20 lbs., though for special purposes higher pressures are employed. To avoid accidental explosions it is best to place the gasoline tank and pump apparatus at a considerable distance from the furnace and join up with 0·25 inch- to 0·375-inch piping. Full and complete instructions always accompany each burner sent out. According to Fulton, a 2-inch Carey burner, under 10 lbs. pressure, will consume from 0·65 to 0·75 gallon of gasoline per hour. A little experience is necessary before these furnaces can be successfully manipulated.

CHAPTER IV.

FURNACE IMPLEMENTS AND APPARATUS USED IN FURNACE OPERATIONS.

I. Furnace Implements.

COMPARATIVELY few tools are required for assay furnaces: the more important are pokers for stoking the fire, tongs for handling crucibles, etc., and iron moulds in which to pour assays, etc.

Pokers.—These are made of stout round bar-iron $\frac{1}{2}$ inch diameter, and are used for stirring the fuel from the top of the furnace and for opening the fire-bars when choked with clinkers and ashes. Two straight pokers, one 30 to 36 inches long and a shorter one 18 inches long, are very suitable. For the purpose

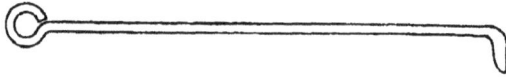

FIG. 20.

of clearing the bars from underneath a poker 36 inches long and bent at the end is advisable (fig. 20).

Tongs.—The features essential in tongs are that they be as light as possible, consistent with strength, grasp the crucible, etc., firmly without danger of tipping, and take up little room in the furnace. Many of the tongs sold for assay purposes are too heavy and clumsy and quite unsuitable; it is usually more satisfactory to have them made by a good blacksmith. The following different kinds of tongs are used, viz. crucible tongs, scorifier tongs, and cupel tongs.

Crucible Tongs.—Several forms of crucible tongs are required. A useful length for these is 24 inches. A very convenient pair of crucible tongs to grasp the sides of the crucible, and operating in little space, is shown in fig. 21. This form of tongs is the one commonly used for furnace work.

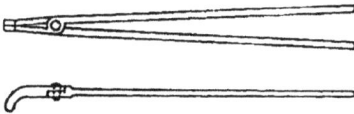

FIG. 21.

Two forms of crucible tongs designed to grasp the body of the crucible are shown in figs. 22 and 23. These tongs are sometimes useful for pouring the contents of a crucible, but owing to the space they take up in opening they cannot be conveniently manipulated when there are several crucibles close together in a large wind furnace or in a muffle.

For lifting large crucibles the basket tongs (fig. 24) are used.

Fig. 25 illustrates a convenient form of crucible tongs for handling the

"Colorado"-shaped crucibles used for the fusion of assays in the muffle (see fig. 59, page 44). They are 24 inches to 30 inches long.

The scissor tongs shown in fig. 26 are very useful for packing coke round a crucible and for handling small crucibles. A useful size is 20 inch.

FIG. 22.   FIG. 23.

Scorifier Tongs.—These are used for handling the scorifiers described on page 45. For this purpose the lower arm of the tongs is divided near the end into two prongs, while the upper arm is made straight throughout. The lower arm fits the bottom of the scorifier, the long upper arm extending across the top.

FIG. 25.

FIG. 24.   FIG. 26.

Scorifier tongs are of different shapes and sizes: those represented in fig. 27 will be found generally suitable. The length is usually about 24 to 28 inches. The guide in the middle of the tongs should be so constructed that the arm moves *inside* the guide as shown. Scorifier tongs and cupel tongs (described below) are frequently made with a guide running through a slit or hole in one

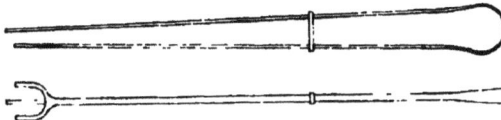

FIG. 27.

arm of the tongs; this arrangement is unsatisfactory, as the guide tends to stick owing to rusting or through being bent accidentally. Two or three pairs of scorifier tongs of different sizes are generally required to accommodate the various sizes of scorifiers. Special forms of scorifier tongs have been introduced from time to time but are not in very extensive use.

Cupel Tongs.—These are used for manipulating cupels (see page 45). Two common forms of cupel tongs are shown in figs. 28A and 28B. They are made of different sizes, but tongs about 28 inches long will be found generally suitable. Some assayers prefer cupel tongs that are curved at the ends so that they fit round the body of the cupel.

Stirring Rod.—Stirrers are used in roasting operations. A round rod of iron about a quarter of an inch in diameter and 24 inches long, flattened out

FIG. 28A.

FIG. 28B.

at one end, is generally used as a stirrer. When required for roasting in a muffle, the flattened end is bent at right angles for the distance of about an inch. Two forms of stirrers are shown in figs. 29 and 30.

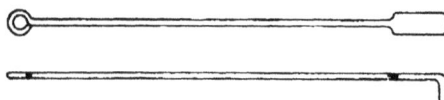

FIG. 29.

Muffle Scraper.—Two forms of scraper in general use are shown in figs. 31 and 32. They are 24 to 30 inches long, and consist of a small iron plate riveted to an iron rod handle. They are used for introducing bone-ash or sand into the muffle and smoothing down the surface of the floor of the muffle. The scraper

FIG. 30.

with the end bent at right angles is employed for removing the pasty mass that forms on the floor of the muffle after frequent use or through the spilling of an assay.

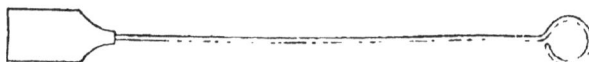

FIG. 31.

FIG. 32.

Scoop.—This is generally made of copper, and is employed chiefly for charging crucibles ; it is shown in fig. 33. A scoop about 9 to 12 inches long, not including the handle, is a convenient size. It should be kept smooth and bright so that the assay charge may flow freely into the crucible.

Cupel Trays.—These are used to hold cupels when carrying them to or from the muffle. Cupel trays are generally made of sheet-iron with or without partitions. Two different forms are shown in figs. 34 and 35.

Moulds.—Moulds for receiving the contents of crucibles, scorifiers, etc., are of various shapes and sizes. They are either open or closed, and are generally made of cast iron: open moulds have either hemi-

Fig. 33.

spherical or conical cavities, as shown in fig. 36 and fig. 37. When large numbers of assays have to be made, a mould of thick sheet-iron, containing

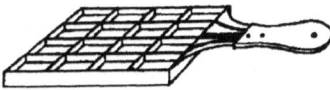

Fig. 34.

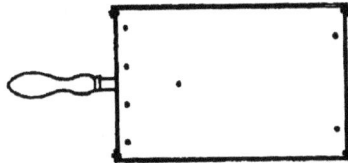

Fig. 35.

9 or 12 or more hemispherical cavities, formed by hammering, is frequently used (fig. 38).

A common form of ingot mould for receiving comparatively large quantities of molten metal is shown in fig. 39.

When thin flat ingots of metal are required, a closed mould (fig. 40) is used.

Fig. 36.

Fig. 37.

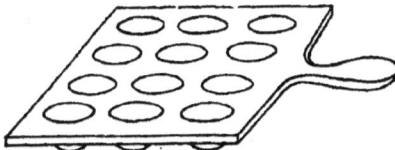

Fig. 38.

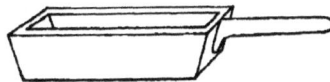

Fig. 39.

The various parts are detachable, and can be adjusted so as to give ingots of different sizes.

All moulds should be cleaned thoroughly inside before being used. Spherical moulds are usually cleaned by rubbing with plumbago powder. This is best done by means of a "plug" consisting of a piece of stick rounded at one

end and covered with a piece of leather or cloth like a drum-stick. The plug
is dipped into plumbago powder and then rotated
several times in the mould. Care must be taken
to remove all powder from the mould before use.

Ingot moulds may be blackleaded inside, although it is more usual to prepare these by wiping
with an oily rag. Small ingot moulds may be
prepared by holding them over a gas or oil flame
so that the inside becomes coated with a thin
deposit of "soot" or "lampblack."

Miscellaneous Appliances.   Anvils. — Two
anvils of hardened steel are usually required, one
for hammering the lead buttons from assays to
free them from slag, etc., and the other for flattening assay beads of gold and silver. An anvil or
stake for cleaning lead buttons is shown in fig. 41.

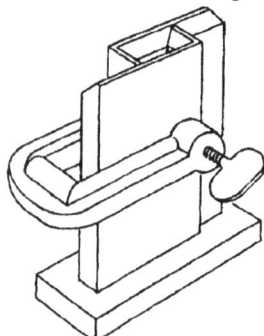

Fig. 40.

It has a hard smooth surface 4 or 5 inches square
and a fang for fixing it securely into a wood block. The anvil should be
mounted on a strong bench in the furnace room and so arranged
that the slag, etc., when hammered off is collected and not
allowed to fall on to the floor and become a nuisance. The
anvil is best mounted on a solid block of wood which extends
to the floor. A suitable bench may be made of strong sheet-
iron, supported on angle iron. The top should measure about
4 feet by 2 feet 6 inches, and should be turned up at right
angles at both ends and at the back to form a flange $2\frac{1}{2}$ to
3 inches high. The legs are quite firm if made of $1\frac{1}{4}$-inch angle
iron braced up with $\frac{3}{4}$- by $\frac{1}{4}$-inch iron strip riveted on.

Two forms of anvil or "slagging" bench are shown in cross-
section in fig. 42 and fig. 43. In fig. 42 the anvil is fixed in the
middle of an oblong trough placed in front of the bench as shown,
so that the slag as it is hammered off falls into the trough. The
top of the bench in this case may conveniently measure 4 feet by 1 foot 6 inches.

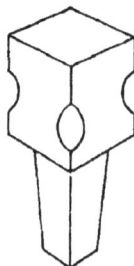

Fig. 41.

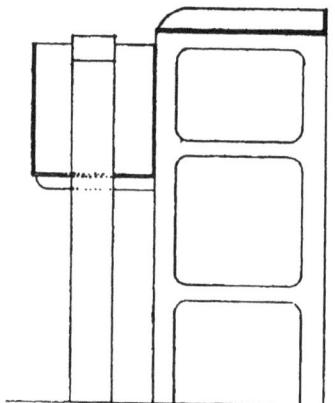

Fig. 42.

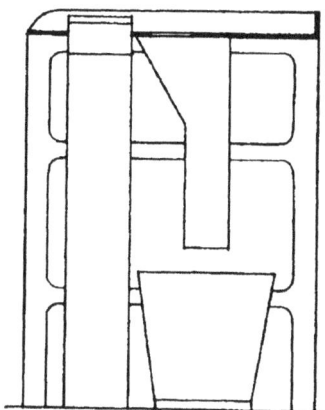

Fig. 43.

A small trap or door is provided at the bottom of the trough to permit of
the ready removal of the slag that accumulates from the day's work.

The anvil shown in fig. 43 should be from $6\frac{1}{2}$ to 7 inches square and fixed into the end of a 10-inch square timber standing on the floor, so that it projects through the bench top about $1\frac{1}{2}$ to 2 inches. As a preventive against slag flying about a flange about $\frac{3}{4}$ inch thick and $\frac{1}{4}$ inch high should be cast on three sides of the anvil except about 3 inches along the back, measuring from one corner, to allow of the broken slag being readily brushed away. Behind the anvil, about 2 inches away from it and in direct line with the opening in the flange, a hole, $2\frac{1}{2}$ or 3 inches square, is cut in the bench top, and through this the slag is brushed so that it falls into a bucket or box placed to receive it under the bench.

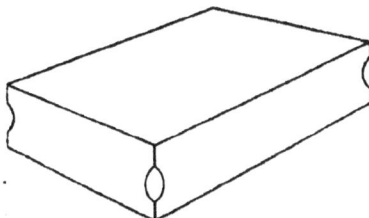

Fig. 44.

The anvil (fig. 44) used for flattening large beads, etc., of gold or silver should be of hardened steel with a face about 6 to 10 inches square. It should be firmly fixed in a block of wood. The face should be highly polished and kept bright by rubbing with an oiled rag and by being covered with a suitable wood cover when not in use.

Hammers.—For use with this anvil it is well to have two hammers of the shape shown in fig. 45, the heads of which weigh 7 and 11 lbs. respectively. The faces of the hammers should be highly polished like the anvil and kept bright: it is a good plan to have an anvil cover sufficiently large to allow of the hammers being kept under it with the anvil.

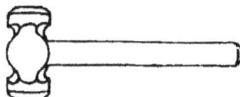

Fig. 45.

To prevent loss when hammering brittle metal it is advisable to have an iron ring about 8 inches in diameter and $1\frac{1}{4}$ inches high to stand on the anvil as occasion requires.

Shears.—A strong pair of shears about 9 inches long with a good cutting edge are generally useful for cutting metals, such as sheet gold, silver, and lead.

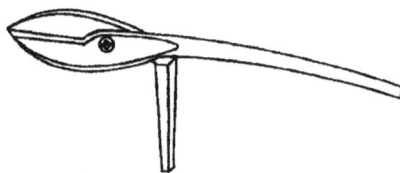

Fig. 46.

For cutting hoop-iron, required in certain assays, a large pair of shears with a spike for mounting on a wood block as shown in fig. 46 is desirable.

Chisels.—The cold chisel (fig. 47) is employed for cutting off portions of metal for assay, etc. It should be 5 or 6 inches long, and about $\frac{1}{2}$ to $\frac{3}{4}$ inch wide, which is the best size for general use. However, for some purposes, as when cutting very tough metals, it is

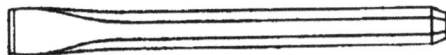

Fig. 47.

convenient to have a chisel only $\frac{1}{4}$ or $\frac{3}{8}$ inch wide, as these metals are so much more difficult to cut, and the small chisel meets with less resistance.

Pliers.—Several forms are used for removing gold and silver buttons from cupels and squeezing them to loosen adhering bone-ash. Half-round nose pliers (fig. 48) are useful for small buttons, while flat-nose pliers (fig. 49) are generally used for large buttons. The nose of these pliers may be conveniently filed to the curvature of the surface of the cupels used.

Cutting Nippers.—These are employed for cutting metal that cannot be cut conveniently with shears. The form of cutters shown in fig. 50 is commonly used. The spring-cutting nipper, fig. 51, is a very suitable form of cutter, as it puts less strain upon the hand. The steel jaws are held in position by screws, and can be adjusted for narrow opening or for wide opening, as illustrated. The jaws can be easily replaced when necessary.

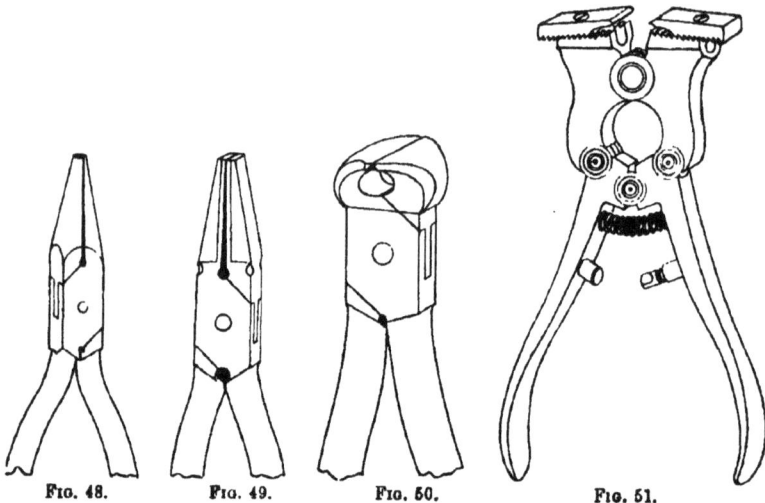

FIG. 48.   FIG. 49.   FIG. 50.   FIG. 51.

Brushes.—Hard stiff brushes are used for cleaning silver and gold buttons from adhering bone-ash. For small buttons a hard tooth-brush is frequently used.

FIG. 52.

A form of brush known as the Freiberg pattern consists of a bundle of stiff bristles fitted into a metal tube (fig. 52). Another form of brush (fig. 53) is made of hog bristles set at a slight angle as shown at A, fig. 53. By mounting the bristles in this way the brush retains its stiffness until worn out, and does not spread out as is usual with brushes in which the bristles are mounted in the ordinary way.

A   FIG. 53.

Punches and Dies.—Good-quality steel punches are used for stamping bullion. They should comprise the numerals from 0 to 9 and the letters of the alphabet, and may vary in size from $\frac{1}{32}$ inch to $\frac{5}{8}$ inch. In addition to these it is usual to have certain word-punches in one piece, such as "Gold," "Silver," "Fine," "Value," "Total," "No.," "Oz.," and "$." Special punches with name or mark of the assayer, mine, or company are also in use and can be procured as desired.

II. Apparatus used in Furnace Operations.

The principal apparatus used in assaying by furnace methods consists of crucibles, scorifiers, roasting dishes, and cupels. The materials from which these are made vary according to the objects for which they are to be used.

Crucibles.—These are vessels made of some refractory or heat-resisting material, for the fusion of assays, melting of metals, etc.

The essential qualities of crucibles to be used for assay purposes are as follows: (1) they must be able to resist a high temperature without softening; (2) they must not be friable when hot; (3) they must be capable of withstanding sudden changes of temperature without cracking, as, for example, when a white-hot crucible is brought out of a furnace into cold air; (4) they are required to resist the corrosive action of metallic oxides that may be contained in the material being fused.

Crucibles are made of fireclay mixed with burnt clay, sand, plumbago, or other infusible material, in order to counteract the tendency to shrink when drying which raw clay possesses. The various substances thus mixed with the raw clay do not contract on drying; by their addition, therefore, contraction on drying is considerably reduced.

The crucibles are moulded, dried, and then burnt in kilns. All crucibles should be cautiously annealed before use by placing them in an *inverted* position over the furnace, otherwise they are liable to crack when placed in a red-hot fire.

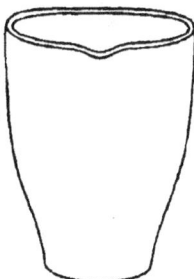  

FIG. 54.　　　　FIG. 55.　　　　FIG. 56.

The crucibles best known in commerce are the varieties designated London, Battersea, Cornish, Hessian, and French. The sizes and shape of crucible best suited for assay purposes have been indicated when giving the charges for crucible work.

London Crucibles (fig. 54).—These are round in shape, and are preferred by many assayers on account of their very refractory nature. They resist the corrosive action of fused metallic oxides better than most clay crucibles.

Battersea Crucibles (fig. 55).—These are similar in shape to the London crucible, but a little taller. They are fairly refractory, and are largely used for the fusion of gold and silver ores. For silver-ore assays the "Battersea Round," E or F sizes, are generally used, while for gold ores the G size is most suitable.

Cornish Crucibles (fig. 56).—These are round, short, and somewhat open crucibles used sometimes for the roasting of ores. They are fairly refractory, though liable to be corroded by metallic oxides. The usual size employed is No. 79, which is 3¾ inches high and 3¼ inches diameter at the top.

Hessian Crucibles (also known as "London Triangle").—These are triangular in shape, and have three spouts; from any of which the contents can be poured. For this reason they are sometimes preferred, but they are not largely employed by assayers. They are fairly refractory, and resist the corrosive action of metallic oxides very well. They are very liable, however, to crack with sudden changes of temperature, and to break when held with the tongs. A crucible known as the " Battersea London Triangle " is another and somewhat more convenient form of triangular crucible (fig. 57).

French Crucibles.—Fig. 58 shows the form of the French clay or " fluxing pot." These crucibles are very carefully made, being smooth, close in grain, and of good quality. They are somewhat brittle, but will withstand high temperatures without softening, and resist the corrosive action of metallic oxides remarkably well. In the assay office they are used occasionally for melting small quantities of gold, silver, and other metals.

Colorado Crucibles (fig. 59).—These are special-shaped crucibles used for the fusion of gold and silver ores in the muffle. They are made of the same material as " Battersea Round Crucibles." The various sizes are specified by their gramme capacity, that is, by the number of grammes of ore with the proper amount of fluxes necessary for fusion which the crucible will hold. The chief sizes have capacities of 5, 10, 12·5, 15, 20, 30, and 40 grammes. The sizes most frequently used are the 20-gramme crucible for 0·5 assay-ton charges, and the 30-gramme for 1·0 assay-ton charges.

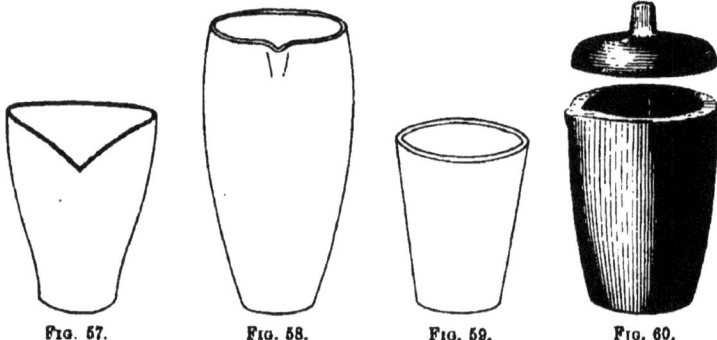

FIG. 57.   FIG. 58.   FIG. 59.   FIG. 60.

Plumbago Crucibles.—The so-called plumbago crucibles are made of raw fire-clay and graphite, in the proportion of 51 of the former to 49 of the latter. The usual shape of these crucibles is shown in fig. 60. Good plumbago crucibles, after a preliminary careful annealing, withstand the greatest changes of temperature without cracking, bear the highest heat without softening, and may be used repeatedly.

They are the least corroded by metallic oxides, but, owing to the reducing property of the carbon they contain, they must not be employed when oxidising actions are required.

The careful annealing of plumbago crucibles is very essential, as they have a very great tendency to split when plunged into a hot fire without proper annealing.

In order to prevent absorption of moisture and salts—as, for instance, in orders for shipment—and to enable rapid alterations of temperature to be better withstood, the finished crucibles are in some cases dipped in milk of clay, dried, glazed, and then dipped in tar. Crucibles made in this way by the Battersea Crucible Company are known as Salamander crucibles ; they will stand forty meltings of gold without sensible deterioration. ·(Roberts-Austen.)

Plumbago crucibles possess many advantages over clay crucibles, and they have almost entirely superseded the latter for melting gold and silver. Their surface is very smooth, and permits of the contents being poured clean.

Scorifiers.—Scorifiers are shallow, circular, cup-like vessels made of fireclay, in which the substances to be subjected to scorification are placed.

Fig. 61 shows a section of the ordinary type of scorifier. They are made of various sizes designated by their outside diameters. The sizes ordinarily made are 1·5 inches, 2 inches, 2·5 inches, 3 inches, and 3·5 inches. The 2·5-inch scorifier is the one most frequently used.

Fig. 61.

The proper use both of scorifiers and of cupels (described below) will be described under the head of Silver Assay.

Good-quality scorifiers are very essential for assay work. A scorifier should resist as much as possible the corrosive action, at high temperatures, of molten lead oxide, either by itself or in admixture with other highly corrosive substances, such as oxide of copper or of iron. It should not crack on exposure to great and sudden changes of temperature. It should be wide across the top, so that a large surface of the materials operated upon may be exposed to oxidation; and so deep in the centre that a considerable proportion of the lead used in the operation may be kept well covered and preserved from oxidation.

Scorifiers should be well annealed before use.

Fig. 62.

Roasting Dishes.—Roasting dishes are flat, shallow, circular vessels of fireclay (fig. 62). They are sometimes employed for the roasting of ores, etc., containing sulphides or arsenides, an operation which may be conducted in a muffle, or by placing the roasting dish in front of the flue of an air-furnace. Roasting dishes 4 inches and 5 inches in diameter are the sizes most generally useful.

Cupels.—These are cup-like vessels in which the operation termed "cupellation" is carried on. In old treatises of assaying they were termed "copples." Both cupel and coppel are derived from the French word "coupelle," which means a little vessel in the form of a cup (Percy). Cupels are generally round, and more or less tapered towards the base, but square cupels are also used. Cupels of the usual form are represented in section in figs. 63 and 64. They must be made of such substances as are not acted upon by certain fused metallic oxides, as those of lead and copper, and their texture has to be sufficiently loose to allow of the molten oxides penetrating their substance readily, and yet be sufficiently strong to bear handling without breaking.

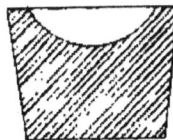

Fig. 63.

Fig. 64.

There are several substances of which cupels are made which fulfil all these conditions, but only one is in general use, viz. the ash of burnt bones (bone-ash). This consists principally of calcium phosphate, with a little calcium carbonate and fluoride.

Other substances which have come into use in recent years for the manufacture of cupels are magnesia (MgO) and, to a limited extent, Portland cement.

Preparation of Bone-Ash.—The best bones for the preparation of bone-ash are stated to be those of sheep and horses, and next, those of calves and oxen (Percy).

They are calcined with free access of air, but before burning them it is advisable to boil them repeatedly in water to dissolve organic matter. If the bones are not quite white after the first ignition, but contain a little carbon, they should be ground up, moulded into small balls, and burned again. Care should be taken not to heat the bone-ash too strongly, or it will have a smooth, glassy fracture, and will not be sufficiently spongy or absorbent to make good cupels. When the bones are burnt white throughout, they must be finely ground, sifted, and washed several times with distilled water till all soluble salts (sodium carbonate and sodium chloride) are removed.

According to Bettel,[1] the bone-ash should be specially treated by washing with an aqueous solution of ammonium chloride, this salt being used to the extent of 2 per cent. of the weight of the bone-ash to be treated. The bone-ash, thus purified, is dried, again pounded, and passed through a sieve of 40 mesh. The fineness of division of the bone-ash is of much importance. If it is too coarse, the cupel will be too porous and permit of a comparatively large absorption of silver and gold. If the bone-ash is too fine, the cupels made from it will be too dense, the cupellation will be prolonged and tend to increase the loss by volatilisation. Cupels too fine in grain are, moreover, very liable to crack when heated, and so cause loss of silver.

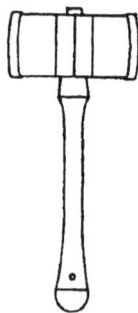

Bone-ash cupels do not bear transport well, hence the assayer has generally to make them, or to supervise their making. They can, however, be purchased: those bearing the well-known name Deleuil, Paris, are exceptionally good and especially suitable for bullion assays. When small quantities only are required they are usually made in moulds by hand, but as cupels should be as uniform as possible as regards density, they are best made by machine as, by machinery, a constant pressure may be obtained. When large quantities are required a machine is generally employed in preference to hand moulds.

FIG. 65A.   FIG. 65B.

Manufacture of Cupels.—The process is conducted as follows :—Finely ground bone-ash is moistened with water and thoroughly kneaded, so that the mass may be of uniform consistence throughout, and quite free from lumps. The material is best worked up by rubbing between the hands. The bone-ash is sufficiently wet when it coheres by gently squeezing it in the hand. If too much water has been used, the cupel will be too dense and compact; if too little, it will be tender and friable.

As a general rule, bone-ash requires the addition of about one-twelfth its weight of water. It is very important, for the reasons stated on page 182, that cupels should not contain carbonaceous matter or carbonates; and therefore, in making them, the bone-ash should not, as sometimes recommended, be mixed with stale beer, or water containing adhesive substances such as molasses, or with alkaline carbonates, etc. Nothing but pure water is necessary, and when the correct proportion is added, the results are perfectly satisfactory.

Cupel moulds (fig. 65A) are generally made of turned iron or gun-metal. They consist of three parts : (1) a hollow cylinder, (2) a disc of metal, and (3) a plunger or plug for compressing the bone-ash and shaping the top of the cupel. The disc (not shown in fig. 65) forms a false bottom for the cylinder.

In forming the cupel, the disc is put in its place, and the cylinder filled with the moistened bone-ash. The plunger is then placed in position and the bone-ash compressed by three or four smart blows from a wooden mallet (fig. 65B).

[1] W. Bettel, *Proc. Chem. and Met. Soc. of S.A.*, xi. 599.

Before removing the plunger it should be turned half-way round upon its axis so as to free it and to smooth the face of the cupel. The cupel is then carefully pushed out by pressing up the disc of metal forming the false bottom.

The removal of the cupels is more easily effected if the mould is somewhat conical, instead of cylindrical, in form. The thickness of the metal disc is varied according to the desired thickness of the cupel. When cupels with a fine smooth surface are required, a thin layer of very fine bone-ash is placed in the mould on top of the coarser bone-ash before inserting the plug. In this way the cavity of the cupel becomes coated with a layer of finer material. The cupels are placed on iron or wooden trays and well air-dried by storing in a warm place. If possible they should be allowed to dry for two or three months before using. It is important to preserve cupels from the access of nitrous or other acid fumes. These are absorbed by the bone-ash and given out in the furnace, so that they may cause "spitting" (see page 181). The mould after use should be oiled to prevent rusting.

Cupels vary in size according to requirements: those upon which buttons of 15, 30, and 40 grammes of lead respectively can be cupelled are most generally useful. A cupel should be capable of absorbing its own weight of molten litharge: this will serve as a guide in the selection of cupels of suitable size.

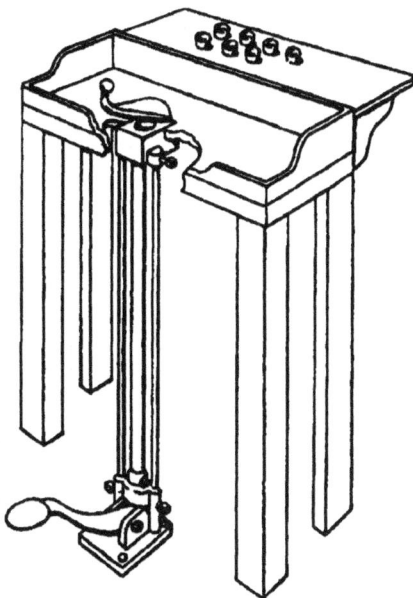

Fig. 66.

The weight of lead absorbed by different-sized cupels of the shape shown in fig. 64 is as follows:—

| Diameter at top . . . | $\frac{3}{4}$, $\frac{7}{8}$, 1, $1\frac{1}{8}$, $1\frac{1}{4}$, $1\frac{3}{8}$, $1\frac{1}{2}$, $1\frac{5}{8}$ inches. |
| Absorption in grammes | 3, 5, 8, 10, 16, 20, 28, 40. |

If the taller cupels of the shape fig. 63 are used, a cupel of a given diameter will absorb a larger weight of lead than that given above for a cupel of the same diameter of the shape shown in fig. 64. When a very large number of cupels are required, a press may be conveniently used.

A simple form of cupel machine is shown in fig. 66. It consists essentially of an ordinary cupel mould as above described, placed in an inverted position with the plug attached to a long rod connected with a pedal fixed to the floor. To make the cupel the mould is filled with bone-ash, then closed by means of the sliding cover shown, and pressure applied to the plug by means of the pedal, thus forming the cupel in an inverted position. When made the cupel is released by sliding the cover on one side and pushing the cupel out by a slight pressure of the pedal.

Considerable pressure may be used with this machine, and the cupels made quite firm.

Another type of machine is Calkin's automatic machine, illustrated in fig. 67.

This machine consists of a hopper to hold the moistened bone-ash and a removable disc with a number of holes which are automatically filled and brought into position under the plunger. To operate the machine, the bone-ash is properly moistened and placed in the hopper, in which there is a small wheel with a rubber rim that keeps the material stirred up and fills the moulds. The lever handle is then raised, the handle on the lower disc being grasped at the same time, and the filled mould is brought beneath the plunger; the lever handle is then forced down. The downward motion of the lever handle compresses the cupel, and by pulling the disc handle in a reverse direction to that formerly given it, the opening in the lower disc is brought beneath the cupel, and further pressure on the lever handle brings a new system of levers into action, ejecting the cupel, which may be caught in the hand. An automatic attachment stops the disc at the proper points, and an adjusting device is arranged for giving different degrees of compression.

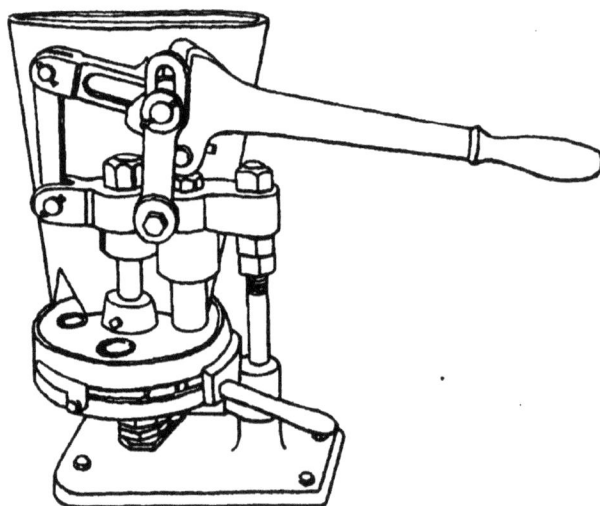

Fig. 67.

Cupels of various sizes may be made by using interchangeable discs and dies, which are easily adjusted to the machine. Perfect edges and a homogeneous cupel can always be obtained. It is claimed that 600 cupels an hour can be made with this machine.

A smaller machine to be bolted to a wall, post, or other convenient support can also be obtained. The same principle of leverage is applied, but it is necessary to feed the bone-ash by hand.

Square cupels are used in some assay offices. The cupels in use at the Royal Mint, London, are square blocks of bone-ash, a little less than 2 inches across, and a little more than three-quarters of an inch deep. Each block carries four circular cavities about 0.7 inch in diameter and 0.3 inch deep. The muffles in use will take six rows of these blocks placed three abreast, thus making provision for seventy-two assays at one time.

A mould for making such cupels is represented in figs. 68 and 69. It will be seen that the mould is in three pieces, and the cupel blocks are made with the cavities downwards. To obtain the necessary pressure the mould is placed in the press shown and the lever pulled down.

A mould for making a block with one large cavity only is also shown.

Magnesia Cupels.—Magnesia was introduced a few years ago, as a substitute for bone-ash, and a large number of brands of so-called "patent" cupels made with a magnesia base are now on the market. The magnesia (MgO) is produced by strongly calcining crude mineral magnesite (magnesium carbonate, MgO,CO_2). The cupels require to be made under high pressure, and should be baked at a high temperature before use. They are then very hard and firm and of a brown colour, resembling fireclay. Owing to the high pressure required (usually hydraulic), magnesia cupels cannot satisfactorily be made in the laboratory like bone-ash cupels. The exact composition of these cupels is generally a trade

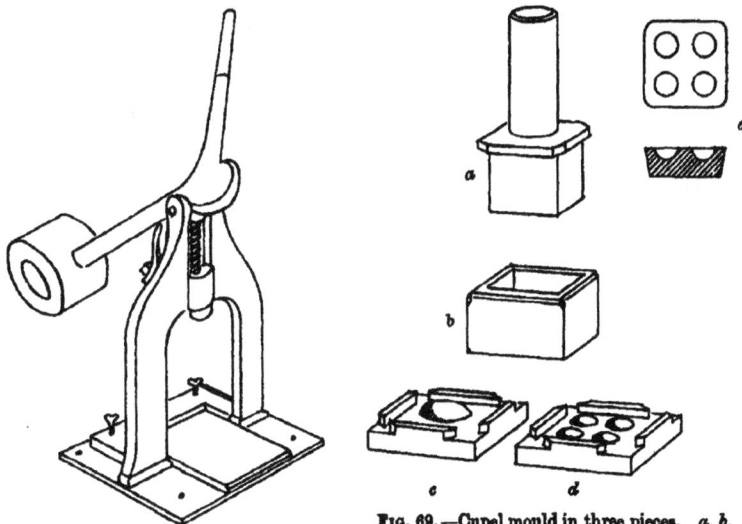

FIG. 68.—Cupel press.

FIG. 69.—Cupel mould in three pieces. *a, b,* and *c,* for cupel with one cavity ; *d,* for cupel with four cavities ; *e,* cupel with four cavities. (Plan and vertical section.)

secret ; several brands contain a certain proportion of bone-ash. The different brands include the Morganite, made by the Morgan Crucible Co., London, the Mabor, Cupellite, Scalite, Velterite, Excelsior, Star, and Elkan brands. The "patent" cupels bear transport well, and on that account have come into wide use in South Africa and other mining centres. Compared with bone-ash cupels their cost is higher in almost all cases. Magnesia cupels absorb and hold less litharge than cupels made of bone-ash.

The properties of magnesia cupels are discussed on page 166.

Portland Cement Cupels.—According to Holt and Christensen,[1] ordinary Portland cement forms a convenient substitute for bone-ash in making cupels, provided the amount of water added is carefully adjusted. The amount of water added should be between 6 to 10 per cent. of the weight of the cement: 8 per

[1] " Experiments with Portland Cement Cupels," T. P. Holt and N. C. Christensen, *Eng. and Min. Journ.*, 1910, xc. 560-561 ; also " Cement *versus* Bone-ash Cupels," J. W. Merritt, *Min. and Sci. Press.*, 1911, c. 649.

4

cent. is the most suitable amount.   If less than 5 per cent. of water is used the cupels are too fragile ; with 20 per cent. of water the cement cannot be readily moulded in the cupel machine.

Cupels with less than 5 per cent. and more than 15 per cent. of water crack at the edges when heated.   Cement cupels are strong, and will absorb their own weight of litharge ; the loss of silver by absorption is stated to be about the same as with bone-ash.   The cupels should be thoroughly dried before use, otherwise they will develop cracks during heating.

Cupels made of a mixture of equal parts of cement and of bone-ash were found to be in some respects superior to either cement or bone-ash cupels. Compared with bone-ash, cement forms a very cheap substitute for making cupels and is frequently more easily procured.   The cupels may be made entirely of cement, or the lower portion from one-half to three-quarters only made of cement and the upper portion of bone-ash or of "patent" cupel material such as "mabor" or "cupellite."

In this case the required quantity of moistened cement is put into the cupel mould and bone-ash added to fill the mould : the whole is then compressed in the ordinary way.

A very smooth surface can be given to the cupels by dusting them over with very finely powdered bone-ash or cupellite (about 200 mesh) just before compression.

CHAPTER V.

BALANCES AND WEIGHTS: WEIGHING.

Construction.—The balance (see figs. 71 and 73) consists essentially of a rigid metal beam suspended slightly above its centre of gravity. The scale-pans are suspended at either end of the beam, and equidistant from its centre. In order to diminish the friction of the working parts at each point of suspension, the beam is suspended at the centre on a knife-edge of agate (or hardened steel), which rests on a polished agate plane fixed to the brass support. At each extremity of the beam a similar knife-edge is fixed in the reverse position, and bearing an agate plate from which, by suitable hooks and light wires, the pans are suspended. The movements of the beam are indicated by a vertical pointer or needle which oscillates in front of an ivory scale, graduated usually into twenty parts, and fixed so that its middle point or zero is exactly behind the needle when the beam is horizontal.

Any inequality in the weight of the arms is compensated by means of a small vane or flag fixed on the top of the beam above the central knife-edge, which may be turned to the right or left as occasion requires, and sometimes by an adjusting screw fixed at each extremity of the beam. The stability of the beam is regulated by the aid of a small ball weight, termed the gravity-bob, which is threaded on a vertical rod attached to the upper edge of the beam over the centre knife-edge on which the beam works.

The beam is divided into an equal number of divisions, frequently 100, to admit of the use of a small weight or rider, described later, and each division is equivalent to a certain fraction of a milligramme according to the weight of the rider used. The rider is manipulated by means of a movable arm or "carrier" fitted to the side of the balance case. To protect the balance from dust, etc., and to prevent disturbance due to air-draughts during the operation of weighing, it is contained in a glass case with counterpoised sliding *sashes*. The case is supported on levelling screws, by which it can be brought into a perfectly horizontal position, as indicated by spirit-levels attached to the base of the balance. When the balance is at rest the beam and the scale-pans are held by mechanical supports which are operated by means of a lever or an eccentric connected with a "milled" head, placed outside the case, by a slight turn of which the supports are simultaneously lowered and the beam and pans set free to swing upon their "knife-edges."

As a precaution against damp it is desirable to keep inside the balance case a small glass beaker or jar about half filled with dry calcium-chloride or small pieces of quicklime which must be renewed as occasion requires.

The main conditions upon which the stability, sensibility, and accuracy of the balance depend may be briefly stated as follows:—

"The *stability* of the balance is ensured by the centre of gravity being situated below the point or axis of suspension.

51

"The *sensibility* of the balance is secured by having the beam as light and as long as is possible consistently with rigidity, and by having the distance between the centre of gravity and the point of suspension as small as possible, also by reducing the friction at all free surfaces to a minimum.

"The *accuracy* of the balance depends upon having the two arms of equal length" (Roscoe).

At least two balances are employed for the assay of gold and silver, one for weighing fluxes and finely pulverised ore or "pulp," and the other for weighing the small "buttons" of precious metal obtained from the assay. This latter—the assay balance proper—is extremely sensitive. In assay offices where general assays are made, a third balance, more sensitive than the pulp balance, is usually necessary : in this case some form of chemical balance is employed.

1. **Pulp Balance.**—The pulp balance for weighing the charges of ore or products and the fluxes for assay should be capable of carrying at least 250 grammes in each pan, and be sensitive to 2 milligrammes. A convenient form of pulp balance is shown in fig. 70.

2. **Assay or "Button" Balance.**—The balance for weighing the minute buttons of precious metal resulting from the assay is the most important part of the assayer's outfit. In purchasing this it is true economy to buy a first-class instrument by some reliable maker. It should be capable of carrying 1 or 2 grammes in each pan, and, for ordinary work, should be sensitive to $\frac{1}{20}$ milligramme or less. For work requiring extreme accuracy balances sensitive to $\frac{1}{100}$ milligramme or even $\frac{1}{300}$ milligramme are in use. In cases where a large amount of general gold and silver assaying has to be done it is usual to employ two button balances, one for the silver buttons and the other for the gold, as the finely adjusted balances used for gold are too slow for rapid silver weighing and are liable to lose their extreme sensitiveness if subjected continually to the strain of weighing large silver buttons.

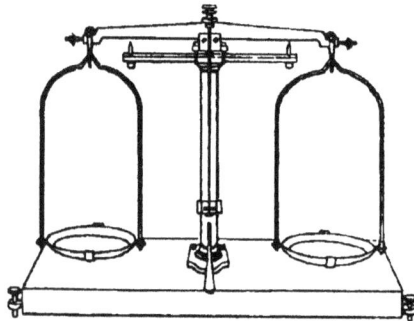

FIG. 70.

The accompanying illustration (fig. 71) shows an ordinary type of assay balance by "Oertling," with 6-inch horizontal beam to carry 2 grammes in each pan and turn with 0·01 milligramme. The beam is constructed with agate knife-edges working on agate planes, and divided into 100 parts.

Another type of button balance (by Oertling) for very exact work is shown in fig. 72. In this "short-beam" type the pan suspenders are decreased in length, thus permitting of very short brass supports for the beam. This arrangement tends to concentrate the movable mass near its central axis, thus giving great stability of poise, without diminishing the sensitiveness. The beam is 4 inches long and divided into 100 parts. The pointer extends upward, and the scale is above the beam. The pointer makes one complete oscillation in 10 seconds, and with 0·1 milligramme shows 10 divisions on the scale. Lenses, which are removable, are provided for reading the divisions on the beam and the scale : they are shown in position in fig. 72. The capacity of the balance is 1 gramme and the sensitiveness $\frac{1}{300}$ milligramme.

In some forms of button balances the pointer is fixed at the end of the beam in a horizontal position and the scale vertical.

A portable lens or telescope to stand inside the balance case is very frequently employed for reading the divisions on the scale and beam of assay and chemical balances.

A very simple arrangement of lenses for this purpose has been devised by

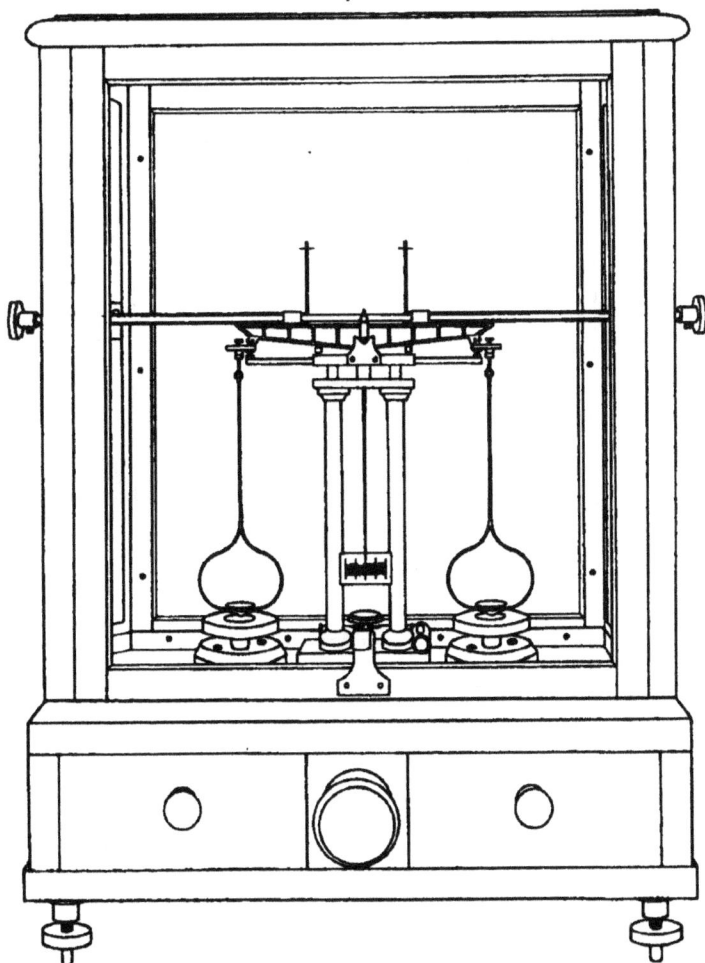

FIG. 71.

G. T. Holloway.[1] The usual telescopic or portable lens arrangements are replaced by lenses cemented on the rising glass front of the balance case. Plano-convex lenses cemented with "seccotine" on the side of the glass are preferred. A 3-inch lens is a convenient size for the pointer reading, but a 2-inch lens is sufficient for reading the beam graduations. The latter is cemented to face the

[1] *The Analyst*, February 1906.

right-hand side of the beam, but a slight movement of the head suffices to bring any part of the beam into the field. The focal length of the lenses is of little importance, but they must not be powerful enough to produce distortion, or thick enough to prevent the rise and fall of the balance-front on which they are fixed. There is always sufficient clearance for any ordinary lens, and as the lenses are fixed inside, they remain clean and require no attention or adjustment.

On account of the great accuracy required in precious-metal assaying it is

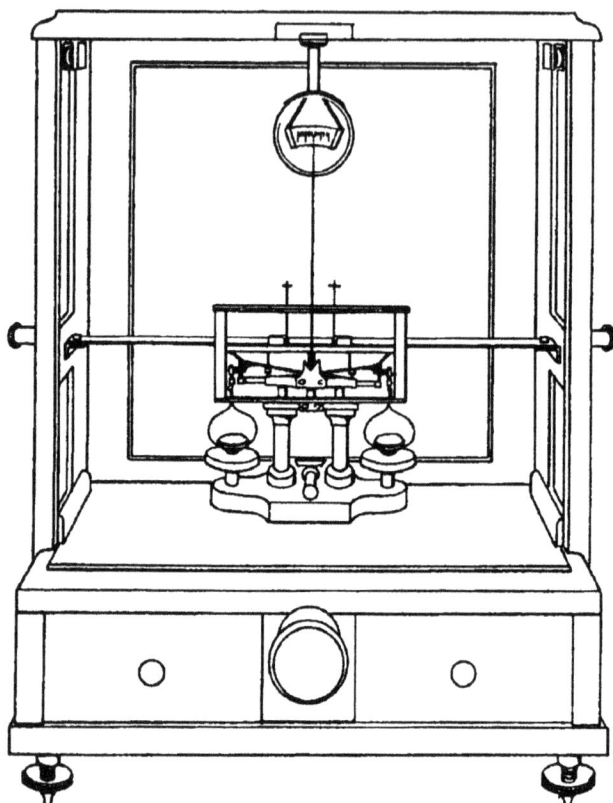

Fig. 72.

very desirable that the assayer should make himself thoroughly acquainted with the capabilities of the balance which he employs.

The degree of sensitiveness may be determined by accurately noting the deviation of the pointer from the zero-point which an overweight of a milligramme and decimals of a milligramme effects under varying conditions as follows:—(1st) when the balance is unloaded, (2nd) when carrying 1 gramme on each pan, and (3rd) when carrying 2 grammes on each pan. The variation in the rates of oscillation as the load increases should also be observed at the same time. When the balance is approaching equilibrium it is possible, after a little experience, readily to estimate the weight required to establish perfect equilibrium from the extent and rapidity of the oscillations as indicated by the pointer. This knowledge is very useful when a large number of weighings have to be made.

Auxiliary Assay Balance.—A balance of special construction for ascertaining the approximate weight of assay buttons before using the fine-assay balance is described on page 274, fig. 143.

3. **Chemical Balance.**—Fig. 73 represents a modern form of short-beam chemical balance suitable for weighing rich ores and products which require to

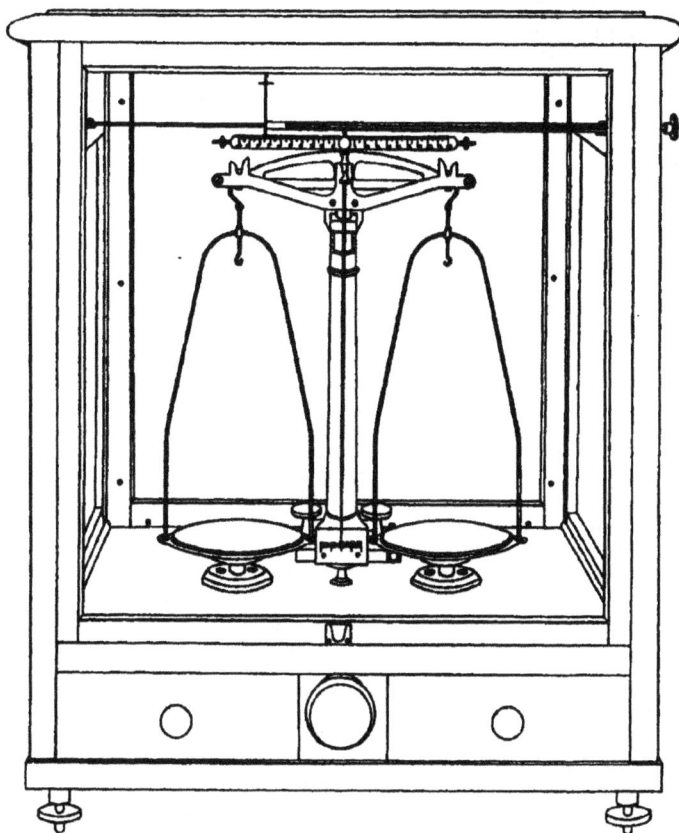

FIG. 73.

be more accurately weighed than is possible with the ordinary pulp balances. This balance will carry 200 grammes in each pan and turn with $\frac{1}{10}$ of a milligramme. The length of the beam is 6 inches.

Setting up Balances.—The balance must stand in a position free from vibration: otherwise it is liable to be frequently thrown out of adjustment. In ordinary buildings it is often difficult to obtain a firm support for the balance: if placed on a table which stands directly on the floor the vibration caused by persons walking in the room is quite sufficient to affect a very sensitive balance. The best foundations are stone or concrete piers set a few feet into the ground and projecting upwards through the floor but without touching it. A simple method of support consists of brackets firmly attached to the wall. This will usually be sufficient to prevent vibrations from the floor. Where slight vibrations occur they may be prevented by placing a rubber block under each levelling screw of the balance. A suitable rubber block (1¼ inches diameter) mounted in two brass caps is shown in fig 74. A small cavity is drilled in the top cap to receive the levelling screw and keep the block in position. An ordinary rubber bung is frequently used, a hole being bored in the bung sufficiently large to admit the leg of the levelling screw so that the milled plate rests on the bung. The bung is placed with the wide end resting on the bench. A perforated bung of the ordinary description will, of course, do as well.

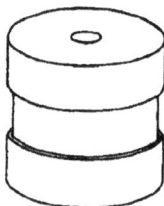

FIG. 74.

The bench top on which the balance stands is best made of some hard, well-seasoned wood : slate is sometimes recommended, but is not desirable. The balance should not stand in a position where it may be exposed to direct sunlight. It should be so arranged in the balance room that in weighing the light falls over the left shoulder of the operator, or from overhead. Facing a strong light while manipulating a sensitive balance is very trying to the eyes. On this account the north aspect, if obtainable, is best for the balance room. The balances should be arranged on the bench with ample room allowed on each side for samples and for the bottles, etc., into which the samples are to be placed after being weighed. It is to be regretted that more attention is not paid to the position of assay balances in mining schools.

To economise space it is sometimes convenient to place the balance bench under the window so that the operator faces a direct light, and the disadvantage of this position may not be felt with the intermittent weighing done by students. This position should, however, be avoided wherever possible, otherwise students may get the impression that it is the most suitable for weighing, while experience proves that it is most injurious to the eyes. In all assay offices where a large amount of continuous weighing has to be done and where the welfare of the staff is considered, the balances are arranged so as to cause as little eye-strain as possible.

Care of Balances.—A balance suffers more from imperfect preservation than from proper usage. When not in use it should be covered with a well-fitting baize or linen bag to keep out dust. A balance in constant use will require cleaning about every three or four months, but this must only be attempted by a person thoroughly acquainted with the construction. The instrument should be carefully taken to pieces and the various parts rubbed with a soft leather or brushed with a camel-hair brush. The movements should be cleaned and oiled with good watch-oil. Occasionally the agate planes will require repolishing by the maker, as the constant working of the knife-edges wears a minute groove in them, easily perceptible by a lens.

The systematic examination of balances should on no account be neglected if the balances are to retain the high degree of accuracy expected from them in precious-metal assaying.

Artificial Lighting of Balances.—In manufacturing towns where fogs are more or less prevalent at certain times of the year the artificial lighting of balances becomes a question of considerable importance when a large amount of weighing has to be done. Unless the balance room is unusually well lighted it is generally necessary to provide a separate light for each balance. These should be arranged, if possible, behind the operator, so that the light falls over his shoulder. When gas or electric light is available, balanced pendants suspended just behind the operator answer very well. Some assayers suspend the lights directly over the balance case or even resting upon it. This is very objectionable, as it heats the beam unequally and puts the balance out of equilibrium, causing the pointer, in the case of very sensitive balances, to oscillate several divisions more in one direction than in the other. The objection may be overcome in most cases by allowing the balance to rest for half an hour after turning on the light so as to allow the case to attain a uniform temperature before the balance is used.

A convenient arrangement, used by the author, consists of a double-arm bracket fixed on the wall or screen behind the balance at a convenient height to the right or left of the balance as shown in fig. 75, front and side views. Ball sockets are used for the joints so that the light can be moved into any desired position and the light thrown from behind the operator.

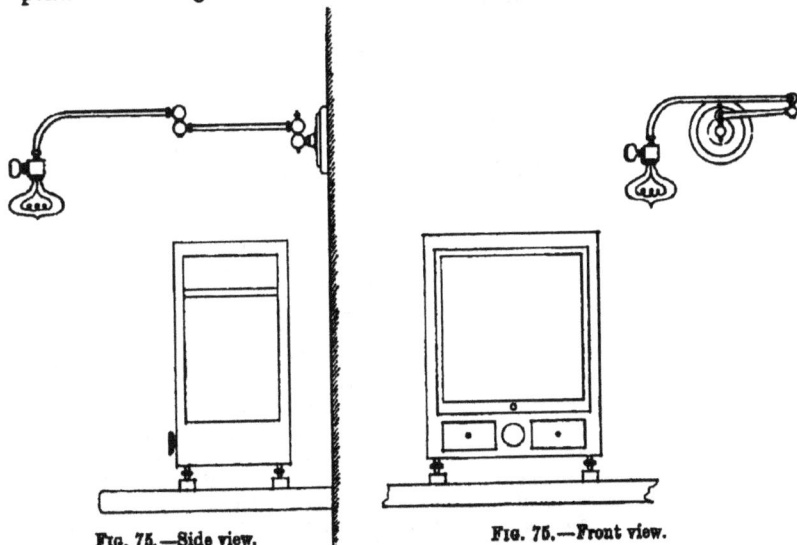

FIG. 75.—Side view.　　　　　FIG. 75.—Front view.

Weights.—The system of weights now very generally adopted by assayers is the French or metric system, in which the gramme (15·432 grains) is taken as the unit. The different denominations in weight of the system are related to the gramme by multiples and submultiples of ten, the fractional weight most frequently referred to being the milligramme, written mg. The denominations are:—

Gramme . . . 1·0
Decigramme . . . 0·1 or $\frac{1}{10}$ part of gramme.
Centigramme . . 0·01 or $\frac{1}{100}$,, ,,
Milligramme . . 0·001 or $\frac{1}{1000}$,, ,,

A set of weights usually extends from 50 grammes to a milligramme, and should contain weights of the following denominations:—

Grammes.	Grammes.	Gramme.	Gramme.	Gramme.
50	5	0·5	0·05	0·005
20	2	0·2	0·02	0·002
10	1	0·1	0·01	0·001
10	1	0·1	0·01	0·001

with several 0·01 (centigramme) riders.

For assay work it is convenient to have a 100 gramme weight included in the set.

The weights from 1 gramme upwards are usually made of brass, and the smaller weights, from 0·5 gramme downwards, of platinum or of aluminium. Although weights of less than 0·01 gramme are generally contained in the set, in practice it is usual to employ only the "deci" and "centi" grammes of these small weights (*i.e.* the first and second places of decimals), the milligrammes and fractions of a milligramme (the third and fourth decimals) being determined far more conveniently and with equal if not greater accuracy by means of the sliding weight or *rider* than by the use of the small weights themselves. According to T. K. Rose, the small weights often differ from their stated value by 5 per cent., or even more, and for this reason riders should be used.

The rider consists of a piece of platinum, gold or aluminium wire weighing (generally) exactly 1 centigramme (0·01 gramme), and bent so that it can be placed astride the beam (see fig. 76) and moved to any desired position on the graduated beam of the balance. It must fit the beam, so that when placed in position it always rests in a plane perpendicular to it. To ensure the rider taking its true position the subdividing marks on the beam should be nicked. The effective weight of the rider decreases as it travels from the end to the centre of the beam.

Each arm of the balance (*i.e.* the distance from the central knife-edge to the extremity of the beam) is divided into ten equal parts, and when a centi-gramme rider is used each part, counting from the centre outwards, equals 0·001 gramme or 1 milli-gramme. If each of these tenths be further sub-divided, the fractions of a milligramme are obtained.

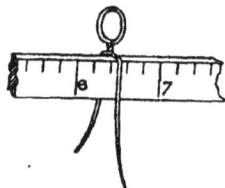

FIG. 76.

The divisions of the beam, therefore, represent milligrammes and fractions of a milligramme, and thus give figures in the fourth place of decimals.

When a rider (weighing 1 centigramme) is placed on the tenth main graduation at the end of the beam its effect is the same as if it were placed in the scale-pan, *i.e.* it is equal to 1 centigramme or 10 milligrammes: if placed on the fifth main graduation it exerts only half this effect, and is equivalent to 5 milligrammes and so on for the other main graduations. The ten main gradu-ations are (generally) subdivided into five equal parts, each of which equals with a centigramme rider $\frac{10}{50}$ or $\frac{1}{5}$ milligramme (0·2 milligramme or 0·0002 gramme). A rider placed midway between two main divisions, say between 6 and 7, would therefore be equivalent to 6·5 milligrammes or 0·0065 gramme, and if placed on the fourth subdivision between the two main divisions 6 and 7 the weight would be 6·8 milligrammes or 0·0068 gramme.

It will be evident that when the rider is placed in the middle of one of the subdivisions it is equivalent to $\frac{1}{10}$ milligramme or 0·0001 gramme. If a rider weighing only 5 milligrammes is used each subdivision will obviously equal $\frac{5}{50}$ or $\frac{1}{10}$ milligramme, and with a 1-milligramme rider each subdivision will be equivalent to $\frac{1}{10}$ milligramme. A thorough knowledge of the use of the rider is very essential to the assayer, as it is almost invariably employed in weighing the very small buttons of precious metal obtained from assays. Its use is not only attended with much economy of time and trouble but also with greater accuracy, as the rider allows the minutest variations to be determined. Riders weighing 5 milligrammes, 1 milligramme, and in special cases 0·5 milligramme are generally used with the most sensitive assay balances, their employment enabling the fractions of a milligramme to be determined with greater accuracy than is possible with a centigramme rider. Needless to say, the balance must always be brought to rest before altering the position of the rider. Further details on the use of riders for weighing gold and silver assay buttons will be found on pages 202 and 210.

It is very necessary to point out that riders are frequently sold which are not of correct weight, and it is essential to test them carefully before using. The errors on riders are reduced according to their position on the beam. At a distance of one-tenth of the half-beam length from the centre the error of the rider is reduced to one-tenth of its full amount. For ordinary work the riders will usually be sufficiently accurate, but when great accuracy is required they must, if necessary, be carefully adjusted by the assayer.

"Platinum riders, if too light, may be gilded in a cyanide bath, but light aluminium riders must be rejected. Any rider may be reduced in weight by light rubbing on a sheet of ground glass, or by other mechanical means. The error on a rider can in this way be made less than 0·005 milligramme without difficulty; but, if greater accuracy is desired, it saves time to determine the error of the weight and to apply it as a correction to the result of each assay." [1]

A good set of weights will be found to be practically unchanged even after many years' use provided they are used with due care. The weights must never be touched with the fingers, but handled always with ivory-tipped or brass forceps: steel forceps should not be used. When not in actual use the box should be kept closed and the greatest care taken to protect the weights from the action of acid fume. A new set of weights should always be tested before being taken into general use to see if they agree among themselves, and the testing should be repeated from time to time, particularly when the weights begin to be tarnished. [2] It is very desirable that an assayer should be able to compare his weights with a standard set, and in many assay offices a set of carefully standardised weights is kept exclusively for the purpose.

In addition to an ordinary box of decimal weights the following special weights are also used by assayers :—

The Assay-Ton System.—In reporting the result of an assay for gold or for silver, the number of ounces troy, per ton avoirdupois, is always returned. In England and Australia the result is reported upon the English or *long ton* of 2240 lbs. (avoirdupois), and in North America and Africa upon the *short ton* of 2000 lbs. (avoir.). To facilitate calculation a system of weights, known as the "assay-ton system" (abbreviated A.T.) has been devised, which is a combination of the troy and avoirdupois systems with the French metrical system. The unit of the system is the "assay ton," which for the English system (2240 lbs. to the ton) is 32·6666 grammes. It is derived as follows :—

There are 2240 lbs. in an English ton, and 7000 grains in 1 lb. avoirdupois; therefore there are 2240 × 7000 = 15,680,000 grains in a ton.

There are 480 grains in 1 ounce troy—consequently there are $\dfrac{15,680,000}{480} =$ 32,666·6 troy ounces in 1 ton avoir.

Now employing the French system. If we take 32,666·6 grammes of ore for an assay, each gramme of precious metal obtained would indicate 1 ounce troy per ton avoirdupois; or, if we take 32·6666 grammes of ore, each milligramme of precious metal obtained will indicate 1 ounce per ton; this quantity, 32·6666 grammes, is the English "assay ton": it contains the same number of milligrammes as there are troy ounces in the ton avoirdupois. As an example of its use, suppose the weight of the resulting "button" of precious metal from one assay ton of ore is 1·2 milligrammes, then the ore contains 1·2 ounces of gold (or of silver) per statute ton.

[1] *Metallurgy of Gold*, T. K. Rose, 1906, 5th edition, page 460.
[2] For the method of testing weights, see Appendix.

The "assay ton" for the American system (2000 lbs. to the ton) is 29·166 grammes, and is derived in the same way. Sets of assay-ton weights for both the English and American systems ranging from ¼ A.T. to 4 A.T. can be purchased or a set can be made up from an ordinary box of gramme weights and its subdivisions. Although in some cases it may be advantageous to employ "assay ton" weights, in general assay practice it is more convenient to work on a purely decimal system, and convert when required into ounces per ton, etc., either by actual calculation or by reference to a set of tables.

Assay Weights.—In assaying silver and gold bullion a special decimal series of weights may be employed in which the unit for silver "assay weights" is 1 gramme, which is stamped "1000," and for gold "assay weights" is ½ gramme, which is stamped "1000." The decimal subsidiary weights are stamped as follows :—

Silver assay weights.		Gold assay weights.	
Number stamped on weights.	Actual weight.	Number stamped on weights.	Actual weight.
1000	1·0 gramme.	1000	0·5 gramme.
500	0·5 ,,	500	0·25 ,,
200	0·2 ,,	200	0·1 ,,
100	0·1 ,,	100	0·05 ,,
100	0·1 ,,	100	0·05 ,,
50	0·05 ,,	50	0·025 ,,
20	0·02 ,,	20	0·010 ,,
10	0·01 ,,	10	0·005 ,,
10	0·01 ,,	10	0·005 ,,
5	0·005 ,,	5	0·0025 ,,
2	0·002 ,,	2	0·001 ,,
2	0·002 ,,	2	0·001 ,,
1	0·001 ,,	1	0·0005 ,,
Riders 0·01 gramme (10 milligrammes)		Riders 0·005 gramme (5 milligrammes)	

The figures stamped on the weights in the silver series denote the number of milligrammes contained in the weight, and in the gold series the number of ½ milligrammes (termed "millièmes") contained in the weight. The advantage of these special "assay weights" is that the report is at once indicated by the figures on the weights without further calculation, provided 1 gramme of silver and 0·5 gramme of gold is used for the assay respectively. They are in use in most bullion assay offices dealing with a large number of assays per diem.

Methods of Weighing.[1]—There are three methods of determining the weight of a substance by means of the balance : (1) the *direct* method, (2) the method of *substitution*, and (3) the method of reversal. Before describing these methods it is well to point out that whatever method of weighing is adopted it is best to work with the balance on the "swing" as it is expressed, that is, the state of adjustment of the balance is indicated by the extent to which the pointer swings or is deflected over the ivory scale. If the balance is in equilibrium the pointer will swing approximately an equal number of divisions on each side of the zero, losing, however, half a division (or other fraction of a division) on each swing, the small loss being due to friction and to a gradual return of the balance to rest.

[1] The author has availed himself of the excellent description of the balance and methods of weighing given in *Quantitative Chemical Analysis*, by T. E. Thorpe.

Equilibrium is established when the number of divisions traversed by the pointer to the left is the mean of the preceding and succeeding number of divisions traversed to the right.   For example, if the pointer swings seven divisions to the right on one swing and six divisions on the next, the intermediate swing to the left should have been 6·5 divisions.   When the swings are thus approximately equal the balance is in equilibrium, and if any substance is being weighed, the weight would be at once recorded without spending the time necessary to bring the pointer to rest at the zero-point.

The most sensitive balances in coming to rest do not usually lose as much as half a division on each swing, and are consequently slower in coming to rest. The first few swings, after releasing the beam, are always disregarded; but if afterwards the pointer swings further in one direction than in the other, either the balance is not level, or is out of adjustment.

In using this method to ascertain the state of adjustment of the balance it is advisable not to let the deflection of the pointer exceed five or six divisions each way.   The method of determining the position of rest or actual zero-point of the balance is fully described on page 62.

Some assayers work by the method of "no deflection" or swing: that is, when equilibrium is established and the beam is lowered gently on its knife-edge, no deflection of the pointer takes place.   "This method, however, is not recommended, as it disregards friction and inertia, and for small weights gives inaccurate results" (Fulton).   If carefully conducted, the method of working by noting the amount of deflection of the pointer to the right and the left of the zero, as described above, is more accurate and rapid than the method of bringing the balance to complete rest, and is the method in very general use.

1. **Direct Weighing.**—To determine the weight of a substance by direct weighing the balance is first adjusted to equilibrium, then the substance placed in one pan and weights added to the other pan until the substance is exactly counterpoised. As this method is the most expeditious it is the one in common use.   If, however, the balance is not in perfect equilibrium, or the arms of the beam are of unequal length (though this is rare in good balances), the true weight is not registered by this method.   This is of no consequence in ordinary operations, since in any one series of weighings the *relative* weights of the different substances only are required.   These will be correctly given, provided the different substances are placed always upon the same scale-pan and the conditions of the balance remain unaltered between the successive weighings of the series.[1]

If the arms are of equal length, and the balance in a perfect state of equilibrium, it is indifferent upon which pan the substance to be weighed is placed.   The right-hand pan is the most convenient for manipulation, and in practice this pan is invariably used for manipulating weights when ascertaining the weight of a substance, but when weighing out a definite quantity of a substance it is usual to put the weight on the left-hand pan and manipulate the substance on the right-hand pan.

2. **Weighing by substitution** gives both the relative and the true weights of a substance, and the results are absolutely accurate, even when the balance is not in perfect equilibrium or the arms not of exactly equal lengths.   In this method the substance to be weighed is first accurately counterpoised; it is then removed, and weights substituted for it until equilibrium is again restored.   It is obvious that the weights thus substituted for the substance will represent the true weight of the substance.   "In practice this method of weighing may be facilitated by using one of the larger weights, heavier than the object to be weighed, as a counterpoise, and adding weights to the object until equilibrium is established.   The object is then removed, and weights substituted until the

[1] Clowes and Coleman, *Quantitative Chemical Analysis.*

balance is again in equilibrium.   The substituted weights express the real weight
of the object" (Thorpe).   Weighing by substitution is in use in some bullion
assay offices where very great accuracy is required.

3. **The Method of Reversal.**—In this method "the substance is weighed
first in one pan, and then in the other.   If the weights are identical, the true
weight of the object is at once given.   If the weights are unequal, their geometric
mean will be the true weight : this is found by multiplying the apparent weights
together and taking their square root.   Practically the common arithmetic mean
of the two weights will be found sufficiently accurate, unless their difference is
considerable " (Thorpe).

"**The Method of Swings.**"—Within certain limits the deviation from the
horizontal in a balance is proportional to the weight causing it.   Advantage may
be taken of this fact to determine the weight of minute quantities (*i.e.* fractions
of a milligramme) of gold or silver with great accuracy.   The position of rest of
the balance, *i.e.* the true zero or equilibrium point, is first found and then the
value of one division of the ivory scale is ascertained.   When the balance is
very nearly in equilibrium it is caused to oscillate, and the position of the pointer
when at rest determined from successive observations of the extreme points
reached by the pointer when swinging.   Disregarding the first swings, so soon as
the deflections of the pointer fall within a certain limit, their extent, as stated
above, commences to decrease at a regular and uniform rate.   Let A_1, A_2, A_3, A_4
be the extreme points (end of swing) consecutively reached by the pointer in its
deflections, A_1 and A_3 being on one side of the middle of the ivory scale and
A_2 and A_4 on the other.   Then the position of rest or equilibrium of the balance
X is given by the formula :—

$$X = \frac{A_1 + 2A_2 + A_3}{4},$$

or with greater exactness by

$$\frac{A_1 + 3A_2 + 3A_3 + A_4}{8}.$$

The following is an example :—

Position of Rest of Empty Balance.—The balance, having been adjusted as
nearly as possible, is made to oscillate, and the extreme positions of the pointer
in its deflections observed, counting swings to the left as minus and to the right
as plus.   (Plus divisions are always reckoned on the side where the weights are
placed, minus divisions on the side where the gold or silver is added.)   The
observations should be made through a lens or telescope.   Suppose the swings,
after the first two, are in succession as follows : $+4\cdot4, -3\cdot8, +4\cdot2$.   The position
of rest or true zero is

$$\tfrac{1}{2}\left(\frac{4\cdot4 + 4\cdot2}{2} - 3\cdot8\right) = +0\cdot25 \text{ division, or one-fourth of a division to the right.}$$

In the actual determination of the position of rest (X) an odd number
(conveniently five or seven) of successive readings are made so soon as the pointer
reaches division 6 on the scale : the arithmetical mean of the half difference
between consecutive pairs of observations gives the position of rest of the pointer
along the graduated scale as shown above.

Value of One Division.—If we know the weight corresponding to a given
deviation from the zero, the estimation of the minute fraction required for exact
equilibrium becomes a simple matter.   We have first to determine the values of
one division of the graduated scale (*i.e.* the weight required to make the pointer
deviate one division) for varying loads.   "Suppose a balance at rest in perfect

equilibrium, with the pointer exactly over the middle point of the scale. Let the scale be a series of points at equal distances along a horizontal line; then, if a small weight be placed on one pan, the pointer will deviate from its vertical position and come to rest opposite some definite part of the scale, which will depend upon the magnitude of the weight added. The law determining this position is a very simple one; the deviation as measured along the points of the scale varies directly as the weight added. For example, with an ordinarily sensitive balance, such as is used for general purposes, $\frac{1}{2}$ milligramme will move the pointer along, say, four divisions of the scale; then 1 milligramme will move it eight divisions; $\frac{1}{4}$ milligramme two divisions; and so on. Of course with a more sensitive balance the deviations will be greater" (Beringer).

To determine the value of one division a weight of, say, 0·3 milligramme (by means of a rider) is placed on the side used for weights, and the number of divisions traversed by the pointer noted as described above. Suppose the swings are − 6·7, + 0·7, − 6·3, the new position of rest or zero is

$$\tfrac{1}{2}\left(+0\cdot7 - \frac{6\cdot7+6\cdot3}{2}\right) = -2\cdot9 \text{ divisions.}$$

The deviation from the middle point due to the overweight of 0·3 milligramme is therefore 2·9 divisions, and one division of the scale would be equivalent to 2·9 ÷ 0·3 = 0·103 milligramme. This value is practically constant from day to day.

If the pointer is not exactly over the middle point when the balance is empty, the slight deviation must be allowed for. Suppose, for instance, that the true zero of the balance was one-fourth of a division to the right (i.e. + 0·25), as in the first example, then the deviation due to the overweight of 0·3 milligramme would be 2·9 − 0·25 = 2·65 divisions; or one division would be equivalent to 0·113 milligramme. A weight of 0·6 milligramme ought to produce double the amount of deviation caused by the 0·3 milligramme: if any difference is observed, the mean of the two observations is taken as representing the true value of one division. In these determinations the greatest care is necessary to preserve the balance under perfectly uniform conditions. The value of one scale division on a very sensitive balance is invariably greater in summer than in winter.

Weight of Minute Quantity of Gold.—The method of determining minute weights by swings or deflections affords a simple means of weighing the minute gold and silver beads obtained from assays, and is sometimes employed by assayers. Suppose the gold is found to require about 1 milligramme to counterbalance it, and the swings are now − 4·6, + 0·2, − 4·4. The new position of rest is

$$\tfrac{1}{2}\left(0\cdot2 - \frac{4\cdot6+4\cdot4}{2}\right) = -2\cdot15.$$

Assuming the position of rest of the empty balance to be + 0·25 and the value of one division to be 0·113 milligramme, the gold weighs

$$1 - \left(2\cdot15 + 0\cdot25 \times \frac{0\cdot113}{1}\right) \text{ milligramme; that is, } 0\cdot73 \text{ milligramme.}$$

In the case of very minute gold beads which, without the addition of weights, only deflect the needle a few divisions, the weight of the bead is the difference in deflection between the two points of rest (i.e. of empty balance and balance with bead) multiplied by the value of one division. This method of working by swings or deflections adds very considerably to the power of a balance in distinguishing small quantities. According to Dr Rose,[1] by weighing in this way results

[1] *Metallurgy of Gold*, T. K. Rose, 5th edition, 1906, p. 460.

correct to 0·002 milligramme are easily obtained by using the best assay balances in which the value of the scale division is 0·005 milligramme: and still greater accuracy is attainable by weighing the gold in each pan in succession, and by taking means of a number of observations.   Unless, however, very sensitive balances are available, the method of weighing by swings is not generally to be recommended: fractional parts of the milligramme should be determined with the "rider."

In working by the method of swings it is a good rule to always place the button to be weighed on the left pan of the balance, the weights on the right; count the divisions of the scale from the middle to right and to left, marking the former plus and the latter minus as before indicated.

Directions for Weighing.—The sensitive balances employed in the assay office must be very carefully handled if their accuracy is to be maintained. Careless manipulation very soon impairs the sensitiveness of balances, and the following general rules may therefore be found of assistance to beginners during the operation of weighing:—

(1) When using a balance sit directly in front of it so as to have the ivory scale in direct line of vision.

(2) Before commencing to weigh, see that the rider hangs on the carrier and not on the beam.

(3) Ascertain that the pans, beam, etc., and the floor of the balance case are perfectly clean; if not, brush carefully with a camel-hair brush, which should be kept solely for this purpose.

(4) Do not take it for granted that the balance is in equilibrium before commencing to weigh.   Test the equilibrium by gently releasing the pans and beam.   If the beam does not oscillate when released, start its motion by gently wafting the air down upon one of the pans.   The pointer should now be seen to oscillate through equal spaces on each side of the zero of the ivory scale.   If not in equilibrium brush the pans, etc., as before, and again test the equilibrium before attempting to adjust the balance by moving the vane over the centre knife-edge.   Should readjustment be necessary, this must be done by an experienced person, as frequent attempts at adjustment by inexperienced hands will certainly disturb the sensitiveness of the balance.

(5) Always have the balance at rest when putting on or removing anything (either substances to be weighed, or the weights) upon the pans.   When arresting the motion of the beam, raise the supports the moment the pointer is opposite the zero of the scale so as to prevent any jerking or sudden vibration of the beam.   Carelessness in arresting the oscillation of the beam greatly interferes with the uniform behaviour of the balance, and with inexperienced operators is the most frequent cause of disarrangement.

(6) Put the weights on the pan systematically.   Try the weights of each denomination in succession, retaining or removing each weight in order according as it is too light or too heavy.   A very little experience is sufficient to tell, roughly speaking, the weight of an object.   Suppose we wish to determine the weight of a sample of gold bullion submitted for assay.   First place on the right-hand pan a weight which by a guess is judged to be approximately equal to the weight of the gold.   Suppose the 20-gramme weight has been selected, and found to be insufficient; one of the 10-gramme weights is added—this is too much; substitute the 5-gramme for the 10-gramme—the weight is again too little; add the 2-gramme—this is too much; remove it, and substitute a 1-gramme —it is too little.   This decides the units.   The gold weighs between 26 and 27 grammes.   Now use the decimals of the gramme in the same systematic order until the first and second decimal places are decided, that is, until the whole number of decigrammes and centigrammes are ascertained.   Add the 0·5-gramme—

this is still too little; add the 0·2-gramme—still too little; so add the 0·1-gramme —this is too much, so the first place of decimal has been ascertained. Remove the 0·1-gramme and substitute 0·05—still too much; remove and try 0·02—the pointer now oscillates much more slowly, but still indicates that the weight is too great, so substitute 0·01 for the 0·02. This determines the second place of decimals: the total weight on the pan is now 26·71 grammes, but it is still insufficient to equipoise the gold. Close the front sash and bring the balance into complete equilibrium by means of the rider. This will give the third and fourth places of decimals, viz. the milligrammes and fractions of the milligramme. Put the rider on the division marked 5; if this is too much, try it at the 4; if this is not enough, the correct weight must be between the 4 and 5— this decides the third decimal place. Now place the rider midway between the 4 and 5; the oscillation of the pointer is now equal in both directions—the balance is therefore in equilibrium. The aggregate value of the weights used is 26·7145 grammes, which represents the weight of the gold. An experienced weigher familiar

FIG. 77.

with the indications of his balance can almost intuitively tell, from the extent and rapidity of the oscillations, what weight is required to establish equilibrium. Until this experience is acquired the weights should be added in the systematic method above indicated. This method will be found much more expeditious than that of selecting the weights at random.

(7) Record the weight in a note-book (not on an odd scrap of paper) immediately the weighing operation is done. The value of the weights should first be read off from the empty spaces in the box, and then checked as the weights are returned one by one, beginning with the highest, to their respective compartments. This double reading should never be neglected: the one method serves to check the other.

(8) Except in the case of pieces of metals, alloys, etc., substances should not, as a general rule, be placed directly upon the pan, but should be contained in a suitable receptacle, which in assay offices is generally a counterpoised metal pan. For weighing materials which are to be transferred to flasks or bottles for wet assay the metal scoop (fig. 77) is sometimes employed.

CHAPTER VI.

GOLD AND SILVER.

Physical and Chemical Properties.

GOLD.—Symbol, Au (aurum); atomic weight, 197·2.
Specific gravity, 19·32; melting-point, 1064° C. (Berthelot).

Gold expands considerably on fusing and contracts again on solidifying. At high temperatures gold is slightly volatile.

Commercial Forms of Gold.—Metallic gold is usually brought into the market in the form of ingots in a more or less refined state and weighing usually 200 or 400 ounces each. The metal sent by the Bank of England to the Royal Mint for coinage is usually in the form of refined ingots, each weighing about 400 ounces and containing over 99 per cent. of gold.

Various forms of gold are used for industrial purposes. The pure gold used by goldsmiths for the preparation of alloys is supplied in four different forms, namely, "brown," "grain," "sheet," and "dull" golds.

Brown Gold is in the form of powder, and is the pure gold left from the refining by the nitric-acid parting processes, simply washed and dried. It is principally used for the preparation of gold solutions for electro-gilding, as gold in this form is easily dissolved.

Grain or Granulated Gold is made by melting the brown powder obtained from refining, and slowly pouring it into water, whereby it is obtained in the form of small granules or grains. Gold in this form is very convenient for weighing out.

Sheet Gold is prepared by casting the metal into a bar and then rolling it out. Sheet gold is used for "anodes" in the electro-gilding process.

Dull Gold is made by heating brown gold to a red heat, when it changes colour and becomes dull yellow. Dull gold is now very little used, but there are still a few jewellers who prefer it to "grain" gold for alloying purposes.

Gold Leaf.—Gold leaf is of various qualities, containing from 90 to 98 per cent. of gold and the rest silver and copper. The different qualities of gold used in the manufacture of gold wares and jewellery are obtained in the form of sheet metal and as wire, etc. Samples of gold in any of the forms described above may be received for assay.

Alloys of Gold.—Although gold is capable of alloying with almost any of the metals, the important alloys of gold are those with copper and silver, and of these the gold-copper alloys are by far the more important, on account of their employment as alloys for coinage, etc. These alloys are dealt with subsequently under Industrial Alloys, Chapter XXI., page 320.

Gold and Silver unite in all proportions when melted together, forming homogeneous alloys, no separation of the constituents taking place on solidification. The colour of gold becomes paler when small quantities of silver are added to it, and is white with a scarcely perceptible tinge of yellow when 50 per cent. of silver is present. Alloys containing more than 60 per cent. of silver are silver-white. This change in the colour of gold, due to the bleaching action of silver, affords an indication of the amount of silver present in the alloy, and is utilised by experienced assayers to guess the approximate proportion of silver in the gold-silver buttons obtained in assays.

A natural alloy of gold and silver, designated "Electrum," is found, and is usually of a pale yellow or amber colour, and contains about 70 to 75 per cent. of gold.

Gold and Copper.—These metals are miscible in all proportions when molten, and on solidification separate only to a slight degree. The first additions of copper to gold cause a rapid lowering of the melting-point. The gold coinage of Great Britain is made from an alloy containing gold (approximately) 91·7 per cent., and copper 8·3 per cent., and termed "Standard Gold." This alloy solidifies at 951° C. or 103° below the melting-point of pure gold. The "Standard" alloy used for the gold coinage of France, America, and most other countries contains 90 per cent. of gold, and solidifies at 946° C. The colour of gold becomes redder when copper is added to it, and if silver is added at the same time in suitable proportions, the two metals counteract the colouring effect of each other, so that triple or "ternary" alloys can be formed having a colour closely resembling that of pure gold. Such triple alloys are largely used by jewellers (see page 321).

Gold and Mercury.—Alloys of these metals are known as *amalgams.* They are formed, although with difficulty, by the direct union of the two metals at the ordinary temperature. Union is more readily effected when finely divided gold is heated with mercury. "The amalgams recovered in gold-mills, in which ores are treated by the amalgamation process, are not true alloys. Before straining, they consist of mercury containing a number of little 'nuggets' or particles of gold into which mercury has penetrated to some extent. On straining, these nuggets, coated with mercury, are separated, and the gold-miner's amalgam is consequently of variable composition, generally containing from 25 to 50 per cent. of gold, the percentage of gold being highest when the average size of the gold particles is greatest."[1]

Gold and Lead.—Lead unites with gold, forming alloys that are readily fusible, very brittle, and hard. About 0·15 per cent. of lead makes pure gold somewhat brittle. If more than 4 per cent. of lead is present, there is marked segregation on solidification.

Gold and Platinum.—These metals can be alloyed in all proportions by fusion, but require a high temperature to effect their union, owing to the high melting-point of platinum. With less than 50 per cent. of platinum the alloys formed are malleable and ductile. When the proportion of platinum exceeds 50 per cent., the alloys are brittle and greyish in colour.

Solubility of Gold.—Gold is not soluble in sulphuric, nitric, or hydrochloric acid alone, but is readily dissolved in aqua-regia, a mixture of one part by measure of nitric acid and three parts of hydrochloric acid, which evolves chlorine, forming a solution of auric chloride, $AuCl_3$. When the solution is evaporated to dryness, and the residue dissolved in water, the concentrated solution deposits reddish crystals of the composition $AuCl_3,2H_2O$. These lose their water when carefully heated, leaving a brown mass of deliquescent crystals; but if strongly heated, the $AuCl_3$ is readily decomposed, leaving a residue of metallic gold. Aqua-regia is the best and ordinary solvent for gold. The mixture is best made when required.

[1] *Precious Metals,* T. K. Rose, p. 45.

The several reactions may be expressed as follows :—

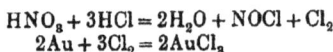

$$HNO_3 + 3HCl = 2H_2O + NOCl + Cl_2$$
$$2Au + 3Cl_2 = 2AuCl_3$$

or

$$2Au + 3HNO_3 + 9HCl = 2AuCl_3 + 3NOCl + 6H_2O$$
$$2AuCl_3 + heat \text{ (about } 205° \text{ C.)} = 2Au + 3Cl_2.$$

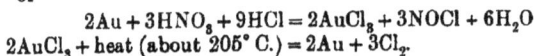

Gold is also dissolved in any other mixture producing nascent chlorine, such as a mixture of bleaching powder and certain acids.

The presence of silver in the gold retards solution, a film of insoluble silver chloride being formed over the metal, and the action may eventually be completely stopped if the percentage of silver present is large ; thus alloys of gold and silver containing over 50 per cent. of gold are difficult to dissolve in acid (see also page 282). Gold in a very thinly laminated or finely divided state is dissolved by aqueous solutions of simple cyanides (such as KCN) in presence of an oxidising agent.[1]

$$4Au + 8KCN + O_2 + 2H_2O = 4AuK(CN)_2 + 4KOH.$$

This is known as Elsner's equation.

Reactions of Gold.—*Sulphuretted Hydrogen*, H_2S, gives from a cold solution of $AuCl_3$ a black precipitate of auro-auric sulphide Au_2S_3, and from a boiling solution a brownish precipitate of aurous sulphide Au_2S. These precipitates are insoluble in either hydrochloric or nitric acids, but dissolve in aqua-regia. The precipitate Au_2S_3 is given with ammonium sulphide $(NH_4)_2S$ and sodium thiosulphate. If nitric acid or any other oxidising agent is present, it must be destroyed before H_2S is passed through the solution.

Caustic Potash (potassium hydroxide), KHO.—Neither the hydroxides nor carbonates of potassium or sodium give any precipitate with *moderately dilute solutions* of $AuCl_3$. From a *concentrated* solution KHO gives a brown precipitate of hydrated auric oxide, $Au_2O_3,3H_2O$, or $Au(HO)_3$. This precipitate is soluble in excess of potash, yielding potassium aurate, K_2O,Au_2O_3, or $KAuO_2$.

Ammonium hydrate, NH_4HO, gives an orange-red precipitate of ammonium aurate or *fulminating gold*, $(NH_3)_2Au_2O_3$, the exact composition of which is not known, thus :—

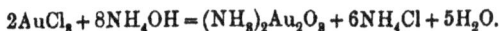

$$2AuCl_3 + 8NH_4OH = (NH_3)_2Au_2O_3 + 6NH_4Cl + 5H_2O.$$

When dry it explodes with violence if struck or gently warmed. It is decomposed without explosion by sulphuretted hydrogen, and by stannous chloride. It is insoluble in water, but soluble in potassium cyanide, with the formation of auricyanide of potassium, $AuCN_3,KCN$.

Potassium Cyanide, KCN.—When potassium cyanide is added to an acid solution of auric chloride, aurocyanide of potassium, $AuCN,KCN$, is formed. Aurocyanides are decomposed by mineral acids, aurous cyanide being precipitated and hydrocyanic acid evolved. By adding potassium cyanide to a perfectly neutral solution of auric chloride, auricyanide of potassium, $AuCN_3,KCN$, is formed, the precipitate first formed being dissolved (T. K. Rose).

It may be noted here that solutions containing aurocyanides of the alkalies and thiosulphates of gold do not give the ordinary reactions of gold, although sulphuretted hydrogen precipitates sulphide of gold from thiosulphates. Solutions containing cyanides or thiosulphates may be tested for gold by heating

---

[1] For further information on the solubility of gold in cyanide solutions, see T. K. Rose, *Metallurgy of Gold.*

them with aqua-regia, evaporating to dryness, taking up with water and a little hydrochloric acid, and testing the solutions in the ordinary way. The decomposition with aqua-regia must be done in a good fume cupboard so as to avoid inhaling the poisonous fumes of hydrocyanic acid.

The most characteristic reactions of gold are based upon the fact that the metal is readily reduced from its compounds and precipitated in the metallic state, hence the detection of gold is attended with no difficulty. Many organic bodies readily precipitate gold from $AuCl_3$. Thus sugar boiled in it gives at first a light red precipitate of metallic gold, which afterwards darkens in colour. The reagents most frequently employed are sulphurous acid, oxalic acid, ferrous sulphate, and stannous chloride. Sulphurous acid and oxalic acid are largely used in the laboratory for the preparation of pure gold (see remarks on page 282).

Sulphurous Acid, H_2SO_3, is a convenient reagent, and is often used in the laboratory, as it acts almost equally well in cold and hot solutions. It gives a brown precipitate of metallic gold.

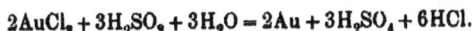

$$2AuCl_3 + 3H_2SO_3 + 3H_2O = 2Au + 3H_2SO_4 + 6HCl.$$

Oxidising agents, if present, must be destroyed before adding sulphurous acid or passing SO_2 gas through the solution.

Oxalic Acid, $C_2H_2O_4$.—This is a very useful precipitant for gold. It does not act very readily in cold solutions, but on being gently warmed with $AuCl_3$ oxalic acid causes the gold to be deposited either as a scaly brown precipitate or as a coherent gold film on the surface of the glass vessel, according to the strength and temperature of the gold solution.

Platinum is not reduced by oxalic acid.

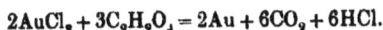

$$2AuCl_3 + 3C_2H_2O_4 = 2Au + 6CO_2 + 6HCl.$$

Potassium Nitrite, KNO_2, reduces $AuCl_3$, being itself converted into nitrate thus :—

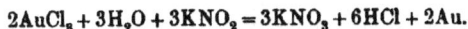

$$2AuCl_3 + 3H_2O + 3KNO_2 = 3KNO_3 + 6HCl + 2Au.$$

Ferrous Sulphate, $FeSO_4$.—In comparatively strong solutions a brown, finely divided precipitate of metallic gold is formed, and in weaker solutions the colour of the solution becomes blue by transmitted light.

$$2AuCl_3 + 6FeSO_4 = 2Fe_2(SO_4)_3 + Fe_2Cl_6 + 2Au.$$

Ferrous sulphate is often used to detect the presence of minute quantities of gold chloride in solution. For this purpose a test-tube filled with the liquid to be tested is held in the hand side by side with a test-tube filled with distilled water, and a few drops of a clear solution of ferrous sulphate added to each. On looking down through the length of the test-tubes from above, with a white surface as background, any slight changes of colour may be detected by comparison, and the liquid may also be compared with the original solution in a test-tube.

In this way, by a little practice, the presence of gold in the proportion of only 1 dwt. per ton of water can be detected (T. K. Rose).

The detection of minute quantities of gold chloride in solution is also effected by means of stannous chloride, which affords a very delicate test.

Stannous Chloride, $SnCl_2, 2H_2O$, gives with $AuCl_3$ a brownish precipitate of variable composition and colour depending on the strength of the solution. The compound is known as *purple of Cassius*, and its composition is not known with certainty. The production of the purple is facilitated by the presence of a small quantity of stannic chloride, such as is always present in a solution of stannous chloride, except when quite freshly made. It is best prepared by dissolving 112·5 grammes SnCl₂ in 200 c.c. of 5N hydrochloric acid and diluting to 1 litre, a little granulated tin being placed in the bottle.

The purple of Cassius test is very sensitive, and by its means a violet coloration by transmitted light can be obtained in a solution containing 1 part of gold in 500,000 parts of water, or even less can be detected.[1] To test the liquid supposed to contain gold in small quantity, it is heated to boiling and poured suddenly into a large beaker containing 5 to 10 c.c. of saturated solution of stannous chloride, and the liquids agitated so as to effect complete mixture. A yellowish-white precipitate of tin hydrate forms, which settles rapidly, and can be readily separated from the bulk of the liquid by decantation. If the solution originally contained at least 1 part of gold in 5,000,000 of water (3½ grains per ton), the precipitate is coloured purplish-red or blackish-purple, according to the nature of the solution and the condition of the precipitant. The colour can be seen without comparing it with other precipitates. The presence of gold in quantities less than 0·1 milligramme may be detected in cyanide solutions[2] with stannous chloride by acidulating the solution, boiling till most of the HCN is expelled, then adding potassium chlorate, KClO₃, and again boiling till most of the chlorous gases are driven off, and finally adding stannous chloride, which gives the purple of Cassius. The final solution should not be too strongly acid, or the colour may not appear. It frequently becomes more marked on allowing to stand for some time. Gold is precipitated from a hydrochloric acid solution of AuCl₃ by most metals, even by silver, mercury, and platinum.

Preparation of Pure Gold.—The purest gold obtainable is required for use as checks or proofs (see page 276) in the assay of gold bullion.

The following method of preparing it is used at the Royal Mint, London,[3] and in many assay offices.

Gold assay cornets from the purest gold which can be obtained are dissolved in nitrohydrochloric acid and the excess of nitric acid expelled by repeated evaporations with additional hydrochloric acid on a water bath. The blackish-red product, smelling of chlorine and consisting chiefly of AuCl₃ . HCl, or HAuCl₄ (chlorauric acid), is then poured in a thin stream into a large glass vessel full of distilled water, and a solution of about 1 ounce of gold in each pint of water (1 gramme of gold in 20 c.c. of water) is formed in this way. After vigorous stirring the solution is left to settle, and at the end of about a week the whole of the precipitated chloride of silver will have subsided to the bottom. The progress of the subsidence is easily watched. The particles fall at the rate of about 3 or 4 inches per day. The clear bright supernatant liquor is now removed by a glass syphon, and diluted to about 1 ounce of gold per gallon of water (1 gramme to 150 c.c.). No further subsidence of silver chloride takes place, even if the diluted liquor is left undisturbed for three months, and the gold may be at once precipitated with sulphurous acid or oxalic acid. Oxalic acid is often used, and is stated by Krüss to be the best precipitant if platinum is present. If platinum is absent, as in good cornet gold, carefully purified sulphurous acid gas is a more convenient precipitant than oxalic acid, and may be substituted for it without any ill effects. Platinum is not precipitated from its solution as tetrachloride by sulphurous acid. Sulphurous acid acts in a few minutes in cold solutions, but if oxalic acid is used, it is better to warm the solution and to leave it to stand for three or four days. If oxalic acid is used, the precipitate of gold will consist largely of scales and plates.

[1] T. K. Rose, in *Chem. News*, vol. lxvi. (1892), p. 271.
[2] P. H. Argall, *Mill and Smelter Methods of Analysis*, 3rd edit., 1908, p. 73.
[3] T. K. Rose, *Metallurgy of Gold*, 5th edit., 1906, p. 488.

After the precipitated gold has settled, the acid solution is syphoned off and the gold transferred to a large flask and repeatedly shaken with cold distilled water, closing the mouth of the flask with a watch-glass. The gold is then repeatedly washed very thoroughly with hot water and turned out into a porcelain basin, covered over, dried, melted in a clay pot, and poured in an iron mould, which is neither smoked nor oiled, but rubbed with powdered graphite, and then brushed clean with a stiff brush. The ingot is cleaned by brushing and heating in hydrochloric acid, dried, and rolled out. The rolls must be clean and bright, and free from grease. The surface of the rolled-gold plate is again cleaned by scrubbing with fine sand and ammonia and also with hydrochloric acid, and is scraped with a clean knife before being used as proofs in bullion assay.

The amount of gold prepared at one time in large bullion-assay offices is usually from 40 to 50 ounces, but only 20 ounces can be washed in a single flask of a capacity of 160 ounces or 4½ litres.

This method of preparing pure gold is, with slight modifications, in very general use. For all practical purposes assayers regard the gold thus prepared as absolutely pure, i.e. 1000 fine, but in cases where a very high decree of accuracy is desired, it would no doubt be a very close approximation to the true purity of the metal to state the fineness as 999·9 per 1000.[1]

Metallic aluminium is also used for precipitating the gold.[2] The gold solution is prepared in the ordinary way and copiously diluted with distilled water, then allowed to flow slowly into a beaker containing a piece of pure aluminium sheet. A perfect precipitation of the gold immediately takes place. The aluminium is removed, the gold well washed in distilled water, dried, and melted, resulting in metal 1000 fine.

SILVER.—Symbol Ag (argentum) ; atomic weight, 107·88.
Specific gravity, 10·53 ; melting-point, 962° (Berthelot).

Silver expands on fusing and contracts again on solidifying. The decrease in volume on solidification is stated by Roberts-Austen and Wrightson[3] to be 11·2 per cent. If silver is kept in a state of fusion for a considerable time in contact with the air or oxygen, it is capable of absorbing about twenty-two times its own volume of oxygen, which it again liberates (with the exception of 0·7 volume) at the moment of solidification. The oxygen evolved often bursts through the outer crust of solidified metal with considerable violence, ejecting portions of the still liquid silver as irregular excrescences constituting the phenomenon known as the "spurting," "spitting," or vegetation of silver. This may be prevented by sprinkling charcoal powder on the melted metal. The presence of small quantities of copper also prevents the spurting of silver. Silver alloyed with as much as one-third of its weight of gold still retains the power of absorbing oxygen and spurting on solidification, but larger proportions of gold prevent the action (Percy).

At very high temperatures silver is slightly volatilised, its vapour in the oxyhydrogen blow-pipe being of a pale blue colour inclining to purple.

Commercial Forms of Silver.—As in the case of gold, metallic silver comes into the market in the form of ingots or bars, but silver bars are made much larger than gold, often weighing 1000 or 1200 ounces.

Mixed bars, containing both gold and silver, are seldom cast of a greater weight than 600 ounces in mills (Rose).

---

[1] See Phelps, "The Accuracy of the Gold Bullion Assay," *Journ. Chem. Soc.*, 1910, vol. xcvii. p. 1275.
[2] J. W. Pack, "Assaying of Gold and Silver in U.S. Mint," in *Min. and Sci. Press*, Nov. 14, 1903.
[3] *Proc. Phys. Soc.*, v., 1884, pp. 97-104.

Refined silver bullion is usually sold in the form of ingots, which weigh from 1000 to 1200 ounces, and often contain over 99·9 per cent. of silver.

For industrial purposes fine silver is supplied in the form of sheet, and of "grain" or granulated metal.

Alloys of Silver.—Silver is capable of combining with most of the common metals by direct fusion of the constituents, but the alloys of silver with copper may alone be said to have any important industrial applications.

Silver and Copper combine when melted together in any proportions, the resulting alloys being comparatively homogeneous, though ingots of the alloys are not absolutely identical in composition throughout.   With alloys containing from 5 to 30 per cent. of copper the interior of the ingot is usually the richest in silver, while with those containing from 30 to 90 per cent. of copper the exterior of the ingot is the most highly argentiferous.   The alloy containing 28·107 per cent. of copper and 71·893 per cent. of silver is almost perfectly homogeneous (Levol).

These alloys are white in colour until the copper amounts to 50 per cent. of the alloy, while the tint becomes more red with the increase in the amount of copper above this limit.

The silver-copper alloy of chief industrial importance is the alloy containing 92·5 per cent. silver and 7·5 per cent. copper, and designated "Standard" or "Sterling" silver.   This alloy is the English standard, fixed by law for the purpose of coinage and for the manufacture of silver wares.   The "Standard" silver alloy for most other countries contains 90 per cent. of silver (see also Industrial Alloys, page 337).

Silver and Lead.—Silver and lead mix together in all proportions while they are molten, but on solidification they separate, so that the alloys are not uniform in composition.   In commerce these alloys constitute what is designated "Base Bullion."   Owing to the segregation which it undergoes, the valuation of argentiferous lead is a matter of difficulty.   When silver-lead alloys are melted in the air the lead is oxidised and the silver remains unaltered.   At a red heat the action is rapid and the litharge (lead oxide) produced becomes molten and is far less viscous than the molten metals.   The cupellation process (see page 157) is based on these properties.

Silver and Gold.—See Gold Alloys.

Silver and Platinum.—These alloys are used in jewellery and dentistry. Silver and platinum do not readily alloy, the platinum having a strong tendency to settle to the bottom in molten mixtures of the metals on account of its high density.

Silver and Tin.—These metals unite readily when fused together to form alloys that are hard and more or less brittle.   The alloys containing between 40 and 60 per cent. of silver are used by dentists.   They are mixed with about an equal weight of mercury, and amalgams are formed which are pasty at first and subsequently become hard.

Solubility of Silver.—Silver is readily soluble in nitric acid even when rather dilute, forming argentic nitrate (commonly known as *lunar caustic*), with liberation of oxides of nitrogen.   Very strong acid does not act so rapidly.   Its best solvent is dilute nitric acid consisting of equal parts of acid and water by measure :

$$6Ag + 8HNO_3 = 6AgNO_3 + 4H_2O + N_2O_2.$$

Hot concentrated sulphuric acid dissolves silver slowly, converting it into argentic sulphate, with the formation of sulphur dioxide.   Dilute sulphuric acid does not act upon the metal.

$$2Ag + 2H_2SO_4 = Ag_2SO_4 + 2H_2O + SO_2.$$

The sulphate is not very soluble in water, and easily crystallises.

Hydrochloric acid has only a very slight surface-action, scarcely noticeable. The metal is dissolved by potassium cyanide in the presence of air or certain oxidising agents. The reaction may be represented as follows :—

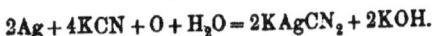

$$2Ag + 4KCN + O + H_2O = 2KAgCN_2 + 2KOH.$$

Alkalies have no action on silver.

Reactions of Silver.[1]—Hydrochloric acid and soluble chlorides give a white curdy precipitate of silver chloride, AgCl, which, on being warmed or stirred, becomes granulated in appearance, and very quickly settles. On exposure to light the white precipitate assumes a slate colour, which gradually deepens to a violet, and finally appears brown or black.

$$AgNO_3 + HCl = AgCl + HNO_3.$$

Silver chloride is quite insoluble in water and dilute acids, but slightly soluble in concentrated nitric and hydrochloric acids : soluble in a concentrated solution of sodium chloride (brine). It readily dissolves in ammonium hydrate, forming the compound $2AgCl, 3NH_4OH$. Nitric acid decomposes this compound, causing the reprecipitation of AgCl, which is practically insoluble in that acid.

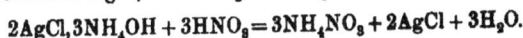

$$2AgCl, 3NH_4OH + 3HNO_3 = 3NH_4NO_3 + 2AgCl + 3H_2O.$$

Silver chloride is soluble also in potassium cyanide, KCN, being first converted into silver cyanide, which dissolves in excess of KCN, forming the double potassium silver cyanide $KAg(CN)_2$.

It is readily soluble in sodium thiosulphate, with the formation of a double thiosulphate : thus,

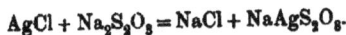

$$AgCl + Na_2S_2O_3 = NaCl + NaAgS_2O_3.$$

Silver chloride melts without decomposition at 451° C., it undergoes a physical change only, and resolidifies to a horn-like mass (horn silver). After fusion it dissolves with much difficulty.

Caustic Alkalies, KHO or NaHO, give a greyish-black precipitate of silver oxide, Ag_2O, insoluble in excess of the caustic alkalies, but readily soluble in ammonium hydrate. Ammonium hydrate when gradually added also precipitates silver oxide, which is readily soluble in excess. If the silver solution is acid, ammonium hydrate gives no precipitate but forms a soluble double salt (see above).

Alkaline Carbonates, K_2CO_3 or Na_2CO_3, give a white precipitate of silver carbonate, Ag_2CO_3, insoluble in excess of the precipitant : soluble in ammonium carbonate, ammonium hydrate, and nitric acid.

Sulphuretted Hydrogen, H_2S, or ammonium sulphide, $(NH_4)_2S$, produces a black precipitate of silver sulphide, Ag_2S, insoluble in dilute acids, except boiling dilute nitric acid, which converts it into nitrate. The H_2S, which by double decomposition is set free, is acted upon by the nitric acid, with the precipitation of sulphur and evolution of nitric oxide, thus :

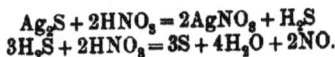

$$Ag_2S + 2HNO_3 = 2AgNO_3 + H_2S$$
$$3H_2S + 2HNO_3 = 3S + 4H_2O + 2NO.$$

Silver sulphide is insoluble in ammonium hydrate, ammonium sulphide, or potassium sulphide.

Potassium Iodide, KI, or hydriodic acid, HI, gives a yellowish precipitate of silver iodide, AgI, insoluble in dilute nitric acid and almost insoluble in ammonium hydrate, which distinguishes it from silver chloride.

[1] Newth, *Chem. Analysis.*

Potassium Cyanide, KCN (or other soluble cyanide other than mercuric cyanide), gives a white curdy precipitate of silver cyanide, AgCN, insoluble in dilute nitric acid, but readily soluble in excess of the precipitant, giving a double cyanide $KAg(CN)_2$. Silver cyanide is soluble in ammonium hydrate, but reprecipitated by dilute nitric acid : it is soluble also in sodium thiosulphate. Silver cyanide closely resembles silver chloride; it is, however, readily distinguished by the fact that when boiled with hydrochloric acid or concentrated nitric acid it is decomposed with evolution of hydrocyanic acid ; and also that, when heated alone, gives off cyanogen, and leaves a black residue of metallic silver and paracyanogen, a non-volatile black compound which, according to Newth, is a polymeride of cyanogen expressed by the formula $(CN)x$.

Potassium Thiocyanate, KCNS, or ammonium thiocyanate, $(NH_4)CNS$, gives a white precipitate of silver thiocyanate, AgCNS, somewhat resembling the chloride in appearance.

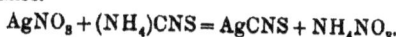

$$AgNO_3 + (NH_4)CNS = AgCNS + NH_3.$$

Silver thiocyanate dissolves in ammonium hydrate, but less easily than either the cyanide or chloride, and is insoluble in dilute nitric acid. It is slowly dissolved by hot concentrated nitric acid. When heated, the thiocyanate is decomposed, and in contact with the air takes fire and burns with a pale blue flame, leaving a black residue.

Reduction of Silver Salts to the Metallic State.—Silver compounds are readily reduced with precipitation of metallic silver by certain organic substances, as sugar, tartrates, aldehydes, etc. Thus a solution of sugar when added to a hot solution of silver nitrate gives a greyish-white powder of metallic silver. Many inorganic salts also, which act as reducing agents, precipitate metallic silver, from solutions of its salts, *e.g.* ferrous sulphate.

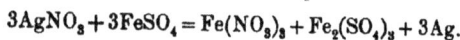

$$3AgNO_3 + 3FeSO_4 = Fe(NO_3)_3 + Fe_2(SO_4)_3 + 3Ag.$$

Many metals are capable of reducing silver compounds. Thus, if a strip of zinc, copper, or aluminium is immersed in a solution of silver nitrate, the silver is thrown down in a crystalline or spongy state according to the condition of the solution.

The reducing action of metals is often employed in the laboratory to convert precipitated silver-chloride into metallic silver. Iron, zinc, and aluminium are frequently employed for the purpose ; the reaction in all cases is similar :—

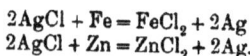

$$2AgCl + Fe = FeCl_2 + 2Ag$$
$$2AgCl + Zn = ZnCl_2 + 2Ag.$$

Mercury in contact with silver chloride similarly reduces it to metallic silver, with the formation of mercurous chloride :—

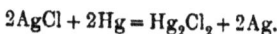

$$2AgCl + 2Hg = Hg_2Cl_2 + 2Ag.$$

Tests for Silver.—Silver can be easily detected when in solution by the precipitation of the white curdy chloride, insoluble in water and nitric acid, and soluble in ammonium hydrate. This test is the one most frequently employed, and serves for the recognition and separation of silver. The best confirmatory test is made by wrapping the precipitate in a small piece of sheet lead, and cupelling, when the silver will be left in the metallic state as a small globule and is easily recognised.

A new delicate test for silver is given by A. W. Gregory.[1] When a solution of a silver salt is added to a mixture of 20 c.c. of aqueous ammonium salicylate

[1] *Chem. Soc. Proc.*, 1908, xxiv. 125.

(20 grammes of salicylic acid neutralised with ammonia, a slight excess of the latter added, and the solution made up to 1 litre) and 20 c.c. of ammonium persulphate solution (50 grammes in 1 litre) an intense brown colour is produced. By this reaction 0·01 milligramme of silver can be detected. As lead does not give this reaction, silver may be tested for in presence of this element; 1·1 milligramme of silver may be detected in this manner in the presence of 0·2 gramme of lead. The brown substance formed in this reaction does not contain silver, and it is probable that the silver salt acts as a catalyst, since on boiling a solution of ammonium salicylate with ammonium persulphate, a similar brown colour is produced.

Preparation of Pure Silver.—The simplest method of preparing pure silver is to dissolve commercial silver in dilute nitric acid (1 to 1), allow the liquid to stand, syphon off from any residual gold, etc., dilute with hot water, precipitate the silver with hydrochloric acid, stir well, allow to settle, and wash by decantation. The silver chloride becomes granular by continuous stirring and settles readily. When the decanted liquid no longer shows a trace of impurity, the silver chloride is allowed to settle and the excess of solution decanted off. Sufficient hydrochloric acid is then added to make the remaining solution distinctly acid, and plates of wrought iron (best Swedish bar) are placed in the silver chloride. The white chloride next the iron at once begins to blacken and is entirely reduced to metallic silver in the course of a few days. When reduction is complete, the grey silver powder, which is contaminated with chloride of iron, is washed by continuous stirring first with dilute hydrochloric acid, then with pure water, renewed many times.

The decanted liquid is tested for iron with ammonium thiocyanate, NH_4CNS, and when no red coloration is given, indicating that the liquid is free from iron, the silver is dried and melted under a little charcoal in a graphite or clay crucible, and poured into a black-leaded iron mould. Silver thus prepared is usually over 999·9 fine.

A sheet of commercial aluminium is sometimes used in place of iron to reduce the chloride. The metallic silver thus produced is washed, heated with hydrochloric acid, washed thoroughly to remove all acid and aluminium chloride, then dried and melted as described.

The washed silver chloride may also be reduced by transferring it to a porous pot which has been previously steeped in hydrochloric acid for some days and thoroughly washed by long, continued standing in clean distilled water frequently renewed. The porous pot is placed in water in an outer vessel, in which a cylinder of wrought iron is dipped as an electrode, and a little hydrochloric acid added to start the action. A cathode of pure silver or platinum is plunged into the silver chloride and connected with the iron anode through an electric battery. The white chloride at once blackens next the cathode, and in the course of a few days the reduction is complete. The chloride of iron which forms diffuses into the inner pot and contaminates the silver, but may be removed by washing. When free from impurity the silver powder is melted as above described. The wrought-iron anode may be replaced by carbon anodes, but the action is slower and the outer liquid becomes strongly impregnated with chlorine gas and should be renewed once or twice a day. The silver in this case is not contaminated with impurity. This method is used in the preparation of pure silver at the Royal Mint, London.[1]

Another convenient method very frequently used is to reduce the silver chloride by fusion with anhydrous sodium carbonate, thus :

$$4AgCl + 2Na_2CO_3 = 4Ag + 4NaCl + 2CO_2 + O_2,$$

[1] T. K. Rose's *Precious Metals*, p. 130.

or in the presence of charcoal powder to combine with the oxygen represented in this equation :

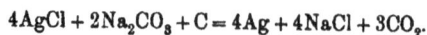

$$4AgCl + 2Na_2CO_3 + C = 4Ag + 4NaCl + 3CO_2.$$

The silver chloride after washing is dried and mixed with an equal *bulk* of anhydrous sodium carbonate, and the mixture fused in a clay crucible first at dull redness, and then gradually raised to bright redness. The reduction is complete in about thirty minutes, and when tranquil fusion has taken place the contents of the crucible are poured into a suitable iron mould and allowed to cool, after which the adhering slag of sodium chloride produced by the reaction is removed from the silver by means of a hammer, and the silver finally cleaned in hot water. As carbonic acid is given off freely during the reduction, care should be taken not to have the crucible more than three parts full, and to avoid a high temperature at the beginning, otherwise the charge may boil over and occasion loss. The silver thus obtained is practically pure, but the fineness is slightly improved by remelting under charcoal powder. The reduction may also be effected by fusion with a mixture of chalk and carbon thus :

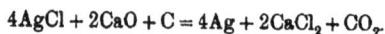

$$4AgCl + 2CaO + C = 4Ag + 2CaCl_2 + CO_2.$$

CHAPTER VII.

PRECIOUS METAL ORES: VALUATION OF ORES.

Ores.—"The term *ore* is applied by the metallurgist only to those minerals from which, on a large scale, metals may be obtained with profit."

Ores are always accompanied by more or less extraneous earthy matter to which the terms *gangue*, vein stuff, or matrix are applied. The gangue may consist of one or more of the following substances :—silica, in the form of quartz ; various silicates, such as felspar and mica, in granite, hornblende, clay-slate, etc. ; carbonates of lime and magnesia, barytes, fluor-spar, iron pyrites, iron oxides, etc.

One of the most striking differences between the ores of the precious metals and those of almost all other metals lies in the extremely small proportion which the gold and silver bears to the worthless gangue with which it is accompanied.

For example : Copper ores are regarded as "rich" when they contain 15 per cent. of metal, whilst ores yielding a few ounces of silver or gold per ton are of commercial value. The *average* value of the ore now treated on the Rand in South Africa is well under 7 dwt. of gold to the ton, but ore of half that value is treated profitably. The gold and silver in ores are usually referred to as the *values.* When ores or metallurgical products of other metals contain gold they are said to be *auriferous,* and when they contain silver they are described as *argentiferous.* Thus a copper ore carrying gold may be described as auriferous copper ore, and one carrying silver as an argentiferous copper ore.

Gold Ores.—Gold occurs very widely distributed in nature, and almost always, but not invariably, in the native, *i.e.* metallic state, usually in veins or lodes in rock formations and in alluvial deposits formed by the disintegration of auriferous rocks.

Native gold is never found pure, but sometimes contains less than 1 per cent. of impurities. It is always alloyed with a varying proportion of silver and smaller quantities of copper and other base metals. It is sometimes associated with platinum and metals of the platinum group ; thus gold from Mexico was found to contain as much as 34 to 43 per cent. of rhodium. The quality of native gold and silver is always expressed in parts per 1000, pure gold and pure silver being described as 1000 fine. Thus an alloy containing 90 per cent. of gold or of silver and 10 per cent. of base metal would be described as 900 fine.

The following analyses will serve to show the general composition of native gold :—

	Australia.	California.	West Africa	Russia.[1]	Wales.
Gold	94·64	89·10	97·81	98·96	89·83
Silver	4·95	10·50	2·19	0·16	9·24
Copper	...	...	...	0·05	...
Iron	0·41	0·20	...	0·35	...
	100·00	99·80	100·00	99·52	99·07

[1] Formerly obtained at Katerineburg in the Urals.

The purest native gold yet found is that from the Pikes Peak Mine at Cripple Creek, Colorado, which contained 99·9 per cent. of pure gold.[1]

The average fineness of gold from different countries is as follows :—

	Gold.	Silver.	Base.
South Africa (banket ore)[2] . .	875	125	...
Western Australia . . .	831	107	62·0
Australia[3]	921	39·0	40·0
,, ...	868	89·0	43·0
South America (Andes) . .	600–700	...	...
Colorado	800	200	...
California	880	...	...
Canada	850–900	...	...

Native gold occurs in the form of minute scales, threads, or grains disseminated through the ore. It is sometimes visible to the eye without magnification, but is more often in too fine a state of division to be seen. Gold also occurs in copper and iron pyrites and with other metallic sulphides and arsenides of base metals, most probably in the metallic state, not as a sulphide. Gold is also met with in combination with tellurium as telluride. The tellurides of gold, containing the compound $AuTe_2$, are very common.[4] They occur in considerable quantities in Western Australia, Colorado, and Transylvania, and have been reported from many other localities. Various mineralogical names have been assigned to the tellurides from different localities, the best known being *Calaverite*, which has the composition $AuTe_2$; *Sylvanite* or graphic tellurium, which appears to be a variable mixture of $AuTe_2$ and Ag_3Te_4 ; *Petzite*, Ag_2Te, in which some silver is replaced by gold ; and *Nagyagite*, or foliated tellurium, which contains a considerable percentage of lead. The tellurides are for the most part dark grey or black in colour, rarely silver-grey. They are often mixed with metallic gold, which sometimes gives them a brassy-yellow colour.

The greater part of the rock material or "gangue" forming the veins or lodes in which gold is found is usually quartz, but various silicates also occur, and sulphides, arsenides, or antimonides of base metals are almost invariably present. The upper portions of these veins near the surface of the ground are often found to have undergone oxidation from the influence of the atmosphere, so that the sulphides of iron, etc., have been converted into oxides. These oxidised portions are termed "gossans," and consist of a loose mixture of silica and iron oxide, and the finely divided gold contained in them is often readily separated.

On the Witwatersrand (Rand), South Africa, the gold ores consist almost entirely of "banket" formation, a conglomerate consisting of water-worn pebbles of translucent quartz set in a matrix composed mainly of iron oxides and silica. The gold exists in the cement, the pebbles being generally barren. The ore contains on an average about 7 dwt. gold per ton.

Veins of iron and copper pyrites, galena, argentite, etc., traversing crystalline rocks are frequently sufficiently auriferous to permit of the profitable extraction of the gold by smelting processes, although the gold may be present in insufficient quantity to render it visible in the ore.

Auriferous Black Sands.—The heavy sands derived from gold-bearing gravel

[1] Furman in *Colliery Manager and Metal Mines*, October 1896, p. 89.
[2] R. B. Young, *Proc. Geo. Soc. of S. Africa*, 1911.
[3] "Gold deposited at Sydney and Melbourne Mints," from *Mint Report*, 1910.
[4] T. K. Rose, *The Precious Metals*, p. 67.

deposits are generally known as black sands by reason of the preponderance of dark-coloured iron minerals they contain. The black sand from some districts consists largely of titaniferous iron ore as well as chromite. These sands are occasionally rich in gold and to a less extent in platinum. Most of the value is found in the particles below ⅛ inch diameter.

Minerals resembling Gold.—Several minerals when disseminated through rock material resemble native gold in appearance and are sometimes mistaken for gold by those not familiar with minerals.

Yellow mica, which is found in some sands on the western coast of America and in other places, looks remarkably like gold ; but when the sand is washed in an iron pan, the mica, being very light, floats off, even before the grains of quartz that accompany it, and can thus be easily distinguished from gold, which on account of its density always remains on the pan. Copper pyrites and iron pyrites also resemble gold, especially when they are slightly oxidised on the surface. When these sulphides are heated they are decomposed and converted into oxides and are thus readily distinguished from gold, which retains its colour on heating and is not altered.

Silver Ores.—Silver occurs in the native state, but, unlike gold, it is generally combined with sulphur, chlorine, bromine, and iodine in its ores. It occurs also in combination with other metals, as arsenic, antimony, etc., and with mercury as a native amalgam, whilst also frequently occurring in ores of other metals, such as lead, copper, zinc, etc., constituting argentiferous ores of these metals. This latter class of ores frequently contains sufficient silver to render its extraction profitable, in which case such minerals are also considered as ores of silver. The ores of silver occur in veins traversing gneiss, granite, clay-slate, quartz, calc-spar, mica-schist, etc., and are usually associated with ores of lead and copper, blende (ZnS), spathic iron ore, brown hæmatite, pyrites, ores of nickel and cobalt, heavy-spar, quartz, earthy carbonates, etc. The ores of silver are more or less abundantly distributed over all parts of the world.

Native Silver occurs in sufficient quantity to constitute one of the chief ores of the metal. Native silver is rarely found perfectly pure, but usually contains also gold, copper, platinum, or other metals in larger or smaller proportions, and is itself always present in small quantities in native gold.

Native silver occurs in laminated or filamentous masses or capillaries and in grains minutely disseminated through silver ores and other minerals.

Amongst the more important natural compounds of silver are the following :—

Argentite, or silver-glance, Ag_2S, is one of the most important ores of silver. It occurs in black soft sectile crystals and masses, which, when pure, contain 87·1 per cent. of silver, but it rarely occurs pure as an ore, being often associated with the sulphides of lead, copper, iron, zinc, antimony, arsenic, and tin, or with the ores of nickel and cobalt: so that the ore containing this mineral as its source of silver yields much smaller proportions of silver than the pure sulphide. Silver-glance constitutes one of the richest, most abundant, and usual forms of occurrence of silver found in veins, and sometimes in large masses. It occurs in great quantities at the Comstock lode, Nevada, Mexico, and in many other localities.

Other widely distributed silver sulphides are :—

Ruby Silver Ores.—Proustite, $3Ag_2S + As_2S_3$; pyrargyrite, $3Ag_2S + Sb_2S_3$; and stephanite or brittle silver ore, $5Ag_2S + Sb_2S_3$.

Fahl ore, or grey silver, another important ore, is a mixture of antimony and arsenic sulphides, with sulphides of silver, mercury, copper, zinc, and iron. Its formula is $4(Ag_2Hg,CuFeZn)S(SbAs)_2S_3$. It may contain any proportion of silver up to 31 per cent., arsenical grey silver being poorer than the antimonial varieties. Ores consisting of fahl ore are sometimes very rich in silver, containing occasionally as much as 1400 ounces of silver per ton of ore, but the more usual content of these ores is from 2 per cent. to 4 per cent. of silver.   Fahl ore occurs in Germany, the Tyrol, Bolivia, Nevada, Mexico, Colorado, etc.

Polybasite, $9(Ag_2Cu)S + (SbAs)_2S_3$, is also a commonly occurring mineral, especially in America.

Stromeyerite is an argentiferous copper-glance, sulphide of silver and copper, Ag_2S,Cu_2S.   The pure mineral would contain 53·08 per cent. of silver, but it is usually more or less mixed with copper-glance (Cu_2S), and hence yields only from 3 per cent. to 30 per cent. of silver.   Other sulphide minerals containing silver also occur, but they are less important than those given above.

In combination with tellurium it occurs as *hessite*, Ag_2Te; and *petzite*, a telluride of gold and silver; and combined with chlorine as *horn silver* or *kerargyrite*, $AgCl$; and *embolite*, $Ag(ClBr)$, all of which are of some importance as ores. Horn silver occurs as a soft, horny-looking mass and accompanies other silver ores somewhat abundantly in Peru, Chili, and Mexico.

Silver also occurs in the ores of lead, copper, zinc, nickel.   Galena, PbS, often contains considerable amounts of silver (up to 5 per cent.), and zinc blende, ZnS, chalcopyrite, CuS,FeS,FeS_2, bournonite, and other minerals also yield more or less silver.

Owing to the comparatively high value of gold and of silver, and the facility with which they are extracted from ores or minerals containing them, their separation is rendered profitable although the ores contain only an exceedingly small percentage of the metal.   Thus gold ore containing only 4 dwt. of gold per ton is profitably treated in South Africa; and in the case of silver, while certain of the silver minerals in their pure state contain a very high percentage of the metal, yet the ores usually treated in the smelting works of Europe and America contain only a comparatively small proportion of silver.

Argentiferous galena is extensively treated for silver, the facility with which the silver may be concentrated in the lead and subsequently separated by the process of cupellation rendering its extraction profitable if it be present to the extent of only 2 ounces or 3 ounces per ton of lead, and a similar remark extends to the extraction of silver from certain copper ores.   The amount of gold and silver in precious metal ores varies considerably, and no exact statement can be made as to what amount of precious metal will constitute a rich or high-grade ore and a poor or low-grade ore.   An ore that is considered rich in one mining district may be regarded as poor in another district.   For assay purposes, however, a gold ore is as a rule considered to be rich if it assays over 3 ounces of gold per ton, and poor if the gold is less than 10 dwt. per ton: between these amounts the ore would be considered to be of average quality.

In the case of silver, an ore containing over 200 ounces of silver to the ton would as a general rule be regarded as rich, while an ore with any amount under 50 ounces would be classed as poor.   Average ores would contain amounts between these.

Valuation and Purchase of Ores.—The assayer should be familiar with the principles of ore valuation.   The value of an ore depends upon the nature of the metal it contains and the ease with which its extraction is attended.   The *gross* value of an ore, based on the market price of the metals contained, is termed the "assay value," as it is calculated from the result of the assay. The "assay value" merely indicates the money value of the ore if the precious metal could be extracted *without cost*.   The "assay value" must, therefore, be distinguished from the real value of the ore, which is the gross value less the total cost of extraction; in other words, the net return to the proprietors.   It is necessary to take note of the distinction, as the term "value" is frequently used indiscriminately as a synonym for "assay value" to indicate the *gross* value of an ore, thus leading to confusion.   For instance:[1] an ore that contains 7·6 dwt. gold per ton has not the "value" of 7·6 dwt., which is about 32s., for from that amount must be subtracted the cost of treatment and realisation. In some cases the "assay value" does not even roughly indicate the actual value of the ore.   Assay values of gold and silver ores and products are usually represented in ounces and decimals or in pennyweight and decimals per ton; the method of representing the assay values in ounces, pennyweight, and grains per ton is now being discontinued.   The values are expressed in terms of fine gold and fine silver respectively, not as "bullion."   In America the assay values of gold and silver ores, etc., are represented in money value in dollars per ton according to the prevailing market price of the metals.   Thus, taking the price of gold at 20·67 dollars per ounce and silver at 55 cents per ounce, the assay value of an ore containing 10 ounces of gold per ton and 150 ounces silver would be—

Gold,   10 ounces at $20·67 per ounce   .   .   . $206·7
Silver, 150   „   at $ 0·55   „   .   .   .   . $85·5

Assay value or gross value of ore in gold and silver $292·2 per ton.

Owing to the higher price of gold and silver compared with other metals, the valuation of precious-metal ores and products is carried to a higher point of accuracy than in the case of ores of lead, copper, zinc, etc.   In the sale of ores both the gold and silver are always paid for when present above a certain amount, which varies with the different classes of ore and in different mining centres.   As a general rule, gold is not paid for if below $\frac{1}{20}$ (·05) ounce per ton, and silver if below 2 ounces per ton.   The minimum amount of silver paid for varies, however, with the richness of the ore or product.   The price of fine or pure gold is definitely fixed at 85s. per troy ounce, and is the same in all civilised countries.   The gold content of ores, etc., determined by assay, when expressed in money values, is calculated on this price.   In the United States currency the price is 20·67 dollars per troy ounce of fine gold, but in the purchase of ores in America for smelting the gold is usually priced at 20·0 dollars per ounce, as this is more convenient for calculation and is the price adopted by smelters.[2]   The price of English "Standard" gold (91·66 per cent. gold) in the London market varies from £3, 17s. 9d. to about £3, 18s.

The price of silver fluctuates considerably, and the value of the silver in ores, etc., is always determined by the prevailing market price of fine silver.   The price of fine silver may be taken at 2d. or 2½d. above the London market quotation for "Standard" silver, which contains 925 parts of silver and 75 parts of copper per 1000.   The average price of silver in the London market for the year 1912 was 28$\frac{1}{16}$d. per ounce standard.   The lowest quotation was 25$\frac{3}{16}$d. and the highest 29$\frac{11}{16}$d.   The American gold and silver standards contain 900 parts of gold or of silver and 100 parts of copper per 1000.

As indicated above, the value of precious-metal ores is not merely dependent on the amount of gold and silver they contain, but also on the cost of treatment, which varies with the nature of the ore and method employed.   A deduction from the gross value of the ore is always made to cover the cost of treatment.

In the case of ores treated by smelting processes the general rule is for the smelting company to buy the ores from the various mines and smelt them in consideration of a smelting charge which is deducted from the gross value of the

[1] Consult note on "Values," *Mining Magazine*, vol. iv., 1911, pp. 251-2.
[2] See "Table of Assay and Coinage Values for Gold," by R. J. Holland, *Min. Mag.*, vol. xi., 1905, pp. 525-530.

ores when paying for the same to the mines. The decline in the price of silver and other metals within recent years has brought down the net value of ores to the miner, but on the other hand the cost of mining and smelting has been largely reduced by improved facilities for transportation, by cheaper labour and by cheaper materials, etc., enabling ores, that formerly were worthless, to be sold at a profit.

Ores in which the precious metals exist in such a state that they can be readily recovered by pulverising and amalgamation are said to be *free-milling*, while ores that require some preliminary treatment before the precious metals can be extracted by ordinary methods of amalgamation or that resist the action of heat and chemical reagents, and are therefore difficult to treat, are said to be *refractory*. Ores that can be smelted readily are sometimes termed "docile" ores.

When the precious metals are to be extracted by smelting processes the value of the ore is considerably affected by the nature of the gangue material, as this largely decides the cost of treatment. In the large majority of cases the predominating gangue material is quartz (silica), which is infusible alone and for smelting purposes requires the addition, in suitable proportions, of fluxes [1] such as iron oxide and lime (calcium oxide), to convert it into a fusible compound or slag; consequently, from the smelter's point of view, the proportion of silica in an ore very largely determines its value. The value of an ore is also influenced by the presence of metalliferous minerals. Lead and copper, if present in any considerable amount, are always paid for by the smelter as they are recovered in the smelting operation, and in addition to the value of the metals themselves one or the other is indispensable to the operation, as they are utilised as agents for the concentration of the precious metals disseminated through the furnace charge. When the ores are smelted the lead or copper they contain is reduced to metal, which sinks slowly through the melting material in the furnace and alloys with the gold and silver, thus collecting and concentrating them in the mass of lead or copper which accumulates at the bottom of the furnace, forming what is termed "base bullion."

Zinc is frequently present in precious metal ores, but is not paid for, as it is not recovered in smelting. When present to the extent of 10 per cent. or more it depreciates the value of the ore, as it gives trouble in the smelting operation, as described below under Treatment Charges.

The following system of valuation, employed on the American continent, where the extraction of silver and gold from their ores by smelting processes is conducted on such a large scale, is, with slight differences in detail, almost universally adopted for the valuation of precious metal ores to be treated by smelting processes.[2] The gross value of the ore is calculated from the amount of the different metals present, and the market price of the respective metals. The assay value of the gold, silver, and lead is always determined by the dry or fire assay. Copper when present is almost invariably determined by volumetric or electrolytic methods.

Silver is always paid for on the assay value, at a given rate per ounce (troy), based on New York quotations, with a reduction of 5 to 10 per cent. to cover loss in smelting. When the price of the metal is high, only 90 per cent. of the contents of medium and low-grade ores is paid for; but when low, 95 per cent. is almost invariably paid, as in the case of high-grade ores. Assays are reported to $\frac{1}{10}$ ounce, and payment is made accordingly.

Gold also, when amounting to $\frac{1}{20}$ ounce and upwards per ton of ore, is paid for at 95 per cent. of its full maket value.

---

[1] The use of fluxes is dealt with in Chapter X., page 127.
[2] H. F. Collins, "Smelting Processes," 1893, vol. cxii., *Proc. Inst. Civil Eng.*

Lead.—The manner of paying for the lead contents is variable, the principle being to pay for what can be recovered, less refining charges and freight to the refinery ; but a sort of sliding scale is adopted according to the richness of the ore, so as to cover increased loss in working poor ores. Thus, for example, the lead in poor ores containing between 10 per cent. and 20 per cent. will be paid for at, say, 20 cents to 30 cents per unit (*i.e.* 1 per cent. of a short ton, or 20 lbs.), whilst between 30 per cent. and 40 per cent. it will be paid for at 40 cents to 45 cents per unit, the price of course varying with the market price of the metal. Under 5 per cent. of lead is not paid for.

Copper is not paid for in lead ores (and this is one of the sources of profit), but when the ores are bought comparatively rich in copper and poor in lead, the former metal is paid for at a settled rate per unit according to the market price of copper, in which case usually nothing is paid for lead, whatever its amount, the practice of not paying for both metals together always leaving a good margin for profit in refining. A sliding scale is usually adopted for the payment of copper, as in the case of lead. For example, a low price is paid per unit for copper under 5 per cent., a little higher price per unit for copper from 5 to 10 per cent., and a still higher price when over 10 per cent., and so on.

Charges for Treatment.—To cover the cost of treatment the smelter makes certain charges which are deducted from the gross value of the ore. The losses of gold and silver in treatment are usually assumed to be 5 per cent. The treatment charges vary enormously in different smelting centres according to the class of ore to be dealt with and the quantity and quality. They are also affected by local conditions such as fuel-supply, cost of labour, and freight. The smelting rate is chiefly fixed according to the fluxing qualities of the ores, and this, as previously stated, is regulated principally by the percentage of silica. Ores containing excess of fluxing constituents such as iron oxide, lime, etc., are termed "fluxing" ores. In districts where fluxing ores are not very plentiful they are in great demand as fluxes for the much more plentiful siliceous ores, and to procure them smelters will make only a nominal charge for treatment, whilst a very siliceous ore may have an almost prohibitory smelting charge put upon it or perhaps be refused altogether. In general it may be stated that when lead ores and "fluxing" ores are plentiful the treatment charges on them rise, whilst those on siliceous and "dry ores" fall, and *vice versa*. As a rule, the charges for smelting are arranged on a sliding scale according to the amount of silica present and other constituents in the ore. As a basis for calculation a price is from time to time fixed for an assumed ore with neutral or self-fluxing gangue ; that is, one in which the slag-forming components balance, leaving no excess of either silica or iron oxide, etc. An extra charge or penalty is made for each unit or per cent. of silica in excess of the neutral basis, and an allowance granted per unit when it is below the iron oxide, lime, etc. Manganese if present is reckoned as iron. A small allowance is made for all lime, magnesia, and baryta present. Ores containing zinc are very intractable, as part of the zinc enters the slag as oxide, and renders it both more pasty and more infusible. Smelters therefore always make a deduction (in America usually 50 cents) from the value of the ore for every unit of zinc in excess of a certain amount, usually 10 per cent., sometimes only 8 per cent. Special deductions are sometimes made in the case of arsenic. The necessity for determining the silica, iron, etc., is judged of according to the apparent mineralogical composition of the ore. From the foregoing remarks it will be evident that whilst only gold, silver, lead, and copper are paid for or mentioned in the bid of a smelting company for any particular lot of ore, every element in its chemical composition is taken into consideration in calculating its *value*, and extra charges (penalties) made or allowances granted on a sliding scale based upon the fixed smelting charge for the assumed neutral ore as stated above.

The method of calculating the value of an ore is well illustrated in the following Tables I. and II. given by Brunton [1] for the valuation of ores in the large smelting district of Denver or Pueblo, Colorado. The rates of course are changed from time to time, and the actual prices paid for silver and lead vary daily with quotations from the New York markets, but the figures serve to illustrate the method. The debit column includes penalties charged for objectionable constituents. The credit column includes the value of the metals present and the allowances granted for "fluxing" constituents. The fixed smelting charge for the assumed neutral ore is 6 dollars per ton.

Table I.—Table of Prices and Rates.

Ore.		Debit.	Credit.
		Dollars.	Dollars.
Au per oz.	. . . .	...	19·00
Ag ,, (per cent. N.Y. quotation)	...	
Pb per unit (with N.Y. quotation 4 cents per lb.) .	...	0·50
Fe ,, ...	...	0·15
CaO ,, ...	...	0·10
Mn ,, ...	...	0·15
SiO₂ ,, ...	0·13	...
BaSO₄ ,, (small percentages occasionally disregarded) .	0·05	...
Zn ,, ,, ,, ,, ,,	0·30	...
S ,, ,, ,, ,, ,,	0·15	...

As an example, an ore of the following composition,

Au.	Ag.	Pb.	SiO₂.	CaO.	Fe.	S.	Zn.	BaSO₄.	Mn.
Oz.	Ozs.	per cent.	per cent.	per cent.	per cent.	per cent.	per cent.	per cent.	per cent.
0·3	30	5	14	8	6	12	7	24	4

would be calculated by the table above given in the following manner, showing that after deducting the total charges, amounting to $12·92, the ore would have a net value in Denver of $14·68 per ton.

Table II.

Prices:—Silver, 60 cents per oz. ; lead, $4·00 per 100 lbs.

Ore.	Debit.	Credit.
	Dollars.	Dollars.
Au, 0 3 oz. at $19·00	...	5·70
Ag 30 ozs. at 95 per cent. of 60 cents ($0·570) . .	...	17·10
Pb, 5 per cent. at 50 cents	...	2·50
SiO₂, 14 ,, at 13 ,, ...	1·82	...
CaO, 8 ,, at 10 ,, ...	...	0·80
Fe, 6 ,, at 15 ,, ...	...	0·90
S, 12 ,, at 15 ,, ...	1·80	...
Zn, 7 ,, at 30 ,, ...	2·10	...
BaSO₄, 24 ,, at 5 ,, ...	1·20	...
Mn, 4 ,, at 15 ,, ...	...	0·60
Smelting charges per ton	6·00	...
Net value per ton in Denver or Pueblo . . .	14·68	...
	27·60	27·60

The following are examples of the valuation of ores containing precious metals in other countries [2] :—

[1] In discussion on paper by H. F. Collins, *ibid.*
[2] *Valuation of Ores*, A. Rzehulka, *Z. Angew. Chem.*, 1910, xxiii. 481-485.

For argentiferous lead ores in Germany the London price of lead (2 marks per 100 kilos of lead being considered equal to £1 per ton) and the Hamburg rate for silver bars are taken.

In the absence of exceptional quantities of objectionable impurities, galena (lead sulphide ore) is usually valued at the combined value of the lead and silver contained, less 4 marks per 100 kilos of ore for smelting, in the case of ores of 55 per cent. of lead and upwards ; 5·5 marks are deducted for ores of 40 per cent. and intermediate percentages are calculated proportionately. Ores of lower grade are subject to special terms. Silver is usually not allowed for if the percentage falls below 0·015, and sometimes if it is below 0·025.

In France the value of the lead and silver present is calculated at the London prices, and 50 to 65 francs per 1000 kilos of ore are deducted.

In Spain the value is calculated according to the following formula :—

$$V = (A - 0\cdot50)a + \frac{P - 4}{100}\beta - 5 \text{ reals.}$$

V being the value of a cwt. of ore at the mine.
A the silver in ounces per cwt. of lead.
a the price of silver per ounce.
P the percentage of lead.
β the value of lead per cwt.

The precious-metal ores smelted in England are exclusively confined to imported ores. Imported silver ores average about 250 to 300 ounces per ton, but run up to 3000 or 4000 ounces and down to 50 ounces per ton. The valuation of the ores and deductions for smelting charges are made in a similar manner to that already described.

Settlement.—The terms of settlement for the purchase of ores, etc., are always arranged between the purchaser and the seller before a lot is dealt with. The terms vary somewhat according to the nature of the ore, size of the lot, and other conditions, but the following description may be taken as representing the general practice. In the ordinary method of assaying ores that are sold the ore is sampled and the final sample thoroughly mixed in the presence of representatives of the buyer and seller. Each representative then usually selects four or more portions (sometimes less), which are each wrapped in a separate packet and sealed. The details of the lot, number and weight, etc., are written on each packet. Some of the packets are delivered to the assayer acting for the purchaser and some to the assayer acting for the seller. A certain number of the packets, agreed upon by buyer and seller, are held under seal in case of dispute, and is called the "referee" or "umpire" sample.

In Custom assay offices (see page 12) in America the final sample, weighing about 2 to 3 lbs., is usually subdivided into four portions, each of which is put into a paper bag and sealed. Two portions are assayed by the seller, one by the purchaser, while the fourth portion is retained under seal for an "umpire" assay, should such become necessary. In America the assays for the purchase and sale of ores are called "control assays" or simply "controls." The control assay, both by the purchaser and seller, is made in duplicate or triplicate and an average of the assays in either case is taken for comparison. If the results of purchaser and seller agree within a certain limit, depending on the value of the ore, settlement is made on the purchaser's assay, or sometimes on the average of the two assays. If these assays do not agree within specified limits, it is the practice of the purchaser and seller to reassay their own samples, or to exchange samples and reassay. If the results of the seller and the purchaser do not agree when the samples are reassayed, the "referee" (or umpire) sample is sent to an assayer mutually agreed upon, and his result is decisive, both parties abiding by his result, on which settlement is made. The party whose result is farthest away from the "umpire's" result has to pay for the assay. The charge made for referee or umpire assays is usually a fee and a half or twice the ordinary fee, and sometimes more, according to the professional position of the referee.

As a matter of fact, in the buying and selling of ores the reference to an "umpire" is not frequent, and usually the seller and the purchaser agree as to the values within prescribed limits on the first attempt. The assays made by assayers representing the buyer and seller of an ore are expected to agree within 0·03 to 0·04 ounce per ton in the case of gold, each 0·01 ounce being termed a *point*, and usually about 2·0 ounces in the case of silver. The amount of the difference to be allowed is mutually agreed between purchaser and seller.

For copper they are expected to agree to within half a unit, that is, 0·5 per cent.

In some cases it is previously agreed that if the umpire's result is outside the assays of the purchaser and seller it shall be rejected, and the average of the buyer's and seller's results taken as the basis of settlement, or the lot may be resampled and the same process gone through.

CHAPTER VIII.

SAMPLING.

Definition. — "By the word 'sample' is meant a small portion of a mass which accurately represents the mean composition of the whole of the mass" (Percy).

Introductory.—It is to be regretted that the word sample is often used in a loose manner as a general term for a small portion of material. It would avoid confusion if its use were restricted to such small portions as are truly representative of the bulk.

It may be well at the outset to mention that there are very few assay offices which do not occasionally receive samples which are not truly representative of the ore-mass from which they are taken. They consist usually of pieces of ore, selected because they are worse or better than the average of what they are meant to represent.

Assays of such individual "specimens" may be accurately performed, but the results are of little value if the assayer has not operated upon an *average* sample. For example, it may be stated, in a prospectus, that the assay of a portion of quartz from a certain mine is 50 ounces of gold to the ton, and this may be certified by an assayer whose name would be sufficient guarantee for a bona-fide *assay*. Yet this report is generally of no practical use as an indication of the value of the property, simply because there is no guarantee that the portion of quartz is an average of what there is in the mine : in other words, the portion is not representative, and cannot therefore be truly labelled a "sample."

An assayer who values his reputation, and wishes his report to be above doubt and suspicion, will not put his name to any report without having first ascertained on good evidence that the samples submitted to him are really average samples of the ore.

The object of an assay is to ascertain the true value of the material represented by the sample ; it is therefore absolutely necessary that the small quantity operated upon should accurately represent the average composition of the ore-heap, etc., from which it is taken ; in other words, it must be a *true* sample, otherwise the careful work of the assayer is worse than useless, because the assay result is misleading.

The selection of a small but truly representative portion may be regarded as the first step in determining the composition and value of a large quantity of material.

To select a *bona-fide* sample may seem an easy matter, but those who have had experience testify that it is not so easy as it looks, and generally regard the selection of the sample as a more difficult matter than the actual assay of it.

The work of sampling is too frequently entrusted to subordinates ; but although the manual labour may be left to them, the principles on which it is conducted and the precautions which need to be taken are not unworthy the study of experts and the work should be conducted under experienced supervision.

The extreme importance of correct sampling, and of the utmost care in assaying, are emphasised when it is borne in mind that the sample submitted for assay may represent many tons of rich ore, etc., and large sums of money may be involved in the settlement made on the assay results; consequently, a want of care in the selection of the sample or in conducting the assay may give rise to errors which may mean considerable monetary loss to the buyer or the seller.

Sampling and assaying are closely related, but they are usually performed by different persons. The selection of the sample is done by men termed "samplers," whose special work it is to select a portion which is truly representative of the mass. An assayer is not as a rule responsible for the accuracy of the sample submitted to him, his duty being simply to ascertain and report its value. But, as pointed out by Beringer, "although 'sampling' is thus distinct from 'assaying,' the assayer should be familiar with the principles of sampling, and rigorous in the application of these principles in obtaining, from the sample sent to him, that smaller portion upon which he performs his operations."

It is also to the advantage of the assayer to have some knowledge of the methods in general use for sampling large lots of ore.

Since the introduction of mechanical sampling into many smelting works, and the establishment of public sampling mills in most of the great mining centres, where the accuracy of the sample is guaranteed by the works, the samples representing large lots of material are in many cases handed to the assayer in sealed paper bags, and need very little preparation for assay purposes.

But in general assay-office practice the samples submitted for assay are more often received in small lots, weighing from 10 to 100 lbs., either in the state of coarse powder or in lumps of various sizes, and the assayer requires a knowledge of sampling to enable him satisfactorily to reduce the bulk of the sample to a convenient size for assay.

Principles of Sampling.[1]—The fundamental principle of good sampling may be said to be "a gradual reduction in bulk simultaneously with a gradual reduction in the size of the pieces"; or, in other words, the bulk should not be reduced a second time without first crushing to a degree of greater fineness. This principle is well illustrated by Beringer[2] in the following example:—

Taking a heap of ore, A, and selecting, say, every twentieth shovelful (i.e. 5 per cent.) as the ore is shovelled into the ore-bins, there is obtained a second heap, B, containing $\frac{1}{20}$ of the heap A. If the ore in the heap B is then crushed until it contains approximately the same number of pieces that A did—which means crushing every piece in B into about twenty pieces,—B will become the counterpart of A. Selecting in the same manner $\frac{1}{20}$ of B, there is obtained a third heap C, which in turn is crushed and $\frac{1}{20}$ selected as before. This alternate reduction and pulverising is carried on until a sample of suitable size is obtained. This may be expressed very clearly thus:—

A, 1000 tons of ore in lumps.

B, 50 tons of large pieces $\frac{1}{20}$ of A or 5 per cent. of A.

C, 2·5 tons of small pieces $\frac{1}{20}$ of B or ·25 per cent. of A.

D, 0·125 tons of coarse powder $\frac{1}{20}$ of C or ·0125 per cent. of A.

The conditions affecting and governing the selection of a true average sample from a large bulk of material are summarised by Clarkson as follows :—

[1] The information on the principles of Sampling is mainly derived from papers on "Sampling" by T. Clarkson, A.R.S.M., see *Journ. Iron and Steel Inst.*, xliv., 1893, 131 ; *Journ. Soc. Chem. Ind.*, xiii., 1894, 214.

[2] *Text-Book of Assaying*, 11th edit., p. 2.

1. The uniform composition or thorough mixing of the bulk.
2. The fineness of division, or the size of the pieces forming the bulk.
3. The mode of occurrence of the metallic constituents.
4. The ratio of weight of sample to the fineness of division.
5. The ratio of weight of first sample to total weight of bulk.
6. The method of cutting out the sample.

These conditions are mutually related and interdependent, and are, of course, quite independent of the question of the relative merits of hand and machine sampling.

1. **The Uniformity of Composition.**—It is evident that, if the bulk of material is thoroughly uniform, a portion taken from any part at random would be truly representative of the whole, quite irrespective of all other considerations. This is the ideal simple case, which seldom, or probably never, occurs in practice.

It is common knowledge to all acquainted with the constitution of ores that the valuable metallic constitutents are not uniformly disseminated throughout the whole mass, but generally occur in separate and distinct minerals that are associated in a greater or lesser degree with non-metallic and valueless minerals, such as quartz, etc. Thus, in a heap of ore, one piece may contain absolutely no metal, another piece may have 60 per cent., and another may be made up partly of valuble mineral and partly of waste rock, so that the average percentage may differ very considerably from that of any particular portion.

To obtain a true sample of such a heap, so as to ascertain how much metal is actually present, it is necessary to take some of the richest, some of the poorest, and some of the intermediate value; and, further, the quantity of each kind taken to make up the sample must be carefully regulated so that the ratio between the different kinds is the same as in the original bulk. For example, if there are twice as many rich pieces as poor ones in the bulk, there must be twice as many in the sample, and so for the pieces of other values.

In general, the composition of ore-heaps is far from uniform throughout the bulk; and in some cases, owing to the difference in the size and in the specific gravity of the various constituents, there is a great tendency for them to separate, the fine and heavy material going to the bottom.

The difference in the size of the pieces forming the bulk is also an important factor, as the fine material is almost invariably different in value to the coarse; sometimes one is richer, and sometimes the other; hence it is very necessary in selecting a sample to keep the fine and the coarse in the same proportion as in the bulk.

To ensure uniformity of composition as far as possible it is very important to thoroughly mix the bulk, especially in the case of "spotted ores" (*i.e.* ores of very irregular composition), for unless the mass is well mixed so that the fine and coarse, and heavy and light, particles are kept together in their proper proportion and relation to each other, it will require a larger percentage to be cut out to obtain a truly representative sample.

Much difficulty is frequently experienced in thoroughly mixing ores that contain the highest values in the finest portions, and to overcome this difficulty it is the practice in some sampling works to damp the ore at the beginning so that the fine and coarse will adhere and thus prevent the coarse from separating out by rolling to the bottom. This practice has been found to give very satisfactory results with rich material.[1]

2 and 3. (2) **The proper fineness of division** or the size of the single pieces forming the bulk before a cut for sample can be safely made is controlled by

[1] See "Sampling of Ores from Cobalt," by F. W. Prigsley, *Eng. and Min. Journ.*, vol. xci., 1911 (April 15).

the size of the bulk, and (3) the **mode of occurrence of the metallic constituents.** Small lots should always be crushed fine before cutting.

If the total bulk is very large and uniform in composition, as in a cargo of pyrites or iron ore, the pieces may be proportionately large, provided that the first sample selected is of fair size, say from 2 to 10 per cent. of the whole, according to the material; but in the most difficult cases of sampling, where the metal is carried by only a comparatively few rich pieces in a large mass of barren material, it is absolutely imperative to reduce everything to small size before taking even the first cut. This has more distinct reference to the ores of the precious metals, especially silver ores which require great care in sampling, owing to the presence of minerals rich in silver and very irregularly distributed. After the first sample has been obtained it is, according to the aforementioned principle of gradual and simultaneous reduction, generally necessary to crush it finer before making a second cut, unless the original material was fine. In that case two or more cuts may be made before re-grinding, according to the original degree of fineness.

The necessity of first crushing to a finer state of division before making a second cut, in cases where the pieces are large, is illustrated by the following example. "Let it be assumed that a lot of ore has been crushed to cubes of 1 inch average size, and that to obtain a correct sample at this size it is necessary to cut out, say, 25 per cent. If the sample, representing 25 per cent. of the original mass, is reduced in bulk by one-half, $i.e.$ 50 per cent., without first recrushing, it is obvious that it simply amounts to taking out $\frac{1}{8}$ or 12·5 per cent. of the original mass in two cuts instead of one; and this proportion, as pointed out, will be 50 per cent. too small to give a correct sample with this particular ore if crushed only to 1-inch cubes.

"It follows, therefore, that if 50,000 lbs. ($i.e.$ 25 per cent.) are taken as the first cut in sampling a 100 short-ton lot of 1-inch average cubes, and this quantity is then crushed to $\frac{1}{4}$-inch average cubes, a weight of 3125 lbs. will give as accurate a sample at $\frac{1}{4}$ inch, as 50,000 lbs. at 1 inch; or, if the original mass is all crushed to $\frac{1}{4}$ inch, 3125 lbs. will do for the first cut; and further, that on a reduction to $\frac{1}{16}$ inch, 195 lbs. bear the same ratio between size of cube and weight of sample as the 50,000 lbs. did to 1-inch cubes, and hence will give a correct sample." (Argall.)

This leads up to the important question of,

4. **The Ratio of the Weight of Sample to Fineness of Division.**—Experience has proved that for successful sampling a definite ratio must exist between the weight of the sample and the size of the ore particles, and, before further reducing the weight of samples, the size of the pieces must be reduced upon the basis that there must be a certain minimum number of particles in a sample below which it is unsafe to go if the representative character of the sample is not to be endangered. What this minimum number is, must be determined experimentally for each class of materials. "If the ore particles are large (10 to 12 inches diameter), a large sample must be taken; if the particles are small (0·10 to 0·20 inch diameter), a small sample will, if properly taken, accurately represent the lot of ore."[1]

As an illustration of this principle, let it be assumed that there are

50 bricks of a material designated by A.
25 ,, another material ,, B.
15 ,, ,, ,, C.
10 ,, ,, ,, D.

[1] Brunton, "The Theory and Practice of Ore Sampling," *Trans. Amer. Inst. Min. Eng.*, xxv. p. 826, quoted by Fulton, *Manual of Fire Assaying.*

Dividing these numbers by their greatest common measure, namely, 5, there would be 10 of A, 5 of B, 3 of C, and 2 of D. These bricks, if taken together, would be a representative sample of the original lot, and would be one-fifth of its weight. To attempt to reduce the bulk of the sample below this point without subdividing the pieces would clearly result in destroying the sample. This also illustrates the difficulty of properly sampling a large bulk of waste material containing only a few particles of value, as, for example, in the case of some gold ores, consisting of a large amount of barren quartz containing a few small nuggets or coarse particles of metal. In such a case it is imperative to crush it all very fine and sift out any flattened nuggets that will not crush. The sifted material may then be sampled and the coarse metallic residue reported separately.

Within recent years considerable attention has been given to the subject of the "safe" weight, *i.e.* the minimum quantity which shall be truly representative of the whole of the ore-mass, and there is general agreement amongst sampling authorities as to the safe weight to be taken for certain classes of ore, but the minimum allowable quantity for any particular ore will depend on the nature of the ore and can only be determined by experiment.

Much time has been devoted to the subject by H. A. Vezin of Denver, Colorado, and as the result of experiments he considers it a safe rule to take 1 ounce (3·1 grammes) as the minimum weight to which a sample should be reduced from material passing a sieve of 1-millimetre (·04 inch) mesh; and, using this as the basis of calculation, he states that the weight to be taken for sample, for material of other sizes (coarser or finer), may be found by "making the minimum weight of the sample proportionate to the cube of the linear dimensions of the average particles." Thus, if the material has passed a sieve of 2-millimetre mesh, by this rule the sample should not be cut down below 8 ounces (249 grammes) without the material being first ground finer. This is equivalent to stating that the "number of particles in a sample should never be below a certain minimum, whatever their size may be."

The number of particles of all sizes in 1 ounce of material (with a specific gravity of, say, 4) that has passed through a 20-mesh sieve is at least 50,000, and on this basis the weight of the smallest sample of the same material passed only through a 3-inch mesh would be about 8 tons. Vezin's experiments were made on the pyritic ores of Gilpin County, Colorado, carrying from 1 to 4 ounces of gold per ton, and the "safe weights" for coarser samples of the same ore, obtained by calculations, are indicated in the following table.

The rule is in use in Colorado as the basis of calculation.

Table III.

Diameter of largest particles.		Minimum weight of sample.	
Inches.		Lbs.	
0·04	¹⁄₂₅	0·0625	¹⁄₁₆ (1 oz.)
0·08	¹⁄₁₂	0·50	½
0·16	⅙	4·00	
0·32	⅓	32·00	
0·64	⅔	256·00	
1·25	1¼	2048·00	1 ton (approx.)
2·50	2½	16348·00	8 tons (approx.)

With regard to this rule, Vezin has remarked[1] that in giving the safe minimum weight of a sample at 1 ounce for material of 1 millimetre diameter he refers to ores containing the precious metals in which these are finely disseminated and not in large quantities.   With rich ores in which the precious metals are very irregularly distributed a larger sample must be taken, while for ores uniform in composition, such as iron ores, much smaller samples will suffice.

Wright has pointed[2] out that Vezin's rule implies that the ratio between the number of particles of mineral and the number of particles of gangue remain the same through all the stages of the crushing of the ore, and that during the grinding the mineral and gangue are crushed at the same rate, whereas in the large majority of commercial ores the valuable mineral—owing to its greater brittleness, etc.—is more readily broken down than the gangue.

From experiments made by Wright upon two samples of the same ore taken respectively at the ¼-inch and the 100-mesh stages of the grinding, it was shown that the ratio of the safe weights—for this ore at least—was very much smaller than that required by Vezin's rule.   From these remarks it will be seen that the weight of the sample is dependent not only on the size of the ore particles but also on the nature of the ore.

As already stated, in cases where the valuable constituents of the ore are uniformly distributed, smaller samples will suffice than are necessary for "spotted" ores in which the values are irregularly distributed.   For example, the rich telluride ores of Colorado, assaying 10 to 15 ounces of gold per ton, are usually very difficult to sample, especially by hand sampling, chiefly on account of the great difficulty in properly mixing the fine and coarse particles.   In sampling such ores with automatic samplers Argall found the following ratio between the average size of the ore cubes and proportional weight of the sample to give accurate results[3] :—

Average size of ore cubes, in inches .	1·0	0·25	0·0625	0·0171
			(8 mesh)	(30 mesh)
Weight of sample (for 100-ton lot) in lbs.	40,000	2,500	157	10
„ „ per cent. . . .	20	1·25	0·0785	0·005

With reference to these figures, Argall remarks that in practical work larger quantities of the finer-sized material would be taken simply as a matter of extra precaution, more especially so in mills where all the ore is ultimately ground fine.

Vezin's rule frequently demands finer crushing than practice indicates to be necessary, and in many sampling works, to save the excessive cost of crushing, sampling rules, based on experience, and deviating from Vezin's rule, have been made which are sufficiently accurate for the purposes of buying and selling.

Prof. R. H. Richards of the Massachusetts Institute of Technology has collected a large amount of data based on actual sampling practice, and as the results of his investigations he has given the following rule for ascertaining the weight of sample to be taken for any given ore :—"The weight taken for sample shall be proportional to the square of the diameter of the largest particles."[4]

By adopting this rule, Prof. Richards has drawn up the set of figures given

[1] In discussion on Clarkson's paper, *Journ. Iron and Steel Inst.*, xliv., 1893, p. 189.
[2] L. T. Wright, *Mining Magazine*, 1910, iii. 353-358.
[3] *Trans. Inst. Min. and Met.*, x., 1902, 284-73.
[4] R. H. Richards, *Ore Dressing* vol. ii. p. 850.

in the accompanying Table IV., which have been found to give results that
conform to good practice.

Table IV.— Weights to be taken in sampling Ore. (R. H. Richards.)

Weight.		Diameter of largest particle (in millimetres).[1]					
		Very low-grade or very uni-form ores. A.	Low-grade or uni-form ores. B.	Medium ores.		Rich or "spotted" ores. E.	Very rich or excessively "spotted" ores. F.
				C.	D.		
grammes.	lbs.	mm.	mm.	mm.	mm.	mm.	mm.
...	20,000·000	207·00	114·00	76·20	50·80	31·60	5·40
...	10,000·000	147·00	80·30	53·90	35·90	22·40	3·80
...	5,000·000	104·00	56·80	38·10	25·40	15·80	2·70
...	2,000·000	65·60	35·90	24·10	16·10	10·00	1·70
...	1,000·000	46·40	25·40	17·00	11·40	7·10	1·20
...	500·000	32·80	18·00	12·00	8·00	5·00	0·85
...	200·000	20·70	11·40	7·60	5·10	3·20	0·54
...	100·000	14·70	8·00	5·40	3·60	2·20	0·38
...	50·000	10·40	5·70	3·80	2·50	1·60	0·27
..	20·000	6·60	3·60	2·40	1·60	1·00	0·17
...	10·000	4·60	2·50	1·70	1·10	0·71	0·12
...	5·000	3·30	1·80	1·20	0·80	0·50	
...	2·000	2·10	1·10	0·76	0·51	0·32	
	1·000	1·50	0·80	0·54	0·36	0·22	
...	0·500	1·00	0·57	0·38	0 25	0·16	
90·0	0·200	0·66	0·36	0·24	0 16	0·10	
45·0	0·100	0·46	0·25	0·17	0·11		
22·5	0·050	0·33	0·18	0·12			
9·0	0·020	0·21	0·11				
4·5	0·010	0·15					
2·25	0·005	0·10					

The figures in each column are based on data obtained in practice, full
details of which are given by Prof. Richards in his *Ore Dressing*. The equivalents in inches and "mesh" may be seen by reference to Table VI., p. 122.

The figures given in columns A and B are applicable to poor or low-grade
ores and to ores in which the valuable constituent is very evenly distributed,
while the figures in columns C and D are applicable to ores of similar character
to A and B but richer or higher grade. Most of the ores of the common metals
would come under one of the heads given in these columns.

Gold ores may also be included under the head of low-grade ores when the
gold is combined with pyritic material which is more or less evenly disseminated
through the ore.

The figures in columns E and F mainly apply to ores of the precious metals
in which the values, as already stated, are very frequently contained in a few
particles of metal (or nuggets), or a few lumps of rich mineral, distributed very
irregularly through a large mass of barren material. Exceptionally rich ores of
the common metals are also included under the heading of "Spotted Ores."

With reference to the table, Richards points out that it is not intended that
the reduction in size shall proceed by all the successive stages shown in any one
column, but whenever any weight given in the first column is taken as a sample
the ore should first be crushed to the corresponding size shown in one of the
other columns.

[1] 25·4 mm. (millimetres) = 1 inch ; 1 ton = 2000 lbs.

For example, with a 20,000-lb. (10 - ton) lot of low-grade ore the entire lot might be crushed to 25·4 millimetres (1 inch) and 1000 lbs. taken for a sample. If this were all reduced to 2·5-millimetre ($\frac{1}{10}$ of an inch) size, the quantity could be reduced to 10 lbs.; and then by grinding to 0·80 millimetre (about 20 mesh) the weight could safely be reduced to 1 lb.

5. **The ratio of weight of first sample to total weight of bulk**, or the smallest percentage to be cut out as a sample at one operation, will depend upon the uniformity of composition and fineness of division of the bulk, and also upon the method of cutting out the sample. In some cases 1 per cent. or even less would be safe, whereas in others 20 per cent. or even more would be desirable. The percentage to be taken for sample for any particular ore is usually agreed upon by buyer and seller before sampling. In general sampling practice it is usual to cut out for first sample from $\frac{1}{5}$ (20 per cent.) to $\frac{1}{20}$ (5 per cent.) of the ore bulk according to the nature of the ore and size of the pieces, experience having shown that if the same quantity of ore is taken each time and at frequent and regular intervals the sample so obtained will truly represent the bulk from which it is taken. Where the ores are of low grade, or very uniform in composition, a small percentage will suffice for first cut. Take iron ore, for example, or ores to be used as fluxes for blast-furnace work, where it is important to keep the ore as coarse as possible : 10 per cent. of ore as coarse as 6-inch cubes can be taken if necessary and, reduced proportionally, will give an accurate sample.

In the case of copper ores sampled automatically, Peters states [1] that the proportion of the ore stream deflected into the sample bin in the first instance may vary from 10 to 50 per cent., the latter amount only being required for coarse ores of enormous and very variable richness, while for ordinary lump ores from 10 to 20 per cent. is the maximum required. Ores of the precious metals that are to be treated in stamps or roller mills, and reduced ultimately to a state of fine division, should preferably be crushed finer for the first cut, and as a matter of precaution, say, 20 per cent. taken out for the first sample.[2]

When the nature and fineness of the ore permit of a very small percentage being taken for first sample it is generally better to obtain this in two cuts. For example, a 1 per cent. sample is obtained by cutting 10 per cent. of 10 per cent. instead of cutting 1 per cent. direct, or a 2 per cent. sample is taken by cutting 10 per cent. of 20 per cent.

6. **The Method of Cutting out the Sample.**—This is obviously of the utmost importance, so that a fair average proportion of all the different qualities and sizes may be obtained. The method to be adopted is dependent on the quantity and nature of the ore and the metallurgical treatment to which it is subsequently to be subjected. "In milling and concentrating plants where fine crushing is a necessary preliminary to the metallurgical treatment of the ore, sampling is a comparatively easy and reasonably accurate operation, but where the ore is to be treated by blast-furnace smelting, crushing of any kind is objectionable and fine sub-division is prohibited": under these conditions sampling becomes more difficult.[3]

When a very large lot has to be dealt with, the chances of error are lessened by dividing it into several smaller lots of about 50 or 100 tons each and sampling each of these lots separately. The commercial value of the ore or product to be sampled is also an important consideration in determining what the allowable limit of error in sampling should be, and this will affect the method of sampling and the number of samples to be taken. In the case of metallurgical products an intimate knowledge of the process by which they are obtained is of great assistance to the sampler in deciding the method of sampling to be adopted. The tendency of modern practice in ore sampling is to employ methods which

[1] Peters, *Modern Copper Smelting*.   [2] Argall, *loc. cit.*
[3] D. W. Brunton, *Trans. Am. Inst. Min. Eng.*, 1909, pp. 837–855.

are as far as possible automatic and independent of the will or judgment of the sampler, and in public sampling works machine methods are very generally employed. Accurate sampling can, however, be performed by hand, and the methods of hand sampling which have been in use for many years are still largely employed for mine sampling and smelting works, as they avoid crushing a large proportion of the ore. When dealing with large lots of ore, the first sample is invariably taken by some method of fractional selection, that is, selecting a certain proportion while the ore is being dumped or deposited in a heap; or, if the ore has already been dumped, trenches are cut through the heap and portions reserved for sample as described below.

If, for instance, an entire cargo of ore has to be sampled, there would be set aside at the time of unloading, according to the importance of the mass or the intrinsic value of the ore, every fifth, tenth, or twentieth skipload or sackful, etc., for an especial heap, which may be considered to have the same composition as the whole mass. The quantity thus set aside is then reduced by successive crushing and division until about 8 ounces is finally obtained, which is crushed to pass a very fine sieve and divided into several portions for assay purposes.

The degree of fineness to which an ore is crushed in the successive stages of the sampling will of course vary with the class of ore and other considerations.

Ores direct from the mines usually range in size from 12-inch to 16-inch cubes down to the finest particles, and contain from 1 to 2 per cent. of moisture in hard ores up to in some cases as much as 25 to 30 per cent. moisture in clayey or talcy ores (i.e. ores containing the mineral talc $((MgO)_6(SiO_2)_4 + H_2O)$.

The first operation is, therefore, weighing the ore and reducing it to a suitable size for sampling.

If the ore is destined for the blast furnace, it is crushed as little as possible and a large sample cut out, not less than 20 per cent., depending of course largely on its value; this sample is then coarsely crushed and reduced as stated above.

As an example of the amount of crushing necessary, the following course of procedure is recommended by Argall in the machine sampling of 8-ounce gold ores to be treated by smelting. The ore is crushed to 3-inch cubes and passed through a Vezin sampler (see page 101), which can be made to cut out pieces as large as 3 inches, the sampler taking, say, 33 tons out of 100. The sample is next crushed to about $\frac{3}{4}$ inch and 4 tons cut out for the sample; these 4 tons, or 4 per cent. of the original weight of the ore, is all that would, in this case, be reduced to a fine state of division in the sampling process, leaving 96 per cent. of the ore in good shape for blast-furnace operations.

The proportion of size of cubes and weight of ore taken as sample in this case is, according to Argall, quite safe for 8-ounce gold ores to be treated by smelting, as already stated.

If the ore is to be ultimately ground fine for chemical processes, it is usually crushed to $\frac{1}{2}$ inch or $\frac{3}{4}$ inch before sampling, and in such cases a smaller sample will answer the requirements of commercial accuracy.

In sampling ores that are to be sold, it is the usual practice to reserve a separate portion, say 200 to 300 lbs., of the finer crushed material obtained towards the end of the sampling, and to utilise this quantity for resampling should this become necessary. By adopting this plan the resampling of the entire lot is avoided.

For crushing the ore, rock-breakers of the Gates or Blake type are invariably used to reduce large rocks to cubes of $1\frac{1}{2}$ inch to 2 inch, but below 2-inch cube rolls are preferable to rock-breakers as a medium of reduction. Rolls are used for any size, from 2-inch to 30- or 40-mesh; below that, some form of laboratory machine for fine grinding is employed.

Moisture Sample (see also page 125).—When it is necessary to determine the amount of moisture in the ore, a separate "moisture" sample, weighing usually from ¼ to 2 lbs., is taken when the ore has been crushed to about 10 mesh (0·05 inch). In sampling mills larger moisture samples, weighing from 15 to 20 lbs., are frequently taken. The whole of the sample is dried, and the percentage of moisture determined from the loss in weight as described on page 125. If the sample taken for moisture is too large, it is reduced in bulk by quartering or by any of the methods given later.

The importance of determining the moisture as soon as possible after the sample is taken will be obvious.

Methods of Sampling.

The methods of sampling are conveniently classified under two heads: (1) hand sampling, and (2) machine sampling.

(1) Hand Sampling.

The following are the methods of hand sampling in general use :—

    (a) Grab sampling.
    (b) Trench and pit sampling.
    (c) Coning and quartering.
    (d) Fractional selection or shovel sampling.
    (e) Split Shovels.

(a) *Grab Sampling.*—This method consists of taking proportionate amounts of the lumps and of the fine ore, at equal distances over the pile, after first removing the surface ore which has been dried by the atmosphere. In sampling the lumps the whole piece should be broken and chips taken from both the outside and the inside; no piece should be larger than about ½-inch cube.

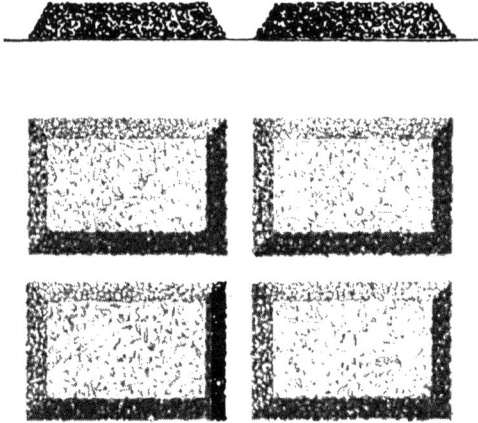

The sample thus selected is crushed finer, well mixed, and then reduced in bulk by coning and quartering (see page 97) or other method.

The method is imperfect, as the sample only represents the upper part of the pile. The accuracy of the sample is largely dependent on the discretion of the sampler: great care and judgment are needed to take the coarse and fine parts in the correct proportion. The method is used chiefly for obtaining an approximate idea of the value or composition of an ore pile, as it is quick and inexpensive.

Fig. 78.

(b) *Trench Sampling.*—In this method the ore is formed into a flattened square or rectangular pile from 2 to 3 feet deep for lots of about 100 tons, and 1 to 2 feet deep for smaller lots.

Trenches about 1 foot wide and crossing each other at right angles are then dug through the centre of the pile with a shovel, as shown in fig. 78, thus converting the large pile into four or more smaller piles. The workman, as he advances in cutting the trench, reserves for sample either all the ore he removes, or a certain proportion, every fifth, tenth, or twentieth shovelful, and throws the rejected portion alternately right and left on to the pile.

The sample obtained in this way may, after crushing, be formed into a new rectangle and reduced in bulk by again trenching or by coning and quartering. Instead of reserving an aliquot portion for sample as just described, it is also the practice when cutting the trenches to reject all the ore removed, each shovelful being thrown to the right and left as before, and when the trenches are completed the sample is taken from the sides of each of the small piles into which the original pile has been divided.

In this case portions are removed from each of the freshly exposed sides of the piles by carefully "rising" the shovel from the bottom to the top, the total quantity taken in this way being about 10 per cent. of the original pile.

Another method, less expensive than trench-cutting, consists in digging a series of holes or pits at equal distances over the surface of the pile and reserving either all or equal amounts, usually a few shovelfuls, of the material excavated from each hole. The reserved portions are then well mixed and reduced in bulk as before.

In digging the pits care must be taken to go right through the layer, otherwise a fair proportion of the finer material, which generally gets to the bottom, will not be obtained.

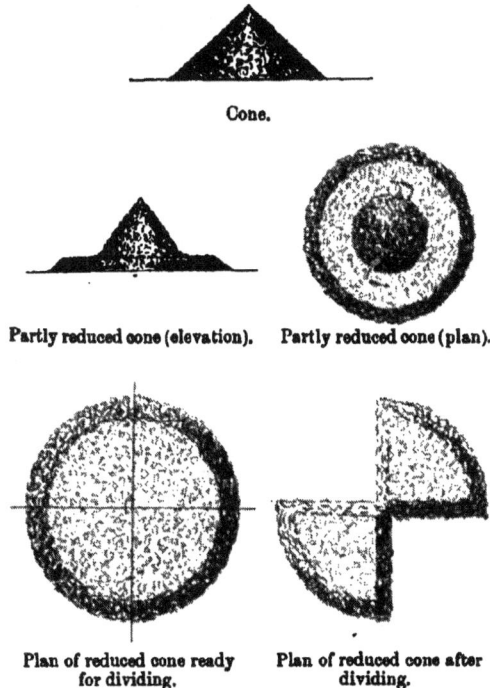

Cone.

Partly reduced cone (elevation). Partly reduced cone (plan).

Plan of reduced cone ready for dividing. Plan of reduced cone after dividing.

Fig. 79.

(c) *Coning and Quartering.*—This old Cornish method of hand sampling has been in use for many years, and is still largely employed for reducing the bulk of large samples or for sampling small lots.

It consists in piling the ore into the form of a cone which is quartered (hence the term quartering), two opposite quarters being reserved for sample and the other two rejected. The various stages of the coning and quartering method of sampling are illustrated in fig. 79. When the coning and quartering are skilfully performed the results are accurate. Before applying this method, and to save unnecessary labour, it is usual with large lots to take an aliquot portion of the ore lot, say every tenth or twentieth truck-load, etc., as previously described; but however the first sample may be selected, the subsequent treatment of it to obtain the final sample is the same.

7

If the selected material is in lumps it must be crushed down to, say, 1½-inch to 2-inch cubes before the bulk is reduced.

After crushing, the ore is well mixed by turning it over several times with a shovel and then piled into the form of a cone, special care being taken that each shovelful is thrown upon the apex of the cone.   To do this successfully requires some skill ; the ore must not be thrown as in ordinary shovel work, but must be tossed vertically into the air, so that it will drop directly upon the apex and fall evenly down the side of the cone.

The even distribution of the material is absolutely necessary in order to obtain a correct sample, and this can only be attained by keeping the cone truly vertical.   Dishonest samplers perform the operation of what is known as 'drawing the cone," i.e. deliberately building up the cone away from the centre of the heap and thus altering the relative proportions of lump and fines in their own favour.

In building up the cone the fine material always forms the apex, while the coarse particles fall towards the base, and as the fine ore is usually richer than the coarse it will be evident that if the cone is not truly vertical when it is divided vertically into four sections, the section towards which the apex leans will be richer than the other sections.   On subsequently levelling and quartering the cone this local richness may be partly corrected or may be intensified according to whether less or more of the fine ore forming the apex is shovelled into one of the sections and not evenly distributed between all four sections.

To ensure the cone being vertical it is a common practice to drive a rod into the ground as a guide to fix the apex.

The object in forming the cone is to thoroughly mix the ore, and when it is completed it is "pulled" down by the men walking around it and shovelling the ore from the apex to the base so that it falls outward, as shown in fig. 79. The lower portion of the cone is not disturbed, and when this process is finished there remains a truncated cone of ore from 6 inches to 12 inches deep which should be uniform in thickness and truly circular in form.

This is divided by diametrical lines into four equal sections by pressing a board or iron plate edgewise into the pile, or by placing a cross, formed of iron plates riveted together, on the pile and driving it into the ore to mark the divisions.

Two opposite quarters are then cut out with the shovel and removed to the ore-bins, and the other two, which represent the sample, are formed into a fresh cone which is drawn down and quartered as before.   This process is repeated until the sample requires recrushing, after which the coning and quartering are continued until the sample has been reduced to a few pounds in weight.

The sample is finally prepared for assay purposes by grinding it sufficiently to pass a fine mesh sieve, but to facilitate this fine grinding it is frequently necessary to dry the sample.

The details of this practice will vary in different works, and the number of recrushings necessary will depend on the quantity and nature of the ore and the size of the pieces forming the original bulk.

In some of the works in Colorado it is the practice first to dump the ore into the form of a ring, leaving a clear space inside of about 5 to 10 feet in diameter according to the quantity of ore.   The men then move around the outer circumference of the ring and shovel the ore into the centre of the clear space to form a cone, which is dealt with as already described.

The coning and quartering method requires considerable floor-space, a 100-ton lot requiring at least 2000 square feet,[1] and when large lots have to be sampled it is somewhat slow and laborious ; but the results are trustworthy, and as good as can be obtained by hand sampling, provided proper care is taken and

[1] Bridgman, *Trans Am. Inst. Min. Eng.*, 1891.

sufficient time allowed. Insufficient crushing and inefficient mixing during the progress of sampling are the chief sources of error.

(d) *Fractional selection or shovel sampling* [1] consists in reserving for sample a certain proportion of the ore as it is being shovelled into the stock-bins or otherwise handled.

For example, if one-fifth is to be cut out for sample, four shovelfuls will be thrown direct into the bins and every fifth shovelful reserved. The proportion to be reserved is dependent on the nature and size of the ore pieces. The selected portion, after crushing, is coned for the purpose of mixing, and then every fifth shovelful again reserved. This is coned as before, and the process repeated until it becomes necessary to recrush, after which the reduction in bulk proceeds in the same way.

To conduct this method successfully proper shaped shovels must be used, and the size of the shovel must accord in capacity with the size of the bulk to be handled, so that smaller shovels must be used towards the end of the sampling.

Sampling by shovel is considered by many to be a more convenient and accurate method of hand sampling a pile of crushed ore than coning and quartering, but the latter method is more generally preferred. The advantages claimed for shovel sampling are (1) it is cheaper in operation; (2) it is quicker, the ore-bulk being reduced at a much quicker rate; (3) that the sample is more accurate for a given reduction as a larger number of cuts are made. This is illustrated by Argall [2] as follows : "Let it be assumed that a heap of 10,000 lbs. has to be divided for sample and reject ; that the shovel used holds 10 lbs., and that the labourer places alternate shovelfuls in the stock-bin for the rejected portion, and on the floor for the sample. In disposing of this heap of ore he will handle 1000 shovelfuls or make 1000 cuts in the heap as against two cuts in the "coning and quartering" method, a manifest advantage in favour of the shovel sampling.

FIG. 80.

"Should the ore, however, be sufficiently fine to warrant a reduction to, say, one-fourth, then every fourth shovelful would go to form the sample and three to the stock-bin. In this case the sample is made up of 250 cuts and reduced to 25 per cent. of the original ore in the operation, whereas the tedious process of coning and quartering would have to be twice performed to give the same reduction, and even then the sample is the result of but four cuts."

An objection to the shovel method is that the selection of the portions to be reserved for sample is left entirely to the discretion of the operator.

(e) *Split Shovel.*—This shovel consists of four long parallel troughs, each 12 inches long and 2 inches wide, with open spaces between them, the spaces being the same width as the troughs (fig. 80). It is sometimes used for sampling small lots of ore weighing not more than about half a ton. The ore must be crushed to ¼-inch diameter. The ore is dug from the pile with an ordinary shovel and then spread over the split shovel, the sampler moving his shovel backwards and forwards across the troughs so that the ore as it slides off is evenly distributed. Half the ore remains in the troughs and the other half drops through the spaces. When the troughs are full the split shovel is lifted and the contents reserved as the sample. When all the ore-heap has been dealt with the portion reserved for sample can be reduced in bulk by again passing it over the split shovel, thus reducing the bulk by one-half.

---

[1] Also called the alternate shovel method when every second shovelful is reserved.
[2] *Trans. Inst. Min. and Met.*, vol. x. (1902), p. 237.

Pipe Sampler or Sampling Iron.—Heaps of finely crushed material, such as dry tailings or concentrates, are sometimes sampled by driving in iron pipes at regular intervals and mixing the samples withdrawn in the pipes. A form of sampling iron (or cheese-taster sampler) for very dry material is described by Richards.[1]

It consists of an iron pipe A, fig. 81, with nearly half its circumference cut away, except for a few inches at the upper end. The pipe is fitted with a "T" piece to hold the handle by means of which the pipe can be rotated. This pipe fits inside a slightly larger pipe B which is similarly cut away, and has its lower end sharpened or drawn out to a point and its upper end provided with a handle. To obtain the sample the pipes are driven into the ore with their openings arranged as shown in the cross-section. The inner pipe is rotated until it reaches the position shown in B', and moved backwards and forwards until it is filled with ore. It is then replaced in its original position A' and both pipes withdrawn with the sample enclosed.

In order to obtain correct samples they should be taken systematically from all parts of the heap and throughout the entire depth of the heap, so as to prevent an undue proportion of the upper layers from being included. In the case of tailings the samples are usually taken from the charge in the vat or from the ore-trucks, and it is also usual to sample the lowest 6 inches to 18 inches separately.

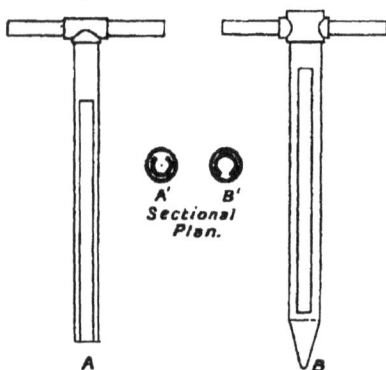

A'   B'
Sectional Plan.

A   B

FIG. 81.

Running Samples.—The term running sample is applied to the sample of any running stream of material, such as the pulp suspended in water, issuing from the screens in a stamp battery used for crushing gold ores, etc.

Dipper or Bucket Samplers.—Running samples are generally taken by catching the whole stream in a dipper or bucket at stated intervals, for example, every fifteen or thirty minutes, for some definite period, usually a few seconds. When a bucketful is collected the sample is allowed to settle for several hours, the clear water is decanted, then the sample dried, well mixed, and reduced in bulk for assay purposes.

Automatic dipper and bucket samplers are now being adopted in many gold stamp mills.[2]

"Salted" Samples.—Samples which have been intentionally or accidentally contaminated with either richer or poorer material so as to alter the true value of their contents are said to be "salted." The intentional salting of samples for deceptive purposes is now seldom attempted.

(2) Machine Sampling.

Mechanical sampling is based upon the principle of intercepting or cutting out a portion of a running stream of ore. This may be done either by taking—

    i. Part of the ore stream continuously, or

    ii. The whole of the ore stream intermittently.

[1] *Ore Dressing*, vol. xi. (1903), p. 844.

[2] Text-books on mining should be consulted for detail of the methods of taking running samples.

Machines for taking a definite portion, usually one-sixteenth, of a falling or sliding stream of ore continuously, were constructed on the assumption that the values are evenly distributed across the ore stream ; but experience has proved that this is not the case, and machines based on this principle have been largely displaced by those which automatically deflect the entire ore stream into the sample division for a varying portion of the time, usually one second in each five seconds. In many machines now in use a narrow scoop passes steadily across the stream of ore at regular intervals of time, and this is considered to be equivalent to the method used in the machines of class ii.

Brunton[1] has pointed out that "it is not practicable to produce a stream of ore which shall be continuous in value through every part of its length, any more than it is possible to produce a stream of ore that is constant in value throughout its width ; but by taking a small sample entirely across a falling stream at very short intervals it is found that, while no single cut would give an exact representation of the composition of the entire lot, the average of a large number of these small samples is so nearly correct that results can be duplicated within very narrow margins, or, in other words, that individual errors are balanced." The accuracy of the sample obtained in this way, provided it is taken sufficiently often, is now well established and recognised.

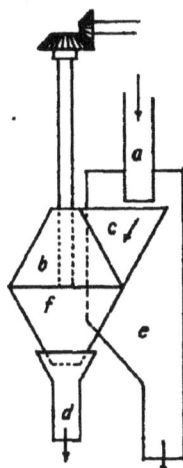

Elevation.

"Numerous different machines working on this principle, such as the Snyder, Vezin, and Brunton samplers, described below, are now in use ; and these types of sampling plants have been perfected to such a degree that where the hopper ore-cars which are now coming into general use are employed, ore may be unloaded, crushed, sampled, and reloaded into the outgoing cars, and the ground samples delivered in a locked steel box without ever having been handled—the entire chain of operations being performed automatically."[2]

The chief drawbacks to machine sampling are the necessity of fine crushing and the difficulty of cleaning the machine.

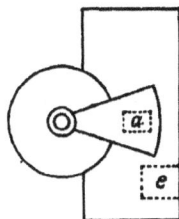

Plan.

The following are types of the sampling machines now in general use :—

Vezin Sampler (fig. 82).—This sampler is accurate, simple of operation, and suitable to handle all grades of ore. It is coming into very general use. According to Argall, it is almost universally employed in Colorado.[3]

6   3   0   6 FEET

Fig. 82.

It consists of two truncated cones b, f, of sheet-steel, with their bases bolted together. A scoop c, made of sheet-steel, with a sector-shaped opening at the top, is riveted to the upper cone b. The angle of the sector may subtend any desired portion of the circumference of a circle, such as one-tenth (36°) or one-sixteenth (22½°). Both cones and the scoop are rotated at about 25 or 30 revolutions per minute by bevelled gearing. The stream of ore falls from a spout a, and when the scoop c passes below the spout, the whole stream of ore falls into the scoop and is delivered into the interior of the lower cone f and thence through d to the

[1] *Trans. Am. Inst. Min. Eng.*, 1909, pp. 837-855.   [2] Brunton, *loc. cit.*
[3] *Trans. Inst. Min. and Met.*, vol. x. (1902), pp. 240-241, with working drawings.

sample bin or to the rolls if it is to be further crushed and reduced in bulk. In this way a sample is taken about every two seconds. The main portion of the ore stream falls into the hopper *e* and passes through the shoot to the storage bins.

A Vezin sampler of about 3 feet in diameter, and requiring a fall of about 6 feet, is trifling in cost, and treats 30 to 40 tons an hour (T. K. Rose). There is no patent on the sampler.

Collom Sampler (fig. 83).—In this machine one or more diverting scoops *b* are mounted on a horizontal arm *a* carried by a revolving vertical shaft. As a scoop passes under the stream of ore from the spout C it cuts out a sample and delivers it through a vertical spout *d* into the sample bin. When the scoop

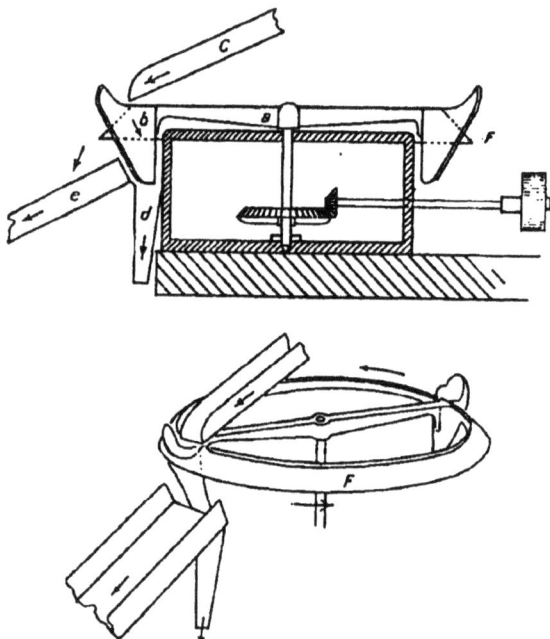

FIG. 83.

has passed, the ore stream falls on to the shoot *e* and thence to the storage bin. To prevent any ore from falling into the sample spout, except when the scoop is passing, the machine is fitted with a sloping metal flange F. This machine is driven at a constant speed in one direction by bevelled gear and pinion.

Snyder Sampler (fig. 84).—This sampler has the form of a circular pan *c* with flaring sides, set on edge on a revolving horizontal shaft. A spout *b* projects through the flaring side, and passes through the ore stream from the feed-spout *a* at each revolution of the machine, thus cutting out a sample which is diverted to the sample spout *d*. During the rest of the revolution the ore falls on the flaring sides and is diverted into the spout *e*.

Brunton Sampler (fig. 85).—This sampler consists of a sheet-iron deflector A, with a sloping side and a scoop B attached to it, slanting in the opposite direction, as shown in section, fig. 85A. The deflector is fastened to a horizontal shaft connected with machinery which causes it to swing backwards and forwards beneath the end of the feed-spout C. As the deflector oscillates the scoop cuts completely across the stream of falling ore at each oscillation, so that a sample is thus deflected to the left by the scoop B and the rest of the ore deflected to the right by the sloping face A. The ratio of rejected ore to

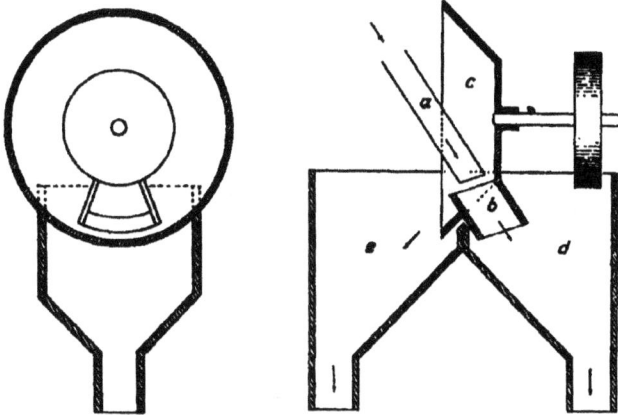

Fig. 84.

sample is governed by the width of the scoop and the width of the ore stream through which the deflector travels. A common ratio is 5 to 1.

Sampling Works or Mills.—The general arrangement of a complete sampling works varies according to the quantity and nature of the ore to be sampled and the method of sampling to be employed.

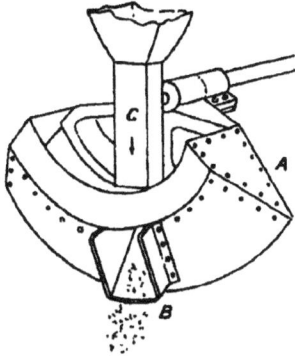

Fig. 85.

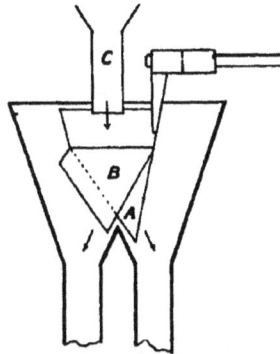

Fig. 85A.

A comparison of hand sampling and machine sampling will be found on page 94.

In the case of ore of high value and of which it is desired to obtain very accurate samples, it is generally desirable to put the entire lot through the sampling machinery.

In cases, however, where it is desired to preserve the ore in lump form, rather than to obtain greater accuracy of sample, only such fractional part of the lot as may have been predetermined is reserved as the first sample and passed to the sampling machinery, while the main bulk of the lot is thrown into the storage bins. In this latter case hand sampling is usually employed in taking the first sample, as it is advisable to limit the maximum size of ore passing through automatic samplers to not more than 2-inch cubes, although ore of larger size is sometimes sampled.

The plant in a sampling works consists of a series of large storage bins for keeping the different lots of ore separate, scales for weighing the ore before sampling, crushers or rock-breakers, crushing rolls or other machinery for fine crushing, one or more automatic sampling machines, elevators for elevating the ore up into bins of sufficient height to discharge into cars or storage bins, etc. If hand sampling is employed, sufficient floor-space must be allowed for manipulating large lots of ore.

Various devices are also in use for sacking the ore when such is necessary.

In some cases revolving screens are used after the first crushing to remove the fine ore from the coarse as soon as possible and thus prevent the production of a large proportion of very fine ore. The fine ore thus removed is mixed with the finely crushed ore that results from the recrushing of the coarse ore.

As a general rule, the sampling machine is placed below the crushing machinery, so that the stream of ore from the crusher is fed directly into the sampler. When this cannot be arranged it is necessary to use an elevator to feed the sampler. The feed-spout above the sampler, whether delivering the ore from the crusher or from an elevator, should be short and should not exceed $3\frac{1}{2}$ inches in diameter. The ore is fed to the sampler in a regular stream while the sampler is running.

The discard spout from the sampler delivers directly into a car, bin, or elevator as required. When, for the reasons stated on page 94, it is desired to pass the ore through two samplers in succession without intermediate recrushing, a bin is interposed between the samplers, in which the ore from the first sampler is allowed to accumulate until there is sufficient to ensure a continuous feed to the second sampler.

Sampling mills are best built on a hill-side when this can be conveniently arranged, as this avoids the necessity of using elevators.

Examples of Sampling Mill Practice.—The following brief descriptions and accompanying flow-sheets will serve to illustrate the arrangement of sampling works and show how the principles of sampling are carried out in practice.

Sampling Mill of Standard Smelting Company, Rapid City, South Dakota.[1]— The method of sampling employed consists of shovel (fractional selection) and riffle sampling (see page 122). The first sample is cut out with the shovel. Assuming 100-ton lot is taken as a unit and every tenth shovelful reserved, a sample of 10 tons is obtained.

This is passed to a Blake crusher with a 9-inch by 15-inch mouth opening, discharging on a Λ-divider which divides the sample into two parts. One half is fed directly to a pair of 24-inch by 12-inch rolls and the discharge from these automatically halved as before, giving a sample of 2·5 tons with no particle larger than 0·375 inch in diameter.

The rolls discharge the crushed ore directly upon a plate iron floor, where it is re-shovelled, every one-fifth or one-tenth (or other fraction) shovelful being reserved as a sample, which now amounts to 500 or 1000 lbs., according to the fraction reserved. This is put through a pair of 12-inch by 12-inch sampling rolls and crushed fine, and then sampled by a large Jones riffle sampler, which takes halves, the operation being repeated until finally a sample of between 15 and 20 lbs. is obtained.

[1] *Manual of Fire Assaying*, C. H. Fulton, 2nd edit., 1911, p. 39.

This is put through a small cone-grinding mill, and after a determination of moisture, is sent to the assay office. Here it is cut down to about 2 lbs. by a small Jones sampler, and then crushed on a buck-board to pass a 120-mesh screen, furnishing the assay sample. This sample is supposed to contain no moisture, but as all settlements are made on dry samples, this final sample is again

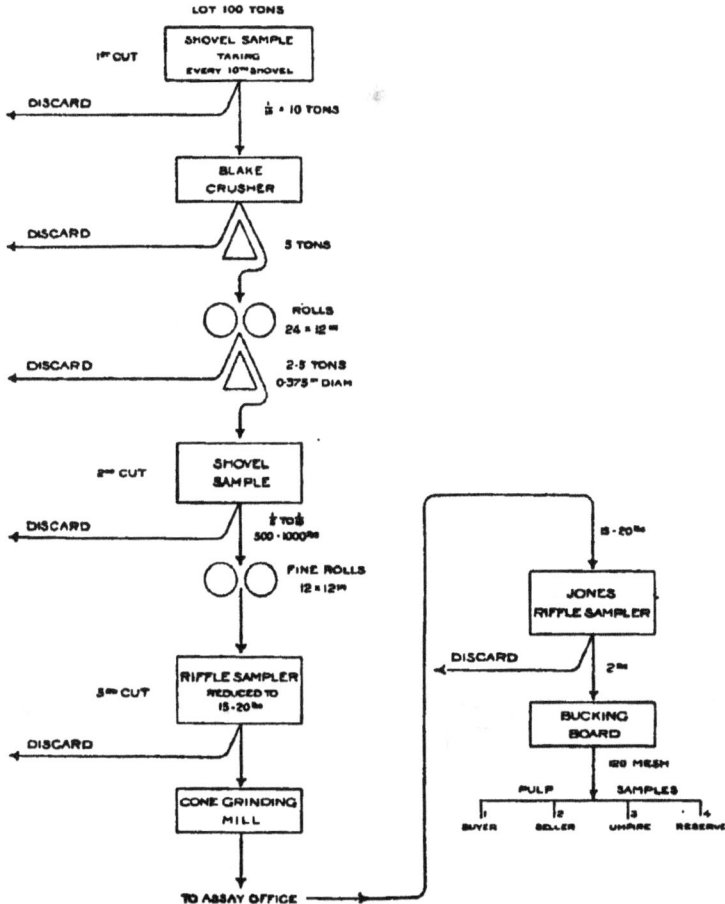

heated at 120° C. for some time in order to expel any moisture which the sample may have absorbed in its passage from the sampling works to the assay office.

The assay sample is divided into four parts and put in paper sacks. One part is assayed by the seller of the ore or product; one part by the purchaser; a third part is kept for emergency; and a fourth part is laid aside for an umpire assay, if such becomes necessary.

The method is illustrated in the flow-sheet given above.

Sampling Plant for Cripple Creek Ores, Colorado.[1]—The following course of

[1] Argall, "Sampling and Dry Crushing in Colorado," *Trans. Inst. Min. and Met.*, vol. x. (1902), pp. 234-299.

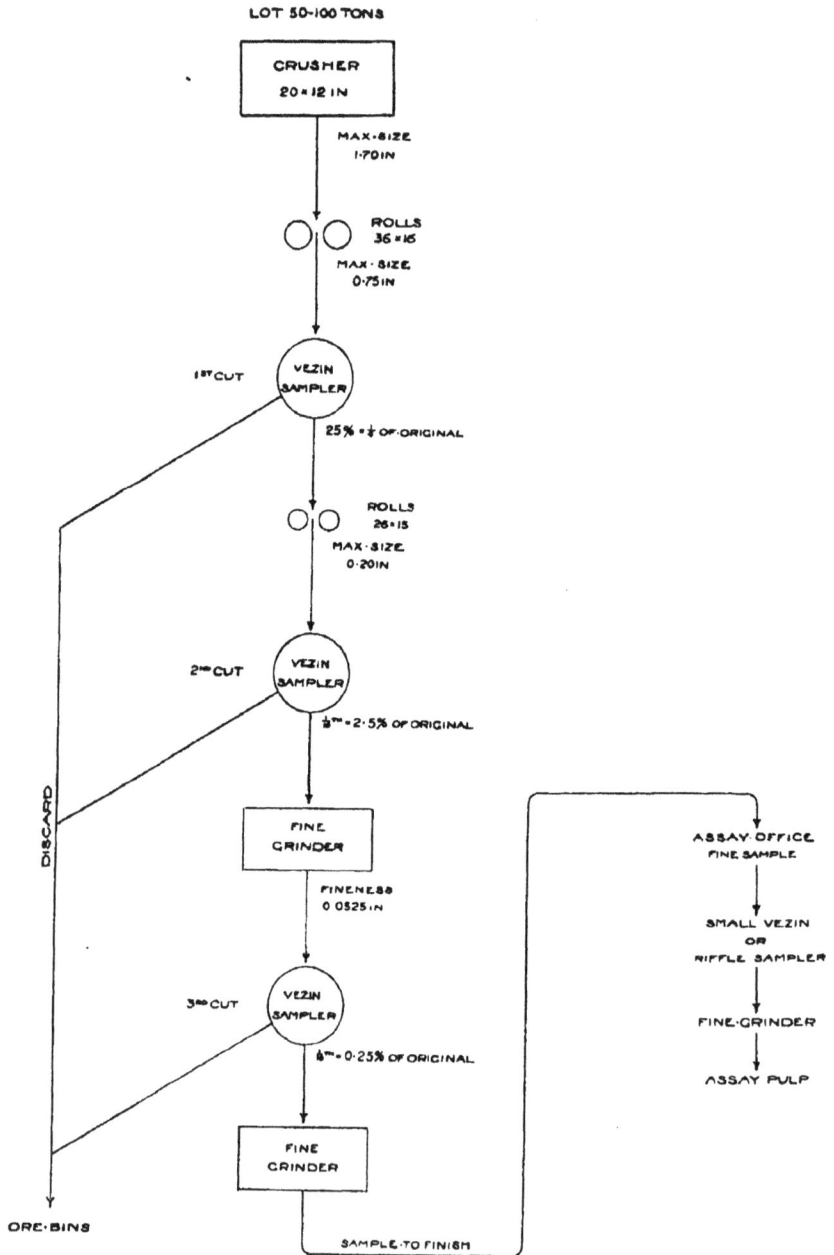

LOT 50-100 TONS

CRUSHER
20 × 12 IN

MAX·SIZE
1·70 IN

ROLLS
36 × 15

MAX·SIZE
0·75 IN

1ST CUT

VEZIN
SAMPLER

25% = ¼ OF ORIGINAL

ROLLS
26 × 15

MAX·SIZE
0·20 IN

2ND CUT

VEZIN
SAMPLER

⅒TH = 2·5% OF ORIGINAL

FINE
GRINDER

FINENESS
0·0525 IN

3RD CUT

VEZIN
SAMPLER

⅒TH = 0·25% OF ORIGINAL

FINE
GRINDER

DISCARD

ORE·BINS

SAMPLE·TO FINISH

ASSAY OFFICE
FINE SAMPLE

SMALL VEZIN
OR
RIFFLE SAMPLER

FINE·GRINDER

ASSAY PULP

procedure is recommended by Argall in machine sampling ordinary quartzose gold ores assaying from 2 to 6 ounces of gold per ton. For lots of 50 to 100 short tons in large lumps:—Crush in 20-inch by 12-inch crushers to a maximum size of 1·70 inch, and crush this in 36-inch by 16-inch rolls (at 35 revolutions per minute) to pass a 0·75-inch mesh screen. The product of the rolls is passed over a Vezin sampler, and 25 per cent. taken for the sample, which should be deflected to a 26-inch by 15-inch rolls to be crushed to 0·20-inch mesh. At this size about 6 per cent. is ample for a correct sample, but to be quite safe one-tenth is cut out; if the ore does not contain over 5 per cent. of moisture, the sample, $2\frac{1}{2}$ per cent. of the original, can be passed direct to a fine grinder, reducing it to 0·0525 inch, and again cut to one-tenth or one-quarter per cent. of the original volume; if damp, the sample is then dried and ground, then passed over a small Vezin sampler, or riffled, as found most desirable, being finally reduced by successive crushing and division until 8 ounces is finally crushed to pass a 120-mesh sieve for assay purposes. The flow-sheet is given on page 106.

In sampling rich telluride ores assaying 10 to 15 ounces of gold per ton, Argall recommends the following method, taking 100 short tons as unit:—With these richer and "spotty" ores finer preliminary crushing is necessary. The ore is crushed to an average of about 1-inch cubes, and 20 tons (i.e. 20 per cent.) taken as the first sample; this sample is then crushed to $\frac{1}{4}$ inch, and 2 tons (10 per cent.) cut out by the second sampler. Crush this to 8 mesh ($\frac{1}{16}$ inch) and reduce by "riffling" to 250 lbs. Dry and crush the sample to 30 mesh ($\frac{1}{50}$ inch), and riffle down to 15 lbs. Pass this through a sample grinder to 90–100 mesh, and riffle down to 1 lb. This is ground on the buck-board to pass 120 mesh ($\frac{1}{210}$ inch), and divided for assay.

Sampling at Cananea, Mexico.[1]—The following plant is in use at Cananea for sampling the domestic smelting ore, which contains a small percentage of copper, with silver and a little gold. The ores are comparatively uniform, soft, and lean. The ore is brought from the mines in train-loads, the cars holding about 30 tons each. These are discharged into receiving bins at the reduction works. One or more train-loads constitute a lot, say 100 tons. The ore is fed on to a screen made of steel rails with a 4-inch clearance. The under-size drops on to a moving belt, and the over-size passes through two 24- by 36-inch Blake crushers set with 4-inch maximum opening, and the ore after crushing immediately joins the under-size on the belt. This delivers the ore to a Vezin sampler, which cuts out one-tenth for sample. This is conveyed to a Blake crusher set to crush to 1-inch maximum size. After passing the latter crusher another sample is cut out automatically by a Vezin machine giving $\frac{1}{100}$th or 1 per cent. of the original lot as a sample. This passes to a small Gates crusher set to crush very fine, and the stream is again cut, $\frac{2}{1000}$th of the original lot being taken. This passes to fine rolls, where it is crushed, after which it is again cut and about $\frac{1}{1000}$th remains as the sample.

This sample, now under 8-mesh maximum size and weighing from 100 to 200 lbs., according to the weight of the lot, is then thoroughly mixed and split in a Jones riffle sampler, until about 4-lb. weight is obtained. The 4-lb. sample is dried and ground in a disc-grinder, split to about 2 lbs., screened and ground to pass a 100-mesh screen. The pulp thus obtained is thoroughly rolled in a mixer and divided into four pulps of about 8 ounces each, which are used for the assays. This process, which is shown in the accompanying flow-sheet (page 108), is stated to be accurate for the class of ore treated. Custom ore is sampled differently. All the discarded ore from this sampling passes to an elevator, from which it is delivered to a belt and thence to the ore beds.

Sampling of Silver Ores containing Cobalt, Nickel, and Arsenic.—The follow-

[1] L. D. Ricketts, *Min. Mag.*, vol. iv., 1911, p. 128.

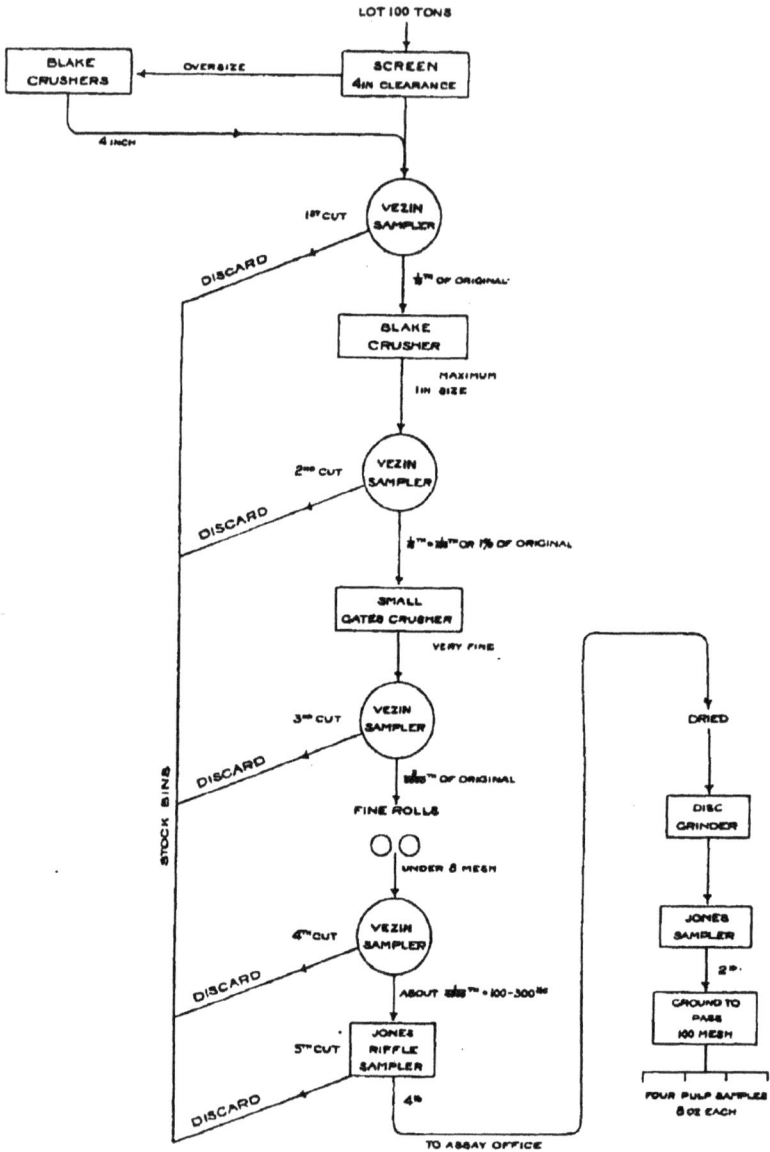

ing details of the sampling of ores containing the arsenides of cobalt, nickel, and silver, and also metallic silver, are given by Handy.[1] The ores are mined in the Cobalt district, Ontario, and are usually very rich in silver, carrying sometimes over 2000 ounces per ton.

Having been first sampled when being unloaded and a determination made of the moisture, the low-grade ore is passed through a jaw-crusher to reduce it to $1\frac{1}{2}$ inches or smaller, and is then divided by shovelling two parts to the storage bin and one to the sampling floor, from which it is again divided, one shovel to bin and one to sample. It is then reduced to $\frac{1}{2}$-inch size and quartered to not less than 1700 lbs. The ore is then crushed again to pass a $\frac{1}{4}$-inch screen, turned over twice, quartered to not less than 300 lbs., crushed to pass a 10-mesh screen, mixed and quartered to not less than 60 lbs., when it is ground to pass a 20-mesh screen and cut down by a Jones sampler to about 4 kilos, which is ground in a pebble mill to pass a 120-mesh screen. The high-grade ore containing over 500 ounces of silver per ton is crushed to 1-inch size, and alternately shovelled to the storage bin and to the rolls, which reduce to $\frac{1}{4}$-inch size. The "metallics" on the screen are collected, weighed, melted, and cast into bars which are assayed. The $\frac{1}{4}$-inch ore is coned and quartered to about 2400 lbs., crushed to pass a $\frac{1}{4}$-inch screen, again coned three times, and quartered to about 400 lbs. It is then crushed in rolls to pass a $\frac{1}{10}$-inch screen, and cut down by a Jones sampler to about 75 lbs.; again ground to pass a $\frac{1}{20}$-inch screen, and reduced to about 5 lbs. It is then finely ground, and screened on a 120-mesh screen to give "metallics" and "pulp."

The samples thus obtained are then assayed by the methods described on page 234.

[1] J. O. Handy, "Eighth Int. Cong. Appl. Chem.," 1912, Sect. IIIa, abstract, *Soc. Chem. Ind.*, 1912, vol. xxxi. p. 880.

CHAPTER IX.

PREPARATION OF THE SAMPLE FOR ASSAY.

General Remarks.—When received by the assayer, samples of ore and metallurgical products are seldom in a fit state for assay purposes : they are generally too large in bulk and too coarse.

To prepare them for assay they have to be crushed finer and reduced systematically to small bulk. If the sample is in lumps, these are broken down to the size of $\frac{1}{4}$-inch cubes or less before reducing the bulk, and preparatory to fine grinding. If the original bulk is small, the entire sample may be pulverised to a fine state and then reduced to the required bulk.

In the assay office the reduction of samples is usually effected by coning and quartering, or by means of a riffle or an automatic machine such as those described later ; but whatever method is employed, the sample is alternately crushed finer and reduced in bulk, until finally the mass is reduced to about 1 or 2 lbs. in weight, the whole of which is pulverised until it is fine enough to pass a sieve of at least 80 holes to the linear inch (*i.e.* 80 mesh). This finely pulverised sample is termed the pulp, and is used for the assay. It is transferred to a sample bag, box, or bottle and carefully labelled.

The number of crushings and reductions to which a sample has to be subjected is dependent on the nature and size of the ore particles and the weight of the sample, and the assayer in obtaining the portion he requires for assay must rigorously apply the principles of sampling already considered.

The preparation of assay samples has been studied by Huntoon,[1] who states "that with ores containing the values in brittle minerals the rich ore is almost entirely separated from the poor portions, if several passes are required in pulverising and the fine material is screened out of each pass. As it is extremely difficult to properly mix pulps varying largely in value, such a procedure often leads to discrepancies in the assay results. Screens should be avoided, the pulverising machinery being adjusted, whenever possible, to deliver a product of the desired degree of fineness, without over-size. The following weights are sufficient to ensure reliable samples :—

Size of ore particle.	Weight of sample.
$\frac{1}{2}$ inch	50 lbs.
$\frac{1}{4}$,,	25 ,,
10 mesh.	6 ,,
20 ,,	3 ,,
40 ,,	24 ozs.
60 ,,	5 to 6 ,,

Pulverising to 100 mesh assists the fusion, but it is not necessary for accuracy of sampling." Experiments on Cripple Creek gold ores are quoted to bear out the statements.

[1] L. D. Huntoon, " Preparation of Assay Samples," *Eng. and Min. Journ.*, 1911, vol. xci. p. 1249.

It is very important that the final reduction of samples in the assay laboratory should be done with the utmost care, as there is little doubt that it is in this last stage of the sampling that errors very frequently occur. The samples received at the assay offices attached to sampling mills are usually in a state of coarse powder from the crushing rolls or grinding machines, and weigh from 12 to 15 lbs. They are reduced in bulk to about 2 lbs., and finally pulverised, then well mixed and divided into four or more portions of about 8 ounces each. These are put into separate sample bags or bottles and sealed. Some of these are assayed by the different assayers representing the buyer and the seller, and some are reserved for a "reference" or "umpire" assay in the case of dispute, as previously described on page 85.

"Metallics" or "Scales."—During the pulverising of both gold and silver ores a residue of small particles is frequently left upon the sieve, and an in-experienced person not understanding that they may be the most valuable portion of the sample is inclined to throw them aside. These particles, which are termed "metallics" or "scales," generally consist of metallic gold or of metallic silver, associated sometimes with silver chloride, bromide, or iodide, and native copper. Owing to their malleability the "metallics" cannot be readily pounded to a fine state of division, and although a part always passes through the sieve, some of the larger particles fail to do so. The particles left on the sieve must be very carefully collected, weighed and put aside to be separately assayed, see page 201.

The sifted sample (without the metallics) is weighed, well mixed, and assayed. The total value of the ore is found by adding the results of these two assays as shown in the example on page 216.

"In pulverising ores and products containing 'metallics' it is of great importance to powder the whole of the selected sample without loss during the process, and of even greater importance to mix well the sifted portion, of which the last particles to come through the sieve are apt to be more than ordinarily rich through the grinding down of some of the metallic particles " (Beringer).

In some assay offices it is the practice to throw the metallics back into the mortar or on to the bucking-plate with a small quantity of the pulverised ore and continue the grinding until everything passes through the sieve. This practice is open to grave objection, as it is impossible to ensure the even distri-bution of the metallic particles throughout the sample ; and, moreover, experience has proved that it gives rise to low results owing to the metallic particles being flattened out and left adhering to the surface of the sample grinder. For example, a sample of rich gold ore in which the gold was present almost entirely as metallic flakes, tested by F. A. Thompson [1] by both methods, gave an average result of 11·71 ounces per ton for the sample ground until all the metallics passed through a 100-mesh sieve, and an average result of 17·92 ounces per ton for the sample in which the metallics were assayed separately and allowed for in the final result.

Rich ores in which the precious metals are present largely in the metallic state are especially difficult to sample and, as indicated above, require very careful attention during the final stage of the sampling.[2] To check any careless-ness in the preparation of samples they should be weighed before and after grinding. In sifting the ore after grinding, the material left on the sieve, which usually consists of the hard gangue, must be again pulverised until the whole of it, with the exception of any metallics, passes the sieve : it is obvious that if this hard portion is rejected the sample will be vitiated, as the sifted portion is thus made proportionately richer.

---

[1] F. A. Thompson, *Western Chemist and Metallurgist*, vol. iii., 1907.

[2] Consult F. White, "Notes on Tests made to ascertain the Effect on Assays of unusually Large Particles of Gold in Ore Samples," *Trans. Inst. Min. and Met.*, vol. xxii., 1912-1913.

Crushing and Pulverising.—The implements employed for crushing and pulverising samples differ according to the number of samples to be dealt with. In assay offices the crushing and reduction of samples is generally done by hand: large lumps are broken down in an iron mortar or by a hammer, and the fine grinding done on a bucking-plate or in some form of laboratory grinding machine such as those described later. At smelting works and sampling mills where large numbers of samples are dealt with daily, small rock-crushers with reciprocating jaws worked by hand or by steam or other power are used for breaking down large lumps, while for fine pulverisation small, steel-faced, high-speed rolls are very generally employed. In selecting a crusher or grinder for laboratory use preference should be given to those machines that can be easily and thoroughly cleaned, and in which the wearing parts can be readily replaced. The wearing parts of all crushing and grinding implements should be made of some tough and hard material such as manganese steel or chromium steel.

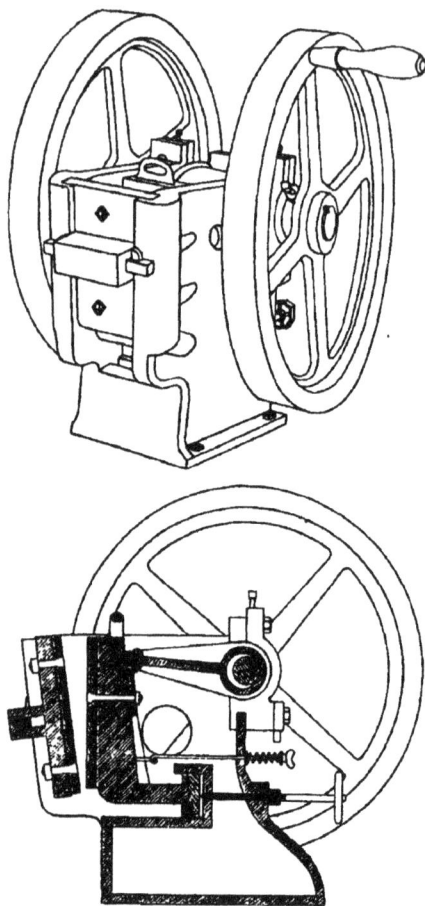

When made of grey cast iron notable amounts of iron are ground off and dilute the finely ground ore. This is especially noticeable in the fine grinding of exceedingly hard and tough ores which may become diluted with iron to the extent of 2 or 3 per cent. The iron should be removed by passing a good magnet through the pulp several times. With soft ores the wear on the grinding surfaces is comparatively small. Whatever form of crushing or grinding implement is used, a supply of barren rock or sand should be kept at hand and a small quantity crushed or ground in the apparatus after each sample has been dealt with in order to thoroughly clean the grinding surfaces before starting a fresh sample. This is very important when dealing with samples of rich ores and products. In assay offices where both rich and poor samples have to be dealt with it is a wise plan to have two grinding machines, one for grinding rich ores and the other for poor ores; the danger of "salting" (i.e. contaminating) the poor sample is thus greatly diminished.

Crushing Implements.—Iron pestles and mortars are still largely used in the laboratory for breaking down lumps when the sample is of small bulk.

FIG. 86.

To prevent fragments of ore from flying out during the crushing the mortar should be covered with a cloth.

Rock-Crushers.—The small crushers used in the laboratory for breaking down the lumps in comparatively large samples are very similar to those employed on a large scale.

The Simplex Crusher shown in elevation and in section in fig. 86 may be taken as a type of hand-crusher for laboratory use. In this machine the crushing is effected by a motion imparted to the vibratory jaw from the eccentric

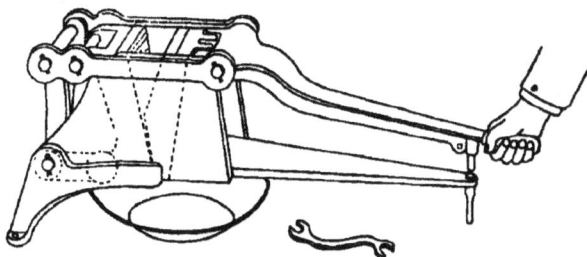

Fig. 87.

on the shaft, and as the heel of this jaw is stationary, the lower portion moves through an arc of a circle, crushing rapidly and ensuring the discharge of the product. Two adjusting screws pressing on the brass box which engages the heel of the vibratory jaw regulate the size of the discharge outlet, and material may be crushed from 2 inches to $\frac{1}{4}$ inch or to a 10-mesh powder.

The crusher is cleaned after pulling out the wedge holding the stationary jaw; removing this jaw gives sufficient access to the interior. By loosening

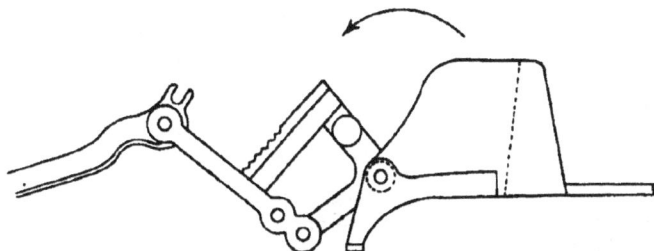

Fig. 88.

the thumb-screw on the spring attachment the vibratory jaw may be thrown back, exposing the entire interior.

When very fine crushing is desired time will be saved by running the ore twice through the machine, set the first time as wide as possible, and the second time at its maximum fineness.

Great care must be taken to clean the crushing jaws before putting a fresh ore through the machine. If necessary, a quantity of barren quartz should be crushed to remove any particles of rich material that may be left adhering to the jaws. For cleaning purposes a hand bellows may be used in place of a brush.

Taylor Rock-Crusher (figs. 87 and 88).—This is a small hand machine of the "Blake" type. The construction of this crusher is such that the size

8

of the discharge opening does not vary when the machine is operated, consequently, when the jaws are set close together, a product of uniform fineness is obtained.   It is possible to pulverise an entire sample to a comparatively fine mesh with one feeding through this machine.   The adjustment for a coarse or fine product is instantly made by turning the knob shown directly under the hand (see fig. 87).

The straps are made of steel, the stationary and vibratory jaw-plates are chilled iron, the frame and hand lever are made of cast iron ; all parts of the crusher are sufficiently strong to withstand hard usage.

The method of opening the machine for cleaning, by throwing the handle back, is illustrated diagrammatically in fig. 88.

Each machine has a cover (not shown in the illustration) to prevent pieces of ore from flying out.

Grinding Implements.—After coarse crushing the sample is finely pulverised on the buck-board or in a grinding machine.   Before the fine pulverising can be done it is frequently necessary to dry the ore, in which case the percentage of moisture is usually determined at the same time by weighing the sample both before and after it is dried, as described on page 125.   The iron pestle and mortar is sometimes used for the finer grinding of small samples, but although useful for breaking down the ore to a coarse sand, the mortar is far inferior to the buck-board for grinding material to pass through a very fine sieve.   If used for this purpose, it should be borne in mind that pulverisation is rendered much easier by operating on a small quantity at a time and removing it at frequent intervals for sifting.   The quantity operated upon

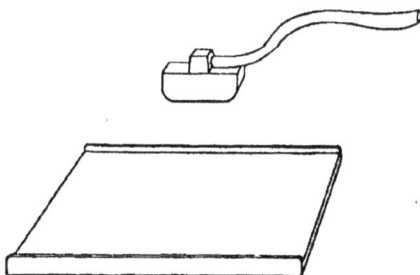

FIG. 89.

at one time is regulated by the hardness or friability of the ore ; the harder it is the smaller the quantity taken.

Bucking-plate.—For general assay-office work the bucking-plate (fig. 89) is most suitable for fine grinding.   It is a smooth iron plate about 18 by 24 inches in size, and 1 inch thick, with a 1-inch rim on two or three sides.   In the latter case the back corners of the plate are usually rounded to facilitate the removal of the fine material.   The muller or grinder to be used with the plate is a heavy piece of iron 4 to 6 inches wide with a smooth, curved surface, weighing about 15 to 20 lbs., and fitted with a handle.   To grind the coarsely crushed ore it is spread on the plate and the muller placed upon it.   The muller is then held with both hands, one holding the handle and the other firmly pressing the iron block downwards, the curved surface being on the ore. It is then moved backwards and forwards while an oscillating movement is imparted by the handle.   After a little practice the method of grinding is soon acquired.

Circular buck-plates from 3 to 4 feet in diameter are also in use.   They are usually placed on a low table so situated that the operator can walk round it and thus alter the position of the muller as he works it backwards and forwards. By working in this way the grinding takes place in different directions across the plate and the wear on the surface of the plate is much more even than in the case of rectangular plates, on the surface of which grooves are liable to be formed owing to the fact that the grinding is always in one direction.   The buck-plate is very effective, and saves considerable time and labour compared with fine grinding in a mortar. It is in very general use for the final pulverising of ores. The pulp is removed from the plate and the plate cleaned by means of the stiff brushes shown in figs. 90A and 90B.

Grinding Machines.—The following machines may be taken as types of the grinding mills, etc., that are used more or less generally as substitutes for the pestle and mortar and the buck-plate for reducing ores and other material to fine pulp for assay purposes.

End-Runner Mills.—These mills are essentially large mortars and pestles worked by hand or other power. The mills vary in size and in detail of construction, but they are all worked on the same principle. Fig. 91 shows a usual type

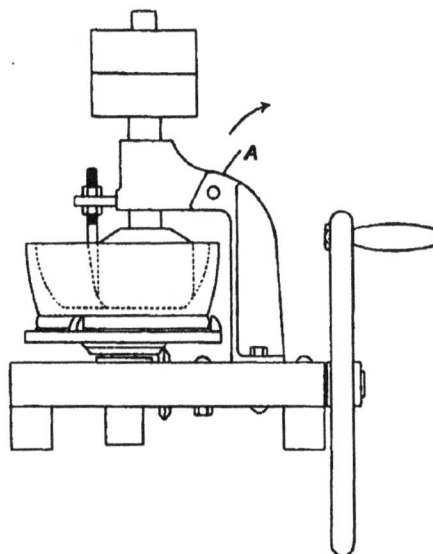

FIG. 90B.

FIG. 90A.

of end-runner mill for either hand or other power. The mill consists of an iron mortar and a runner or pestle, the runner being turned true to fit the inside of the mortar. The mill is shown in position

FIG. 91.

for operating. The mortar is mounted either on a table which runs on ball-bearings, power being applied to the runner, or the table is made to revolve by hand or by gearing below the mortar, the runner being held down by two heavy weights as shown in the illustration. A scraper is attached to the runner frame and so arranged that it removes any material that may adhere to the surface of the runner. The runner is hinged at A so that it can be swung back easily for cleaning purposes and the mortar lifted off and emptied.

When the runner is raised, it rests directly over the mortar, and may be cleaned with a brush and palette knife. All ore that is cleaned off the runner falls directly into the mortar. The usual size of the mortar is 15¼ inches diameter and 5 inches deep. The size of the runner or pestle is 7½ by 6 inches. The mill is run at a speed of 90 to 120 revolutions per minute. These mills are very suitable for grinding ore samples which have previously been coarsely crushed in a rock breaker.

In some of the smaller machines of this type attempts have been made to make the motions of the machine identical with those given to the pestle when worked by hand, and in addition the mortar is, in some cases, made to rotate with the table to which it is attached.

Radial Bucking-Plate (Calkins).—This machine has been devised as a substitute for the buck-plate. The motions of the machine are very similar to those given to the muller when working a buck-plate by hand. Fig. 92A shows a perspective view. Fig. 92B is a vertical section showing the automatic refeeding of the material. Fig. 92C shows the machine open for cleaning.

The appliance is designed to reduce crushed ore samples quickly to a pulp of any desired fineness. It consists of a semicircular iron trough the thick bottom of which forms the bucking-plate, and to the sides of which is journalled an axle carrying a shoe or muller which is pressed against the plate by two spiral springs. More or less compression is obtained by tightening or loosening these springs, and both ends of the plate depart somewhat tangentially from the true radial line of the grinding surface, which prevents the material being thrown out while in operation. At one end a screen of any desired mesh may be inserted, and from time to time the muller should be swung clear of the plate as shown in fig. 92C and the material brushed against the screen, causing that which is fine enough to drop through. Both ends of the shoe terminate in a sharp edge which gathers up the coarse particles as it works, and as it approaches the extremity of the plate these particles roll or slide over the top of the surface of the shoe or muller and enter the crosswise aperture in the centre of the shoe, and are thus automatically re-fed between the grinding surfaces, as shown in fig. 92B.

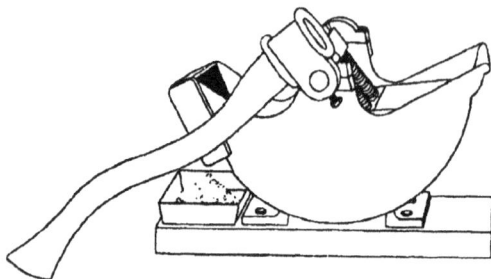

FIG. 92A.

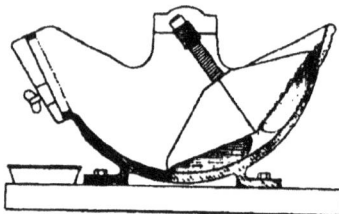

FIG. 92B.

FIG. 92D.

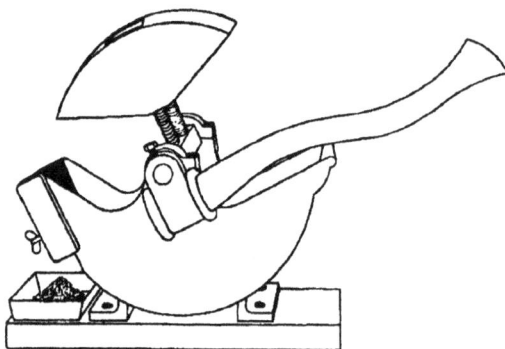

FIG. 92C.

The face of the shoe with aperture is shown in the illustration (fig. 92D) on a reduced scale.

The importance of this construction is apparent, as by gathering up the coarser particles it prevents their being surrounded and protected by the fine powder. The shoe and muller also wear to place, and this prevents their becoming ineffective through use. The screen is held in place by means of a thumb-screw, and as it may be removed easily, various meshes may be used.

The length of the machine is 18 inches and the width 6 inches.

Disc Grinders.—This type of grinding machine was introduced a few years

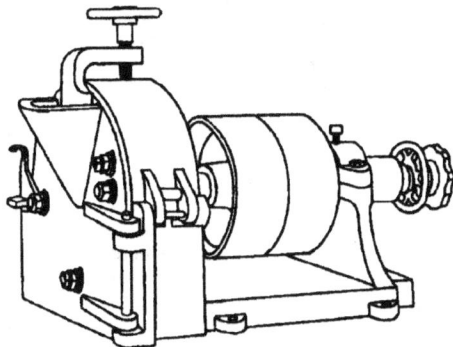

FIG. 93A.

ago and is gradually gaining favour. Figs. 93A and 93B show a grinder of this type made by the Sturtevant Engineering Co., Ltd. The construction of the machine is shown clearly in the illustrations. The grinding is done between two iron or steel discs or grinding plates one of which is stationary and the other revolves at a speed of 750 to 800 revolutions per minute. The machine is fed through the spout or hopper in the door to which the stationary disc is attached.

FIG. 93B.

The door and cover are hinged, and can be swung open as shown in fig. 93B to give full access to the interior for cleaning. This arrangement enables all the pulp to be brushed into a pan, for which provision is made when the machine is mounted.

The space between the discs is regulated by the hand wheel shown, so that the machine will deliver a coarse or fine product as desired. The stationary disc is so arranged that it can be moved frequently from its centre, thus changing the circle of the grinding surfaces. By this means any cuts or scores made on the surfaces when coarse or very hard cutting ores are being treated will grind themselves out. This simple arrangement also enables the plates, when grinding most ores, to be kept true, so that the output does not become uneven in character.

To give the best results the hand wheel must be turned slightly as the grinding proceeds and very frequently when the ore is cutting.

The capacity of pulverisers of this type depends on the fineness to which the material has been previously crushed. If the product fed into the pulveriser is ¼ mesh, it will be slower than if this product were crushed finer before being put through the machine. If large quantities are to be pulverised, it will usually be accomplished in less time if the product is first ground to about 10 mesh and then re-fed through the machine and ground to 100 mesh or other desired mesh. The life of the grinding plates is naturally dependent on the character of the material treated, but when worn out they are easily replaceable. As a general rule, one set of discs is capable of grinding about 2000 to 3000 lbs. of quartz ore of medium size to a pulp of very fine mesh. The grinding plates in machines of this class must be made of steel of very hard quality, otherwise the product will be contaminated with notable amounts of iron. With very fine pulverising of hard and tough ores the product may contain as much as 2 or 3 per cent. of iron even when grinding plates of good quality are used.[1] This fact constitutes one of the chief drawbacks of disc-grinding machines. With softer ores the dilution with metallic iron may be nearly or quite negligible. The amount of metallic iron in a sample may be determined by digesting a weighed portion of the pulp overnight in a solution of copper sulphate and determining the iron dissolved in the solution, or the ore may be treated with dilute sulphuric acid and the iron that passes into solution determined by titration with permanganate or bichromate.

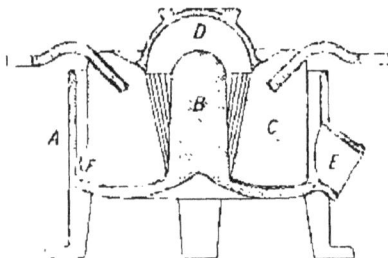

FIG. 94A.

When cleaning these machines before starting a new sample a hand bellows may be used in place of a brush with advantage. The machine described requires about 3 H.P. to run to full capacity, but smaller machines requiring only 1 H.P. are obtainable.

Combined Crusher and Pulveriser (Weatherhead).—This consists of a large iron mortar with a central post, round which a heavy pestle rotates. The inner surface of the pestle is corrugated and acts as a crusher, while the lower surface is smooth and is utilised for grinding the crushed material. Three illustrations of the machine are given. Fig. 94A shows a section of the crusher and pulveriser. The description is as follows :—A, iron casing or mortar with crushing post B, which is corrugated and slightly oval ; C, rotating pestle with two handles fitted into sockets ; D, loose cover ; E, spout through which the pulverised material is discharged ; F is one of four lugs on the side of the rotating pestle, which carry the pulverised material to the discharge spout.

Fig. 94B illustrates the machine with both handles in position, making it very easy to lift out the pestle to clean the mortar. The cover is removed to show the conical corrugated opening or hopper in the centre of the rotating pestle, where the material to be crushed is introduced.

In fig. 94c the machine is shown with the rotating pestle lifted out ready for

[1] Consult V. Lenher, "Contamination of Laboratory Samples by Iron derived from Crushing Machinery," *Chem. News*, vol. cvi., October 11, 1912.

cleaning. For convenience in cleaning the pestle is placed sideways on a wooden stand (not shown in the figure) with the edge overlapping the edge of the mortar as shown, so that any material remaining on the pestle can be brushed off into

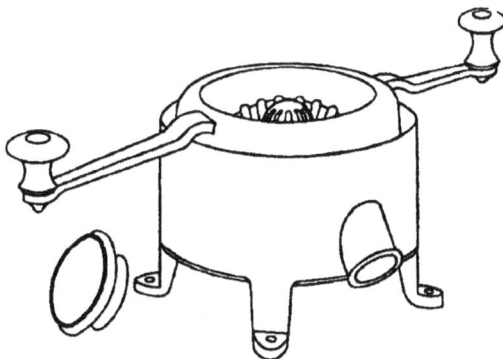

Fig. 94b.

the mortar and then through the outlet. The material to be pulverised must be dry, and should be broken in pieces about $\frac{1}{4}$ to $\frac{3}{4}$ inch diameter: the harder the material the smaller it should be broken. To operate the machine the cover

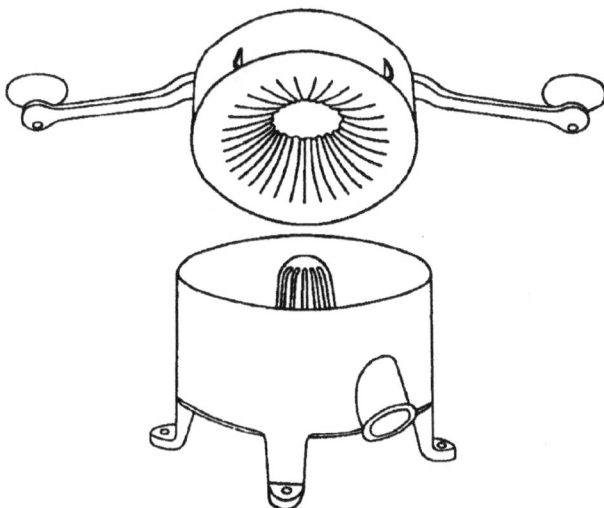

Fig. 94c.

is removed, the proper amount of material, according to its hardness, is fed into the hopper, the cover replaced, and both handles adjusted. The pestle is then worked backwards and forwards until the larger pieces are crushed, and when this is accomplished one handle is removed and the grinding continues with the regular rotary motion. As the material is gradually crushed it works down to the bottom of the conical opening or hopper and passes under the smooth face of the pestle, where it is finely pulverised. The centrifugal force carries the product to the sides, where it is caught by the lugs F, which carry it to the spout and discharge it. When nearly all the material has passed out of the hopper, another lot of ore is fed in and the operation repeated. It is customary with a great many users of this machine to return the material to the hopper a second time, and repeat the pulverising process until a product of the desired degree of fineness is obtained. The diameter of the machine is $10\frac{1}{4}$ inches and the height $8\frac{1}{2}$ inches.

Fineness of Samples.—Opinions differ as to the degree of fineness to which assay samples should be pulverised, but recent experience has proved that better results are generally obtained when the samples are crushed finer than was formerly thought to be necessary. There is now general agreement amongst assayers that samples should never be coarser than 80 mesh, and many assayers recommend 100 mesh and some 200 mesh. The author's experience is in agreement with that of Dr Rose[1] and others that "as a general rule all ores require crushing to 100 mesh, and that finer crushing may be dispensed with in the case of simple ores, but that complex ores, such as tellurides, should be crushed as finely as possible." The want of close agreement in assay results may frequently be traced to the use of samples that have not been ground sufficiently fine.

The influence of fine crushing on the assay value in the case of South African gold ores is shown in the following Table V., which gives the results of experiments by A. Whitby.[2] The figures show that very finely crushed material gives higher results on assay by fusion than does the same material when less finely crushed. Whitby recommends the 100-linear mesh as a sort of standard.

Table V.—Showing the Influence of Fine Crushing on the Assay Value. (A. Whitby.)

Assay No.	Fineness of pulp.			
	30·60 mesh.	60·90 mesh.	150 mesh.	150 mesh.
1 2 3	ozs. per ton. 4·10 5·00 5 10	ozs. per ton. 4·20 5·00 4·40	ozs. per ton. 4·52 5·55 5·64	ozs. per ton. 4·88 5·70 5·28
Average result	4·73	4·53	5·23	5·28

In mine-assay offices where the assaying is largely confined to one class of ore, a few tests should be made on the ore to ascertain what degree of fineness is necessary to give accurate results: if 80 mesh gives correct results, finer grinding is obviously unnecessary. However, in the case of ores with which the assayer is not familiar, it is better not to omit grinding the ore sufficiently fine to pass a 100 mesh.

Screens or Sieves.—To ensure, as far as possible, uniformity in the size of the particles, the finely pulverised ore must be sifted. For this purpose sieves

[1] *Metallurgy of Gold*, 5th edit., 1906, p. 438.
[2] "Gold Assaying: Note on the Influence of Fine Crushing on the Assay Value," A. Whitby, *Journ. Chem. Metall. and Min. Soc. of S. Africa*, 1904, v. 95.

of various sizes and different degrees of fineness are necessary. Two kinds of sieves are in general use, open or drum sieves and closed or box sieves. The open sieve is a round wooden frame with a meshwork of brass or iron wire gauze stretched over one end. For finely pulverised gold and silver ores, where it is very desirable to prevent loss, the box sieve of tinned iron is frequently employed. It consists of three parts : the sieve, a cover which fits tightly on the top to retain the dust, and a tightly fitting box or receptacle to receive the sifted pulp. Iron wire gauze is used for sieves coarser than, say, 20 mesh, while brass wire cloth is invariably used in the laboratory sieves of finer mesh. The size to which ore is pulverised is generally stated in terms of "mesh," that is, the number of holes per linear inch measuring from centre to centre of the wires. Thus a 40-mesh sieve would have 40 holes to the linear inch, or 1600 to the square inch ; and a 90-mesh sieve 90 holes to the linear inch, or 8100 to the square inch. In view of the tendency of modern assaying practice towards very fine grinding, it may be advisable, here, to make brief reference to wire screens. To those familiar with laboratory wire screens the mesh number, such as 40, 80, or 100 mesh, represents a fairly definite maximum size of ore particle. It is, however, generally known that the number of mesh to a linear inch does not clearly define the size of opening through which the ore particle passes, as this is dependent on the thickness of the wire used. For example, a 26-mesh sieve made from No. 26 wire (·0181 inch) gives an opening or mesh of ·0203 inch, while a 26-mesh sieve made from No. 36 wire (·0090 inch) gives an opening of ·0294 inch, a variation of over 30 per cent.[1] With coarse sieves the error is not great and often unimportant, but with the finer sieves the wire itself occupies so much space that the size of particle passed by the sieve may be anything from about one-half to about four-fifths of the size indicated by the word "mesh." It is well, therefore, in using fine sieves for the preparation of assay samples to bear in mind that the stated mesh is not always an indication of the degree of fineness of the pulp obtained. A sieve of 120 mesh, if made with No. 43 gauge wire, will pass particles not exceeding $\frac{1}{500}$ inch in diameter ; but this size particle will also pass a 130-mesh sieve if made of No. 45 wire, and a 140-mesh sieve if made from No. 46 wire. Owing to great variations in the thickness of wire in different screens it is always necessary, in referring to wire screen mesh (if anything approaching accurate definition is required), to give either the gauge of the wire or the actual measurement of the opening between the wires. The importance of knowing the actual dimensions of sifted ore particles both in laboratory work and in making grading tests with dressing plant, etc., has long been recognised, and has resulted in the adoption of several series of standard screens. As the result of careful investigation by a special committee, the following Table VI., of standard laboratory screens, has been established, and adopted in 1907 by the Institute of Mining and Metallurgy, London.[2] It is intended for use in making grading tests and for the correlation of screens used in commercial or other work. In this series, which is being very generally adopted, the diameter of the wire is equivalent to the linear width of the opening, i.e. the open spaces total 25 per cent. of the entire screen surface.

---

[1] Argall, *Trans. Inst. Min. and Met.*, vol. x., 1902, p. 244.
[2] *Bulletin No.* 38, Institute of Mining and Metallurgy, Sept. 1907. For other standard series and bibliography of the subject see paper by T. J. Hoover on "A Standard Series of Screens for Laboratory Testing." *Trans. Inst. Min. and Met.*, 1910, xix. p. 486.

Table VI.—I.M.M. Standard Laboratory Screens.

Mesh or apertures per linear inch.	Diameter of wire.		Aperture.		Screening area.
	in.	mm.	in.	mm.	per cent.
5	0·1	2·540	0·1	2·540	25·00
8	0·063	1·600	0·062	1·574	24·60
10	0·05	1·270	0·05	1·270	25·00
12	0·0417	1·059	0·0416	1·056	24·92
16	0·0313	0·795	0·0312	0·792	24·92
20	0·025	0·635	0·025	0·635	25·00
30	0·0167	0·424	0·0166	0·421	24·80
40	0·0125	0·317	0·0125	0·317	25·00
50	0·01	0·254	0·01	0·254	25·00
60	0·0083	0·211	0·0083	0·211	24·80
70	0·0071	0·180	0·0071	0·180	24·70
80	0·0063	0·160	0·0062	0·157	24·60
90	0·0055	0·139	0·0055	0·139	24·50
100	0·005	0·127	0·005	0·127	25·00
120	0·0041	0·104	0·0042	0·107	25·40
150	0·0033	0·084	0·0033	0·084	24·50
200	0·0025	0·063	0·0025	0·063	25·00

Reducing the Bulk of Samples.—In the laboratory, large samples are reduced in bulk by the method of coning and quartering fully described on page 97, or one of the following methods may be employed.

Riffle "Sampler" (or sampling tin).—This consists of a frame containing a number of narrow troughs or riffles placed parallel to one another, with spaces between, equal in width to the width of the riffles (see fig. 95). In operation an

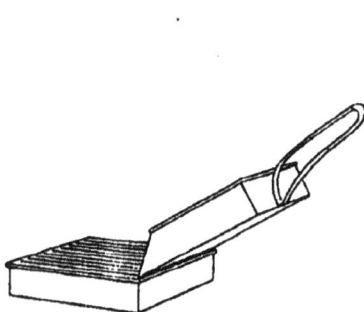

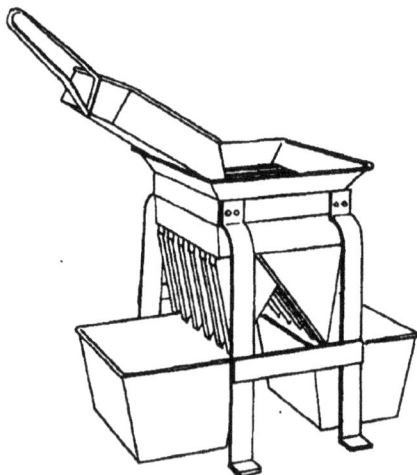

Fig. 95.

Fig. 96.

even stream of ore is allowed to fall on to the riffles from a scoop the same width as the series of riffles. Half the ore falls into the riffles, and is reserved for sample, and the other half falls through the open spaces and is rejected or reserved as a duplicate sample. A repetition of the process, with recrushing when necessary, reduces the sample to any required extent. The width of the riffles varies according to the size of the ore particles, but they should never be less than $3\frac{1}{2}$ times the diameter of the largest ore particle to be passed through them. For material with particles not more than a $\frac{1}{4}$ to $\frac{1}{2}$ of an inch in diameter a sampler with riffles 1 inch wide is suitable, while for finely ground materials riffles about $\frac{3}{8}$ inch wide are convenient. From six to twelve riffles are usually fitted to each sampler. Care must be taken not to overload the riffles, otherwise the samples will not be accurate —one half will contain a larger percentage of coarse material, and the other half a larger percentage of fine material. To prevent errors from overloading, riffle samplers of the Jones type, shown in fig. 96, were introduced. In the samplers of this type, instead of alternate riffles and spaces, two sets of riffles with inclined bottoms sloping in opposite directions are placed side by side, and so arranged that the stream of ore as it falls upon the riffles is deflected alternately to the right and to the left, thus yielding two samples which fall into separate pans as shown. For large samples this type is preferable to the flat-bottomed riffles. To obtain accurate results the riffle samplers of both types must have an even number of divisions, and the scoop used in feeding the ore on to the riffles must be the same width as the sampler, so that the ore is evenly distributed over the whole series of riffles.[1] Careful experiments have proved that, under these conditions, each half of the sample is a representation of the whole. Riffle samplers are clean, convenient, and rapid in operation, and are now extensively used for reducing the bulk of laboratory samples.

Umpire Sampler.—A sectional view of this machine is shown in fig. 97. The pulp is charged into a hopper, which is continually agitated, causing the material to fall in a steady stream into two buckets, which are revolved in opposite directions by means of gearwheels. Each bucket is divided into four divisions, two being closed and two

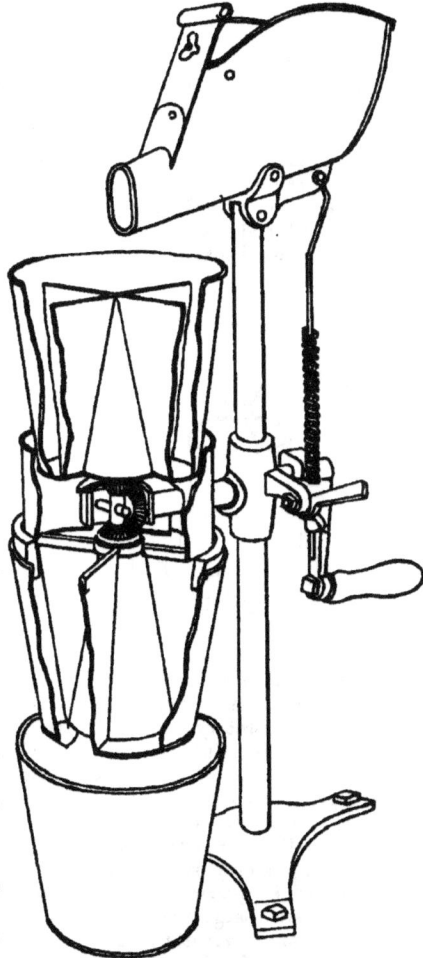

Fig. 97.

---

[1] See "Accuracy of Riffle Ore Samplers," L. D. Huntoon, *Eng. and Min. Journ.*, vol. xc. (1910), 62–65. A description, with working drawings, of a complete set of riffle samplers arranged by W. A. Brown for reducing the bulk of mine and mill samples is given in the *Eng. and Min. Journ.*, vol. lxxxvii., 1907, p. 232.

open; therefore, if 16 lbs. of ore be fed into the machine from the hopper, 8 lbs. will remain in the upper bucket, 4 lbs. in the lower bucket, and 4 lbs. or one-quarter of the original amount will pass into the receptacle at the bottom. By repeating the operation, the original sample may be reduced to $\frac{1}{16}, \frac{1}{64}, \frac{1}{256}$, etc., as desired.

A cam on the crank shaft agitates the hopper by striking a strap, which is attached to the hopper with a coiled spring, and by means of an eccentric lever the force of the blows may be varied at will. The upper and lower buckets may be quickly removed, and all portions are readily accessible for cleaning, making it impossible for subsequent samples to be contaminated.

Sample Dividers.—After the pulp sample has been obtained it usually happens that several persons wish to have a portion of it for separate assay; thus the sample may be divided into three parts—one for the buyer, another for the seller, and a third for the referee. Sometimes six or more distinct samples are wanted, and in order that the results of the separate assays may agree, it is clearly essential that all the samples should be identical—that each should have

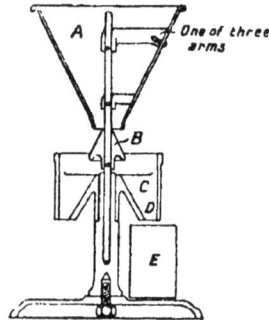

FIG. 98A.   FIG. 98B.

a fair proportion of the fine and the coarse, the rich and the poor, the heavy and the light particles. A very usual method of dividing small lots of pulverised ore in the laboratory consists in spreading the pulp out in a thin layer with a spatula on a sheet of glazed paper or oilcloth and digging out small portions at regular intervals. A small scoop made of sheet brass is preferable to a spatula for digging out the portions. Before spreading the ore it must be thoroughly mixed by drawing up first one corner and then another of the paper or oilcloth so as to roll the ore over on itself. After one sample has been taken as described, other samples may be obtained by repeating the process. The pulp may also be divided by quartering in the usual way after being well mixed. In assay offices where a large number of samples have to be dealt with, special "sample dividers" are frequently used.

Dividing Machine (Clarkson).—The divider is shown in elevation in fig. 98A, and in section in fig. 98B. It consists of a rotating hopper A, a cone B, and a dividing cup C. The material to be divided is formed into a rotating annular stream by means of the rotating hopper and cone, and at the same time thoroughly mixed by three arms in the upper part of the hopper. The stream of ore splits on the cone B, and is divided into six samples by means of the dividing cup C, which is made with six equal segments, delivering the samples by the spouts D into the tins (or bottles) E, one of which only is shown. To use the divider the hopper is set spinning by a touch of the hand, then the sample is thrown in, and it is divided into six equal and exactly similar portions in less than the same number of seconds. The speed of the hopper should be from 60 to 70 revolutions per minute. It must be set turning before the material is charged in. The height of the dividing cup can be adjusted to suit the size of bottle or tin, and the hopper and cup can be removed in two or three seconds for thorough cleansing. The material put through the divider should not be coarser than $\frac{1}{8}$ of an inch. This machine may be also used as a mixer for mixing fluxes, etc. The machine can be arranged for more or less than six samples according to requirements.

Mixer and Divider (Bridgman's).—These are used for the final mixing and distribution of assay samples. The pulverised material is introduced into the large covered funnel or mixer (fig. 99), the outlet being first closed by the thumb or finger as may be most convenient. The mixer and contents are then well shaken for a few moments, and then, with opened outlet, passed to and fro over the set of distributing funnels (divider) and bottles as shown. With very finely pulverised or very light material the flow may be assisted by a slight shaking or tapping with the hand. The little skill necessary to use the apparatus successfully is readily acquired. The mixer will also be found very useful for the rapid and thorough mixing of crucible assay charges, and all other work of similar character.

Labelling Samples. — The prepared pulp samples are stored in proper sample bottles, bags, or boxes, and clearly labelled by numbers for easy reference. Many assayers and smelting works use linen or strong manilla paper bags, printed with the name of the assayer or company, and provided with a tag for convenience in noting details of the samples or in sending them by post.

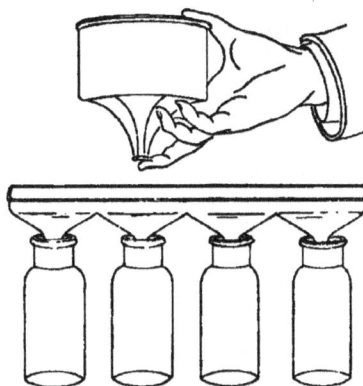

FIG. 99.

Drying: Determination of Moisture.—The term "moisture" is applied to all water lost on heating at 100° C., and ore which has been dried at this temperature is termed "dry ore." As settlements in the sale and purchase of ores are made on the "dry ore," it is very necessary that the samples received by the assayer should be carefully dried before being assayed, as they frequently contain more or less moisture even though they may have been previously dried. "The reason for reporting assay results on the 'dry ore' will be apparent when it is borne in mind that 'dry ore' has a constant composition, and the results of all assays of it will be the same, no matter when made. On the other hand, the percentage of moisture may vary from day to day, and it is best to limit this variability of the moisture by considering it separately and thus avoid having the percentage of the gold or silver rising and falling under the influence of the weather" (Beringer). The determination of the moisture present in any given sample of ore is quite as important as the determination of the metallic contents, for the moisture sample decides what percentage of the weight is to be excluded in the settlement between buyer and seller. This apparently simple process must, therefore, be conducted with care. In practice the ore is generally dried during sampling and the moisture determined at the same time by the samplers, the percentage of moisture found being stated on the label attached to the sample sent to the assay office. It is frequently assumed that the ore dried by the sampler contains no moisture: this may or may not be the case according to the nature of the ore. If the sample contains hygroscopic minerals it may reabsorb moisture, especially if a considerable time has elapsed between the drying of the ore and the assay of it. It is a safe rule, therefore, to dry all ores received for assay and to expel any moisture which may have been absorbed during their passage from the sampling floor to the assay office, unless the assayer knows from the source or nature of the sample that this step is unnecessary. The method usually adopted is to take from one-half to one pound of the coarsely crushed ore and dry it in a shallow iron pan, fig. 100 (such as a frying-pan), heated over a gas flame, or in an ordinary oven, with frequent stirring until no deposition of moisture takes place, when a cold piece of glass is held over it. The ore is then weighed, heated again for one half hour, and reweighed : when two successive weighings agree, the ore is considered dry. The percentage of moisture is calculated from the loss of weight. Thus the difference between the weight of the original ore and the dried ore is the loss by driving off the water; this difference, divided by the amount of ore taken, and multiplied by 100, gives the percentage of moisture in the ore. The heating should be carefully conducted, and the temperature kept as near 100° C. as possible, especially in the case of ores or products that oxidise readily, such as those containing finely divided sulphides, etc. Samples of ore and products that are hygroscopic should only be dried shortly before the assay is made. In the case of small samples the whole sample is dried, and frequently the drying is done in a porcelain basin heated over a sand-bath or water-bath; but if smaller quantities than those stated above are used, extra precaution must be taken, especially in the weighing.

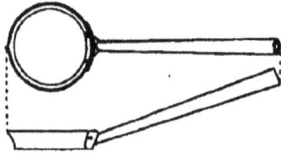

FIG. 100.

CHAPTER X.

FLUXES AND OTHER MATERIALS EMPLOYED.

THE dry, or furnace, method of assay is almost invariably employed for the determination of the precious metals in ores, etc., the ore being brought into a liquid state by fusion at a high temperature with suitable reagents or "fluxes." Very few of the constituents of ores are fusible when heated alone at the temperature attainable in an assay furnace, but if certain solid reagents are mixed with the pulverised ore, compounds are formed which fuse readily at a moderate temperature, producing a molten mass sufficiently liquid to enable a heavy metal such as lead, that may be diffused through it, to sink and unite into one mass at the bottom of the crucible.

Fluxes.—The reagents added for the purpose of forming fusible compounds with infusible or difficultly fusible bodies are termed fluxes.

Slag.—The fusible compounds resulting from the fusion of fluxes with infusible or difficultly fusible bodies are termed slags.

For example, in auriferous quartz the gold exists in the metallic state and is diffused through the mass in particles and in extremely small proportion in comparison with the quartz particles. When such powdered quartz is heated to a temperature far above the melting point of gold (1064° C.) no separation of the metal takes place, owing to the infusibility of the quartz. If, however, sodium carbonate is mixed with the quartz a fusible sodium silicate will be formed, and the melted gold, by virtue of its high density, will then separate out easily. The sodium carbonate in this case would be designated a flux, and the resulting sodium silicate a slag. The quantity

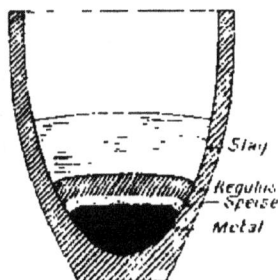

FIG. 100A.

of gold (and silver) present in ores is generally so small that it tends to remain diffused through the molten mass, and in assaying it is necessary to collect it by means of metallic lead, which alloys with the gold and silver and sinks to the bottom of the crucible. The metallic lead is obtained by the reduction of lead oxide added for the purpose.

In addition to slag and metallic lead other products may be obtained as the result of the fusion of an ore according to the constituents present. If metallic sulphides are present, an artificial sulphide or regulus (sometimes termed *matte*) may be formed, and if the ore contains arsenical minerals, a compound of a metal or metals with arsenic, termed a *speise*, may result. Assuming these substances to be present in a charge, they would separate according to their densities, when the fused mass solidified, in the order shown in fig. 100A. At the bottom of the crucible a button of lead would be found; above this a thin layer of speise; then a regulus, next a slag, and, in special cases, on the top of this a layer of more fluid slag consisting usually of fusible-alkaline chlorides and sulphates. A regulus, which is a compound of one or more metals with sulphur, is a yellowish-grey, metallic-looking mass, usually brittle, and often crystalline. A speise is usually hard, brittle, and greyish-white in appearance.

The correct use of fluxing materials is a subject of considerable importance to the assayer, and ought to be regulated according to chemical principles. It will be conducive to the thorough appreciation of the principles which should underlie the fluxing of ores, etc., if reference is first made to the action of the various reagents usually employed for this purpose. The principal fluxes employed in assaying are as follows :—

Basic Fluxes.
　Sodium carbonate (Na_2CO_3).
　Sodium bicarbonate ($NaHCO_3$).
　Litharge (lead oxide) (PbO).
　Red lead (lead oxide) (Pb_3O_4), or ($PbO_1Pb_2O_3$).
　Hæmatite (iron oxide) (Fe_2O_4).
　Lime (CaO).
Acid Fluxes.
　Borax (sodium pyroborate) ($Na_2B_4O_7$).
　Borax glass ($Na_2B_4O_7$).
　Silica (SiO_2).
　Glass.
Neutral Fluxes.
　Fluor-spar (calcium fluoride).
　Common salt (sodium chloride).

Sodium Carbonate (commonly called soda), Na_2CO_3.—The powdered anhydrous sodium carbonate is the most convenient for fluxing purposes. It is a readily fusible basic flux, and when heated with silica (quartz) forms a fusible sodium silicate and liberates carbon dioxide (carbonic acid).

$$Na_2CO_3 + SiO_2 = Na_2SiO_3 + CO_2.$$

Sodium Bicarbonate, $NaHCO_3$, may be used, but it is much less convenient, as the same bulk contains a smaller quantity of soda, the fluxing constituent. When heated, water and carbonic acid are given off at a comparatively low temperature, and sodium carbonate left.

$$2NaHCO_3 = Na_2CO_3 + CO_2 + H_2O.$$

The fluxing power of the bicarbonate is about two-thirds that of the carbonate. The fusion of charges containing sodium carbonate or bicarbonate must not be performed rapidly, otherwise the escaping gas will cause frothing and loss through the boiling over of the charge. This is very liable to occur when the bicarbonate is used, and on that account its employment is to be deprecated. Potassium carbonate may be wholly or partially substituted for sodium carbonate, but it has the disadvantage of deliquescing and of being more costly. A mixture of equal parts of potassium and sodium carbonates forms a very suitable flux and is sometimes employed. Owing to their great fusibility, the alkaline-carbonates can retain in suspension, without losing their fluidity, a large proportion of pulverised infusible materials such as an earth, charcoal, graphite, etc.

Lead Oxide.—This is a valuable basic flux : it is used in the form of red lead (PbO,Pb_2O_3) or of litharge (PbO). With silica it forms a very fusible lead silicate.

$$PbO + SiO_2 = PbSiO_3.$$

With silicates, also, it forms very fusible double silicates, and in the presence of silicates or of borax it acts as a flux for lime and magnesia. Metallic oxides that are infusible or difficultly fusible alone are easily dissolved by lead oxide, with which they form a strongly basic slag which has a very corrosive action on the crucible. This is particularly the case with fused mixtures of the oxides of lead and copper, which readily unite with the silica of the crucible. Red lead when fused is decomposed into litharge and oxygen, so that whether the lead oxide used as a flux is added in the form of red lead or of litharge the lead will exist in the slag as monoxide (litharge).

$$2Pb_3O_4 = 6PbO + O_2.$$

The excess of oxygen in red lead is thus available for oxidising purposes, and on that account red lead is in some cases to be preferred to litharge. Red lead of good quality is often more readily obtained than good-quality litharge, and on this account the former is the form of lead oxide used by many assayers. Lead oxide, in addition to its use as a flux, is used for obtaining the lead necessary for collecting the precious metals. For this purpose it is acted upon in the crucible by a reducing agent, see page 131, such as charcoal, whereby metallic lead is obtained as follows :—

$$2PbO + C = 2Pb + CO_2.$$

As red lead and litharge always contain silver in sensible quantity, this must be determined as described on page 145, and taken into account in returning the assay result. The following more or less frequent constituents of ores also act as fluxes when present.

Iron Oxides are of frequent occurrence in ores either as hæmatite (natural ferric oxide, Fe_2O_3) or more commonly as limonite (hydrated ferric oxide, $Fe_2O_3 . H_2O$). Ferric oxide is very infusible, but when converted by reducing agents into the lower oxide it unites readily with silica to form ferrous silicate, which is easily fusible. The reactions are expressed as follows :—

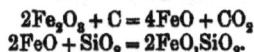

$$2Fe_2O_3 + C = 4FeO + CO_2$$
$$2FeO + SiO_2 = 2FeO,SiO_2.$$

Manganese oxides, which are also found in ores, act as a flux in a similar manner.

Lime, present in ores as limestone (calcium carbonate, $CaCO_3$), acts as a powerful base for fluxing silica. When heated, limestone is converted into calcium oxide and the carbonic acid driven off. Lime is very infusible, but in combination with silica it forms a fusible silicate the fusibility of which is increased by the addition of another base.

Carbonate of lime also occurs in ores in combination with magnesium in the mineral dolomite, a double carbonate of lime and magnesia ($CaO,CO_2 + MgO,CO_2$). As a flux magnesia acts in a similar manner to lime.

Alumina (Al_2O_3) is often found in ores, usually in combination with silica. It acts as a basic flux. It is a common constituent of rocks and materials such as clays and slates.

Borax, $Na_2B_4O_7$.—This may also be written $Na_2O,2B_2O_3$. Borax is a hydrated sodium biborate containing nearly half its weight of water. When heated the water is expelled, causing the borax to swell and finally fuse into a clear glass known as borax glass. The swelling up may cause loss in the fusion of an assay by forcing part of the charge out of the crucible. On this account the borax should be freed from water by carefully heating it at a temperature a little above the boiling point of water, or it may be fused in a crucible and the resulting glass poured upon a clean surface, and then powdered for use.

Borax acts as an acid flux, and is the chief acid flux employed by assayers. It

9

contains an excess of boron trioxide (B_2O_3) which is capable of dissolving many metallic oxides to form very fluid slags. It is especially valuable in fluxing infusible substances such as lime, zinc oxide, etc. Borax increases the fluidity of almost any charge, and on that account is frequently added. In the form of borax glass it is largely used as a " cover " for scorification assays.

Silica, SiO_2.—This is sometimes used, in the form of very finely pulverised quartz or siliceous sand, as a flux for extremely basic ores to lessen the corrosive action on the crucible. It combines with the metallic oxides or bases present in the charge to form a slag mainly composed of silicates. It is only very occasionally that an ore requires to be fluxed with silica, as most basic ores contain a sufficient amount of silica for fluxing purposes. Powdered glass is a very convenient substitute for silica : broken beakers, etc., cleaned, dried, and powdered, are very suitable for the purpose. Glass is a mixture of sodium and potassium silicates with small amounts of other silicates. It contains generally from 60 to 80 per cent. of silica.

Fluor-spar (calcium fluoride, CaF) should be selected as free as possible from other minerals (such as galena) and powdered. It is occasionally used in assaying, as it is a suitable flux for infusible substances like calcium phosphate (bone-ash) and infusible silicates. It also fuses easily with barytes and gypsum. Fluor-spar requires a comparatively high temperature to fuse it, and melts without decomposition. When melted it is very fluid, and is capable of holding in suspension unfused particles of mineral matter without seriously diminishing the fluidity of the slag. The greatest portion of the fluor-spar is found unchanged in the slag merely in a state of mechanical mixture. Its extreme fluidity constitutes its chief value as a flux ; it facilitates the fusion, and increases the fluidity of the mass. Slags containing fluor-spar have a stony appearance.

Salt (sodium chloride, NaCl).—Ordinary table salt. Fuses easily into a very thin fluid and is not decomposed during fusion. It volatilises at a red heat. It is occasionally used as a " cover " on the top of assay charges, as it helps to preserve certain slags from oxidation, and also by reason of its fluidity washes down particles of metal adhering to the sides of the crucible. Sodium chloride decrepitates when heated, it should therefore be dried and powdered before being used as a flux.

The following metals are employed in dry assaying :—

Iron.—Wrought-iron, in the form of hoop-iron, rods about half an inch in diameter, or nails, is used in the crucible assay of ores containing sulphur, arsenic, and lead. It acts as a desulphurising agent, as described on page 137. Wrought-iron crucibles are employed with advantage for ores consisting largely of galena (PbS).

Metallic Lead.—This is used in assaying in the form of (*a*) sheet-lead, and (*b*) granulated lead. Ordinary commercial lead generally contains too much silver to be suitable for assay purposes ; lead refined by desilverisation processes and sold as assay lead or test-lead is therefore employed.

(*a*) *Sheet-Lead.*—Sheet-lead is chiefly used in the direct cupellation of pieces of metal, which are wrapped up in the lead. The thickness of the sheet should be about equal to that of an ordinary visiting card. It is usually purchased, but may be prepared if necessary by casting the lead in the mould shown in fig. 40 to obtain a thin flat ingot, which is rolled out to the desired thickness. Very pure lead for this purpose may be obtained by the reduction of white lead with charcoal : 100 parts of white lead require about 3 parts of charcoal powder to reduce it to metallic lead.

(*b*) *Granulated Lead.*—Granulated lead is used for assays by the scorification method. It may be prepared by pouring the molten metal into a strong wooden box and rapidly shaking the box until the lead has wholly solidified. A rectangular oak box about 13 inch by 9 inch by 9 inch, provided with handles at the ends, is very suitable for the purpose. The inside is blackleaded, and the lead, when just on the point of solidifying, is quickly poured in and subjected to a vigorous shaking, by which means the metal is broken up into a fine state of division. The temperature of the molten metal may be tested by inserting a piece of stick, and when this is no longer sensibly charred the lead is ready for pouring. If the lead is poured too hot, the box is burnt and the resulting fume causes inconvenience in shaking. After being granulated the lead is thrown on to a sieve of about 13 mesh, and any metal that fails to pass the sieve is remelted and again granulated. Small quantities of granulated lead may be prepared by pouring the molten metal into a large crucible, throwing a little charcoal powder upon it, and stirring the mixture rapidly with a stout stick or iron rod until the lead has solidified. The charcoal powder is afterwards removed by winnowing or by washing.

Lead-shot is sometimes used as a substitute for granulated lead, but its use is to be deprecated, since it contains arsenic.

Reducing and Oxidising Agents.

In assaying ores by the dry method reactions frequently take place which involve reduction or oxidation, and it is essential that the assayer should thoroughly understand these reactions.

Reduction.—When a metal is separated or set free from a state of chemical combination it is said to be reduced, and the process of separation is termed reduction. Also, when a compound is changed from a higher to a lower state of oxidation, it is said to be reduced.

Reducing Agent.—The agent by which reduction is effected is termed the reducing agent. In ores the metals exist, in many cases, in the state of oxides or of sulphides, and the term reducing agent is most frequently used to signify "a substance that removes oxygen or sulphur (or analogous element) from a compound and sets free the metal," or "that changes a compound from a higher to a lower state of oxidation." Thus when lead oxide is heated with carbon, carbonic acid is formed and the lead is reduced to the metallic state; or when iron is heated with lead sulphide, iron sulphide is formed and the lead set free in the metallic state. Also, when ferric oxide (the higher oxide) is heated with a suitable proportion of carbon, carbonic acid is formed and the iron reduced to ferrous oxide (the lower oxide).

In these cases the carbon and iron are reducing agents. As a general rule, reducing agents consist mainly of carbon and hydrogen, or their compounds, but in some instances the common metals and their compounds are employed. Although the above definition of a reducing agent is generally applicable, it must be remembered that under suitable conditions sulphur itself, either alone or in combination, will act as a reducing agent. For example, when sulphur is heated with lead oxide sulphur dioxide is formed, and the lead is reduced to the metallic state as in the case with carbon: thus, $S + 2PbO = 2Pb + SO_2$. Also, when the sulphides and oxides of certain metals are heated together, complete reduction takes place, and in such cases the sulphides are regarded as reducing agents. The following reactions between the sulphides and oxides of lead and of copper are examples :—

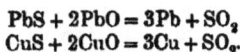

$$PbS + 2PbO = 3Pb + SO_2$$
$$CuS + 2CuO = 3Cu + SO_2.$$

A reducing agent that removes oxygen from a compound is also termed a *deoxidiser*, and one that removes sulphur is termed a *desulphuriser*.

From the assayer's standpoint the subject of reduction and of reducing agents is mainly confined to the consideration of the substances that will reduce metallic lead from lead oxide.   For this purpose the assayer chiefly employs some form of carbonaceous matter such as charcoal, argol, or flour, but he also makes use of the reducing power of the sulphur present in the metallic sulphides frequently found in ores.   He also employs iron as a reducing agent for the reduction of ·certain metallic sulphides such as galena (lead sulphide), as stated above.   The chief reducing agents are as follows :—

Charcoal.—Powdered wood charcoal is used.   It contains about 90 per cent. carbon, the rest being mainly ash and moisture.   The advantage of carbon as a reducing agent consists in its great affinity for oxygen, and at a red heat it surpasses most other substances in this respect.   When heated with metallic oxides, carbon removes the oxygen and liberates the metal with the production of, at low temperatures, carbon dioxide, and, at higher temperatures, carbon-monoxide. The reducing effect of charcoal varies with the quality and the conditions under which it is used, the actual reducing effect of any given sample should therefore be determined under the conditions in which it is to be employed.   Finely powdered coal or coke may be used as substitutes for charcoal, but are not so suitable.   Coal, especially soft coal, swells on heating and is inconvenient, but as the quantity required in an assay is usually small, it is sometimes employed. Coke is not very suitable as a reducing agent, as it burns very slowly.   It should never be used when it is possible to avoid it.   When employed it must be very finely powdered.

Flour.—Ordinary wheat flour is frequently employed as a substitute for charcoal and has many advantages.   It can usually be readily obtained, and being in such a fine state of division it mixes very intimately with the substance to be reduced.   It contains moisture which varies in amount with different samples.   When heated, flour gives off inflammable reducing gases and leaves a residue of finely divided carbon.   The reducing effect of flour is about one-half that of charcoal.   When flour is employed it is useful to remember that 1 gramme reduces about 11 grammes of lead from litharge.   Starch, which has nearly the same reducing power as flour, is sometimes used.   It should be well dried before use.   Sugar also possesses similar reducing properties.   Its decomposition by heat gives a fair proportion of carbon, but it undergoes a great increase in volume when heated, and on that account is unsuitable for assay purposes.

Tartar or Argol ($2KHC_4H_4O_6$).—This is a crude potassium bitartrate.   Cream of tartar, the purified salt, may be used.   When heated, tartar is easily decomposed; it gives off inflammable gases, and agglomerates without fusing or boiling up, leaving a carbonaceous residue.   The residue contains about 15 per cent. of carbon when produced from crude tartar, and 7 per cent. when from cream of tartar.

The decomposition may be represented thus :—

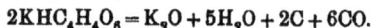

$$2KHC_4H_4O_6 = K_2O + 5H_2O + 2C + 6CO.$$

The carbon and carbon monoxide thus set free act as the reducing agents and the K_2O is available as a basic flux.   The reducing effect of tartar varies with the purity of the sample, but may be taken as equal to about one-fifth that of charcoal.

Black Flux.—This consists of a mixture of carbon and potassium carbonate obtained by deflagrating crude tartar with potassium nitrate (nitre).   For this purpose the mixture of tartar and nitre is poured into a red-hot crucible placed under a chimney with a good draught.   A good plan is to stand the hot crucible in an empty wind furnace and add the mixture, in small lots at a time, from a metal scoop.   The mixture deflagrates and gives off nitrous vapours, leaving as a residue potassium carbonate and carbon, the proportion of which varies with the quantity of nitre added.   When nitre is deflagrated with tartar the carbon is oxidised by the oxygen of the nitre and potassium carbonate formed.

By varying the proportion of nitre, black flux with different reducing effects may be obtained Black flux with a vigorous reducing action is made from 3 parts of tartar and 1 part of nitre; for less vigorous reducing power, from 2 to 2½ parts of tartar are used for 1 part of nitre. As black flux is hygroscopic, it must be frequently prepared fresh and kept in a dry place. On this account a substitute consisting of a mixture of potassium (or sodium) carbonate and flour is used in preference. Usually from 20 to 25 per cent. of wheat flour is added to the alkaline carbonate. Owing to the trouble of preparation and its hygroscopic nature black flux is only used to a limited extent in assaying.

Potassium Cyanide (KCN).—This salt is easily fusible, and when melted is somewhat volatile. When fused it acts as a powerful reducing and desulphurising agent readily removing oxygen and sulphur from metallic compounds. It combines with oxygen, forming potassium cyanate,

$$PbO + KCN = Pb + KCNO,$$

and with sulphur, forming potassium thiocyanate,

$$PbS + KCN = Pb + KCNS.$$

Commercial samples of potassium cyanide vary much in purity, some containing less than 50 per cent. of pure cyanide. The better qualities only should be used in assaying. Potassium cyanide is usually sold in cakes and is powdered for use, but when powdering it great care must be taken not to inhale it, as it is a powerful poison. During the operation the mortar should be covered with a cloth. The powdered salt should be kept from the air as far as possible, as it absorbs moisture. It is only occasionally used in gold and silver assaying.

Reducing Power.—The actual effect produced by a reducing agent is termed the *reducing power*, and is always given in terms of the lead reduced. It may be defined as follows:—By reducing power is meant the weight of lead yielded (or reduced) by 1 gramme of the reducing agent when fused with litharge. For example, one gramme of pure carbon will (theoretically) reduce 34·5 grammes of lead from litharge; the reducing power of carbon is therefore said to be 34·5. But charcoal contains, in addition to carbon, more or less inert ash which has no reducing effect; the actual amount of lead reduced will consequently be materially less. In practice it is usually found to be between 20 and 30 grammes for 1 gramme of charcoal according to the quality, and these figures express the reducing power of charcoal. The reducing power of the common reducing agents is given in Table VII., but the figures must be taken as approximate only, as the results will vary somewhat according to the quality and dryness of the sample. For all practical purposes, however, they are sufficiently near to indicate the reducing effect of the different agents most frequently employed in assaying.

Table VII.—Approximate Reducing Effect of Various Reducing Agents.

Reducing agent.	Quantity of lead in grammes reduced from litharge by 1 gramme of reducing agent.
Wood charcoal	22–30 grammes.
Powdered hard coal	25 ,,
,, soft coal	22 ,,
,, coke	34 ,,
Argol (crude tartar)	5–9·5 ,,
Cream of tartar	4·5–6·5 ,,
Wheat flour	10–12 ,,
Starch	11·5–13 ,,
Sugar	12–14·5 ,,
Potassium cyanide	

It will be seen from these figures that the reducing power of flour on litharge is about one-half, and of argol about one-fourth, that of charcoal.   As red lead (Pb_3O_4) is frequently employed in assaying instead of litharge as a source of metallic lead to collect the precious metals, it will be well to remember that it contains an excess of oxygen which will use up part of the reducing agent before any lead separates out.   It is obvious therefore that 1 gramme of reducing agent will reduce less lead from red lead than from litharge, a fact which has to be taken into consideration when red lead is used in making up assay charges.   According to Beringer, the reducing power of 1 part of charcoal, flour, and argol on litharge and on red lead is as follows :—

Reducing agent.	From litharge.	From red lead.
	grammes.	grammes.
1 gramme of charcoal reduced . .	22·6 lead.	12·0 lead.
,, flour	11·2 ,,	6·0 ,,
,, argol	5·0 ,,	3·0 ,,

Comparing these figures, it will be seen that the reducing effect on red lead is only about one-half of the effect on litharge.

Reducing Effect of Metallic Sulphides.—The metallic sulphides act as reducing agents and will reduce lead from litharge in the same way that carbonaceous reducing agents do.

The following are the sulphides most frequently found in ores :—

Iron sulphides . . .	{ Pyrite (iron pyrites) . .	FeS_2.
	Pyrrhotite	Fe_7S_8.
	Arsenopyrite (mispickel) . .	FeAsS.
Copper sulphides . .	{ Chalcopyrite (copper pyrites) .	$CuFeS_2$.
	Copper glance	Cu_2S.
Lead sulphide	Galena	PbS.
Antimony sulphide . . .	Stibnite	Sb_2S_3.
Copper and antimony sulphide	Tetrahedrite fahlerz (fahl ore) .	$4Cu_2S,Sb_2S_3$.
Zinc sulphide	Sphalerite (zinc blende) . .	ZnS.

Iron pyrites usually forms by far the largest proportion of the sulphides found in ores.   The weight of lead reduced from litharge by 1 gramme of mineral sulphide varies with the different sulphides, and varies also for the same sulphide according to a combination of conditions.   In discussing the reducing power of sulphides it is important to remember that sulphur, and in some cases the metals, are capable of two degrees of oxidation, and, whether the lower or the higher oxide is formed, will depend very largely on the conditions under which reduction takes place.   In an assay as ordinarily performed, other substances beside litharge are usually present, and in this connection the character of the slag resulting from the fusion is an important factor in modifying the reducing effect of sulphides. When the slag is acid owing to the presence of much siliceous material in the charge the sulphur will, in most cases, be eliminated as sulphur dioxide (SO_2). On the other hand, if the slag contains a large proportion of a strong base like soda, the tendency will be for the sulphur to be converted into sulphur trioxide (SO_3) with the formation of sulphates.   When a sulphide is fused with litharge alone, all the sulphur will be liberated as the lower oxide (SO_2).   Taking iron pyrites as an example, the following equation expresses the reaction when it is fused with litharge—

$$FeS_2 + 5PbO = FeO + 5Pb + 2SO_2 ;$$

i.e. 1 gramme of pure pyrites reduces 8·6 grammes of lead. In this case the iron and the sulphur are only oxidised to the lower states, the iron to ferrous oxide and the sulphur to sulphur dioxide (SO_2), as no alkaline bases are present to induce the formation of SO_3, and consequently of sulphates. If the reaction takes place in the presence of much soda, the sulphur is converted into sulphur trioxide (SO_3), which combines with sodium oxide to form sodium sulphate (Na_2SO_4), the reactions being expressed by the two following equations :—

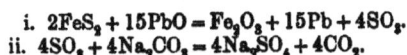

i. $2FeS_2 + 15PbO = Fe_2O_3 + 15Pb + 4SO_3.$
ii. $4SO_3 + 4Na_2CO_3 = 4Na_2SO_4 + 4CO_2.$

In this case 1 gramme of pyrites exerts its greatest reducing effect and reduces 12·9 grammes of lead. In other words, both the sulphur and the iron have been fully oxidised.

It will be seen from the above equations that the amount of lead reduced from litharge by a sulphide is influenced by the presence of soda, a fact of some importance to assayers which is more fully discussed on page 199. By decreasing the amount of soda or otherwise altering the conditions of the fusion, the sulphur may be converted partly into sulphur dioxide and partly to sulphur trioxide, and it will be obvious that the amount of lead reduced from litharge by 1 gramme of iron pyrites may theoretically vary from 8·6 grammes to 12·9 grammes. In actual practice it may range from about 9 to 11·2 according to whether the charge is acid and contains little or no soda, or is basic from the presence of much soda. The approximate reducing power of some of the common sulphides on litharge in the presence of one-fourth of its weight of soda is given in Table VIII. The amount of lead theoretically obtained from 1 gramme of sulphide for the two degrees of oxidation is also given.

Table VIII.—Reducing Power of Metallic Sulphides.

Metallic sulphide.	Weight of lead reduced from litharge by 1 gramme of sulphide.[1]	Theoretical weight of lead reduced from litharge by 1 gramme of sulphide when—	
		(a) Sulphur and metal are oxidised to the lowest degree.	(b) Sulphur and metal are oxidised to the highest degree.
Antimonite . . .	6 grammes.	5·5 grammes.	8 grammes.
Blende	7–8 ,,	6·4 ,,	8·5 ,,
Copper pyrites . .	7–8 ,,	7·5 ,,	11·6 ,,
Fahlerz	7–8 ,,	...	...
Galena . . .	3 ,,	2·6 ,,	3·4 ,,
Iron pyrites . . .	11 ,,	8·6 ,,	12·9 ,,
Mispickel . . .	7–8 ,,	...	...

The figures given in the table represent the reducing power of the different sulphides in a pure state, but it must be pointed out that it is more usual to find several sulphides associated in an ore than to find one sulphide only, and due consideration is given to this fact in dealing later with the assay of sulphide ores. The reducing power of a few metallic sulphides on red lead is given in Table IX. from experiments by Beringer. The results obtained with litharge under the same conditions are inserted for comparison.

[1] From experiments by Beringer (*Text-Book of Assaying*).

Table IX.—Reducing Power of Metallic Sulphides.

Sulphide.	Lead reduced from litharge.	Lead reduced from red lead.
1 gramme of antimonite reduced .	6·2 grammes.	3·6 grammes.
,, copper pyrites . .	8·9 ,,	5·9 ,,
,, galena . . .	8·4 ,,	1·7 ,,
,, iron pyrites . .	11·6 ,,	8·5 ,,

Determination of Reducing Power.—The reducing power is determined from the quantity of lead which is yielded by fusing 1 or 2 grammes of the reducing agent with 60 grammes of litharge and 15 grammes of sodium carbonate. The button of lead obtained is weighed, and the weight divided by the weight of reducing agent used, the result being the reducing power. The presence of soda does not influence the amount of lead reduced from litharge with carbonaceous reducing agents, but it has already been pointed out that soda has considerable influence on the reducing effect of metallic sulphides. In assaying, it is sometimes necessary to determine the reducing power of ores containing more or less pyritic material, and in this case the flux used in determining the reducing power should be the same, and in the same proportion as that to be employed in the charge for the assay of the ore, so that the conditions in both cases may be the same. It is only by working in this way that the true reducing effect of an ore can be ascertained.

Oxidising Agents.

Oxidation.—When oxygen combines with an element the compound formed is termed an oxide, and the act of combination is known as oxidation. In assaying, oxidation usually has reference to the conversion of a metal or element to the state of oxide, or the changing of a compound already in an oxidised state to one of higher oxidation; for example: lead to lead oxide (PbO), sulphur to sulphur dioxide (SO_2), or ferrous oxide (FeO) to ferric oxide (Fe_2O_3). The process of oxidation is taken advantage of in the operations of roasting, scorifying, cupelling, etc., described subsequently. When the amount of pyritic material (*i.e.* sulphides, arsenides, etc.) present in an ore is in excess of that required to act as a reducing agent, the excess is usually oxidised and passed into the slag. To effect oxidation, special substances termed oxidising agents are employed. They may be defined as follows :—

An oxidising agent is a substance that imparts oxygen to an element or compound.

The following oxidising agents are employed in assaying :—

Nitre (potassium nitrate, KNO_3) is a powerful oxidising and also desulphurising agent. Nitre fuses very easily to a watery liquid. It contains nearly half its weight of oxygen, which it readily loses, in two stages, when heated. At a low temperature (a red heat) it is decomposed into potassium nitrite (KNO_2) and free oxygen, and at a higher temperature more oxygen is set free and potassium monoxide (K_2O) left.

The decomposition is shown thus :—

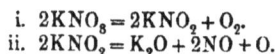

i. $2KNO_3 = 2KNO_2 + O_2.$
ii. $2KNO_2 = K_2O + 2NO + O.$

The action of melted nitre is very energetic, sulphides being readily oxidised to sulphates, arsenides to arsenates, and most metals converted into oxides. The oxidising action of nitre on iron pyrites may be expressed thus :—

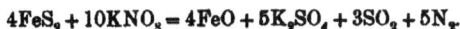

$$4FeS_2 + 10KNO_3 = 4FeO + 5K_2SO_4 + 3SO_2 + 5N_2.$$

In the presence of a strong alkaline base such as soda, alkaline sulphates are formed as the result of oxidation, and float as a watery liquid on the surface of the less fluid slag. In this case the oxidising action of nitre on pyrites is expressed by the following equations :—

i. $6KNO_3 + 2FeS_2 = Fe_2O_3 + SO_3 + 3K_2SO_4 + 3N_2.$
ii. $3SO_3 + 3Na_2CO_3 = 3Na_2SO_4 + 3CO_2.$

It is for the conversion of metallic sulphides into oxides that nitre is most frequently used in assaying.

Red Lead (Pb_3O_4) is an oxide of lead, frequently used instead of litharge as a source of lead, especially in cases where an oxidising action is required in the fusion. When heated it parts readily with about one-fourth of its oxygen, which is available for oxidising purposes without any separation of metallic lead.

$$2Pb_3O_4 = 6PbO + O_2.$$

The lead oxide (litharge, PbO) that is formed can be used as a flux, or the oxygen it contains can be used for oxidising purposes, but in this case metallic lead will be liberated to the extent of about 13 parts of lead for each part of oxygen made use of.

Red lead can only be used to a limited extent as an oxidising agent for assay purposes, as the employment of a comparatively large quantity will give too much lead to be dealt with satisfactorily. The oxidation of iron pyrites by means of litharge is expressed in the following equations :—

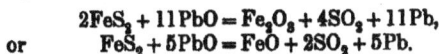

$$2FeS_2 + 11PbO = Fe_2O_3 + 4SO_2 + 11Pb,$$
or $$FeS_2 + 5PbO = FeO + 2SO_2 + 5Pb.$$

Hot Air.—The oxygen of the air is used as an oxidising agent in all roasting operations. The sulphur of metallic sulphides is oxidised and passes off as sulphur dioxide, and the arsenic of minerals like mispickel, etc., is oxidised to "white arsenic," As_2O_3. The metals present are converted into oxides, mixed with more or less sulphate or arsenate. Thus iron pyrites heated in air is oxidised to ferric oxide and the sulphur to sulphur dioxide.

$$2FeS_2 + 11O = Fe_2O_3 + 4SO_2.$$

A certain proportion of ferroso-ferric oxide (magnetic oxide), Fe_3O_4, or Fe_2O_3, FeO, is also formed thus :—

$$3FeS_2 + 5O_2 = Fe_3O_4 + 3SO_2 + 3S.$$

Desulphurising Agents.—When an oxidising agent is employed to remove sulphur by oxidation, it is also termed a desulphurising agent. Most oxidising agents act as desulphurisers, and the substances described above are used for this purpose.

Iron is also a desulphurising agent frequently employed in assaying. When heated with metallic sulphides the iron combines with the sulphur to form ferrous sulphide, which will be dissolved in the slag or separate out as a regulus according to the conditions of the fusion. Iron is used in the form of hoop-iron, iron rods about ¼-inch in diameter, or of nails.

Examples.—(1) Iron decomposes lead sulphide, forming iron sulphide and metallic lead thus :—

$$PbS + Fe = FeS + Pb.$$

(2) Iron pyrites, when heated with iron, parts with half its sulphur, forming ferrous sulphide.

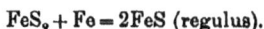

$$FeS_2 + Fe = 2FeS \text{ (regulus).}$$

Alkaline Carbonates.—Sodium and potassium carbonates act not only as basic fluxes but also as desulphurising agents. The alkaline oxides Na_2O and K_2O react on certain metallic sulphides, separating the metal, with the formation of sodium or potassium sulphate and sulphide. Thus sodium carbonate partially decomposes galena (lead sulphide), setting free some lead and forming sodium sulphide, which combines with the remaining lead sulphide to form a double compound thus :—

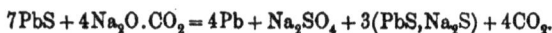

$$7PbS + 4Na_2O.CO_2 = 4Pb + Na_2SO_4 + 3(PbS,Na_2S) + 4CO_2.$$

The double sulphide of lead and sodium is decomposed by iron, and metallic lead set free, thus :—

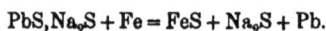

$$PbS,Na_2S + Fe = FeS + Na_2S + Pb.$$

Sodium oxide (Na_2O) in the presence of carbon (which removes the oxygen) acts as a desulphuriser by combining with sulphur to form sodium sulphide. In this case carbon is employed as an auxiliary to the desulphurising action of the alkali.

CHAPTER XI.

PRINCIPLES OF FLUXING: ASSAY SLAGS.

To understand the principles of fluxing it is necessary that the student should have a thorough knowledge of the nature of the extraneous earthy or stony matter found associated with precious-metal ores. Such matter is termed gangue, vein stuff, or matrix. The nature of the gangue is of importance in determining the nature of the flux necessary to convert it into a suitable slag. For convenience, gangue-forming materials are classified according to their chemical character into *acid* and *basic*, silica, as a general rule, being the acid and the metallic oxides the basic constituent. The gangue as a whole is classed as acid or basic according to whether acid or basic constituents are in excess. If both acid and basic materials are present in about the same proportion the gangue is said to be neutral. In the comparatively rare cases in which the proportion of basic material is such that it neutralises the silica present and combines with it to form a readily fusible compound, the ore is said to be self-fluxing.

The principal gangue materials, classified according to their chemical character, are given in Table X.

Table X.—Gangue-forming Materials.

Acid gangues.	Basic gangues.
Silica, uncombined ; as quartz crystals, quartz rock, quartzite, sandstone, etc.	*Metallic oxides*, notably those of iron and manganese.
Siliceous rocks, rocks in which silica, either free or combined, predominates ; as granites, porphyry, etc.	*Metallic carbonates*, as those of calcium (limestone), magnesium, iron, etc.
Silicates, or silica, combined with a base ; as felspars, mica, hornblende, etc.	*Barytes*, or heavy spar (barium-sulphate).
	Fluor-spar (calcium fluoride).

Quartz gangues will contain from about 80 to 100 per cent. of silica. Siliceous rocks such as granite will contain from 70 to 75 per cent. of silica, and from 13 to 20 per cent. alumina.

The silicates present in the gangues are mainly silicates of alumina, lime, and magnesia, and contain, roughly speaking, from 45 to 65 per cent. of silica, 15 to 20 per cent. of alumina, 5 to 15 per cent. of iron oxide, and 5 to 15 per cent. of lime and magnesia, with a small percentage of alkalies. In the majority of cases the gangue of an ore will consist of rock composed mainly of silica or of siliceous material, and will therefore be acid in character owing to the preponderance of silica.   In some cases the gangue may consist of a mixture of acid and basic material and be neutral in character.   In comparatively few cases the gangue may consist entirely of basic material and consequently be basic in character.   But in addition to the gangue, metalliferous minerals, notably iron pyrites, are found associated with precious-metal ores, and these, when present in any large quantity, play an important part in determining the character of an ore for fluxing purposes, as their oxidation produces basic oxides.   For example, an ore, consisting mainly of iron pyrites, with a little quartz would, if roasted, be converted into iron oxides and become strongly basic.

To facilitate description, ores may be classified as follows :—

1. Ores with siliceous gangue, as ordinary quartz.
2. Ores with basic gangue, as iron oxide, limestone.
3. Pyritic ores, as iron pyrites, with or without other metallic sulphides.
4. Complex ores, as antimonial and telluride ores.

The nature and quantity of the flux to be used depend upon the character and composition of the ore.   When smelting ores on a large scale the metallurgist always determines the composition of the ore, and then carefully calculates the quantity of flux to be added to produce a suitable slag.   In assay practice, however, it is impracticable to make analyses of ores to determine the composition for fluxing purposes, but with a knowledge of mineralogy the assayer can, by a few simple tests, ascertain sufficiently accurately the nature of the ore which he wishes to flux.   Much valuable information may be obtained by a careful examination of a few characteristic lumps of the ore when these are available.   If the ore is in the state of powder, a small quantity should be washed or "panned" in a small sheet-iron pan, and the washed material carefully examined with a lens, and, if necessary, with a blowpipe.   By this means the gangue is in most cases · readily separated from any metalliferous minerals, and the nature of the ore and the approximate proportions of the various constituents are ascertained.   For example, it will be comparatively easy to tell at once whether metallic sulphides such as iron pyrites or galena are present, or whether the ore contains limestone, limonite, baryta, or other mineral, and in what general proportions.   The information thus gained invariably repays the time spent in obtaining it.   The nature of the gangue having been ascertained, the best flux for producing an easily fusible slag is added in accordance with the following simple general rule for fluxing :—

To flux an acid gangue use a basic flux.
To flux a basic gangue use an acid flux.

1. **Ores with Siliceous Gangue.**—Taking first the simplest case of a gangue composed almost entirely of quartz or siliceous material, which constitutes the gangue of a large proportion of precious-metal ores.   The best and most frequently used flux for silica is soda.   It combines with silica to form sodium bisilicate, which is very fusible and makes a good glassy slag.

With sodium carbonate the reaction is :—

$$SiO_2 + Na_2CO_3 = CO_2 + Na_2SiO_3.$$

With sodium bicarbonate :—

$$SiO_2 + 2NaHCO_3 = 2CO_2 + Na_2SiO_3 + H_2O.$$

Theoretically, 1 part of silica will require 1·77 parts of carbonate or 2·8 of the bicarbonate to form sodium bisilicate.   In practice, from $1\frac{1}{2}$ to 2 parts of sodium carbonate are usually employed to flux 1 part of quartz or sand.

An acid gangue may also be composed largely of the natural silicates, which may be considered as combinations of silica and metallic oxides, with usually more or less silica in excess. As previously stated, the most common silicates found associated with ores are the silicates of alumina, lime, and magnesia. The silica in silicates is already partially neutralised by combination with the metallic base present, but as the natural silicates are, in most cases, fusible only with great difficulty, they are, in assaying, generally fluxed in the same way as uncombined silica. Although soda is the flux most generally employed for quartz and siliceous material, it is sometimes advantageous to use litharge as a partial or complete substitute for soda. Litharge, when fused with quartz, forms lead silicate, which is a yellow glass, easily fusible, and more fluid than sodium silicate. The reaction is quieter than with soda, as there is no evolution of gas.

$$PbO + SiO_2 = PbSiO_3.$$

Theoretically, 1 part of silica requires 3·7 parts of litharge. In this reaction the heating must be carefully regulated, otherwise the litharge will fuse and drain away from the unattacked quartz, which is left as a pasty mass on the surface. In using litharge it must be remembered that it is only the lead oxide in excess of that required to give the necessary weight of metallic lead for collecting the precious metals in the ore that is available as a flux.

Borax is also used as a partial substitute for soda to flux silica. Its fluxing action is about equal to that of litharge. With siliceous material, borax forms borate silicates which increase the fluidity of the slag. When borax is used it should be borne in mind that it is an acid flux, consequently the acid character of the slag (which is a bi-silicate or acid slag) is increased by increasing the proportion of borax.

As an example of the use of two fluxes let us suppose we wish to flux 50 grammes of siliceous ore, which would require theoretically 88·5 grammes of sodium carbonate if this flux alone be used. If we add 71 grammes of sodium carbonate (which is only sufficient to flux 40 grammes of silica) it leaves 10 grammes of silica to be fluxed by other reagents. To flux this we can add 37·2 grammes of litharge or the same weight of borax, since the fluxing action of these reagents is about equal. Or, instead of using either litharge or borax separately, we could use both and flux the 10 grammes of silica with 18·6 grammes of each reagent. A sufficiently fluid slag would still be obtained even if the proportions of the three fluxes were altered by decreasing the quantity of any of them and proportionally increasing either or both the others. In practice it is more usual to rely mainly on soda as the flux for silica and silicates.

2. Ores with Basic Gangue.—The chief ores which come under this head are those which consist mainly of limonite (hydrated iron oxide), and in most cases are believed to be the result of the decomposition of pyritic material by atmospheric agencies.

Ores containing excess of hæmatite (Fe_2O_3) magnetite (Fe_3O_4), calcite ($CaCO_3$) dolomite ($CaCO_3 + MgCO_3$) or other basic minerals are also included under this head. The bases to be fluxed off in ores of this class are therefore the oxides of iron, lime, alumina, and, usually in smaller proportion, the oxides of magnesium, manganese, barium, copper, zinc, etc. The gangue may consist of one of these oxides only, but it is much more general to find several oxides present, and silica, in greater or lesser quantity, is present in practically every ore. The chief flux for basic material is borax, which forms fusible compounds with nearly all bases, being especially useful in uniting with metallic oxides. It also forms fusible compounds with silica. When a basic gangue contains a fair proportion of silica it may be partially self-fluxing, as the silica will combine with part of the basic material present to form fusible silicates, leaving only the excess of basic material to be fluxed. In the comparatively rare cases in which an ore is extremely basic and deficient in silica, sand or powdered glass is added. As an exception, the addition of sand is frequently necessary in fluxing the excess of ferric oxide which results from the roasting of very pyritic ores. Sand in suitable proportion forms fusible silicates with alumina, lime, iron oxides, etc., and protects the crucible from corrosion. Ordinary crucibles consist of quartz and clay, and are rapidly attacked by slags that are too basic. Even when silica is added as a flux it is a common practice to add also a little sodium carbonate, as slags consisting of the double silicates of sodium and other bases are much more fusible, and more suitable for the requirements of the assayer, than slags formed of silicates free from sodium. The proportion of soda should be increased, and that of the borax diminished, as the quantity of silica becomes greater. For average basic ores a mixture of equal weights of soda and borax answers very well. An important consideration in the fluxing of basic material consisting mainly of ferric oxide (Fe_2O_3) is the reduction of this higher oxide to the ferrous state (FeO), in which state alone it unites with silica to form a readily fusible silicate. Ferric oxide being infusible, its presence in the slag tends to stiffen it and render it pasty. "When fluxed with as much as six times its weight of borax, ferric oxide yields an unsatisfactory slag, while on the other hand the lower oxide (ferrous oxide) is easy to deal with ; fused borax will dissolve about its own weight of it, and sodium silicate, such as makes up the bulk of most assay slags, will take up at least half as much " (Beringer).

Experience has shown that slags containing much ferric oxide are very liable to retain notable amounts of gold and silver. The reduction of the iron to the state of the lower oxide is effected by the addition of charcoal powder or other reducing agent ($2Fe_2O_3 + C = 4FeO + CO_2$). For ores consisting mainly of ferric oxide such as pyritic ores after roasting the quantity of charcoal powder required will be about one-twelfth of the weight of ore taken for assay ; the exact amount needed for any particular ore must, however, be left to the experience of the assayer. In the case of ores oxidised by roasting, experience will frequently enable the adjustment of the charcoal to be made from the colour of the roasted material. It may be pointed out that brown hæmatites (hydrated ferric oxide) contain in some cases as much as 15 per cent. of water (which is expelled on heating), while red hæmatite (ferric oxide) contains none; hence more charcoal is needed for the latter than for the former. Manganese peroxide (MnO_2) also acts in the same way as ferric oxide, and when present must be reduced to the lower oxide (MnO). It requires about twice the amount of charcoal for its reduction as compared with ferric oxide. Manganese is not a common constituent of precious-metal ores, and when present it is usually in small proportion only. Basic ores containing easily reducible oxides, such as copper and antimony oxides, should be partially fluxed with lead oxide, as it is advantageous in such cases to have an excess of lead oxide in the slag. The lead oxide dissolves the metallic oxides present and prevents any copper, or other metal, from contaminating the lead present for collecting the values. Where there is no such reason for the presence of lead oxide it is generally advisable to use only an amount sufficient to give the desired weight of lead. Ores containing any appreciable amount of copper or antimony belong to the class of complex ores which are more fully dealt with in Chapter XV. page 221.

3. **Pyritic Ores.**—Mainly iron pyrites. In assaying ores of this class it is necessary to eliminate the sulphur, otherwise it will pass into the lead, making it more or less brittle, or the lead will be accompanied by a layer of regulus or "matte," *i.e.* an artificial sulphide of the metals, usually lead and iron. The

reducing action of metallic sulphides has already been dealt with on page 134. When only a moderate amount of pyritic material is present, say under 10 per cent., its effect is simply to act as a reducing agent and increase the weight of the lead reduced, provided there is an excess of litharge in the charge. This effect is easily counteracted by using less charcoal in the assay charge or by omitting it. When the charge contains only sufficient litharge to yield the desired weight of lead for collecting the values, any excess of pyritic material beyond that necessary to reduce the litharge may render the lead sulphury, or even in some cases give a button of regulus instead of lead. If, on the other hand, an excess of litharge is present, the amount of lead reduced may be considerably in excess of that required to collect the values.

Metallic sulphides are partially decomposed by alkaline carbonates, but carbonaceous matter must be present. In these decompositions alkaline sulphides are formed, and combine with and retain a certain quantity of the metallic sulphides present (see page 138). The proportion of the sulphides which remains in combination with the alkaline sulphide depends on several circumstances. It is always less when a large proportion of alkaline carbonate has been added to the charge, and when a high temperature has been employed.

According to Beringer, the slag resulting from the fusion of 50 grammes of quartz, with 100 grammes of sodium bicarbonate and a little borax (which represents an ordinary assay charge), will dissolve some 10 or 12 grammes of iron sulphide. In assaying, the decomposition of the metallic sulphides in ores is effected either by means of metallic iron, or some method of oxidation is employed to oxidise the sulphur and thus eliminate it.

Metallic iron acts as a desulphuriser by combining with sulphur to form, in the case of iron pyrites, ferrous sulphide ($FeS_2 + Fe = 2FeS$), which will remain diffused through the slag when it contains sodium sulphide (formed during the fusion) instead of separating as a layer of regulus. This is also true of other metallic sulphides if not present in too large a quantity and if the charge is not fused at too high a temperature. A moderate amount of sulphur may be successfully converted into ferrous sulphide by metallic iron and dissolved in the slag without a separate layer of regulus forming, but with ores containing excess of pyritic material it is difficult to effect complete reduction by this method, even when large quantities of metallic iron are added to the charge; the lead button obtained is always more or less brittle, owing to the presence of sulphur, and is frequently accompanied by a layer of regulus.

The presence of a large excess of ferrous sulphide in the slag is objectionable, as it is readily oxidised and causes the slag to become pasty, and frequently to retain appreciable quantities of the precious metals, especially silver. Ferrous sulphide in moderate quantities in a slag is not injurious, and hence it is a common and beneficial practice to insert metallic iron in all crucible charges of moderately pyritic ores to decompose the sulphides. It also tends to reduce the metals present in the slag to the lower states of oxidation. It is important to remember that when iron is used as a desulphurising agent the amount of lead oxide added to the charge should only be sufficient to give a button of lead of the required weight, as any excess will be reduced to metallic lead by the iron. The slag obtained from the fusion should be strongly alkaline, and to ensure this an excess of sodium carbonate should be added to the charge. The desulphurising action of iron is taken advantage of by assayers as far as possible, as it is a very convenient method of decomposing sulphides, but if the first fusion of a pyritic ore by this method gives a layer of regulus or speise, it is best, in most cases, to repeat the assay by eliminating the sulphur by other methods. Ores containing a large proportion of pyritic material (say over 50 per cent.) are generally submitted to some method of oxidation either before or during fusion. In the first case the ore is roasted as described on page 150, and in the second case the oxidation is effected by the addition of lead oxide or of nitre to the charge.

As the roasting of an ore occupies a considerable amount of time and attention, other methods of oxidation are generally preferred by assayers whenever they are applicable. Roasting can generally be dispensed with in the case of silver ores, but it is sometimes desirable with gold ores where larger quantities have to be used for assay. Fuller details of the methods of oxidation by means of lead oxide and nitre are given on page 196. When the pyritic material in an ore has been oxidised by a preliminary roasting, it consists of metallic oxides and is consequently dealt with as a basic ore, to which class it now belongs.

4. Complex Ores.—Under this head are included those ores which contain large proportions of metallic sulphides, other than iron pyrites, such as copper pyrites, arsenical iron pyrites, sulphides of antimony, zinc, lead, bismuth, etc.

Those ores are also included in which the precious metals exist in chemical combination with another element, such as tellurium.

Complex ores require special treatment, and are dealt with later (see page 221).

"Universal" Fluxes.—As the selection of the fluxes and the quantity of each to be employed is dependent on the nature of the gangue material and the presence or absence of pyritic material, it is impossible to give definite charges to meet the requirements of all ores: the best fluxes to be used in each particular case must, therefore, necessarily be left to the judgment and experience of the assayer. Typical charges for each class of ores will be found in Chapters XIII. and XIV. Universal fluxing mixtures suitable for fluxing all classes of ores are frequently recommended, but although these mixtures may be suitable in many cases for certain classes of ore, they sooner or later lead to failure if used for all ores. It must be emphasised that to obtain the best results ores must be assayed on scientific principles, each ore being treated with the fluxes and reagents best suited to its composition. The variation in the chemical character of ores makes it impossible to obtain a perfect mixture that will satisfactorily flux them all. Silver ores especially are often so complex that with a fresh ore it is sometimes necessary to make a trial fusion. In cases where an assayer's work is mainly confined to ores of any particular type or district, experiment should be made with various fluxes to ascertain the best mixture for the ores and this mixture made up as a stock flux. As previously stated, it is usually possible, by varying the proportions of the different fluxes used, to make up several mixtures, all of which are equally suitable for fluxing a particular ore, but other points being duly considered, preference should be given to the cheapest mixture.

As a general rule, from 2 to 3 parts of flux are quite sufficient to yield a fluid slag with 1 part of ore; in some cases less will suffice, and in other cases more may be required, but usually if it is found that 4 parts of flux are not sufficient, the remedy lies in using a different flux rather than in taking a larger quantity. An excess of flux will sometimes be a distinct disadvantage, as it will tend to stiffen the resulting slag and render it pasty. For example, soda, which is a very strong base, when used to flux silicates of lime and magnesia, may act prejudicially if it is in sufficient excess to set free notable quantities of lime and magnesia, which but for that excess would exist in combination as complex fusible silicates. Also, "There are many minerals which with but little soda form a glass, but with more yield a lumpy, scoriaceous mass. There are many minerals, too, which are already basic (for example, calcite, $CaCO_3$) and, when present, demand either a less basic or an acid flux according to the proportions in which they are present in the ore."[1] When dealing with an unknown ore the first assay may be more or less unsatisfactory, but the result of the fusion will indicate what modification of the charge is necessary to give a good result.

The Assay of Reagents.—Litharge and red lead invariably contain silver in appreciable amounts, and as both reagents are used in the assay of ores, etc., the silver must be determined and taken into account in returning the assay result. If this is not done, it is obvious that the silver will be ascribed to the ore and make the return too high. Litharge is sometimes sold as "silver free," but it is advisable to test the samples before taking them into use. Gold also is not infrequently present, but the quantity is usually too minute to be taken into consideration in returning the value of the ore.

To determine the silver in litharge or red lead, the following charge is weighed out and intimately mixed :—

Litharge or red lead . . . 100 grammes.
Charcoal . . . . . { 1 gramme for litharge.
{ 2 grammes for red lead.
Sand . . . . . 7 grammes
Borax (or soda) . . . 20 „

The charge is fused at a red heat in a fire-clay crucible for about fifteen to twenty minutes, whereby the lead is reduced. When fusion is completed, the charge is poured into an iron mould and allowed to cool. The slag is afterwards detached from the button of lead, which should weigh about 20 grammes. The lead is then carefully cupelled, and the resulting small button or "prill" of silver weighed. From the weight of this button the quantity of silver in the litharge or red lead used for each assay of an ore, etc., must be calculated, and deducted from the total weight of the silver obtained in the assay. Lead oxides usually contain on an average from 1·0 to 2·0 milligrammes of silver in 100 grammes, which is equivalent to 6·5 dwt. to 13 dwt. per ton of 2240 lbs. The common varieties of litharge usually contain more silver than red lead. To test the button for gold it is flattened out, transferred to a porcelain crucible, and treated with a small quantity of dilute nitric acid. On gently warming, the silver is dissolved, leaving the gold (if any) as a small black speck. The acid is carefully poured off, the gold washed with water, and then heated to redness : when cold, it is, if possible, weighed, but, as stated above, the gold usually amounts to "traces" only and is disregarded.

Granulated lead and sheet-lead (test lead) always contain more or less silver, which must be determined and deducted from the assay result as in the case of litharge, etc. The silver is determined by weighing out 100 grammes and scorifying it down to about one-fourth of its weight and then cupelling. Test lead can be obtained very pure and containing only traces of silver, but it is not safe to assume that a sample is silver free. The silver usually amounts to 1 or 2 milligrammes in 100 grammes. It is tested for gold in the manner described above.

In cupelling the lead buttons it is very necessary in all cases to remove the cupels from the muffle immediately the cupellation is finished, as the very small silver buttons obtained are very liable to sink into the cupel. When the amount of silver in a new lot of litharge, granulated lead, etc., has been determined, the result should at once be recorded in some convenient place for reference.

The fluxes and other reagents commonly employed in dry assaying do not contain silver or gold, but if they are left exposed in the laboratory they some-

[1] Beringer, *Text-Book of Assaying.*

10

times become accidentally (and perhaps purposely) contaminated or "salted" with material containing gold and silver values. A blank assay (that is, a charge in which the ore is omitted) of the reagents will readily determine whether any precious metal is present. The precaution of assaying fluxes is very seldom, if ever, necessary in well-conducted assay laboratories, but it is very desirable in cases where irresponsible assistants are employed to do the grinding and fluxing of ores and also when these operations are conducted in the same room. It is a wise precaution also in laboratories where rich ores are being dealt with continually.

Assay Slags.

The main object of the fluxing of an ore is to convert the infusible gangue materials into a slag. but it is necessary to consider the nature and quality of the slag produced. When fused it must be liquid and homogeneous and not too corrosive on the crucible. If the slag is stiff or not sufficiently liquid when fused, it may retain some of the precious metals in the charge and, if it fuses readily and is very liquid, the particles of reduced lead disseminated in the fused mass may fall to the bottom of the crucible before they come into contact with all the gold and silver present. Generally, the slag resulting from an assay is glassy, and varies much in colour according to the nature of the ore : it should, however, be uniform in colour and structure ; not streaked or variegated, as that generally indicates that the mixture has been imperfectly fused. Slags usually consist of artificial silicates formed by the combination, in suitable proportions, of silica with the common bases (i.e. metallic oxides, etc.). The silicates are classified according to the ratio of the oxygen in the base to the oxygen in the acid (i.e. silica). One molecule of silica consists of óne atom of silicon and two atoms of oxygen, so that the various silicates have the composition given below. The metallurgical classification is shown in Table XI., in which the formula RO represents oxides such as PbO, FeO, Na_2O, etc., and the formula R_2O_3 oxides such as Al_2O_3, etc.

Table XI.—Series of Formulæ for Silicates.

Formula.	Name.	Oxygen ratio in base : in acid.	Example.
$4RO + SiO_2 = R_4SiO_6$	subsilicate	2 to 1	$4PbO, SiO_2.$
$4R_2O_3 + 3SiO_2 = R_8Si_3O_{18}$	,,	2 to 1	
$2RO + SiO_2 = R_2SiO_4$	monosilicate	1 to 1	$2Na_2O, SiO_2.$
$2R_2O_3 + 3SiO_2 = R_4Si_3O_{12}$	,,	1 to 1	$2Al_2O_3, 3SiO_2.$
$RO + SiO_2 = RSiO_3$	bisilicate	1 to 2	$CaO, SiO_2.$
$R_2O_3 + 3SiO_2 = R_2Si_3O_9$	,,	1 to 2	
$2RO + 3SiO_2 = R_2Si_3O_8$	trisilicate	1 to 3	$2FeO, 3SiO_2.$
$2R_2O_3 + 9SiO_2 = R_4Si_9O_{24}$	,,	1 to 3	
$4RO + 3SiO_2 = R_4Si_3O_{10}$	sesquisilicate	2 to 3	$4MgO, 3SiO_2.$
$4R_2O_3 + 9SiO_2 = R_8Si_9O_{30}$	,,	2 to 3	

In this classification the monosilicate is the neutral silicate : it contains in base and acid equal amounts of oxygen. If a silicate contains but one base it is termed a monobasic, or simple, silicate ; but if two or more bases are present it is termed a multibasic, or double, silicate, and in writing the formula for a double silicate the silicates of the various bases are united by a plus sign. Thus a monosilicate slag containing the three bases Na_2O, FeO, CaO would be written :—

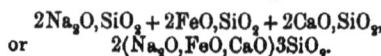

$$2Na_2O,SiO_2 + 2FeO,SiO_2 + 2CaO,SiO_2,$$
$$\text{or} \quad 2(Na_2O,FeO,CaO)3SiO_2.$$

The fusibility of silicates is dependent on the proportions of silica and base they contain. The most fusible are the subsilicates or basic silicates, which contain least silica, and, in the silicates classified, the fusibility decreases as the proportion of silica increases. Within certain limits the fusibility of silicates is also greatly dependent on the nature of the particular base present.

The strong bases Na_2O, K_2O, and PbO give easily fusible silicates; FeO, MnO, and CuO give silicates that fuse less readily; MgO, CaO, and Al_2O_3 give silicates that fuse with difficulty. The silicates of lime, alumina, and magnesia are the least fusible, and are almost or entirely infusible at the temperature of the assay furnace, so that in order to obtain a sufficiently fluid slag for assay purposes the proportions of these more infusible bases must be materially smaller than the proportions of the more fusible bases Na_2O, PbO, and FeO. This is effected in practice by the addition of suitable proportions of fluxes containing the more fusible bases. The multibasic silicates are usually more fusible than the simple silicates, and this fact is taken advantage of by the assayer. A slag may be regarded either as a mixture of several silicates or as a solution of one silicate in another. The alkaline bases Na_2O and K_2O and all the common bases FeO, CuO, MgO, Al_2O_3, MnO, ZnO, etc., with which the assayer has to deal form silicates which are soluble in one another.

Borax and the boron trioxide set free by its decomposition act in a similar manner to silica, combining with the bases present to form borates, and as borax is commonly added as a flux, the assay slag in most cases consists of a solution of silicates and borates of various bases.

The following are examples of borate slags :—

$$i.\ Na_2O, FeO, 4SiO_2, 2B_2O_3 = FeO, 4SiO_2, Na_2B_4O_7.$$
$$ii.\ Na_2O, PbO, Al_2O_3, 6SiO_2, 2B_2O_3 = PbO, Al_2O_3, 6SiO_2, Na_2B_4O_7.$$

The assay slag is usually a double silicate, as almost all ores contain one or more of the common bases in greater or lesser quantity. In addition to silicates and borates the slag may contain dissolved infusible material and also certain metallic sulphides, notably those of sodium and iron. According to Fulton,[1] "the most desirable constitution for an assay slag in general is that of a mono-silicate or a sesquisilicate—sometimes, but more rarely, a bisilicate. If the ore is basic a bisilicate may be approached, if acid a monosilicate, or even a subsilicate, in order to ensure complete decomposition." The various assay charges given subsequently will give slags that approximate to the above constitution. The assay slag may be regarded as having a more or less constant constitution, and the assayer having determined approximately the composition of the ore, adds basic or acid fluxes in correct proportion to produce the desired slag.

It has been pointed out previously that litharge possesses the property of dissolving or holding in suspension certain proportions of other metallic oxides, and when an ore is fluxed with litharge the resulting slag will consist of a mixture of the oxides of lead and other metals with, in some cases, a certain proportion of lead silicate. In the scorification process litharge is produced during the operation, so that the resulting slag is of the nature described.

Calculation of Assay Charge.—The following example will serve to illustrate the method of calculating a charge to produce a desired slag from an ore of known composition.

Let it be assumed that the ore contains the following constituents to be fluxed:—

Silica (SiO_2)	60 per cent.	
Ferrous oxide (FeO) . . .	16	,,
Magnesia (MgO)	3	,,

[1] *Manual of Fire Assaying*, p. 64.

In order to simplify the calculations whole numbers are taken, and the percentage of carbonic acid, etc., combined with the bases is disregarded. Let it be assumed also that the ore is to be fluxed so as to produce a monosilicate slag represented by the formula

$$2Na_2O, SiO_2 + 2FeO, SiO_2 + 2MgO, SiO_2.$$

The amount of silica slagged by the bases present is first ascertained by proportion with the aid of the atomic weights, thus,

$$\begin{array}{c|c} 2FeO & SiO_2 \\ 2(56+16)=144 & 28+(16\times 2)=60 \end{array} \left\{ \begin{array}{l} 144 \text{ parts of FeO require 60 parts of silica} \\ \text{to form a monosilicate :---} \end{array} \right.$$

therefore 16 parts of FeO will require $\dfrac{60 \times 16}{144} = 6{\cdot}66$ parts.

The amount of silica required by the 3 parts MgO is found in the same way :---

$$2MgO, SiO_2 = 80 : 60 \ \frac{60 \times 3}{80} = 2{\cdot}25.$$

Thus giving a total $(6{\cdot}66 + 2{\cdot}25)$ of $8{\cdot}91$ parts by weight of silica slagged by the bases, so that $60 - 8{\cdot}91 = 51{\cdot}09$ parts of silica remain to be converted into slag as sodium monosilicate. According to the same proportion,

$$2Na_2O, SiO_2 = 124 : 60 \ \frac{124 \times 51{\cdot}09}{60} = 105{\cdot}6.$$

Hence to flux $51{\cdot}09$ parts of silica $105{\cdot}6$ parts of Na_2O are required, but as this is added in the form of, say, sodium bicarbonate, which contains approximately 40 per cent. of Na_2O, the total weight of this flux required would be $\dfrac{105{\cdot}6 \times 100}{40} = 264$ parts.

Therefore, if 50 grammes of ore are taken for assay, $\dfrac{264}{2} = 132$ grammes of sodium bicarbonate must be added to obtain the monosilicate slag desired. If other bases are present or different fluxes are used the quantities required are calculated in the same way after determining what slag is to be produced. When the iron in an ore consists of ferric oxide it is essential to remember that it must be converted into ferrous oxide by the addition of carbon, to form a satisfactory slag with silica.

The amount of carbon required is calculated from the following reaction :---

$$2Fe_2O_3 + C = 4FeO + CO_2.$$

One gramme of Fe_2O_3 requires $\dfrac{12}{320} = 0{\cdot}037$ gramme of carbon. If charcoal powder, which contains only about 90 per cent. of carbon, is used, the quantity required to reduce 1 gramme of Fe_2O_3 will be $\dfrac{{\cdot}037 \times 100}{90} = 0{\cdot}041$ gramme charcoal.

This amount of charcoal would of course be added in addition to that required to reduce the PbO to give the lead necessary for collecting the values.

Magnetite (Fe_3O_4) and manganese dioxide (MnO_2) require reduction in a similar manner.

The accompanying Tables XIIA. and B, based on tables computed by Professor Balling, will simplify slag calculations.[1]

---

[1] Quoted by Roberts-Austen, *Introduction to Study of Metallurgy.*

*Table XII.—For calculating the Amounts of Fluxes required
to obtain given Slags.*

Table A.				Table B.			
For ascertaining the Necessary Amounts of Bases to convert Given Amounts of Silica into Slag.				*For ascertaining the Necessary Amounts of Silica to convert Given Amounts of Bases into Slag.*			
Base.	Parts by weight of bases required to convert one part by weight of silica into a—			Base.	Parts by weight of silica required to convert one part by weight of the base into a—		
	Mono-silicate.	Bi-silicate.	Sesqui-silicate.		Mono-silicate.	Bi-silicate.	Sesqui-silicate.
Na₂O	2·07	1·02	1·37	Na₂O	0·486	0·972	0·729
PbO	7·36	3·69	4·92	PbO	0·136	0·271	0·203
FeO	2·40	1·20	1·60	FeO	0·416	0·833	0·625
CaO	1·86	0·93	1·24	CaO	0·535	1·070	0·803
MgO	1·33	0·66	0·88	MgO	0·750	1·500	1·125
Al₂O₃	1·14	0·57	0·76	Al₂O₃	0·873	1·747	1·310
MnO	2·36	1·18	1·57	MnO	0·422	0·845	0·633

In order to show the use of this table the above example may be worked with its aid. For ascertaining the amount of silica converted into slag by the bases in the ore, Table B is used. In the section for the monosilicates, the corresponding figures are found for the bases present, and these figures multiplied together give the following products :—

$$FeO \qquad . \qquad . \qquad . \qquad . \qquad . \qquad 16 \times 0\text{·}416 = 6\text{·}66$$
$$MgO \qquad . \qquad . \qquad . \qquad . \qquad . \qquad 3 \times 0\text{·}750 = 2\text{·}25$$
$$Total = 8\text{·}91$$

Thus 8·91 parts of silica are converted into slag, and of the total amount (60 parts) of silica present there remain 60 − 8·91 = 51·09 parts to be converted into slag by the addition of sodium oxide. On referring to Table A, it will be seen that the corresponding amount of sodium oxide is 2·07, the product being 51·09 × 2·07 = 105·75.

Several graphic methods for facilitating the calculation of charges are given in *An Introduction to the Study of Metallurgy*, by Sir W. Roberts-Austen.

CHAPTER XII.

ASSAY OPERATIONS.

THE dry assay of ores involves the operations of roasting, fusion, scorification, and cupellation.

I. Roasting.

The oxidation of pyritic material in ores is sometimes accomplished by roasting, in which operation advantage is taken of the oxygen of the air as the oxidising agent. It may be well to point out that the operation is sometimes termed calcination, the terms roasting and calcination being frequently used indiscriminately to indicate the same operation. But, in a metallurgical sense, the term calcination should be confined to the application of heat for the expulsion of carbon dioxide or volatile matter. In roasting, the sulphur and arsenic in such minerals as pyrites and mispickel are oxidised by the hot air and pass off principally as sulphur dioxide and arsenious oxide (As_2O_3), while the metals generally remain in the final product in the form of oxides. The quantity of ore, 20 to 50 grammes, to be used for the assay must be carefully weighed out before roasting. The roasting is most conveniently effected in a roasting dish (fig. 62) sufficiently large to allow of the ore being stirred without loss by spilling. The dish is placed in a muffle heated to a dull red heat (that is, at a temperature of about 600° C.), and the door removed to allow of free access of air. The ore is frequently stirred, though not violently, to expose fresh surfaces to the air and to prevent clotting. A suitable stirrer for the purpose consists of a piece of stout iron wire flattened at one end, and bent at a right angle about 1 inch from the end. The temperature at which the roasting is started is of importance, as it is during the first few minutes of the operation that the formation of lumps is most to be feared. The employment of too high a temperature at the outset, or neglect in stirring, will cause partial fusion of the more fusible sulphides, such as Sb_2S_3 and PbS, that may be present, the fused matter so formed being troublesome to roast efficiently.

The ore is maintained at dull redness until the blue flame of burning sulphur has ceased, and no incandescent particles are visible on stirring, after which the temperature is raised to redness to decompose sulphates. The operation, which occupies from thirty to sixty minutes for 50 grammes of ore, is complete when the ore remains of a uniform colour on stirring and no further smell of SO_2 is discernible. When finished, the ore is said to be "dead" or "sweet" roasted. In the case of an ordinary sulphide ore, such as pyrite (FeS_2) with quartz, the roasting takes place in two stages. Pyrite, when heated, parts readily with one of the two atoms of sulphur, thus :—

$$FeS_2 + Heat = FeS + S.$$

The first stage of the roasting consists, therefore, in liberating and oxidising one-half of the sulphur, which burns with a blue flame, forming sulphur dioxide, which is volatilised. The heat generated by the oxidation of the sulphur causes the ore particles to glow. If a low temperature be employed, a certain proportion of the sulphur is oxidised to sulphur trioxide, which unites with metallic oxides to form sulphates ($FeO + SO_3 = FeSO_4$). The second stage consists in oxidising the remaining sulphur to sulphur dioxide and the iron to ferrous oxide, thus:—

$$FeS + 3O = FeO + SO_2.$$

Owing to the oxidising conditions of the roasting, a large proportion of the ferrous oxide is converted into ferric oxide ($2FeO + O = Fe_2O_3$). With prolonged heating ferroso-ferric oxide (magnetic oxide) Fe_3O_4, or Fe_2O_3,FeO, is formed. Any sulphate formed in the first stage will be decomposed as the temperature increases, forming ferrous oxide and sulphur trioxide.

$$\text{i. } FeSO_4 + Heat = FeO + SO_3.$$

The decomposition of sulphates is essential, otherwise they may be reduced to sulphides in the subsequent fusion of the ore in the crucible, thus:—

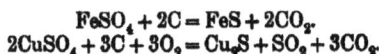

$$FeSO_4 + 2C = FeS + 2CO_2.$$
$$2CuSO_4 + 3C + 3O_2 = Cu_2S + SO_2 + 3CO_2.$$

Sulphates are decomposed by a higher temperature towards the end of the operation. Lead and zinc sulphates require a comparatively high temperature to effect their decomposition.

The roasting operation is more difficult and complex when ores contain copper, arsenic, antimony, and lead.

When copper is present, the copper sulphate formed during roasting requires a higher temperature to effect its decomposition than ferrous sulphate. The decomposition is facilitated by the addition of ammonium carbonate, which converts the copper sulphate into ammonium sulphate, which is volatilised, thus:—

$$CuSO_4 + (NH_4)_2CO_3 = CuO + (NH_4)_2SO_4 + CO_2.$$

For this purpose the ore is first roasted until "sweet," then cooled, mixed with about one-fourth of its weight of dry powdered ammonium carbonate, and re-roasted at a low heat until fumes cease to be given off.

Ores containing arsenic and antimony require considerable care and attention during roasting to successfully eliminate these two volatile elements. Both arsenic and antimony, during roasting, are first oxidised to the state of the lower oxides As_2O_3 and Sb_2O_3, which are volatile and largely expelled, but a certain proportion is further oxidised to the higher oxides As_2O_5 and Sb_2O_5, which, in the presence of certain metals, notably silver, copper, and iron, form arsenates and antimonates of these metals. These compounds are stable even at high temperatures, thus causing the arsenic and antimony to be retained in the ore instead of being eliminated. Arsenates and antimonates will carry silver into the slag in the subsequent fusion of the roasted material unless previously decomposed. The decomposition of the arsenates and antimonates is best effected by mixing about an equal volume of powdered charcoal or coal with the ore after it has been roasted " sweet," and roasting again at a red heat until all the carbon is burnt off. During this second roasting the air should be excluded in order to secure a reducing atmosphere, and thus facilitate the reduction of the arsenates and antimonates (see also page 221).

Iron arsenate is readily decomposed when heated with charcoal powder, and practically the whole of the arsenic eliminated, but the reducing action of charcoal on copper arsenate is very slight, only a small elimination of arsenic taking place. The reduction of iron arsenate is shown in the following reaction :—

$$Fe_2O_3,As_2O_5 + 2C = Fe_2O_3 + As_2O_3 + 2CO.$$

Antimonates are also decomposed by charcoal, but less easily than most arsenates, and the antimony eliminated. The presence of sulphur facilitates the elimination of arsenic and antimony by combining with them to form volatile sulphides.

Owing to the readiness with which antimony sulphide fuses, the roasting of ores containing this mineral must be performed at a low temperature to avoid clotting.

Ores containing galena (PbS) are difficult to roast on account of the ready formation of lead sulphate, which is only decomposed at a comparatively high temperature. Under these conditions there is a great tendency for the ore to clot or even fuse, as lead sulphate, and any lead oxide formed may combine to form, in the presence of silica, a readily fusible lead silicate. As a general rule, ores containing large proportions of galena are treated by other methods of desulphurisation described subsequently.

The mechanical loss during roasting may be considerable if proper precautions are not taken. The chief causes of loss are spilling through excessive agitation of the ore, and decrepitation of the pyritic material, due in many cases to the coarseness of the sample. Certain varieties of pyrites, and some metallic sulphides, however, have a great tendency to decrepitate when heated, even when finely powdered, but loss may be prevented by covering the ore with an inverted roasting dish until all danger of decrepitation is over. It is somewhat difficult to prevent partial fusion or clotting of the ore containing sulphides of antimony or lead, and, in the event of the ore fusing, the only satisfactory procedure is to weigh out a fresh quantity of ore and roast again with more care. In the absence of fusible minerals the temperature at the commencement of the operation should not be below a dull red heat, as a low temperature favours the formation of sulphates, which should, as far as possible, be avoided for the reason already stated. The roasting may also be performed in the crucible in which the fusion is afterwards made. In this case the crucible is usually placed in a wind furnace and inclined to the operator, so that the draught of air passing to the furnace flue may impinge as much as possible on the surface of the ore. The top of the furnace must be open, so that air may have free access to the crucible. The ore is stirred with the straight stirrer shown in fig. 30. The crucible employed should be shallow and wide-mouthed, not deep and narrow, and sufficiently large to allow of the subsequent fusion of the ore being made in it. When the roasting is finished, the ore is turned out into an iron mortar and mixed with the fluxes, taking care to clean the stirrer in the mixture. The charge is then replaced in the crucible used for the roasting and fused in the furnace.

Ores may be conveniently roasted also in an iron pan (fig. 100) over the crucible furnace. The pan should be coated with a thin layer of chalk or red ochre to protect it from injury through the sulphides attacking the iron.

As previously stated, the roasting of an ore occupies a considerable amount of time and attention, and, when possible, it is better to take advantage of the desulphurising action of lead oxide or nitre. The roasting of silver ores is seldom attempted, as it is very liable to give low results, especially with high-grade ores. The roasting method of oxidation is frequently used for gold ores, consisting mainly of pyritic material, as it permits of a larger quantity of ore being taken than is permissible with other methods of oxidation.

The tendency of modern assay practice, however, appears to be to obviate the roasting of ores as far as possible.

II. Fusion.

The object of this operation is to concentrate the precious metals in a button of lead, while the gangue, etc., forms, with suitable fluxes, a fusible slag in which the lead sinks. The fusion of an assay charge is effected in a fire-clay crucible in either a wind or muffle furnace. A deep and narrow crucible is usually best for the purpose. The crucible is carefully annealed in the ashpit of the furnace before using, otherwise it is very liable to crack when placed in the furnace. The ore and fluxes (excepting a small part of the borax) are well mixed and charged into the crucible, which must not be more than half full. The mixture is then covered with the remainder of the borax, reserved for that purpose, and if iron is required the nails or hoop-iron are inserted in the charge. The crucible is put into the wind furnace at a low red heat (about 600° C.) and packed round with fuel. For this purpose both ordinary and basket tongs are useful (see figs. 21 and 24). A plan adopted by some assayers is to place an old crucible into the furnace and pile the coke round it; then carefully withdraw it, and lower the charged crucible into the hollow thus formed. If coke is used, all fuel touching the pot should be in the first instance at a black heat. The temperature should not be urged at first, but slowly raised, and after ten or fifteen minutes the charge should have just reached a dull red heat.

This exposure of the charge to a dull red heat before fusing is known as "fritting," and is of considerable importance. At this time the lead is all reduced and remains entangled in the charge in the form of minute liquid beads. Chemical action now begins between the fluxes and the ore. Carbonic acid is given off from the sodium carbonate as it unites with the silica, and the charge is continually stirred up by the disengaged gas. The temperature is now gradually raised to a full red heat, and the lead beads unite and sink through the liquefying charge. After a period of thirty or forty minutes from the time of charging in, the charge is in a state of tranquil fusion, and the lead containing the precious metals is collected at the bottom. The nails (or hoop-iron) are then withdrawn, and all adhering lead shaken off them into the crucible. The crucible is now lifted out of the furnace and tapped on the top of the furnace to assist the lead to settle in the slag. The charge is poured into a warm iron mould and allowed to solidify. It is important that the mould should be clean and quite free from any loose plumbago powder, as this would cause the lead to spurt and occasion loss. The lead is found at the bottom of the mould, and, when cold, is detached from the slag by hammering the edges of the mass. It should be soft, malleable, and free from any regulus, and should weigh from 20 to 30 grammes. When freed from the slag, the lead button should be hammered into a disc with rounded edges, or into a cube with rounded corners. If the edges or corners are left sharp, they are very liable to injure the face of the cupel when the button is placed on it. After pouring the charge the crucible may be employed a second time, or even more; but the practice is to be deprecated, unless the ore being assayed is low grade. It should on no account be done with rich ores. The same crucible is used for "cleaning the slag" when this is necessary.

The fusion of the charge too frequently receives but scant attention at the hands of the assayer, but it may be pointed out that the success or failure in collecting all the gold or silver in the charge is to a large extent dependent on the care bestowed on the fusion. If the heat should be too strong at first, the quickly escaping gases may cause part of the charge to be ejected from the crucible, or the flux might melt and sink to the bottom of the crucible, leaving the quartz, etc., as a pasty mass above. The more slowly the charge is melted the better the chance afforded of bringing the lead into intimate contact with all parts of the pasty mass, and of collecting the precious metals from the ore. With a gentler heat, also, the combination between ore and flux is more complete, and, when the heat is subsequently raised, the fusible compound is simply melted into homogeneous slag. As stated above, the temperature at charging in should be about 600° C., while "the final temperature" need not, as a rule, be above 1100° C., and may be lower in certain cases. The influence of the rate of heating on the result is very great, as shown by experiments by Dr Rose.[1]

Four charges, consisting of equal weights of pyritic ore, litharge, and sodium carbonate with a little borax powder and the necessary charcoal, were made.

Two charges were heated very slowly, and remained in the furnace for fifty minutes before fusion was complete. The other two charges were heated as rapidly as possible, and were in the furnace for only twenty minutes. The results were as follows :—

	Gold.	Silver.
	ozs. per ton (2240 lbs.) (1) 0·130 (2) 0·125	ozs. per ton (2240 lbs.) (1) 4·42 (2) 4·28
Heated slowly . {		
Heated rapidly . {	(1) 0·075 (2) 0·092	(1) 2·33 (2) 2·71

The mean deficiencies in the charges heated rapidly were thus :—Gold, 34 per cent. ; silver, 42 per cent. ; but the results were also less regular than when slow heating was used. Another example is given on page 228.

Fusion in the Muffle (see fig. 101). — Crucible fusions can be made satisfactorily in a muffle furnace if it is large enough, and in Colorado, and in America generally, muffle furnaces have largely replaced crucible furnaces for general assaying. The temperature required is about the same as that used in scorification, and the conditions of the fusion are similar to those already described. Crucibles of the "Colorado" shape (fig. 59) are generally used. The advantages claimed for the muffle are that it is cleaner, easier to manipulate, and more uniform in working, the temperature of a muffle being better controlled, and more easily kept constant and uniform than that of an ordinary fusion furnace. When large-size muffles are employed, six or eight, or even more fusions can be made at one time, and on that account muffle furnaces are favoured in smelter and mine assay offices where frequently a large number of assays have to be dealt with daily. The chief disadvantage of the muffle for fusions is that the weight of the

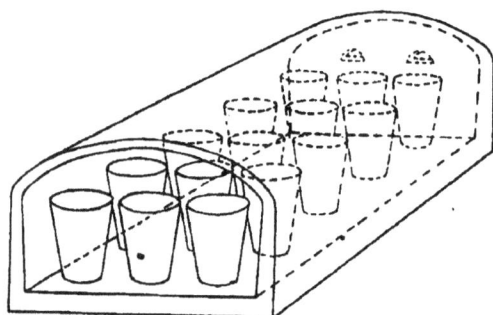

FIG. 101.

[1] *Metallurgy of Gold*, 5th edit., 1906, p. 444.

ore taken for assay is limited to about 20 grammes; though this objection may be obviated by making several fusions and scorifying together the resulting lead buttons so as to concentrate the precious metals into one button of lead.

Cleaning or Washing the Slag.—The slags of very rich ores may retain enough precious metal to necessitate further treatment. The slag is collected, roughly crushed, and "cleaned" by mixing with the following charge:—

Litharge	30	grammes.
Charcoal powder	1·5	,,
Sodium carbonate . . .	10	,,

If any regulus or speise forms during the fusion of the ore, and has not been scorified with the lead button should this have been necessary, it must be preserved, carefully crushed, and mixed in with the slag charge and a little more sodium carbonate added. The charge is placed in the crucible previously used for the fusion, and kept melted for ten to fifteen minutes. If the fused mass is thick and pasty, a small quantity of borax may be added with advantage. When tranquil fusion takes place the contents of the crucible are poured into a mould, the slag detached from the button of lead, and thrown away. The lead button from this second fusion is cupelled either by itself or with the first lead button. It is also a common practice in cleaning slags to fuse the slag alone, and then add a mixture of litharge and charcoal. In this case the crucible is replaced in the furnace as soon as the first charge has been poured, and the slag put back into it, without crushing, after being detached from the lead button. When the slag is fluid, a mixture of 20 grammes of litharge and 1 gramme of charcoal is charged on to the surface with a copper scoop. When the charge is again in a state of tranquil fusion, it is poured and the lead button extracted.

Some assayers consider that it is not necessary to pour the slag resulting from the first fusion before cleaning it, and adopt the following method: after the ordinary fusion of the ore has been made and the crucible contents have become tranquil, a mixture of litharge and charcoal is added, and the charge poured as usual when again tranquil. In this way the "cleaning" or "washing" of the slag forms part of the ordinary fusion and not a separate operation, and is therefore more expeditious (see Balling's method, page 200).

The above methods of cleaning slags are applicable to all ores and products which are assayed for precious metals. The richer the ore or product, the more needful is it to clean the slag; in the case of poor ores it may generally be omitted. However, in the case of ores with which the assayer is not familiar, it is better not to omit the examination of the slag until experience has shown that this may be safely done. As a general rule, slags from rich ores, slags containing many metallic oxides or metallic sulphides, and slag containing much ferric oxide should be cleaned.

III. Scorification.

This operation consists of an oxidising fusion of an ore with metallic lead in the muffle furnace, and has for its object the concentration of the precious metals in a button of metallic lead, and the removal of every other constituent of the ore by the solvent action of molten litharge formed during the operation by the oxidation of part of the lead added.

Scorification is conducted in a scorifier (fig. 61, p. 45), which must be well dried at a gentle heat for ten or fifteen minutes before use. If this precaution is not taken, "spurting" of the lead may occur during the operation. Sometimes the scorifier is coated inside with ferric oxide to protect it as far as possible from the corrosive action of the molten litharge.   A thin paste made by mixing powdered hæmatite with water is used for the purpose.   The amount of ore and lead used varies according to the nature of the ore, being, on the average, about ten to twenty parts of lead to one part of ore.   The charges for the scorification of different classes of argentiferous and auriferous materials are given subsequently in Chapters XIII. and XIV.

Granulated lead is generally employed for the purpose.   The ore is mixed in the scorifier with about half the lead, the mixture smoothed down, the rest of the lead spread evenly over it, and a small quantity of borax glass (0·5 to 1 gramme) placed on the top: the object of the borax being to lessen the corrosion of the scorifier, and to render the slag more liquid.   Instead of mixing the ore and the lead as described, it is equally satisfactory, according to Beringer, "to wrap the ore up in thin sheet-lead of the required weight; and if the ore contains much sulphur, the borax may with advantage be added (wrapped in tissue paper) some five or ten minutes after the operation has started."

The scorifier is then put into the muffle at a full red heat with the scorifier tongs (fig. 27), and the door closed until fusion has taken place.   As soon as the lead is melted the door is partly opened to allow a current of air to pass over the scorifier and oxidise the contents.   At this stage some of the ore may be seen to be floating on the surface of the molten lead, and any sulphides present are rapidly oxidised, chiefly by the air, but partly by the litharge which immediately begins to form.   The sulphur, arsenic, antimony, etc., in sulphides and arsenides are thus soon eliminated as volatile oxides, while the base metals are converted into oxide, and combine with the litharge and borax to form a fusible slag.   Silica also forms a fusible compound with the litharge.   Thus the litharge, as fast as it is produced, reacts on the various substances contained in the ore, forming with them a clean slag, while the precious metals pass into the unoxidised lead at the bottom of the scorifier.

The slag as it is produced forms a ring, leaving the centre of the lead bath or "eye," as it is termed, exposed to oxidation.   As oxidation of the lead continues, the litharge flows to the sides and increases the quantity of slag, until at length the whole of the lead is covered with slag and the eye has disappeared, leaving a flat, uniform surface.   The scorification is then considered finished, but it is usual before pouring to clean the slag.   This is done by placing a pinch of charcoal or coal powder (about 0·2 gramme), wrapped in tissue paper, on the surface of the slag, with a pair of tongs, and closing the muffle door.   A small quantity of litharge is thus reduced, forming a number of globules of lead which sink in the slag, alloying with any silver and gold which the slag may contain, and concentrating them in the molten lead below.   The temperature at the finish should be sufficiently high to completely liquefy the slag.   A temperature of 1050° to 1100° C. is usually enough.   When the slag has again become tranquil, the scorifier is withdrawn, and the contents poured into an iron mould (figs. 36 and 38).   The contents should pour clean and not leave any imperfectly fused material attached to the surface of the scorifier.   The assay should be poured as soon as finished, and not left in the muffle, as oxidation of the lead will continue although none is exposed to the air, owing to the interchange of oxygen between litharge and other oxides present.   When cold, the slag is detached from the button of lead as already described.   The lead button should be comparatively soft and malleable, and weigh from 15 to 20 grammes.   In most cases the button may be cupelled direct, but if it weighs much more than 20 grammes, its weight is reduced by another scorification.   If the scorification is prolonged to produce buttons weighing less than 15 grammes, appreciable amounts of precious metals are liable to be retained in the slags, especially with rich ores.   The lead buttons from the scorification of ores containing copper, antimony, etc., are very liable to be hard, owing to contamination with these metals, especially if insufficient lead has been used, and in order to get a pure button for cupellation it is generally necessary to rescorify them with the addition of fresh lead. When the temperature has been too low, the lead is very liable to retain lead oxide and be more or less brittle. Scorification usually occupies from thirty to forty minutes. The temperature must be well maintained during the operation, and must be high enough to melt litharge when contaminated by silica and oxides of copper, iron, manganese, etc. It ranges usually from 1000° C. to 1100° C., although a higher temperature may be employed for highly siliceous ores.

The slag from scorification should be homogeneous and glassy, while in colour it varies according to the nature of the ore. Scorification slags are essentially oxide slags consisting of metallic oxides dissolved in excess of litharge, and exercise a very corrosive action on the material of the scorifier, especially if they contain copper oxide. If the slag be stony in appearance and white patches of lead sulphate appear on the surface, it generally indicates that too low a temperature has been employed.

The slags from the scorification of average ores may generally be thrown away, as the button of lead will contain practically all the silver; but with rich ores and products it is advisable to save the slags and subsequently "clean" them by fusion in a small crucible, as previously described. In this case the "cleaning" by charcoal powder during scorification is omitted.

The slags from several scorifications are crushed and mixed with about 20 grammes of lead oxide, 1 gramme of charcoal, and 5 grammes of borax. The resulting lead button is cupelled.

The slags may also be fused with the cupels used for the cupellation of the lead button obtained from the scorification of the ore. In this way the silver lost in the slag and that absorbed by the cupel is recovered at one operation (see page 349).

The losses incurred in scorification are chiefly due to the employment of too low a temperature. If the muffle is too cold at first, the scorification is always more or less unsatisfactory, and the silver and gold retained in the slag will be by no means inconsiderable. If the slag appears pasty even when the temperature is sufficiently high, borax may be added with advantage at short intervals in quantities of 0·5 to 1·5 grammes at a time. The borax, wrapped in tissue paper, is placed on the surface of the slag. Only a small quantity of borax should be added at the commencement of the operation, as it would cover the surface of the lead and retard its oxidation. If the ore contains copper, powdered glass or silica may be used with advantage in the place of borax. Scorification is a very convenient and quick method of assay, and needs very little variation for the different classes of ores. It is largely employed for the determination of silver in ores and products. The small quantity of material that can be treated constitutes the chief disadvantage of the operation; consequently, gold ores, as a general rule, are better assayed by the fusion method, so that a comparatively large quantity of material may be operated upon.

IV. Cupellation.

Cupellation has for its object the removal, as oxides, of base metals such as lead and copper from gold and silver, which are non-oxidisable and are left in the metallic state. The same remarks apply to platinum and metals of the platinum group. The operation is performed in a muffle furnace in a porous vessel, called a cupel (fig. 63), the nature and manufacture of which has been previously described. Cupellation is based on the fact that "when lead is exposed to the action of air at a temperature considerably above redness, it combines with the oxygen of the air to form litharge (PbO), which at the temperature of its formation is a *liquid*. Consequently, if the lead is contained in a porous vessel, which allows the fused litharge to drain away as fast as it is formed, a fresh surface of the lead will be continually exposed to the action of the air, and the operation goes on until the whole of the lead has been removed. Silver or gold exposed to similar treatment does not oxidise, but retains its metallic condition ; so that an alloy of lead and silver or gold similarly treated would yield its lead as oxide, which would sink into the porous vessel, while the silver or gold would remain as a button of metal."[1]

There is a considerable difference between the surface tension of molten lead and molten litharge, and while litharge can "wet" the bone-ash surface and hence be absorbed, molten lead either cannot do so, or does so only to a very slight extent, and hence is not absorbed.

For the same reason the high-surface tension of the molten silver and gold, left on the cupel after the oxidation of the lead, prevents them from being carried into the pores of the cupel, although, as stated subsequently, there is always a small loss of precious metal during cupellation, which is caused by absorption by the cupel.

The oxides of lead and bismuth, in a state of purity, are the only oxides which possess the property of being absorbed by the cupel ; but by the aid of either of these, various oxides which by themselves form infusible scoriæ on the cupel acquire the property of being absorbed. For example, copper, antimony, tin, and most of the base metals form powdery oxides, which are not of themselves easily fusible, and it is necessary when they are present to add some solvent, which is invariably lead oxide (litharge), to render the oxide sufficiently fluid to be absorbed by the cupel.

Advantage is taken of this fact in cupelling material containing a base metal like copper, etc., such as an alloy of silver and copper, and sufficient lead added to supply the proper proportion of litharge to dissolve the copper oxide formed in the operation and carry it into the cupel. The proportion of litharge necessary varies with the nature of the substances cupelled and other circumstances. The quantity of lead required to supply the correct proportion of litharge in ordinary cases will be given subsequently.

Cupels.—It is extremely important that good cupels should be used for cupellation. They must, obviously, have a sufficiently loose texture to allow the fused oxides to be absorbed readily, and at the same time be sufficiently firm to bear handling without breaking. They must also be made of material that will not enter into fusion with molten litharge. The method of making cupels has already been described on page 45.

Cupels should be made many months before they are used, and stored in a dry place so that they may be dried very slowly, but completely, as otherwise cracks appear when they are heated and loss is thereby occasioned. "The cupels (bone-ash) used at the Royal Mint, London (see page 49), are made two years before being used, and are dried slowly on shelves at some distance from the furnaces. More rapid drying has been tried on many occasions with less satisfactory results. Nevertheless, fair results, but not the best, can be obtained when using cupels only a few days old" (Rose). It is well to remember that a bone-ash cupel will absorb its own weight of litharge. "About 98·5 per cent. of the litharge formed during the operation is absorbed by the cupel and the remaining 1·5 per cent. volatilised" (Fulton). It is important that the cupel used should be of suitable size for the lead to be cupelled upon it, as it is unsatisfactory to have the cupel too strongly saturated with litharge. As a general

[1] Beringer, *Text-Book of Assaying.*

rule the cupel (bone-ash) should be at least one-fourth as heavy again as the lead button.

The number of cupellations made at one time is dependent on the size of the muffle, but it is not usual to cupel more than fifty in one batch or "fire" as it is sometimes termed. The cupels must be carefully arranged in a definite order in the muffle when more than one lead button has to be cupelled.

The operation of cupellation is conducted as follows :—The cupels are first cleaned by blowing off the loose dust and by gently rubbing the cavity with the finger ; they are then placed in a muffle at a red heat and allowed to remain there for about ten to fifteen minutes until they attain the temperature of the muffle, and in order to expel any moisture or burn off any organic matter that may be present. If this precaution is not taken and the lead placed direct into a cold cupel, the lead will not clear, and if all the moisture is not expelled the steam will escape and bubble violently through the molten lead, causing what is termed "spurting," i.e. the ejection of lead particles and consequent loss. A similar result will be produced if the cupels contain particles of carbonaceous matter owing to the escape of bubbles of carbon dioxide resulting from the reduction of the litharge by the carbon (see also page 181).

When the cupels have attained the temperature of the muffle the lead buttons are placed in them, each in its proper cupel, by means of the cupellation tongs (fig. 28, page 38), and the door of the muffle closed. Some assayers heat the cupels for another ten minutes or longer after they have attained the temperature of the muffle before charging in the lead buttons. The "charging in" should be done promptly, so as not to unduly cool the muffle.

When a large number of assays have to be "charged in" great care must be taken to ensure that the temperature is sufficiently high before charging in, that the chilling which necessarily takes place during the operation shall not cool the muffle below the requisite temperature. The start of the cupellation requires a fairly high temperature, and is a critical part of the operation : the general experience is, however, that beginners are more likely to have the furnace too hot. The exact temperature suitable for cupellation will be discussed later. If the temperature is correct the lead quickly melts and a grey-black scum of lead oxide forms at once on its surface, but soon fuses. This is called the "clearing," "uncovering," or "opening up" of the lead button. The molten lead then appears much brighter than the cupel and rapid oxidation commences. The lead buttons should clear as soon as possible, and if this does not take place the buttons are said to be "frozen," in which case the temperature must be raised. The risk of "freezing" at the start is much greater with a cupel which has not been properly heated before introducing the lead. If freezing takes place, the lead may be made to clear by placing a piece of charcoal near the cupel so as to raise the temperature locally, or a little charcoal powder may be dropped on to the surface of the lead. Freezing may also take place during the cupellation owing to the temperature falling. In this case the assays should be repeated. The closing of the muffle door at first is simply to increase the heat until the lead buttons have "cleared," and when this has taken place the door is partly opened to admit air and to cool the muffle, as the temperature needed to clear the buttons is higher than that required to keep the cupellation going. When cupellation has started satisfactorily the muffle is cooled to as low a temperature as is compatible with the continuation of the operation. A higher temperature is again required at the finish : the muffle door is therefore closed and the furnace urged towards the end of the operation.

It will be observed that as soon as the lead has cleared the surface of the metal becomes covered with oily-looking drops of litharge, which are rapidly absorbed by the porous cupel and replaced by others. They pass over the surface slowly at first, but as the operation continues they move with greater rapidity. Towards the end of the operation the metal suddenly appears uniformly dull and glowing, and iridescent bands, produced by extremely thin films of fluid litharge, resulting from the oxidation of the last traces of lead, immediately pass over it. On the disappearance of these coloured bands a bright liquid globule or button of gold or silver or an alloy of both is left. The appearance of the molten globule of metal left on the cupel is practically the same whether it consists of pure gold or silver or an alloy of both. If the cupel is withdrawn from the furnace, the button becomes very bright at the moment of solidification, and is said to "flash," "brighten," or "blick." This "flashing" of the buttons of gold or silver is due to the evolution of the latent heat of fusion or "recalescence," which momentarily reheats the cooling globule to its melting-point. The flashing is much more noticeable with gold than with silver buttons. Molten gold has a peculiar green colour which is easy to recognise, and just after solidifying it glows beautifully, even with very small buttons. With very small silver buttons the flashing is rarely noticeable. The flashing is prevented if the buttons contain metals of the platinum group. Silver buttons require careful attention at the finish of the cupellation. Silver when molten absorbs oxygen from the air and gives it off suddenly when solidifying, causing a cauliflower-like growth on the surface of the button, and particles of silver may even be ejected out of the cupel and cause serious loss. This "spitting," "spurting," or "vegetation" as it is termed does not take place readily with silver buttons weighing less than about 5 milligrammes, but all buttons above this weight are very liable to spit. Spitting does not take place if gold is present in the silver button to the extent of 33 per cent. or more (Levol), as the solubility of oxygen in silver is lowered by alloying with gold. If the button is small the cupel may be withdrawn from the muffle as soon as cupellation is finished without risk of "spitting," but with large buttons precautions must be taken. Spitting can be prevented by slow cooling, and in practice the following methods are usually adopted. The cupels are allowed to cool in the muffle itself, the door being closed; or another cupel, previously heated to redness, is inverted over the cupel containing the button, and both carefully withdrawn from the muffle, thus cooling the button slowly between two hot cupels; or the cupel may be withdrawn very gradually towards the door of the muffle. Cooling in the muffle is invariably adopted with a large batch of assays, as, in addition to the danger of upsetting a cupel, it is inconvenient to handle a large number of hot cupels (see page 184). Buttons that have "spitted" must be rejected and the assays repeated. The button of precious metal when cold is detached from the cupel with a pair of pliers and the under-surface distorted by squeezing or hammering so as to loosen the adhering bone-ash. It is then cleaned by brushing with a hard brush and weighed. The silver button from cupellation should be bright, round, and slightly crystalline on the upper surface, and adhere slightly to the cupel. When it retains base metals it adheres more firmly to the cupel. Gold may be present in the silver button to the extent of 50 per cent. without showing any yellow colour. Platinum when present destroys the characteristic silver lustre and renders the button dull and grey but more crystalline. Gold, with little or no silver, is easily recognised by its characteristic colour, and the "flash" of gold is so marked that it is a valuable indication of its presence. Gold buttons are more globular than silver buttons of similar size and adhere less to the cupel.

Temperature of Cupellation.—The expression "temperature of cupellation" is usually understood by assayers to mean the temperature of the muffle in which cupellation is conducted, and this meaning of the expression is adhered to in the present volume. It must, however, be pointed out that the true temperature of cupellation is the temperature of the lead being cupelled.

Owing to the heat developed by the active oxidation of lead, the actual temperature of the molten lead in the cupel is always higher than the temperature of
the muffle where the cupel stands. The muffle temperature always bears a
distinct relation to the temperature of the lead being cupelled, and for all
practical purposes the muffle temperature affords sufficient indication of the
correct temperature at which to start and conduct cupellation. Experienced
assayers, however, always closely observe the temperature of the lead on the
cupel, and carefully regulate the supply of air through the muffle and the
temperature of the muffle so as to keep the lead at the correct temperature.
The regulation of the amount of air passing into the muffle during cupellation
is of considerable importance, and is a matter which does not always receive
sufficient attention. The exact temperature suitable for cupellation can only
be ascertained by practice, and the varying light of the day may occasion error
in judging of the degree of heat. As previously stated, in general it is well
to employ a comparatively high temperature at the commencement so as to
well uncover the lead, then to cool down, and increase the heat at the end of
the operation for a few minutes, in order to aid the brightening.

There is a German adage many centuries old which runs :—

> "Kühl getrieben, heisser Blick,
> Ist des Probierer's Meisterstück."

This may be translated :—

> "Cool for the working, hot for the 'blick' (*i.e.* brightening),
> This is the assayer's master trick."

With lead buttons poor in silver or gold, the lowering of the temperature
after "clearing" is a matter of importance, especially when the temperature
at the start has been excessive ; but when dealing with material such as
bullion containing large proportions of the precious metals, it is of more
consequence that the muffle should be uniformly hot throughout than that
any absolute degree of heat should be attained, as it is usual in cupelling such
material to use check or proof assays (see page 181) to eliminate uniform errors
due to high temperature. The time occupied by the cupellation when small
quantities of lead only are used, as in bullion assays, is frequently too short to
allow of much variation of temperature being made. The temperature of
cupellation has to be varied somewhat according to the nature of the material
to be cupelled; for example, gold, which solidifies at 1064° C., requires a higher
temperature than silver, which solidifies at 962° C. A higher temperature is
also required for lead containing base metals and a large percentage of the
precious metal than for lead containing little gold or silver and free from base
metals. Most lead buttons from ore assays will be comparatively pure and
contain very much less than 1 per cent. of precious metal, and will melt
practically at the melting-point of lead, while in bullion assays the proportion
of precious metals is much greater ; and base metals such as copper are frequently
present and raise the melting-point of the lead alloy on the cupel, so that the
temperature of cupellation must be higher in order to prevent freezing. As
cupellation proceeds the percentage of lead in the alloy decreases, and that of
silver and gold increases. The lead button from the assay of ores, etc., may
be regarded, therefore, as an alloy of lead and silver (or gold) which, during
cupellation, undergoes enrichment from practically pure lead melting at 327° C.
to pure silver melting at 962° C., or pure gold melting at 1064° C., and, owing
to the higher melting-points of the precious metals, the finishing temperature
must be above that used during the operation. It will be evident that, as the
lead is gradually removed by oxidation, the alloy remaining on the cupel will
become gradually enriched in precious metal and will consequently tend to

11

solidify or freeze if the temperature is not raised towards the end, and the solidified metal will contain lead.

This tendency to freeze is greater with lead containing base metals and comparatively large amounts of precious metal such as in bullion assays, than it is with lead with smaller silver or gold content, such as is ordinarily obtained from ore assays.

"In the cupellation for silver it would seem at first sight that a final temperature of 962° C. (i.e. the melting-point of silver) is necessary in order to prevent freezing and to obtain a silver bead free from lead. However, the phenomenon of the 'surfusion' of the silver, i.e. silver in a molten state below its true melting-point, due probably to its separation from its lead alloy by the oxidation of the lead, appears to indicate that this temperature is not necessary. It is true, nevertheless, that the finishing temperature, depending somewhat upon the amount of silver present, may not fall much below 910° C."[1]

The remarks on the cupellation of silver apply also to that of gold, except that, owing to the somewhat higher melting-point of gold (1064° C.), the finishing temperature should be higher. In the case of alloys of silver and gold the finishing temperatures will be intermediate between those for the pure metals, and will be higher as the proportion of gold increases.

During cupellation there is always a slight loss due to absorption by the cupel and volatilisation (see page 178), but if the temperature employed is much above that actually required for the operation, the loss of silver and gold is very much increased.

Diversity of opinion exists between assayers as to the correct temperature for cupellation, especially for silver, which is subject to greater loss than gold during the operation. As a general rule, the temperature is considered to be properly adjusted when the muffle is at full red heat, the cupel dull red, and the melted lead much more luminous than the cupel; the fumes should rise slowly to the crown of the muffle, and no crystals of litharge (termed "flakes" or "feathers") should form on the cupel.

"Feathers" are crystals of solid litharge sublimed from the vapour and deposited on the rim of the cupel, which is invariably cooler than the molten lead on the cupel. It is stated by Fulton that they will not form above 820° C.[2] When the temperature is too high, the outline of the cupels becomes indistinct, the melted metal is seen with difficulty, and the fumes are almost invisible; when it is too low, the fumes do not reach the top of the muffle, and a ring of indistinct crystals of litharge forms on the top and outer surface of the cupel. Many assayers prefer to conduct the cupellation of silver at a temperature low enough to enable crystals of litharge to form on the cupel. "When cupellation of silver is carried on under these conditions the temperature should not, according to Fulton, be above 820° C., in which case crystals of litharge (feathers) form on the side of the cupel towards the muffle mouth. If the temperature is too low for the cupel successfully to absorb practically all the litharge, these feathers form low down in the cupel. When the temperature is correct, they form near the upper rim of the cupel. It is, however, to be noted that the draught through the muffle influences the formation of feather litharge; i.e. if the draught is strong, feathers will form, although the temperature is somewhat above 820° C. During cupellation the door of the muffle should never be left wide open, but should be set slightly ajar, so that the cold air will not strike directly upon the cupels."[3]

This description applies more particularly to the cupellation of large lead buttons containing comparatively small amounts of silver such as are obtained from the assays of ores.

[1] Fulton, *Manual of Fire Assaying*, 2nd edit., 1911, p. 86.
[2] *Loc. cit.*, 82.   [3] *Loc. cit.*, 82.

The formation of "feathers" is generally accompanied by a sluggish, heavy movement of the fumes, which fall in the muffle. Beyond the fact that it shows that the muffle has not been too hot, it is doubtful whether there is any advantage in cupelling at a temperature sufficiently low to permit of the formation of feathers. It is open to the objection that the silver buttons are very liable to retain lead and other base metals. In any case, the degree of heat should not be below that required to produce feathers in small quantity only. The author is in agreement with Dr Rose and others that "it is probably better that the litharge should be completely absorbed."

The practice almost universally adopted by American assayers is to cupel at a temperature sufficiently low to leave a distinct ring of feathery litharge crystals round the edge of the cupel, marking the area covered by the original lead button.

Beads of silver cupelled at this heat always retain traces of lead, but it is assumed that the *plus* error thus introduced is much less than the *minus* error or loss due to volatilisation and to absorption by the cupel, so that a rough approximation to accuracy is obtained.[1]

Since the introduction of accurate pyrometers, such as the thermo-couple, many attempts have been made to determine the true temperature of cupellation, *i.e.* the temperature of the lead undergoing oxidation in the cupel. In the early experiments many muffle temperatures were taken on the assumption that these would give the temperature of cupellation, but the discrepancies between the different observations clearly indicate that the true temperature of cupellation cannot be ascertained in this way.

The discrepancies between the different observations of muffle temperature are due, among other causes, to the variation of the air current passing through the muffle, and to the fact that the temperature of the air above the cupel cannot be taken by a thermo-couple on account of the radiation of heat from the cupel and from the walls of the muffle. The radiated heat is absorbed by the thermo-couple much more than by the air passing through the muffle, with the result that the thermo-couple always shows a temperature far above that of the air immediately surrounding it.[2]

As previously stated, the temperature of the lead in the cupel is always greater than that of the muffle, and in the most recent experiments attempts have been made to ascertain the actual temperature of the molten lead during cupellation. This determination presents many experimental difficulties, as the protective tube of the thermo-couple is rapidly destroyed by the corrosive action of the litharge.

Another important consideration in determining the temperature of cupellation, is the exact freezing-point of litharge, which is most difficult to determine. A cooling curve gives no definite point of arrest. In the most recent and accurate determinations the freezing-point of litharge is given by Doeltz and Mostowitsch[3] as 906° C., and by Mostowitsch[4] alone later as 884° C., this latter figure being in all probability the more accurate. There is no doubt that at 906° C. litharge is completely molten. At 640° C. it is completely solid. Between these temperatures there is a long pasty stage. The exact point above which the pasty litharge is to be called liquid, and below which it is to be called solid, is very difficult to decide. According to Dr Rose,[5] litharge is not absorbed by a bone-ash cupel maintained at 750° C., but is absorbed by one kept at 825° C.; and Fulton[6] observes that absorption does not take place at 815° C., but is rapid at 883° C. If absorption is to be taken as the criterion of liquefaction, then the melting-

[1] Collins, *Metallurgy of Lead*, 2nd edit., 1910, p. 468.
[2] T. K. Rose, *Trans. Inst. Min. and Met.*, 1909, vol. xviii. p. 463.
[3] *Metallurgie*, iv. 1907, p. 290.   [4] *Metallurgie*, iv., 1907, p. 468.
[5] *Trans. Inst. Min. and Met.*, 1909, vol. xviii. p. 463.
[6] *Western Chemist and Metallurgist*, vol. iv. p. 47.

point of litharge must apparently be placed at about 825° C.   In cupellation, therefore, the cupel itself must be above this, but the temperature of the air in the muffle near the cupel may differ considerably from this for the reasons stated. The results of the most recent attempts that have been made to measure the temperature of muffles during cupellation and also the actual temperature of cupellation are briefly summarised as follows.

The most complete and probably the most reliable experiments to determine the temperature of cupellation were made in 1908 by C. H. Fulton,[1] assisted by O. A. Anderson, I. E. Goodner, and J. D. Ossa.   The following is a summary of the results obtained by these authors.   The temperatures apply to the cupella-tion of silver on bone-ash cupels.

1. In the case of the lead buttons not containing any appreciable amount of copper or iron, etc., a *muffle temperature* of at least 800° C., and, better, one of 850° C., is required to "uncover" or start cupellation.

2. This temperature may be lowered to about 770° C. during the oxidation of the greater part of the lead.

3. Toward the end of the cupellation, in the case of silver, it must again be raised to about 830° C. in order to get a pure silver button.

4. The *actual* temperature of the cupelling lead is always appreciably higher than the *muffle* temperature.

5. The *actual* finishing temperature of the cupellation cannot be safely carried below about 910° C.

6. The greatest observed surfusion of silver was 77° C., and this is probably very near the maximum.

7. Silver beads finishing with surfusion are free from lead.

8. "Feathers" or crystals of sublimed litharge on the cupel are an indication of the proper cupellation temperature, provided the air-draught is not excessive.

9. It is just as essential to regulate the air-draught of the muffle as its temperature.

Another series of experiments on the temperature of cupellation were made in 1909 by R. H. Bradford,[2] who determined the temperature of the lead button and of the interior of the cupel during cupellation by means of the Le Chatelier pyrometer, the couple at the hot junction being protected by a thin coating of fireclay.   It was found that before "driving" (*i.e.* rapid oxidation of the lead) commenced, the temperature of the molten lead rose to 900° C. or above. Temperature determinations at different stages showed that when once the lead commenced to oxidise rapidly, the temperature of the button rose owing to the heat of oxidation of the lead, this heat being sufficient to keep the litharge melted, even when the cupel was drawn forward to a cooler position in the muffle. Cupellation proceeds as long as the heat developed is sufficient to keep the litharge melted, and this can be attained by maintaining the temperature at 650°–750° C., about ¼ inch above and near the front of the cupel, and working with a gentle draught through the muffle.

In determining the temperature of cupellation for silver, Lodge obtained a number of results by means of a Le Chatelier pyrometer, the junction being kept about ¼ inch above the lead buttons.   The results show that at 700° C. crystals of litharge form all about the lead button, at 775° C. crystals of litharge form on the cooler side of the cupel only, while at 850° no more crystals of litharge are formed.

With regard to the temperature for the cupellation of gold, very few experi-ments have been made.   The figures recorded are mainly muffle temperatures.

[1] Fulton, *Fire Assaying*, 2nd edit., 1911, p. 97 ; also West, *Chem. and Met.*, vol. iv., 1908, pp. 31–54 ; and *Eng. and Min. Journ.*, vol. lxxxvi., 1908, p. 326.
[2] R. H. Bradford, *Journ. Ind. and Eng. Chem.*, 1909, i. 181–184.

As an example, the temperature of a muffle during the cupellation of standard gold (91·66 per cent. gold and 8·34 per cent. copper) at the Royal Mint, London, was determined some years ago by Dr Rose.[1] The muffle charge consisted of seventy-two assays, arranged in twelve rows of six, the amount of metal taken for each assay being 0·5 gramme wrapped in eight times its weight of lead. The Le Chatelier pyrometer was used, the position of the junction being about 1 inch above the cupel (bone-ash) in all cases.

The muffle was 15 inches long and 6½ inches wide, fired by coke, and it was found that the temperature gradually rose from about 1050° C. to 1080° C. in passing from front to back, whilst along the sides the temperature was 1 or 2 degrees higher than in the middle line. The mean temperature of the muffle was about 1070° C.

It is believed that with gas furnaces and the present arrangements of the muffle the variation in temperature is much less. The temperatures given are very probably too high, for the reasons stated on page 163.

From the various figures quoted it will be apparent that, notwithstanding the amount of work done by different experimenters, the subject of cupellation still offers a field for investigation, and it is to be hoped that, by the introduction of new methods of investigation, more exact knowledge respecting this most important of assay operations will soon be forthcoming.

As the correct temperature of cupellation can only be ascertained by practice, beginners should cupel weighed quantities of pure silver with lead and note the loss, and if this is excessive the experiment should be repeated until the results show that the correct temperature has been employed. For this purpose six pieces of pure silver, each exactly 0·5 gramme, are carefully weighed and each wrapped in 2 grammes of sheet lead. These are charged into separate cupels that have been heated at least ten minutes previously in the muffle. The cupels should be arranged in two rows of three in the middle of the muffle. When cupellation is finished, the muffle is cooled slowly until the silver has become solid. The cupels are removed, the buttons cleaned by squeezing sideways and brushing, and then carefully weighed on an assay balance. They should agree closely in weight, and the loss should not exceed 0·5 per cent. The experiment is repeated with fresh quantities of silver until the desired result is obtained.

It is also well to experiment with larger proportions of lead. Thus 0·20 gramme of silver may be cupelled with 20 grammes of lead in sets of six arranged as before. In this case the loss should not exceed, say, 1·8 per cent. Other series, in which a small addition of copper is made, may also be tried with advantage.

As a guide in cupellation the following scale of colour temperatures is given by Fulton [2]:—

	Degrees centigrade.
Lowest red visible in the dark	470
Dark blood red or black red	532
Dark red, blood red, low red	566
Dark cherry red	635
Cherry red, full red	746
Light cherry, light red	843
Orange	900
Light orange	941
Yellow	1000
Light yellow	1080
White	1205

[1] *Journ. Chem. Soc.*, vol. lxiii., 1893, p. 707.

[2] White and Taylor, in *Trans. Am. Soc. Mech. Eng.*, xxi. p. 628; H. M. Howe, in *Eng. and Min. Journ.*, lxix. p. 75; quoted by Fulton, *Manual of Fire Assaying*, p. 80.

The student may obtain a practical knowledge as to the meaning of these figures by placing in the muffle small pieces of metals or alloys of known melting-points and carefully noting the temperature "colour" of the muffle when they melt.

The metals should be in the form of thin sheet and triangular in shape to give a sharp point. The pieces are placed on clean cupels in the muffle almost entirely closed; the temperature should be very slowly raised, and the appearance of the muffle when each metal begins to melt carefully noted. The following metals and alloys are convenient for the purpose :—

		Melts at
Lead		327° C.
Zinc		419°
Aluminium		657°
Silver eutectic alloy { silver, 72 per cent. / copper, 28 per cent. } . . .		770°
Pure silver		962°
Silver-gold alloy { gold, 20 per cent. / silver, 80 per cent. } . .		1005°
Pure gold		1064°
Copper		1084°

Cupellation on Cupels of Material other than Bone-ash.—Since, as stated on page 49, a large number of cupels made of materials other than bone-ash (mainly magnesia) are now supplied for assay purposes, it is well to point out that during cupellation there is a considerable difference in the behaviour of these so-called "patent" cupels as compared with bone-ash cupels, due to differences in the thermal properties of the materials used.[1]

The diffusivity of heat and the specific heat of magnesia cupels are greater than those of bone-ash cupels. If similar cupellations be conducted in bone-ash and in magnesia cupels side by side, a marked difference will be seen in the behaviour of the lead. The lead on the bone-ash cupel during the cupellation is very bright, whereas the lead on the magnesia cupel is comparatively dull during a considerable part of the operation, and is not so hot, although the muffle temperature is the same for both. "This is due to the fact that the extra heat generated by the oxidation of the lead is diffused as soon as it is generated, owing to the superior diffusivity of the magnesia cupel, and hence cannot serve to raise the temperature of the lead, as is the case in the bone-ash cupel. Hence for the same 'muffle temperature' the actual cupellation temperature of the lead in the magnesia cupels is 50° to 60° C. lower than in the bone-ash cupels."[2]

It has been pointed out on page 164 that the heat generated by the oxidation of the lead is sufficient to carry the cupellation to a finish with bone-ash cupels, provided the muffle temperature is not lowered at the end of the operation; but, for the reasons stated above, it is necessary in the case of magnesia cupels to employ a slightly higher temperature during the cupellation, and to raise the temperature towards the end of the operation.

The difference between the temperatures of the lead during cupellation on magnesia and on bone-ash cupels, which can be noticed by observation, is much more marked at the beginning of the cupellation, and, in fact, is hardly discernible at the very end of the operation.

It will be noted also that magnesia cupels retain heat longer than bone-ash cupels, consequently silver beads take longer to solidify and to spit on

[1] Bannister and Stanley, "Thermal Properties of Cupels," *Trans. Inst. Min. and Met.*, vol. xviii., 1909, pp. 439-465.
[2] Fulton, *loc. cit.*, p. 103.

magnesia than on bone-ash cupels, after being withdrawn from the same muffle temperature. Silver beads are also much less liable to spit on magnesia than on bone-ash cupels, and the nature of the spit is different in the two cases, the spit in the former case generally taking the form of a frosty appearance only instead of the well-known "vegetation" obtained with bone-ash. These important differences in the properties of magnesia and bone-ash cupels are not always recognised by assayers.

An assayer, when asked to test magnesia cupels, usually puts half a dozen in the muffle with half a dozen bone-ash, and cupels under conditions suitable for the bone-ash, with the result that he forms an unfavourable opinion of the magnesia cupels. When using magnesia cupels under these conditions the lead is very liable to "freeze," and the results are unsatisfactory; but if the proper conditions for magnesia cupels are employed, the results are quite as satisfactory as those obtained with bone-ash. Although a somewhat higher temperature is required for cupellation on magnesia cupels, the tendency is to have a very much higher temperature than is necessary, in order to make the cupelling lead look like that on bone-ash cupels; and this is a great mistake, whereby one of the most important advantages of the magnesia cupels is lost. The loss of silver due to absorption is usually less with magnesia than with bone-ash cupels (see page 177).

Portland Cement Cupels.—The behaviour of Portland cement cupels during cupellation is very similar to that of bone-ash cupels, although, as shown on page 177, the loss by absorption is slightly higher. When using cupels made entirely of Portland cement for gold assays, especial care must be taken to thoroughly clean the buttons, otherwise when subsequently parting in nitric acid insoluble silica will remain adhering to the cornet and be weighed as gold. This difficulty may be overcome by "facing" the cement cupel with bone-ash. In this case the cupel mould is filled about two-thirds full with cement, and bone-ash added to fill the mould, the cupel then being finished in the ordinary way.

Cement cupels are also made from mixtures of Portland cement and bone-ash, the proportions being usually equal parts of each.

Influence of Base Metals on Cupellation.—The materials which are submitted to cupellation may be divided into two main classes, viz. :—

1. Material containing a very small proportion of precious metal and a large excess of lead, such as lead buttons from the assay of ores, etc.

2. Material containing a large proportion of precious metal and requiring the addition of a suitable quantity of lead for cupellation, such as bullion.

Various base metals may be present in greater or smaller quantity in any of the materials that can be submitted to cupellation, and may exert considerable influence on the results. When metals such as tin, antimony, and iron are present, the oxides of which dissolve only sparingly in molten litharge and produce a "scoria" or corrode the cupel, they should be previously removed by scorification. The lead buttons from the assays of ores are, as a general rule, practically free from base metals, as it is the usual practice to scorify lead buttons which may unavoidably contain an appreciable quantity of such metals. The materials of the second class may contain comparatively large proportions of base metals, notably copper and zinc, and the satisfactory removal of these during cupellation requires a careful adjustment of the lead. The behaviour of the lead button when hammered usually affords an indication of the presence of base metals and other impurities. Copper, iron, and tin tend to make lead hard, while antimony, zinc, sulphur, and arsenic tend to make it brittle. Lead buttons may also be brittle from the presence of lead oxide due to the employment of too low a temperature for the fusion of the ore. The appearance of the cupel will also frequently indicate the presence of base metals as described below, and when the cupel shows signs of these in objectionable quantity it is usually more satisfactory to repeat the assay and scorify, so as to remove them before cupellation. When lead is the only oxidisable metal present the stained portion of the cupel is yellow when cold except for a small, dark grey patch where the button has rested. The influence of various base metals, etc., on cupellation is summarised below.[1]   It should be noted that copper exerts its greatest influence towards the finish of the operation, while the influence of the more easily oxidised metals, such as antimony, tin, etc., is exercised at the commencement of cupellation.

1. *Antimony* does not interfere if less than 2 per cent. is present.   If 4 per cent. is present a slight yellow scoria of lead antimonate is formed, and the cupel is stained dark brown.   The margin of the cupel is sometimes cracked when the antimony exceeds the amount stated.

2. *Arsenic* at first burns with a blue flame and causes spurting ; after a few minutes the flame turns greenish-white and becomes very brilliant.   The cupel is stained pale brown, and if 4 per cent. of arsenic is present, there is much pale brown scoria.

3. *Bismuth* is oxidised and absorbed by the cupel in the same way as lead. Argentiferous bismuth may be cupelled alone, but towards the end of the operation the bismuth appears to act in a similar manner to tellurium, although to less extent, the alloy of bismuth and silver wetting the surface of the cupel . and being partially absorbed owing to the breaking down of the surface tension of the metal.   This action does not appear to take place with comparatively large buttons of silver.

4. *Cadmium* causes a black sooty ring to form inside the cupel near its margin.   When cupellation begins, the action is very violent, a yellowish-red flame is seen, followed by much spurting and the appearance of large, red-hot scales floating on the lead.   Cadmium oxide is not absorbed by the cupel.

5. *Copper.*—The copper oxide formed during cupellation is dissolved in the molten litharge, and the mixed oxides absorbed by the cupel, provided the lead is present in proper proportion.   If, however, the lead is not present in sufficient quantity, all the copper will not be removed, and the button of silver (or gold) still retaining copper will be found covered with a black coating of copper oxide. The proportion of copper carried into the cupel by litharge varies not only with the temperature, but even for the same temperature, according to the ratio of copper to lead in the alloy.   Copper does not oxidise as readily as lead, and tends to concentrate in the button ; consequently the lead, which may originally have contained only a small percentage of copper, becomes relatively richer in copper as the operation proceeds.   This is an important fact, as experience shows that most of the copper is eliminated during the last stage of cupellation, just before the lead is worked off.   It is on this account that the ill-effects of the copper are most marked at the close of the operation, and that copper oxide is found accumulated around the button of precious metal when there is insufficient litharge to remove it.   Copper requires at least twenty parts of lead to carry it into the cupel, and in practice a much larger ratio of lead to copper is employed to eliminate the copper in gold and silver alloys (see Chapter XXI.).   Since, as shown above, the removal of copper takes place quicker when the ratio of copper to lead is not too great, it is evident that two or three successive cupellations with small quantities of lead will be more effectual than a single cupellation with a larger quantity, and this method of procedure is sometimes adopted by assayers.   Copper, if not completely removed, has a very marked effect on

[1] See "Notes on Cupellation and Parting," by T. K. Rose, *Chem. Met. and Min. Soc. of S. Africa*, Jan. 1905.

the appearance of the silver button : the metal is spread out, is ragged at the edges, and adheres firmly to the cupel. Even when the cupellation has been satisfactory the lower surface of the silver button is flatter, and adheres more firmly to the cupel than buttons from the cupellation of silver free from copper. A small quantity of copper is persistently retained by both silver and gold at the end of the cupellation, especially by gold buttons from which the last traces of copper cannot be entirely eliminated. Copper colours the cupel a dirty green or blackish-green according to the amount of copper present. Small quantities of copper give a greenish tint. The intensity of the green coloration affords a useful indication of the amount of copper present, and is utilised by experienced assayers to get an idea of the approximate percentage of copper in the material being assayed.

6. *Iron.*—Lead buttons containing iron melt slowly. Iron oxide is not readily fusible with lead oxide, and a brown scoria is sometimes left on the cupel, which may entangle lead globules and so contain precious metal. The lead may contain about 4 per cent. of iron without the formation of any scoria. The cupel is stained dark red, and is moderately corroded.

7. *Nickel* forms a dark green scoria and greenish stain. Its effect on cupellation is similar to that of iron.

8. *Platinum* and the so-called platinum metals give a greenish stain and very crystalline buttons. The surface of the button is dull after the cupellation has terminated, and is crystalline even when very small quantities of platinum are present. The effect of platinum and of the so-called platinum metals on cupellation is more fully discussed in Chapter XXIV.

9. *Tellurium.*[1]—Tellurium has a great affinity for gold and silver, and exerts considerable influence in cupellation when there is insufficient lead to remove it. The effect of tellurium is to very greatly decrease the surface tension of the lead-gold-silver alloy.

Tellurium is removed comparatively slowly during cupellation, and if towards the end there is sufficient left to amount to anything approaching equality of the gold, or gold plus silver, then the surface tension of the globule breaks down completely and the alloy spreads over a wider area, "wets" the cupel, and is completely absorbed. This occurs in the case of bone-ash cupels when the tellurium in the button equals the gold, or gold plus silver, and the lead does not exceed ten times the tellurium.

When the lead exceeds this amount the behaviour is intermediate between the extreme cases of complete absorption and perfect cupellation. Partial subdivision of the lead on the cupel then takes place. This has long been recognised as a possible occurrence when tellurium is present.

It is brought about by portions of the alloy becoming detached at the line of contact between the cupel and the surface of the molten lead during cupellation. The presence of tellurium causes the surface tension of the lead to be weakened, with the result that it tends to adhere to or "wet" the face of the cupel. Portions of the alloy become detached and are either absorbed or leave minute beads on the cupel.

When the gold and tellurium are equal in amount and the lead amounts to 80 times the tellurium, very little loss occurs on a bone-ash cupel, and this loss is not appreciably decreased by additional lead. It is evident, therefore, that in all ordinary cases the button of lead from a crucible fusion of telluride ore weighing 30 grammes or more contains ample lead for cupellation. In the case of silver, more lead is required (from 100 to 120 times instead of 80) to remove the same amount of tellurium as from gold and leave a rounded bead.

[1] S. W. Smith, "Behaviour of Tellurium in Assaying," *Trans. Inst. Min. and Met.*, vol. xvii. 1908, p. 463.

Tellurium appears to be held more tenaciously by silver, and the fact that it comes out to a considerable extent in the later stage of cupellation is shown by the white patch of tellurium dioxide which underlies the silver bead.

Silver undoubtedly protects gold from possible losses due to the presence of tellurium, and even when present in amount only equal to that of the gold a considerable saving is effected. With an increase in the proportion of silver the protection is more complete. Tellurium being less easily oxidisable than lead is removed comparatively slowly, and tends to concentrate during cupellation, as shown by the following results by S. W. Smith, obtained by analysing a series of partially cupelled buttons originally consisting of,

$$
\begin{aligned}
&\text{Lead} \quad . \qquad . \qquad . \qquad . \qquad . \qquad 20 \ \text{grammes.} \\
&\text{Tellurium} \quad . \qquad . \qquad . \qquad . \qquad 0\text{·}05 \ \text{gramme.} \\
&\text{Gold} \quad . \qquad . \qquad . \qquad . \qquad . \qquad 0\text{·}05 \quad \text{,,}
\end{aligned}
$$

(1) Cupelled to 11·860 grammes, *i.e.* 59 per cent. of its weight.
Amount of tellurium still remaining, 77 per cent.
In this case a very distinct concentration of tellurium has taken place.
(2) Cupelled to 4·205 grammes, *i.e.* 21 per cent. of its weight.
Amount of tellurium still remaining, 21 per cent.
At this stage the percentage of lead and tellurium which have been removed are equal.
(3) Cupelled to 1·635 grammes, *i.e.* 8·15 per cent. of its weight.
Amount of tellurium remaining, 5·88 per cent.
At this stage the tellurium is at last leaving the button at a greater relative rate than the lead.

Cupellation of lead buttons containing tellurium should be conducted at a low temperature to obtain the highest results, as it would seem that the tendency towards "wetting" and absorption is greater at higher temperatures, since the effect of tellurium on the cupellation loss is accentuated under these conditions.

Widely different results may be obtained with similar buttons cupelled under different conditions of temperature and draught. Bone-ash cupels show a greater tendency to become "wetted," giving rise to subdivision, than patent cupels. During cupellation tellurium is oxidised to the dioxide TeO_2.

10. *Tin* gives a brown scoria if it constitutes more than 3 per cent. of the lead button. In all cases it rises to the surface, oxidises, and forms a scoria which floats and delays cupellation, but is subsequently carried off by the litharge. The cupel is stained pale brown.

11. *Zinc* burns with a brilliant green flame at first, and is partly volatilised; it forms a voluminous white or pale yellow scoria which may or may not be subsequently removed by the litharge according to the amount of zinc present and the proportion of lead used. The cupel is deeply corroded.

Losses in Cupellation.—The weight of the button of silver or gold left on the cupel does not represent the actual amount of precious metal in the material submitted to cupellation, as certain losses, incidental to the operation, always take place. These losses may be divided into (1) losses by volatilisation, and (2) losses by absorption, the latter being by far the more important.

1. **Volatilisation.**—Although it has been proved by Makins[1] and others that loss by volatilisation does occur during cupellation, the loss is altogether inconsiderable, unless the temperature at which the operation is performed is much too high.

[1] *Journ. Chem. Soc.*, vol. xiii., 1860, p. 77.

2. **Absorption by the Cupel.**—The loss during cupellation by absorption by the cupel is by no means always inconsiderable. The amount of this loss varies under different conditions and is influenced by :—

(a) The temperature of cupellation.
(b) The amount of lead employed.
(c) The amount of precious metal present.
(d) The influence of gold and silver on one another.
(e) The nature and amount of the impurities present.
(f) The physical condition of the cupel.

Although much experimental data relating to the losses in cupellation, especially of silver, has been published, the results are frequently of little value to the assayer owing to the lack of essential details. Such data is of value only when it gives the amounts of lead and gold or silver, etc., used and describes the exact conditions under which the experiments were made. But even when these details are given the results must be regarded only as affording some indication of the losses that take place during cupellation, and not as constants that can be utilised by the assayer for the purpose of correcting the results of an assay. It must be emphasised that the losses during cupellation are dependent on so many conditions that each assayer should determine for himself the exact losses that take place with the cupels he uses and the conditions of working that he adopts.

Very frequently in assay practice the amount of the cupellation loss has to be carefully determined by the methods described subsequently and the loss duly allowed for in returning the assay results. The figures given in the accompanying tables have been selected from the results of more modern experimental work as representing the average losses which occur during cupellation under what may be regarded as normal conditions, e.g. a temperature of about 750°–775° C.

(a) **Effect of Varying Temperature.**—The loss of silver and of gold both by absorption and volatilisation is very much increased by the employment of high temperatures, as shown in the case of silver in Table XIII. and gold in Table XIV.

Table XIII.—Effect of Varying Temperature on Cupellation of Pure Silver. (Eager and Welsh.) [1]

Amount of silver.	Amount of lead.	Ratio of lead to silver.	Temperature.	Total losses.
milligrammes.	grammes.	approx.		per cent.
204·62	10	50 to 1	700° C.	1·02
205 00	10	50 to 1	775° C.	1·28
203·00	10	50 to 1	850° C.	1·73
203 00	10	50 to 1	925° C.	3·65
203·00	10	50 to 1	1000° C.	4·87

Results of Similar Experiments by Beringer. [2]				
400	20	50 to 1	About 750° C.	1·75
400	20	50 to 1	,, 1000° C.	4·32

[1] Lodge, *Notes on Assaying.*   [2] *Text-Book of Assaying.*

Table XIV.—Effect of Varying Temperature on Cupellation of Pure Gold.
(Eager and Welsh.) [1]

Amount of gold.	Amount of lead.	Ratio of lead to gold.	Temperature.	Total losses.
milligrammes.	grammes.	approx.		per cent.
201	10	50 to 1	775° C.	0·155
201	10	50 to 1	850° C.	0·430
204	10	50 to 1	925° C.	0·460
201	10	50 to 1	1000° C.	1·430
201	10	50 to 1	1075° C.	3·000
Results of Similar Experiments by Beringer. [2]				
1000	20	20 to 1	775° C. ?	0·137 [3]
1000	20	20 to 1	775° C.	0·192 [3]
1000	20	20 to 1	1000° C. ?	0·604
1000	20	20 to 1	1000° C.	0·620
1000	20	20 to 1	1000° C.	0·645

(*b*) **Effect of Varying Lead.**—The loss of silver increases as the quantity of lead employed is increased, as shown in Table XV., but if the quantities of silver and lead employed are increased in the same proportions, the loss slightly decreases, as shown in Table XVI. The experiments by the author were in all cases made on hand-made bone-ash cupels in a gas-fired muffle at a temperature of about 750° C.

Table XV.—Showing Loss on Cupellation of Pure Silver with Varying Amounts of Lead and Silver. (W. H. Kauffmann). [4]

(Temperature not stated, but probably about 750° C.)

Amount of silver, milligrammes.	Total loss of silver per cent.			
	With 5 grammes lead.	With 10 grammes lead.	With 15 grammes lead.	With 25 grammes lead.
25	2·14	2·63	2·69	2·09
50	1·43	2·23	2·14	2·16
100	1·30	1·61	1·68	2·12
200	0·86	1·24	1·40	1·74
Results obtained by the Author.				
Amount of silver, milligrammes.	With 3 grammes lead.	With 6 grammes lead.	With 9 grammes lead.	With 12 grammes lead.
500	0·875	1 140	1·240	1·625
750	0·803	1·000	1·093	1·136

[1] Lodge, *Notes on Assaying.*   [2] *Text-Book of Assaying*, 12th edit., 1910, p. 145.
[3] Highest and lowest results in ten cupellations.
[4] *Eng. and Min. Journ.*, 1902, lxiii. p. 829.

Table XVI.—*Showing gradually Decreasing Loss on Cupellation when Lead and Silver are increased in the same Ratio.* (Hambly.) [1]

Amount of silver in grammes.	Amount of lead in grammes.	Ratio of silver to lead.	Loss of silver per cent.
0·065	0·65	1 to 10	1·225
0·650	6·50	1 to 10	1·135
1·625	16·25	1 to 10	1·067
Results obtained by the Author.			
0·125	1·5	1 to 12	1·40
0·250	3·0	1 to 12	1·32
0·500	6·0	1 to 12	1·20
1·000	12·0	1 to 12	1·14

In the case of gold, Rossler [2] has shown that the loss during cupellation slightly increases with the amount of lead used, but is much less than in the case of silver, and the extent of the loss is dependent upon the quantity of gold cupelled. Some results obtained by the author are given in Table XVII.

Table XVII.—*Showing Effect of Varying Lead on the Cupellation of Pure Gold.* (E. A. Smith.)

Amount of gold in grammes.	Total loss of gold per cent.			
	With 3 grammes lead.	With 6 grammes lead.	With 9 grammes lead.	With 12 grammes lead.
0·250	0 113	0·123	0·137	0·145

(c) **Effect of Varying Silver.**—The figures in Table XV. also show the cupellation loss with a constant weight of lead and increasing quantity of silver. It will be seen that although the *actual* loss increases with the quantity of silver present, the *percentage* loss is greater on the smaller buttons than on the larger buttons.

Hambly's original figures (Table XVI.) are given in grains, and have been converted into grammes by the author.

(d) **Effect of Silver on the Cupellation of Gold.**—When gold is alloyed with silver the cupellation loss is diminished, the total loss of gold decreasing as the amount of silver present increases. This protective action of silver is shown by the results of experiments by Rose, [3] given in Table XVIII.

[1] *Chem. Gazette*, 1856, xiv. pp. 185-6.
[2] *Dingl. Polytech. Journ.*, ccvi. p. 185.
[3] *Eng. and Min. Journ.*, lxxix. p. 708.

Table XVIII.—Loss on the Cupellation of Gold in the Presence of Silver. (*T. K. Rose.*)

Amount of gold.	Amount of silver.	Amount of lead.	Temperature of cupellation.	Total loss of gold.
milligrammes.	milligrammes.	grammes.		per cent.
1	4	25	900° C.	1·20
1	6	25	900° C.	1·05
1	8	10	900° C. ·	0·90
1	10	25	900° C.	0·80
1	6	25	700° C.	0·45
1	10	25	700° C.	0·39
500	1250	10	900° C.	0·055
		Results obtained by Beringer.		
300	300	10	900° C. ?	0·157
300	600	10	900° C.	0·107
300	900	10	900° C.	0·057

(*e*) **Effect of Base Metals on Cupellation Loss** : *Copper.*—The absorption of precious metals by the cupel is always greater when copper is present. The reason of this is not thoroughly understood, but it is usually stated that the copper oxide (Cu_2O) acts as an oxygen carrier (see page 179) in some special manner at the end of the cupellation and facilitates the formation of a small quantity of silver oxide (or gold oxide) which is dissolved in the molten mixture of lead and copper oxides and is thus carried into the cupel.

The influence of small amounts of copper on the cupellation of silver and gold is shown in the following results obtained by Lodge, Tables XIX. and XX. The temperature of cupellation was 775° C.

Table XIX.—Influence of Copper on the Cupellation of Silver. (*Lodge.*) [1]

Silver.	Lead.	Copper.	Percentage of copper in lead.	Ratio of lead to copper.	Loss of silver.
milligrammes.	grammes.	grammes.			per cent.
202	10	0·0101	0·1	1000 to 1	1·05
203	10	0·0202	0·2	500 to 1	1 08
202	10	0·0303	0·3	333 to 1	1·29
202	10	0·0404	0·4	250 to 1	1·45
204	10	0·0500	0·5	200 to 1	Retained copper. .

Results obtained by the Author. (*The cupellation loss with the same quantities in the absence of copper is inserted for comparison*)

Silver.	Lead.	Copper.	Percentage of copper in lead.	Ratio of lead to copper.	Loss of silver.
grammes.	grammes.	grammes.			per cent.
0·5	3	None	...	...	0·875
0·5	3	0·04	1·333	75 to 1	1·005
0·5	6	None	...	...	1·140
0·5	6	0·04	0·666	150 to 1	1·380
0·5	9	None	...	...	1·240
0·5	9	0·04	0·444	225 to 1	1·335
0·5	12	None	...	...	1·625
0·5	12	0·04	0·333	300 to 1	1·700

[1] *Notes on Assaying.*

Table XX.—Influence of Copper on the Cupellation of Gold. (Lodge.) [1]

Gold.	Lead.	Copper.	Percentage of copper in lead.	Ratio of lead to copper.	Loss of gold.
milligrammes.	grammes.	grammes.			per cent.
202	10	None	...	...	0·155
202	10	0·0101	0·1	1000 to 1	0·19 [2]
201	10	0·0202	0·2	500 to 1	0·20 [2]
200	10	0·0303	0·3	333 to 1	0·13 [2]
201	10	0·0404	0·4	250 to 1	0·165
202	10	0·0500	0·5	200 to 1	0·250

Tellurium.—The important influence of tellurium during cupellation has already been dealt with on page 169. The absorption loss due to the presence of tellurium is a function of the temperature of cupellation. As previously stated, the presence of silver protects gold from possible losses due to the presence of tellurium.

Very few figures appear to have been published showing the loss on the cupellation of pure gold or of pure silver with tellurium. The influence of tellurium on the cupellation of gold in the *presence of silver* is shown in Table XXI.

Table XXI.—Influence of Tellurium on Cupellation of Alloys of Gold and Silver. (F. C. Smith.) Cupelled with 12 grammes of lead in every case.

Milligrammes.		Total alloy. Milli-grammes.	Tellurium added. Milli-grammes.	Loss by absorption.		Loss by volatilisation.	
Gold.	Silver.			Gold.	Silver.	Gold.	Silver.
				per cent.	per cent.	per cent.	per cent.
24·76	5·04	29·8	5·0	13·44	27·08	5·65	0·69
23·64	4·81	28·45	15·0	34·22	35·78	5·28	1·75
18·42	3·75	22·17	15·0	29·85	32·01	11·92	17·95

Antimony, Tin, etc.—Antimony and other easily oxidisable metals that tend to form a scoria during cupellation may cause a mechanical loss of silver or gold. When large amounts of scoria form, the gold and silver may be retained as minute beads, which cannot be collected and weighed. The formation of a scoria is, however, not necessarily accompanied by a loss of precious metal. During the cupellation, antimony is oxidised and combines with litharge to form lead antimonate, which is partly absorbed by the cupel and partly remains as a scoria according to the amount of antimony present. The exact losses in cupellation due to the presence of base materials were determined by Rose.[3] In each case 25 grammes of lead were cupelled with 1 milligramme of gold, 4 milligrammes of silver, and 1 gramme of the impurity. Bone-ash cupels were used. The results were as follows :—

---

[1] *Notes on Assaying*, p. 143.
[2] These are actual losses ; the buttons retained 0·16 per cent. copper.   [3] *Loc. cit.*

Impurity added.	Loss of gold. per cent.	Loss of silver. per cent.	Remarks.
No impurity . . .	1·2	12·8	
Antimony . . .	5·3	13·2	No scoria.
Arsenic . . .	3·9	16·3	Much scoria.
Bismuth	21·8	27·9	No scoria.
Cadmium . . .	3·5	13·1	Ring of black feathery scoria.
Copper	10·0	32·6	No scoria.
Iron	4·0	16·6	No scoria.
Nickel	...	...	Total loss due to scoria.
Tin	2·0	13·9	Slight ring of scoria.
Tellurium . . .	55·8	67·9	No scoria.
Selenium	54·1	64·5	No scoria.

In the case of tellurium and selenium practically the whole of the missing gold and silver was recovered by fusing the cupels. The percentage losses observed in these experiments appear very large, but it must be remembered that the absolute losses were very small. The percentage losses are much smaller when the precious metal beads are larger, as shown by the following results by Beringer :—

Impurity added.	Silver added.	Lead used.	Loss of silver. per cent [1]
	grammes.	grammes.	
No impurity added . . .	0·1	20	2·95 } 2·9
,, ,, . . .	0·1	20	2·85
With 0·5 gramme antimony .	0·1	20	3·30 } 3·2
,, 0·5 ,, ,, .	0·1	20	3·10
,, 0·5 ,, copper . .	0·1	20	5·75 } 4·9
,, 0·5 ,, ,, .	0·1	20	4·05

(f) **Porosity of Cupels.**—The amount of precious metal absorbed by a cupel is influenced by the porosity, and this is regulated by the differences in the relative cohesion of the materials employed in making cupels. The porosity of bone-ash cupels varies considerably owing to the differences in the composition and screen analysis of commercial bone-ash and in the amount of pressure used in making them, some assayers preferring to have them well compressed, while others prefer to have them more porous. Cupels made by machine, in which a constant pressure may be obtained, are always more uniform as regards density than cupels made by hand. Magnesia cupels are made by hydraulic pressure and are much denser than those made of bone-ash, consequently the absorption is, as a general rule, less on magnesia cupels. The nature of the surface of the cupels is also of importance, and no doubt has a considerable influence on the losses due to absorption. At present there is little evidence as to the comparative variations in the absorption of silver and gold by bone-ash and magnesia cupels respectively, but it would appear from the data published that the absorption is less with magnesia cupels than with bone-ash. Whatever cupels are used, the absorption should be tested frequently.

Table XXII. gives the results obtained by the cupellation of silver on bone-ash and magnesia (morganite) cupels. Pure silver of the amount given was in each case cupelled with 10 grammes of sheet lead. The temperature employed was sufficiently low to give "feathers."

[1] The highest and lowest results of three experiments in each case.

*Table XXII.—Comparison of Cupellation Losses with Bone-ash and
Magnesia Cupels. (Anderson and Fulton.)* [1]

Amount of silver taken.	Bone-ash cupels. Weight of silver bead obtained.	Loss per cent.	Magnesia cupels (morganite). Weight of silver bead obtained.	Loss per cent.
milligrammes.	milligrammes.	per cent.	milligrammes.	per cent.
5	4·85	3·0	4·80	4·0
5	4·94	1·2	4·89	2·2
10	10·00	Nil	10·00	Nil
10	9·68	3·2	9·86	1·4
15	14·36	4·3	14·50	3·3
15	14·70	2·0	14·80	1·3
20	18·92	5·4	19·52	2·4
20	19·98	0·1	19·68	1·6
25	24·60	1·6	24·84	0·64

Benner and Hartmann [2] compared different kinds of cupels with regard to
loss of silver by cupelling from 35 to 45 milligrammes of silver with 10 grammes
of lead, and found that cupels made of bone-ash of different grades and fineness
gave identical results within the limits of experimental error. The average
percentage loss of silver with different cupels under similar conditions was
as follows:—Morganite, 1·99; bone-ash, 2·36; brownite, 2·89; cement and
bone-ash (1 : 1), 2·95; casseite, 3·09; cement, 3·38.

The absorption of silver by cement cupels has been ascertained by Holt and
Christensen, [3] who tested cement cupels, bone-ash cupels, and cupels made of
a mixture of equal parts of cement and bone-ash.

In each case 100 milligrammes (0·10 gramme) of fine silver were cupelled
with about 20 grammes of lead. The cupellation was performed at an
"average" muffle temperature, and the actual temperature of the cupels
measured by inserting a thermo-couple into a hole bored beneath the cavity
of the cupel.

The results are tabulated in Table XXIII.

*Table XXIII.—Cupellation Losses with Cement Cupels.
(Holt and Christensen.)*

Average temperature of cupel.	Portland cement (V.S.). Loss per cent.	Portland cement (R.D.). Loss per cent.	Equal parts cement and bone-ash. Loss per cent.	Bone-ash. Loss per cent.
915° C.	1·30	1·34	1·21	1·26
925° C.	1·81	1·72	1·54	1·70
945° C.	2·53	2·56	2·42	2·42
965° C.	3·37	3·42	3·05	2·96

Table XXIV. gives the results of absorption loss with bone-ash and cement
cupels obtained by J. W. Merritt [4] on cupelling 10 milligrammes of fine silver
with 15 grammes of lead.

[1] O. A. Anderson and C. H. Fulton, see Fulton, *Manual of Fire Assaying*, 2nd edit.,
1911, p. 104.
[2] *Indust. Eng. Chem.*, 1911, iii. 805–807.
[3] "Experiments with Portland Cement Cupels," *Eng. and Min. Journ.*, 1910, vol. xc. p. 560.
[4] J. W. Merritt, "Cement versus Bone-Ash Cupels," *Min. and Sci. Press*, 1910, vol. c. p. 649.

Table XXIV.—Comparison of Cupellation Loss with Bone-Ash
and Cement Cupels   (Merritt.)

Temperature.	Absorption per cent.	
	Bone-ash.	Cement.
Light cherry heat (800° C. ?) .	4·62	4 91
Orange heat (900° C. ?) . .	6·38	6·64

Very few determinations of the absorption loss of gold during cupellation have
been recorded, but it is always much less than silver.

As shown in Table XIV., page 172, Beringer found in three cupellations of
1 gramme of gold with 20 grammes of lead made purposely at a very high
temperature, that the bone-ash cupels absorbed 6·04, 6·20, and 6·45 milligrammes
of gold respectively.   Hence at a high temperature there may easily be an
absorption loss of more than 0·5 per cent. of gold when cupelled with the
proportion of lead given.   In ten cupellations with the same quantities of gold
and lead, but at an ordinary temperature, the gold recovered from the cupels
varied from 1·37 to 1·92 milligrammes, and gave an average of 1·59 milligrammes,
or 0·159 per cent.

With ordinary gold-bullion assays the absorption of gold is stated by
S. Smith[1] to be about 0·5 part in 1000 for bone-ash cupels, and about 0·3
parts in 1000 in the case of morganite cupels.   It was found that this absorption
difference for cupels of these materials was practically constant.

Theory of Cupel Absorption.—Although the absorption of precious metals by
the cupel is the predominant cause of loss during cupellation, very little has
been published as to the nature of this loss.   There seems little doubt, however,
that the absorption is due partly to the infiltration into the cupel of minute
particles of lead alloy, and partly to the formation of small quantities of the
oxides of silver and gold, which are dissolved in the molten litharge and thus
absorbed.   The loss by absorption occurs mainly towards the end of the
cupellation.

The figures in Table XXII., page 177, showing the absorption with cupels of
different materials, clearly indicate that the physical condition of the cupel as
regards porosity influences absorption ; but that the absorption is not entirely due
to the infiltration of the molten alloy, is evident from the following experiment
by Beringer.[2]

A cupel on which an alloy consisting of 0·80 gramme of silver, 0·47 gramme
of gold, and 25 grammes of lead had been cupelled, was found to contain 7·5
milligrammes of silver, and rather less than half a milligramme of gold.

Assuming, for the sake of argument, that the half-milligramme of gold had
filtered into the cupel in the form of minute drops of alloy, it would have been
accompanied by less than a milligramme of silver, and the presence of the extra
6 or 7 milligrammes of silver must have been due to a different cause.   Similar
differences between the ratio of silver to gold in the alloy submitted to cupellation, and the ratio of silver to gold recovered from the cupel, will be noted in the
experiments by F. C. Smith, page 175.   There can thus be little doubt that
the greater part of the absorption loss must be due to some cause other than
the mere filtration of the fused alloy into the cupel, as it is difficult to ascribe
the great change in the proportion of the metals to this cause.

[1] Sydney Smith, Trans. Inst. Min. and Met., 1909, vol. xviii. p. 460.
[2] Text-Book of Assaying, 11th edit., 1910, p. 101.

This greater part of the loss is ascribed to the formation of oxides of silver and gold, and from the above results it would appear that silver is oxidised to a greater extent than gold.

Both metallic silver and gold are converted into oxides when heated with the higher oxides of lead, copper, and some other metals, and under the conditions existing during cupellation oxides of the precious metals are undoubtedly formed and carried into the cupel by the molten litharge.

As, however, according to H. Rose and other authorities, silver oxide is decomposed between 250° C. and 300° C. into silver and oxygen,[1] and as the litharge formed during cupellation is subjected to a temperature considerably above this, the existence of silver oxide might be called in question. But although silver oxide is decomposed at a temperature considerably below redness, there is evidence that it is not reduced to metal by heat alone, when mixed with an excess of lead oxide.

In favour of the view that silver is capable of existing as oxide in litharge are the observations of St Clair-Deville and Debray, and those of Troost and Haute-Feuille,[2] that silver oxide can be formed at high temperatures. In addition, Wait[3] has extracted 18·67 and 19·25 per cent. of the silver from litharge containing 2·94 per cent. of silver by means of acetic acid, and as metallic silver is insoluble in acetic acid it may be assumed that the metal must have been present as oxide. The limiting temperature below which silver oxide is stable must consequently be regarded as not yet definitely settled.

In the case of gold, Rivot states that it is oxidised to some extent at a red heat in the presence of litharge, cupric oxide, or antimonic oxide, and that it is the oxidised part which is absorbed by the cupel. It is to be noted that the absorption loss of both silver and gold is greater in the presence of copper oxide.

The oxidation of the precious metals during cupellation may be attributed either to the direct action of atmospheric oxygen or to the action of oxides of base metals, which act as "oxygen carriers" and convey oxygen to the precious metals.

The latter is the explanation usually given to account for the formation of precious-metal oxides during cupellation; but although this is a very possible explanation, further research is necessary before finally accepting the view that the oxidation is due to the action of base-metal oxides rather than to free oxygen.

This conclusion is hardly borne out by the results of the experiments by Dr Rose[4] on the oxidation of base metals in argentiferous gold bullion. The question of oxygen carriers is discussed by Rose, but he remarks that "the whole trend of the results of the work is to show that it is silver that carries oxygen to base metals, not the other way round, and that the oxidation of silver does not take place to any extent until the base metals have been almost entirely removed. This has not been previously recognised, but nevertheless seems to be a perfectly sound generalisation, and is of importance both to the assayers and refiners."

As already pointed out, the amount of precious metal absorbed by the cupel is greatly influenced by the temperature at which cupellation is conducted, a comparatively slight increase of temperature causing a marked increase in absorption. Whether this increased absorption is due to an increased oxidation of the precious metal, or to a decrease in the surface tension of the lead alloy, is difficult to determine. That the presence of certain base metals in a lead button

[1] Schnabel, *Metallurgy*, vol. i., 1905 edit.
[2] Graham-Otto-Michaelis, *Anorg. Chemie*, 1884, p. 985.
[3] *Trans. Am. Inst. Min. Eng.*, vol. xv., 1886, p. 423.
[4] "Refining Gold Bullion with Oxygen Gas," T. K. Rose, *Trans. Inst. Min. and Met.*, vol. xiv., 1905.

carrying gold and silver is accompanied during cupellation by a weakening of
the surface tension of the molten metal and corresponding increase in absorption
is proved by the experiments of S. W. Smith, described on page 169.

Determination of Cupellation Loss.—In assaying rich ores and alloys of
gold and silver it is necessary to make a correction for the loss during cupellation, and this is determined either by fusing the cupel and actually recovering
the precious metal absorbed or by the use of "check" or "proof" assays worked
off under the same conditions as the material being assayed. The quantity of
silver and gold absorbed by the cupel in the case of ore assays is comparatively very
small, and is not generally determined except when great exactness is required.

Assay of the Cupel.—All clean bone-ash is broken off the cupel and rejected,
and the stained portion crushed so as to pass through an 80-mesh or 100-mesh
sieve. Fine crushing is necessary, as bone-ash, owing to its infusible nature,
does not combine with fluxes to form a slag, but remains suspended in the
fused mass. The charge is made up as follows for bone-ash cupels:—

Cupel (bone-ash) . . . . 40 to 60 grammes.
Sodium carbonate . . . . 30 „
Borax . . . . . . 50 „
Litharge . . . . . 50 „
Argol . . . . . . 2·5 „

For magnesia cupels the following charge may be used:—

Cupel (magnesite) . . . 40 to 60 grammes.
Sodium carbonate . . . . 20 „
Borax . . . . . . 20 „
Litharge . . . . . 40 „
Silica . . . . . . 15 „
Argol . . . . . . 2·5 „

Cement cupels may be fluxed with the following charge:—

Cupel (cement) . . . . 40 to 60 grammes.
Sodium carbonate . . . . 20 „
Borax . . . . . . 40 „
Silica . . . . . . 15 „
Argol . . . . . . 2·5 „

If other reducing agents are used instead of argol, the amount added should be
sufficient to give a lead button weighing from 30 to 35 grammes.

The addition of fluor-spar is frequently recommended as a flux in fusing
cupels and is advantageous in producing a fluid slag, but it is not always easily
procured. The litharge in the bone-ash acts as a flux. The mixture is fused in
a crucible and the resulting button of lead cupelled. The weight of the silver
button thus obtained is added to that of the silver obtained from the assay of
the ore. To save time and labour, the powdered stained portion of the cupel
may be added to the slag from the ore fusion, with any regulus or scoria, and
the whole assayed for silver and gold with the addition of fluxes as circumstances
may require. The "cleaning" of slags is described on page 155.

The following method of determining the cupellation loss, by which a fusion
of the cupel is obviated, is given by Beringer.[1] It may be used for determining
the cupellation loss in assaying argentiferous lead and rich ores and products.
The following example is given.

Suppose we have an alloy of silver and lead in unknown proportions, and that
by cupelling two lots of 10 grammes each we get silver weighing (1) 0·1226

[1] Beringer, *Text-Book of Assaying*, 11th edit., 1910, p. 104.

gramme, and (2) 0·1229 gramme. We should know from general experience that the actual quantity of silver present was from 2 to 4 milligrammes more than this. To determine more exactly what the loss is the two silver buttons are wrapped up each in 10 grammes of sheet lead and cupelled side by side with two other lots of 10 grammes of the original alloy. If now the buttons (1) and (2) weigh 0·1202 gramme and 0·1203 gramme respectively, they will have suffered in this second cupellation an average loss of 2·5 milligrammes. Suppose the two fresh lots of alloy gave 0·1233 gramme and 0·1235 gramme of silver, the average loss on these would also be 2·5 milligrammes. Add this loss to each result and take the mean, which is in this case 0·1259 gramme, thus giving 1·259 per cent. as the correct result for the silver in the alloy.

A similar method is also recommended by S. Smith[1] for determining the cupellation loss in bullion assays.

Considering the simplest case of the gold-bullion assay, where the gold is nearly pure and the weight of the gold, silver, copper, and lead taken for the assay is known exactly.

A certain number of such assays are cupelled, and after cupellation the buttons weighed and the loss made up by the addition of more copper and lead. The assays are then recupelled side by side with a fresh set of similar assays and the resulting buttons parted in nitric acid in the ordinary way. The only difference between the two sets of cornets will be that one set has been cupelled twice and the other set only once. The difference is the cupellation loss.

In the case of bullion assays this is, according to S. Smith, the simplest method of determining the cupellation loss.

Checks or Proofs.—The method adopted in the assay of gold and silver alloys is to assay "check" or "proof" pieces of pure silver or gold side by side with the metal being assayed, so that they are subjected to exactly the same conditions, and to assume that the loss of weight on the "checks" is the same as that experienced by all the assays worked with them. This method of working is termed "assaying by checks or proofs," and may be illustrated by the following example. Suppose we wish to determine the actual amount of silver in a sample of comparatively pure commercial silver. Two lots of the sample of 1 gramme each are weighed and each wrapped in 5 grammes of sheet lead. One check consisting of 1 gramme of pure ("proof") silver is also accurately weighed and wrapped in 5 grammes of sheet lead. The two samples and check are then cupelled as nearly as possible under the same conditions, the cupels being placed in one row of three across the muffle and the check charged into the middle cupel. Supposing the resulting silver buttons weighed as follows :—

Sample.	Check.	Sample.
0·9968.	0·9992.	0·9964.

The loss on the check is therefore 1 − 0·9992 = 0·0008 gramme.

The average weight of silver from the sample is 0·9966, and assuming that the loss on the sample is the same as that on the check, the corrected result will give 0·9966 + 0·0008 = 0·9974 silver in the sample, or 99·74 per cent. When the alloy submitted to assay contains copper as well as silver it is necessary to add about the same quantity of copper to the checks as is supposed or known to be present in the samples. When the composition of the sample is quite unknown a preliminary assay must be made to determine this, as directed in Chapter XVIII., page 276, where the subject of checks for gold and silver alloys is more fully discussed.

Remarks and Further Details of Manipulation.—*Spurting of Cupels.*—As previously stated, "spurting" of the molten lead may take place at the commencement of cupellation if the cupels contain carbonaceous matter as an impurity in the bone-ash. Little fountains of metal are thrown up, and some part may be ejected from the cupel and occasion serious loss if the lead contains much silver or gold. This "spurting," or "bumping," as it is sometimes termed, is, however, only occasionally due to the escape of carbon-dioxide produced by the oxidation of the carbon remaining in the bone-ash. "Spurting" more frequently occurs when, during the heating up of the muffle into which the cupels have been placed, the draught through the muffle has been shut off by tightly closing the door. Under these conditions the atmosphere of the muffle becomes reducing, and the reducing gases are absorbed by the bone-ash cupels. When the lead is put in and cupellation commences in an oxidising atmosphere, the absorbed gases escape and bubble through the molten lead and cause the spurting. It may also be caused partly by the escape of carbon-dioxide resulting from the reduction of litharge by the reducing gases when cupellation commences. When cupels are allowed to heat up in the muffle and an oxidising draught is kept going all the time, the spurting does not occur. The author has observed that spurting occurs much more frequently with gas muffles than with coke- or coal-fired muffles, especially if excess of gas is turned on in heating up the muffle. Sperting seldom occurs if the door of the muffle is left slightly open when heating up, so as to allow of sufficient draught of air to keep the atmosphere of the muffle oxidising as stated. The presence of calcium carbonate ($CaCO_3$) in bone-ash also causes "spurting," as it begins to give off CO_2 at the temperature of cupellation. Firth[1] observed considerable spurting with cupels made from bone-ash containing nearly 8·0 per cent. of calcium carbonate.

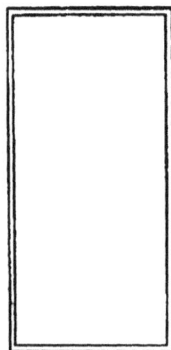

At the finish of the cupellation, when the temperature is raised for the brightening of the buttons, the door of the muffle should not be closed too tightly, otherwise the atmosphere becomes reducing and the last traces of lead are not eliminated and the brightening is delayed. This is especially liable to occur with gas muffles if the gas-supply is not properly regulated.

Within recent years special apparatus has been introduced for the quick manipulation of a large number of assays. The following are used in connection with cupellation.

FIG. 102.

Cupel Tray.—To facilitate the charging in and withdrawal of a large number of cupels, a cupel tray is sometimes used. A cupel tray used at the Royal Mint, London, is shown in plan and section in fig. 102. It is 11¾ inches long and 6 inches wide, and holds cupels for 72 assays, the cupels fitting tightly inside the rim. The cupels are in sets of four, as described on page 48. The tray is made of "Salamander" graphite, and lasts for some weeks. Iron trays were found to interfere with cupellation, and to be rapidly destroyed. Fireclay trays soon break. The tray is sprinkled with bone-ash, and the cupels placed on it before it is charged into the muffle. It is charged in and withdrawn by an iron tool provided with two flat prongs, 1¼ inches wide, and of the same length as the tray, which slide into the grooves underneath the tray. At some assay offices, instead of using a tray, a large number of cupels are made in one block, but the difficulty of maturing the cupels, and the risk of breakage, increases with the size of the block. The advantage of the cupel tray is that single cupels can be rapidly arranged on the tray by hand, before it is put into the muffle. This is an important advantage if the "charging" device described below is used

[1] A. T. Firth, *Journ. Chem. and Met. Soc. of S. Africa*, 1903–4, p. 176.

for charging in the assays, as it is necessary in this case to have the cupels carefully placed in position.

Charging Tray.—A device called a "charging tray," for charging a number of assays on to the cupels at one time, is now in use in many assay offices. It is best made of nickel, and consists of a top plate with holes corresponding exactly to the position of the cupels in the muffle. Underneath is a sliding plate with holes corresponding exactly with those of the upper plate. The lead buttons or assay pieces to be cupelled are placed in order in the holes of the upper plate,

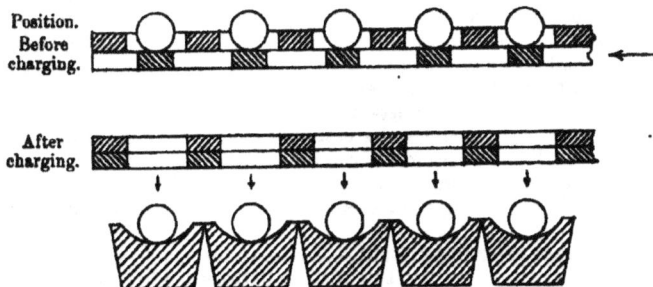

Fig. 103.—Cupel-charging tray.

and rest on the lower plate before introducing the instrument into the muffle. When it is placed in position over the cupels, which have been properly arranged in the muffle, the lower plate is pushed forward to a stop point, thus bringing the apertures of the two plates into register, and allowing the lead buttons to drop into the cupels. The action of the charging tray is diagrammatically shown in section in fig. 103. The instrument is provided with a long handle and arrangement for working the sliding plate, and is generally fitted with two side pieces which rest on the floor of the muffle, and thus ensure the plates being at the proper distance above the cupels, and also give some help in placing the instrument into position. When block cupels are used, the lower plate of the charging tray has no holes, and is drawn right out when charging the assays.

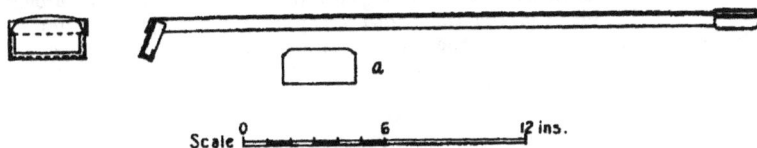

Fig. 104.

Cupellation Mirror.—When cupelling a large number of assays a little difficulty is experienced in ascertaining when all the cupellations have finished, and to obviate this difficulty the simple device shown in fig. 104 was introduced some years ago by Mr Henry Westwood, the Assay Master of the Birmingham assay office, and is now in use in several assay offices It consists of a small mirror of plate glass held in a sheet-iron frame fitted to a long handle at an angle of 30° to the perpendicular, as shown in fig. 104. The mirror is kept in position by a piece of asbestos card at the back, which also protects it from being blistered by the heat. The mirror is held in the muffle and passed quickly from left to right over each row of cupels in succession, the angle of the mirror allowing a reflection of the buttons to be seen, and the progress of the cupellation ascertained.   With a little experience the observation can be made very rapidly, and excessive heating and consequent blistering of the mirror prevented.   The mirrors, cut to shape (a, fig.104), should be kept in stock ready for use, but with care one mirror can be used daily for many weeks before requiring to be replaced.   The mirrors should be of plate glass, as thinner glass is very liable to crack when heated.

Uniformity of Temperature.—To secure uniformity of temperature of the muffle during cupellation as far as possible, the heating of the gas muffles at the Royal Mint,[1] London, has recently been improved by enlarging the combustion chamber round the muffle, and filling the entire space with clay fire-balls resembling those used in ordinary gas stoves.   These act as baffles to the flame, ensuring more complete combustion of the gas before it leaves the furnace.   As a result it has been found that, apart from a saving in gas, the furnace is heated more rapidly and is ready for use in less time after being lighted, and that the temperature is more uniform and under better control, so that the quality of the work is improved.   When a new muffle has been fitted in a furnace it should be tested for uniformity of temperature by cupelling a batch of proof assays of pure silver and carefully comparing the losses that take place in different parts of the muffle.

Cooling of Assays after Cupellation.—In cooling the muffle to prevent spitting of silver buttons due to dissolved oxygen,

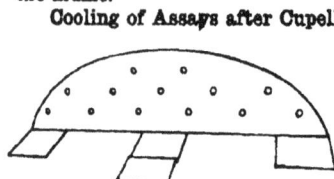

the rate of cooling should be such that the cupels cool at the same rate as the muffle. The cupels retain the heat longer than the muffle, and if the cooling is too rapid, the buttons will solidify on the exposed surface while the interior is kept molten by contact with the hot cupel, and when the button solidifies, spitting is very liable to take place.

FIG. 105.

If, however, the cupel is slowly cooled from below, the under surface of the button will solidify first and the dissolved oxygen escape before the silver solidifies as a whole.   The rate of cooling is of much more importance than the avoidance of draughts, although, of course, excessive draughts should be avoided.   For many years the author has cooled down in a gas muffle batches of thirty and forty 1-gramme silver bullion assays with a gentle draught of air passing through the muffle the whole of the time, with exceptionally few spits.   After all the buttons have brightened, the top half of the muffle door is removed, and the small perforated sheet-iron "grid," shown in fig. 105, inserted in its place and left in position until all the buttons have solidified.   The gas is turned down gradually, at about ten-minute intervals, to certain points ascertained by experiment.   The silver buttons cooled under these conditions are always brighter than those cooled in a muffle from which air-currents are excluded.

Systems of Working.—The importance of system in conducting an assay laboratory cannot be too strongly emphasised.   If assays are not made in some definite order, and a regular system adopted, confusion is sure to result sooner or later, and when one assay gets displaced it throws doubt on all the rest.   It does not follow that a system that is satisfactory for the particular work of one laboratory is suitable for all laboratories.   Each assayer must adopt a system best suited to his particular work, but when once the method of procedure is determined it should be strictly adhered to by all who work in the laboratory. With a good system of working, an experienced assayer has little difficulty in following a large number of assays through all the various operations without confusion, but beginners may find it an advantage to mark the scorifiers, etc.,

[1] *Fortieth Report of Royal Mint*, 1909, p. 53.

so as to identify them. As soon, however, as the system has been learned, the marking should be discontinued. Each day's work should be systematically numbered. The general practice is to work always from left to right in the handling of the assays in the various operations. This systematic order of arrangement of scorifiers, cupels, etc., is kept up, both in and out of the furnace, and the routine of working is never varied, so that a sample can always be identified by its relative position. When a large number of assays of the same kind have to be made, the samples are first placed in order on the balance bench to the left hand of the assayer, ready for weighing. He then takes the first sample, weighs up the quantity required, transfers it to a scorifier or crucible arranged on a tray on his right hand, and duly notes its position. As the weighings are finished, the samples are placed in the same order on his right hand out of the way. The order in which the samples are weighed for assay, and the amount taken, with any other details that may be necessary, should be recorded in the assayer's book as soon as each sample is weighed. These details should never be entrusted to memory alone.

One of the most frequently adopted systems of handling a number of assays from the weighing of the charge to the weighing of the gold or silver buttons is illustrated in the accompanying sketches. The illustrations apply to the handling of fifteen assays, but the system is equally applicable to a larger number. Let it be assumed that seven ore samples, marked A to G, and some " metallics " belonging to sample G, are to be assayed in duplicate by scorification, and that (for convenience of description) the samples are taken in alphabetical order. The charges are weighed and transferred to their respective scorifiers arranged on a tray, the first charge weighed being placed at the left-hand back corner, then the duplicate assay and the rest in succession. Sometimes more than two assays are made of the same sample, but however many assays are made, all the quantities required for that sample should be weighed out before another ore is proceeded with. With duplicate assays the record in the assayer's book would read as follows :--

Assay Number.	Sample.	Assay Number.	Sample.
1	A.	9	E.
2	A. Duplicate of A.	10	E. Duplicate of E.
3	B.	11	F.
4	B. Duplicate of B.	12	F. Duplicate of F.
5	C.	13	G.
6	C. Duplicate of C.	14	G. Duplicate of G.
7	D.	15	G. " Metallics of G."
8	D. Duplicate of D.		

When all the samples are weighed they will occupy the positions on the tray shown in fig. 106. The tray is carried to the furnace room and the assays put into the muffle in the order shown in fig. 107, assay No. 13 being put in first and placed at the left-hand back corner of the muffle, then No. 14, and the rest as shown, working always from left to right. When scorification is finished the assays in the front row are naturally withdrawn first and poured into the ingot mould in the order shown in fig. 108, thus bringing the assays into the original order again. After cleaning the buttons from slag they are returned to the same position in the tray, and the same method of procedure adopted for cupellation as for scorification. During cupellation the buttons will be in the reversed order, as in fig. 107, and when withdrawn from the furnace they are placed again in their proper sequence, as in fig. 108, and are ready for weighing. If the muffle is too

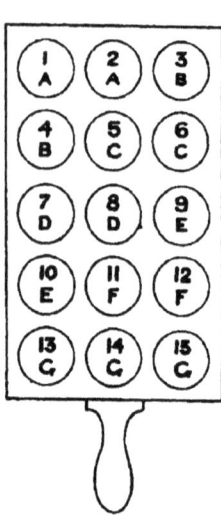

Fig. 106.

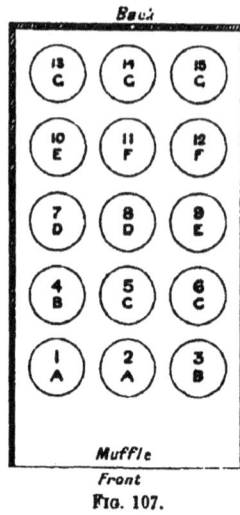

Fig. 107.

Fig. 108.

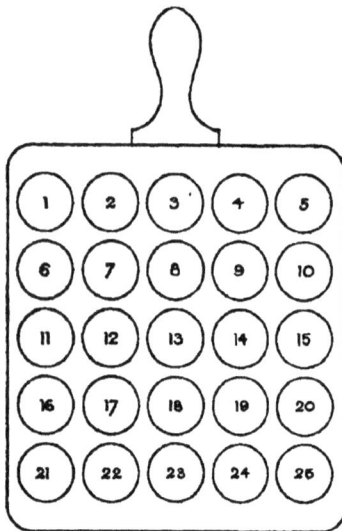

Fig. 109.

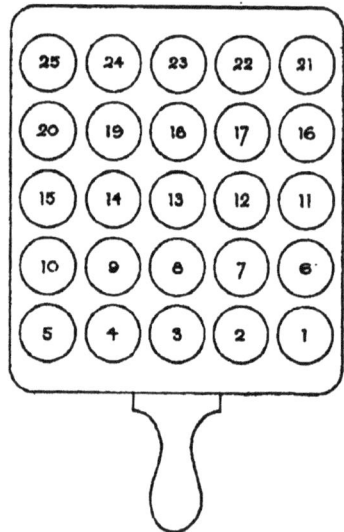

Fig. 110.

small to take all the scorifiers or crucibles, they are divided between two muffles, but the same system of loading and unloading the muffle is adopted.

Another method adopted at a large smelting works visited by the author is illustrated in figs. 109, 110, 111, 112. Large muffles capable of holding twenty-five 3-inch scorifiers are used. The method differs from that described in the fact that the order of the assays is reversed by turning the tray before placing the scorifiers into the muffle, so that in the muffle they occupy the same position as when weighed out. In this case the charges as they are weighed out are placed on a tray in the position shown in fig. 109, *the handle of the tray being away from the operator as shown.* When carried to the furnace *the tray is turned round,* so that the assays occupy the positions shown in fig. 110. Assay No. 1 is then put into the muffle first and placed in the left-hand back corner, then No. 2, and so on, until all the assays have been transferred, and occupy the positions shown

MUFFLE

Fig. 111.

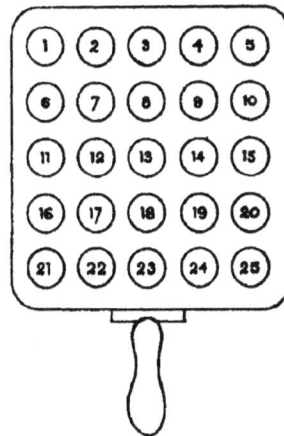

Fig. 112.

in fig. 111. When the assays are withdrawn, No. 21 is poured first into the mould in the left-hand front corner of the mould, and the rest in succession, as shown in fig. 112. By this arrangement the assays occupy the original order of sequence in the muffle, both for scorification and for cupellation. This is a great advantage, since it enables the exact position occupied in the muffle by any particular assay to be easily ascertained, which is very desirable, especially in the case of the cupellations where temperature variations are of such importance.

For cupellation assays only many assayers charge the assays into the muffle in the same order as that which they occupy in the tray after weighing, as in fig. 106. In this case assay No. 1 would be put into the muffle first, and placed in the cupel in the left-hand back corner, and then the other assays in succession. When withdrawn from the muffle the assays are returned to their respective places, the cupels first withdrawn being placed in the front row of the tray, and so on. This method obviates the reversing of the order in the muffle, which is necessary in the case of scorifiers, as in fig. 107, and thus facilitates the calculation for cupellation loss, as the order of the cupels is the same as the silver buttons.

CHAPTER XIII.

THE ASSAY OF SILVER ORES.

Introduction.—The quantity of silver and gold contained in ores is too minute to allow of its being accurately determined by wet methods, and the assay of precious metal ores is, therefore, almost universally conducted in the dry way, *i.e.* by furnace methods.

The usual method of assay is to fuse the ore with materials yielding metallic lead, which alloys with the gold and silver and sinks to the bottom of the molten charge. Suitable fluxes are added to make the ore readily fusible at a moderate temperature. The lead obtained by this operation is detached from the slag and subjected to cupellation to separate the precious metals. The preparation of the ore for assay by sampling and fine crushing has been described in Chapters VIII. and IX.

As already pointed out, opinions differ as to the fineness to which the sample should be crushed for assay purposes. It is usually considered fine enough if all passes through an 80-mesh sieve ; but some ores, such as those containing tellurides, must be crushed through 100-mesh or even a 200-mesh sieve. It is a good rule to crush all ore samples to pass a 100-mesh sieve, unless experience proves that a coarser sample will give satisfactory results with the particular ore undergoing assay.

If metallic particles (*i.e.* metallics) are caught on the sieve, they are collected and assayed separately. The precaution previously enjoined with respect to the metallics is very essential (see page 111).

As the subsequent treatment of an ore is dependent on its nature, a sample should be very carefully examined, as described on page 140, to ascertain as far as possible the nature and proportion of each mineral constituent. Much may be learnt as to the nature of the ore by the examination of unbroken lumps and, after panning, of pulverised ore with a lens. The minerals usually associated with precious metal ores are described on page 139. The presence of sulphides may be confirmed by the characteristic odour of sulphurous oxide (SO_2) which they evolve when roasted on a spatula.

Before weighing out the portion of ore required for the assay, the sample must be thoroughly mixed. It is very necessary to emphasise the extreme importance of thoroughly mixing the sample, as it is obvious that if the mixing is performed carelessly, the portion selected for assay will not be truly representative of the mass. The value of an assay depends very largely on the care with which the portion of ore used for the assay is taken from the sample submitted. Neglect or deficient skill in this selection renders practically useless all the care that has previously been expended in obtaining a truly representative sample of the material. The amount of ore taken for assay is weighed in grammes or in "assay tons" (A.T.). The "assay ton" system of weights, which is fully explained on page 59, is very frequently adopted in practice where a large number of assays of one kind are made, as it saves calculation.

188

For general assay-office practice it is more convenient to work on a purely decimal system, and convert when required into ounces per ton, etc., by reference to a set of tables. In this case it is usual first to calculate from the assay result the percentage of precious metal in the ore, and then refer to tables to ascertain the equivalent in ounces per ton, etc. Therefore, to avoid a needlessly troublesome calculation, it is well to take such a quantity of ore for each assay as by a simple multiplication will yield the percentage.

In the assay of ores there are certain losses incidental to the methods used. The chief losses are retention of precious metals in the slag during fusion, and absorption of precious metals during cupellation. As a general rule, the slag losses are considerably higher than the cupel losses. The amount of these losses varies according to the nature of the ore and the conditions under which the assay is conducted; they are by no means constant, and present no regularity even for the same material. With ordinary gold and silver ores of average grade assayed under normal conditions, both the slag loss and cupel loss are very small and are usually disregarded unless very great accuracy is required. In the case of rich ores and complex ores it is frequently necessary to determine the losses and correct the assay result accordingly. An assay in which the slag loss and cupellation loss has been determined and taken into account in reporting the assay value of the material is termed a "corrected assay."

The assays made at metallurgical works for ascertaining the value of purchased ores, except those of very high values (and generally for all other purposes), are almost invariably direct uncorrected assays.[1] The presence of copper, bismuth, and especially of tellurium and selenium in the lead button, increases the losses of gold and silver during cupellation. Care must therefore be taken in the scorification or the crucible fusion of the ore to prevent these metals from entering the lead as far as possible.

Although the dry assay of gold ores resembles in its main particulars the dry assay for silver ores, it is convenient to describe the assay of each metal separately.

The systems of working off large numbers of assays have been discussed on page 184.

Methods of Assay for Silver Ores.

The various ores of silver are described on page 79.

For assay purposes a silver ore may be regarded as rich when yielding over 200 ounces of silver per ton, and poor if yielding under 50 ounces per ton : ores yielding intermediate amounts may be said to be average ores.

The assaying of a sample may comprise the following operations :—

(1) Concentration of the silver in a button of lead by (A) scorification, or by (B) fusion in a crucible.

(2) Cupellation of the resulting button of argentiferous lead.

(3) When great accuracy is required, the determination of the silver absorbed by the cupel.

(4) Determination of the silver in the slag produced in the process of scorification or fusion.

(5) Assay of the "metallics."

(6) Weighing the silver.

(7) Examination of the silver for gold.

Two assays at least should be made of each sample.

1. (A) Scorification Method.—This is a simple and convenient method of assaying silver ores. It consists in exposing the ore, mixed with granulated lead

---

[1] Collins, loc. cit., p. 471.

and placed in a scorifier, to the action of a bright red heat, in an ordinary assay muffle.

The lead is oxidised by the air, and the silica and other mineral constituents in the ore are fluxed by the litharge, borax, etc.   The precious metals are taken up by the lead, which diminishes in amount as oxidation proceeds, until the slag forms a continuous layer over the lead.   The charge is then poured into an iron mould, the lead detached, and cupelled to obtain the silver.   The operation of scorification is fully described on page 155.

The amount of ore taken for scorification varies from 3 grammes to 5 grammes; but 3 grammes is the amount most frequently taken.

If the ore contains less than 1 per cent. of silver, 5 grammes or about $\frac{1}{6}$ A.T. are taken for a charge; if more than 1 per cent., 2·5 or 3 grammes (from $\frac{1}{10}$ to $\frac{1}{12}$ A.T.); and of very rich ores 1 to 0·5 gramme, or about $\frac{1}{30}$ to $\frac{1}{60}$ A.T.

As a general rule, if more than 5 grammes of ore must be taken, the crucible assay should be adopted.   The ore is weighed accurately to 0·001 gramme unless very poor, when less accurate weighing will suffice.   The proportion of lead added varies from 40 to 100 grammes according to the nature of the ore.   A very usual charge for ordinary silver ores is 3 grammes (0·10 A.T.) of ore and 40 to 50 grammes of lead, with a small quantity of borax as a cover.

Table XXV. gives examples of the proportions of lead to be used for different classes of ores :—

Table XXV.—Proportion of Lead for Scorification of Ores.

Character of ore.	For one part of ore.	
	Parts of granulated lead.	Parts of borax glass.
Arsenical .	16	0·10 to 0·50
Antimonial	16	0·10 to 1·00
Basic	8 to 10	0·25 to 1·00
Blende	10 to 15	0·10 to 0·20
Cobaltiferous	10 to 20	0·10 to 0·20
Cupriferous	10 to 20	0·10 to 0·15
Fahlerz	12 to 16	0·10 to 0·15
Galena, pure	6	0·15
,, siliceous	8 to 11	0·2
,, zinciferous	8 to 11	0·15 to 0·30
Iron pyrites	10 to 15	0·10 to 0·20
Nickeliferous ores	15 to 20	0·20 to 0·30
Siliceous (quartzose)	8 to 10	0
Tin ores	20 to 30	0·15 to 0·20
Zinciferous ores	15 to 20	0·20 to 0·30

The borax renders the slag more liquid and lessens the corrosion of the scorifier, but its quantity is kept as low as possible to prevent the slag from completely covering the surface of the molten lead too soon.   If the ore consists largely of quartz, sodium carbonate should be added instead of borax.   In the case of highly cupriferous ores, it is advantageous to add about 0·5 gramme of finely powdered sand to lessen the corrosive action of copper oxide on the scorifier. Where a large number of assays have to be made, the granulated lead may be measured by the copper ladle or measure, fig. 113, instead of weighing.

A scorifier, 2·5 inches in diameter, is the size most commonly employed for ordinary scorifications.   The scorifiers being charged with the proper proportion of ore, lead, and flux, and the muffle heated to a full red heat, they are removed to the furnace, and as many are introduced as the muffle will accommodate. The introduction of the scorifiers at first greatly reduces the temperature of the muffle, the door of the muffle should, therefore, be closed as soon as the charging in is complete, and remain closed until the lead melts, and the assay is thoroughly heated, after which the door is removed to allow air to enter.

When scorification is complete, a little powdered anthracite may be added to clean the slag (see page 156). When the surface is again tranquil the scorifiers are withdrawn from the muffle by means of proper tongs (fig. 27, page 37), and their contents rapidly poured into suitable moulds. When cold, the buttons of lead are readily separated from the adhering slags by a few blows with a hammer. The lead is cleaned from all slag and then cupelled, and from the weight of the resulting silver buttons the assay value of the material is calculated, as described on page 212. Should the lead exceed 25 grammes in weight, or should it be hard, it must be re-scorified until it is reduced to the required weight and is soft. The slags from very rich ores should always be kept and separately re-treated, as described on page 155.

It is very essential that the slags should be perfectly and uniformly liquid at the time of pouring from the scorifier; if they are very pasty and contain lumps of partially fused material, it usually indicates that part of the ore has been left unacted upon. This may be due to the employment of too low a temperature or to insufficient lead or borax. If the slags should not appear perfectly liquid before pouring, when a sufficiently high tempera-ture is maintained in the muffle, and the other conditions of the operation have been attended to, it will be necessary to add more borax, and, in some instances, even a little nitre. An infusible scoria is very liable to form on the surface of the slag in the case of ores containing large quantities of basic oxides, such as ferric oxide (Fe_2O_3), manganese oxide, and copper oxides. The addition to the charge of borax glass up to 3 grammes greatly facilitates the formation of fluid slags free from infusible scoria. If from any cause the slag is unsatisfactory, and contains unfused portions of ore, it is best to make a fresh assay, using less ore or more lead.

Fig. 113.

Most ores and metallurgical products containing silver may be assayed by scorification; but more experience is required to obtain correct results than by some of the fusion methods. Scorification is, however, not so well adapted for assaying substances poor in silver as the quantity of material operated upon is comparatively small: though this objection may be obviated by combining the lead buttons obtained from several assays and re-scorifying them so as to con-centrate the silver into one button of lead, which is then cupelled.

For the assay of ordinary and rich silver ores, and for products rich in the precious metals, scorification is a very desirable method. It is also a very convenient method for concentrating the precious metals in residues resulting from the treatment of material by combined wet and dry methods. From its convenience and the short time required for the operation, it is very generally employed in establishments where a great number of assays have to be made daily. The small charges used for scorification necessitate the employment of good balances and care in weighing. The silver buttons from duplicate assays of material of moderate richness should agree closely. It should be borne in mind that a difference of 1 milligramme in the weight of the silver button represents a difference of 6 to 10 ounces per ton (2240 lbs.), according to the amount of material taken for assay. The assay should be repeated if the difference between the weights of the buttons exceeds 0·5 milligramme, equivalent to 5·4 ounces per ton when 3 grammes are taken for assay. In the case of

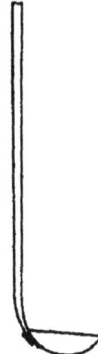

"argentiferous lead ores assayed for silver by scorification in triplicate on one-tenth assay ton of ore, the maximum difference between the three silver buttons should not be more than one-tenth milligramme on buttons weighing 1 centigramme each (1 ounce on 100 ounces). When the difference between the heaviest and lightest of the three does not exceed this amount, the mean may be taken, but if the difference is greater, two more check assays are made."[1]

With rich material, variations in the weights of the silver buttons are not infrequent, even when every care has been exercised, and the only safe plan is to make several assays, and take the mean of the results obtained. The number of assays to be made to control the accuracy of results will vary from 2 to 5 or more, according to the richness and nature of the ore.

1. (B) Crucible or Fusion Method.—The object of the fusion in a crucible, as in scorification, is to concentrate the silver in a button of lead which is subsequently cupelled : and to retain the gangue material in the slag. The charge for the fusion of any given ore varies somewhat according to its nature, and is made up by the assayer according to his judgment and experience.

The various fluxes employed and the principles of fluxing have been fully dealt with in Chapters X. and XI.

The amount and kind of flux required is judged in the first instance from the appearance of the ore and the preliminary tests, and is subsequently modified if necessary according to the result of the fusion. It is well in this connection to bear in mind the general principle that for a siliceous or acid gangue a basic flux is needed, and for basic material an acid flux. Sodium carbonate being a flux for silica and silicates, and borax for lime, iron oxide, and other bases, the relative quantities of the two must be varied according to the nature of the ore. As a general rule, borax should not be used unless the nature of the ore requires it. As previously stated, from two to three parts of flux are usually quite sufficient to yield a satisfactory slag with one part of ore. Lead oxide in the form of litharge or red lead, and charcoal, or other reducing agent are added to supply the lead necessary for collecting the precious metals.

A mixture of 30 grammes of litharge and 1 to 1·5 grammes of charcoal powder will give a button of lead weighing about 20 grammes, which is a satisfactory quantity for most silver ores when 20 to 25 grammes of ore are used for assay.

Sometimes a comparatively large quantity of lead oxide is added to the charge, and a portion of it allowed to remain unreduced to serve as a flux for the oxides of such metals as copper, antimony, iron, etc. The quantity of litharge or of red lead used must be fairly accurately weighed or measured out, so that the amount of silver it contains may be correctly deducted in accordance with the result of the preliminary assay of the reagent. The charcoal or other reducing agent should also be weighed carefully, and the quantity adjusted so as to produce a lead button of suitable size for cupellation on one cupel. Although it may be desirable, by way of scientific training, for students to weigh out all fluxes, an assayer who possesses the requisite aptitude for manipulation soon acquires the skill which enables him to dispense with the tedious process of weighing out all his reagents. Many assayers either estimate the needed quantity with the eye, or make use of tin or copper measures, which, when filled, contain about the required quantity of flux : this latter is the better of the two methods.

When the same class of ore is being assayed repeatedly, it is usual to make up a "stock" flux, and use a measured quantity for each assay.

When the charge has been prepared, it is well mixed, transferred to a crucible of suitable size, and fused as previously described on page 153. An

[1] Collins, *Metallurgy of Lead*, 2nd edit., 1910, p. 463.

E crucible, Battersea round, is very suitable for silver-ore assays in a wind furnace, but if the fusion is to be made in a muffle or furnace of reverberatory type, a crucible of the Colorado shape is generally preferred.

The fusion must be preceded by an exposure to a dull red heat for ten minutes, so as to ensure the complete "fritting" of the charge. The temperature is then raised and fusion effected, and when all action has ceased the charge is poured into a round mould. When cold the lead is detached from the slag, which on account of its brittleness readily separates from the lead on hammering. If the button of lead weighs less than 20 grammes, the quantity of charcoal must be increased, remembering .that charcoal will reduce about twelve times its own weight of lead from red lead and 22·5 from litharge. The lead is cupelled and the assay value of the ore calculated from the weight of the resulting silver button as described on page 212.

In all cases the slag should be kept until it is ascertained from the result of the cupellation whether it is desirable to "clean" it. The slag is generally glassy, but varies with the nature of the ore, and should be uniform in colour and composition. For convenience of description the various ores are divided into four classes :—

     I. Ores with siliceous gangue.
     II. Ores with basic gangue.
     III. Pyritic ores (mainly iron pyrites).
     IV. Complex ores.

The following are typical charges for the several classes of ores :—

I. *Ores with Siliceous Gangue* (see page 140).—Ores of this class present little difficulty. For average ores take :—

Ore	20 to 25 grammes.
Litharge or red lead	30 ,,
Charcoal	1 to 2 ,,
Sodium carbonate	30 ,,

If the ore contains a small amount of pyritic material, metallic iron in the form of nails or hoop-iron should be added to decompose the sulphides and yield a soft and malleable lead button.

II. *Ores with Basic Gangue* (see page 141).—In ores of this class the chief bases to be fluxed are oxides of iron, with alumina, lime, and oxides of manganese, barium, magnesium, and other metals in smaller quantity. The principal fluxes for basic material are borax and lead oxides. More or less quartz or siliceous material is, however, almost invariably present in all basic ores, and a certain proportion of sodium carbonate is usually added to the charge. When the siliceous material in the ore predominates, the charge given above for siliceous ores will be satisfactory, but the proportion of sodium carbonate should be diminished, and that of borax, and if necessary lead oxide increased as the quantity of metallic oxides in the ore increases. The addition of excessive amounts of borax should be avoided, otherwise the charge may fuse too readily. A mixture of equal weights of sodium carbonate and borax usually answers very well for the majority of basic ores. As previously pointed out, both basic and acid materials are generally present in " basic " ores, and as these combine to form more or less fusible compounds, the quantity of additional flux required is frequently comparatively small.

Ores containing large quantities of iron oxide (or manganese oxide) require more charcoal or other reducing agent to be added to the charge for the reasons already given on page 142. As an example, 25 grammes of ore, consisting of practically pure ferric-oxide, will require 3 grammes of charcoal. The addition of fine sand is necessary to flux basic ores in which silica is deficient.

13

For average basic ores the following charge may be taken :—

Ore	20–25 grammes.	
Litharge or red lead	40	„
Charcoal	2–3	„
Sodium carbonate	20	„
Borax	10	„
Sand or powdered glass.	0–10	„

For ores consisting mainly of earthy bases, such as lime, alumina, etc., or of zinc oxide, the quantity of alkaline flux is advantageously increased to 40 grammes.

The quantity of litharge should be increased to 50 to 60 grammes for ores containing oxides of metals such as copper, antimony, zinc, etc., in moderate amount only. The excess of lead oxide passes into the slag and carries the copper oxide, etc., with it, thus preventing the production of hard cupriferous or antimonial lead. Large proportions of lead oxide should, however, be avoided, as slags with a large excess of this oxide are very liable to retain silver and gold. Ores containing large quantities of the oxides of copper, antimony, etc , are more satisfactorily treated by the special methods described subsequently.

III. *Pyritic Ores* (see page 142).—The ores included in this class consist mainly of iron pyrites with small amounts of other metallic sulphides. The sulphides most commonly present, in addition to the sulphurised minerals of silver, are iron pyrites, copper pyrites, galena (lead sulphide), blende (zinc sulphide), and mispickel (arsenical iron pyrites). Ores containing excess of the sulphides, other than iron pyrites, are dealt with under the head of " Complex Ores." Ores containing much pyritic material are easily recognised by the ready separation of the heavy sulphides when the powdered ore is washed, or by the characteristic odour of sulphurous oxide (SO_2) which they evolve when a small quantity is roasted on a spatula.

The presence of mispickel is indicated by the odour of arsenious oxide.

As previously stated on page 134, metallic sulphides act as reducing agents, and it is of importance that the whole of the sulphur should be either removed or oxidised during the process of assaying, as otherwise if present in large amounts it will yield large lead buttons and render the lead sulphury and somewhat brittle, or the lead will be accompanied by a regulus or matte. The following are the methods in general use for the removal of the sulphur :—

(a) Desulphurisation with metallic iron.
(b) Oxidation by roasting.
(c) Oxidation with lead oxide.
(d) Oxidation with nitre.

(a) *Desulphurisation with Metallic Iron.*—In this method metallic iron is added to the charge to take up the sulphur forming iron sulphide which dissolves in the slag in moderate quantity and does not form a separate layer of matte. Slags that are strongly alkaline have considerable solvent action on metallic sulphides. The iron added to the charge should be in the form of thick hoop-iron (two or three pieces), in strips about 6 inches long, or small rods of wrought iron. Stout iron nails are frequently employed for the purpose, but they present less surface, and are more troublesome to manipulate. If they are used, from four to six should be added to the charge with the heads downwards, in order to present as large a surface as possible. Sodium carbonate is the chief flux used, as the slags must be strongly alkaline. As a general rule, the amount of sodium carbonate should be twice that of the ore taken. Only as much lead oxide should be added as will give a button of lead of the weight required, since practically the whole of the lead oxide in the charge is reduced to the metallic state by the iron. It is advisable to add sufficient lead oxide to give a comparatively large button of lead, say about 25–30 grammes. The amount of lead oxide may be diminished for ores containing appreciable quantities of galena.

The following is a suitable charge for moderately pyritic ores :—

Ore .	.	.	.	.	.	.	20–25 grammes.
Litharge or red lead		.	.	.	40	"
Sodium carbonate	.	.	.	.	30–40	"
Borax	.	.	.	.	.	.	10	"

With the addition of metallic iron.

The fusion should be performed at a comparatively low temperature, and the crucible should remain in the furnace for about five minutes after the contents are in tranquil fusion. The iron is then taken out and examined while hot, and should any shots of lead be found adhering to it, they must be washed off by immersing it in the molten slag, after which it is withdrawn, and the crucible left in the furnace for a few minutes longer. The crucible is then removed, and its contents poured into a mould in the ordinary way. The fusion should be conducted in a reducing atmosphere as far as possible, as the large amount of ferrous sulphide in the slag is readily oxidised, causing it to become pasty, and give trouble in pouring. To preserve the slag from oxidation, many assayers add a layer of salt on top of the charge; but in the author's opinion it is preferable to use a cover to the crucible, leaving a small opening to permit of the escape of the gases. It is usually advisable to "clean" the slag unless the ore is low grade.

When arsenic is present the iron will decompose the arsenical compound with the formation of a speise, which separates as a hard, greyish-white layer on the surface of the button of lead. The separation of a small quantity of speise may be disregarded, as it seldom retains an appreciable quantity of precious metal; but if large quantities of arsenic are present, it must be eliminated by other methods. The employment of a low temperature tends to prevent the formation of speise.

Hard and somewhat brittle lead buttons will be obtained from ores containing copper and antimony sulphides, as these compounds are partially reduced to the metallic state by iron, the metals passing into the lead button and causing trouble in the cupellation, unless previously removed by scorification. In this case a cupriferous regulus, which retains appreciable amounts of gold and silver, is also frequently formed. The presence of antimony compounds will produce brittle lead buttons, and give rise to the formation of a scoria on the cupel, which tends to retain part of the precious metals and leads to unsatisfactory results. Ores containing much copper or antimony are best treated by special methods, described subsequently.

This method of assaying pyritic ores is in very general use, as it can be successfully applied to most ordinary sulphide ores. But, as previously stated, only a moderate amount of sulphur can be carried into the slag by the use of iron, and with ores containing a large proportion of metallic sulphide it is very difficult to effect complete decomposition by this method, even when a large quantity of metallic iron is used: the lead button is always more or less brittle owing to the presence of sulphur, and the production of matte is inevitable.

This objection may be overcome by using smaller quantities of ore for assay, but this involves risk of multiplying errors of assay.

Many assayers use metallic iron for ores rich in sulphides even when its use necessarily involves the production of matte, but in this case the matte is saved and scorified together with the lead button, prior to cupellation, or the matte is crushed and mixed with the slag so that both are cleaned at the same time. The use of iron is not applicable to ores containing large quantities of arsenic, and metals such as copper and antimony, for the reasons stated above. Ores rich in arsenic, copper, or antimony are best treated by the methods described subsequently.

(b) *Oxidation by Roasting.*—An oxidising roasting of the ore previous to fusion is a satisfactory method of eliminating the sulphur in pyritic ores, and is sometimes conveniently employed for gold ores ; but as smaller quantities of ore are permissible for the assay of silver ores, unless very poor, roasting may generally be dispensed with, and advantage taken of the oxidising power of red lead or nitre, or the desulphurising power of iron.

When silver ores are roasted, especially such as consist of argentiferous minerals containing copper, it is generally necessary to employ a very low temperature, as, from their great fusibility, they would otherwise be liable to "clot" or partially fuse, and the further expulsion of sulphur would be rendered difficult.

The method of roasting ores is described on page 150. The quantity of ore used for assay must be weighed before roasting. The roasting of silver ores tends to give low results, especially if the ores are high grade. After roasting, the metals exist in the state of oxides, and the ore is therefore basic in character and is fluxed accordingly.

(c) *Oxidation with Lead Oxide.*—In this old method of assay, lead oxide, either litharge or red lead, is employed as the oxidising agent. Both these oxides readily attack all the sulphides, arsenio-sulphides, etc., and oxidise their constituents, whilst a proportionate quantity of lead is set free. An excess of lead oxide must be used, as it not only acts as an oxidising agent, but also as a flux for the gangue of the ore and the metallic oxides resulting from the oxidation of the pyritic material. Silica requires nearly 5 parts of litharge, and the metallic oxides from 2 to 10 parts, to form fusible compounds. The proportion of litharge required to form fusible compounds with the principal metallic oxides is as follows [1] :—

Proportion of Litharge to Flux Metallic Oxides.

One part of	As_2O_3.	Cu_2O.	CuO.	Fe_3O_4.	Sb_2O_3.	ZnO.	Fe_2O_3.	MnO.	SnO_2.
Requires parts of PbO	1	1·5	1·8	4	5	8	10	10	13

Charcoal powder is added in cases where the proportion of sulphides in the ore is very small, and not sufficient to give the required weight of lead.

Red lead is used in preference to litharge, on account of the larger amount of oxygen it contains. As a flux, lead oxide has the advantage of being heavy, and consequently occupies very little space in the crucible. It may be remarked, however, that care is needed in its employment as a flux by itself, as it has a strong corrosive action on the crucible, and if the fusion of the assay is performed too quickly, it sinks to the bottom of the crucible, owing to the readiness with which it fuses, leaving the lighter earthy matters of the charge as a pasty, semi-fused mass in the upper portion of the crucible, where they escape the fluxing action of the litharge. The fusion should be effected at a comparatively low temperature. It is not desirable to prolong the fusion, on account of the corrosive action litharge has upon the substance of the crucible, which it rapidly destroys.

[1] Hofman, *Metallurgy of Lead.*

To lessen this action, it is usual to add fine sand or glass equal to twice the weight of the ore taken, or less according to the proportion of silica in the ore. The proportion of lead oxide required to effect complete oxidation varies according to the nature of the sulphide, but in all cases is large. As a general rule, 1 part of a metallic sulphide requires from 20 to 30 parts of red lead to yield a button free from sulphur. " When less than the requisite quantity of lead oxide is used, only a portion of the sulphide is decomposed, and a corresponding quantity only of lead reduced, whilst the remainder of the sulphide forms, in combination with the litharge and metallic oxide which is produced, a compound belonging to the class of oxysulphides which are generally very fusible " (Mitchell). If oxysulphides are present to a large extent in the slag, they cause the retention of appreciable quantities of the precious metals, notably silver.

An objection to this method of assay is the large quantity of lead produced unless the ore is sufficiently rich to permit of small quantities being taken for assay or contains only a small proportion of metallic sulphides.

The amount of lead reduced by 1 gramme of different sulphides varies from 6 to 11 grammes, as shown by the table on page 135. It is evident, therefore, that if an ore contains a large proportion of these sulphides, and 20 grammes is used for the assay, the quantity of lead reduced will be very much larger than that actually required for an assay. For example :—With an ore containing 75 per cent. of iron pyrites, 20 grammes would yield 165 grammes of lead, and if there were not sufficient lead oxide to yield this weight of metal, the button would be sulphury. This inconvenience may be obviated by oxidising a part of the pyritic material with nitre, by the judicious employment of which buttons of almost any required weight may be obtained as described below. One gramme of nitre is equivalent in oxidising effect to 20 grammes of red lead or 26 grammes of litharge. When lead oxide alone is used as the oxidising agent, the following charge may be used for moderately rich and highly pyritic ores :—

Ore	5–10	grammes.
Red lead (or litharge)	100–200	,,
Sand	0–10	,,

Borax to cover, 10 grammes.

The slags are very fluid and pour readily, and the lead button is generally clean, and separates easily from the slag.

The weight of the button should not be less than 30 grammes, and is usually much above this. If necessary, it should be scorified down to about 15 grammes, and cupelled.

This method can be conveniently applied to highly pyritic ores when sufficiently rich to allow of comparatively small quantities being taken for assay, but is not suitable for low-grade ores, unless the amount of pyritic material present is small.

Sulphide ores rich in copper and antimony are satisfactorily assayed by this method, as these metals are oxidised and dissolved in the slag and not reduced to metal, so that they do not contaminate the lead button as in the case with metallic iron. Except in cases where an excess of lead oxide is required in the slag to flux metallic oxides such as copper, it is the general practice to limit the quantity of lead oxide to that necessary to oxidise the pyritic material, and to add borax and sodium carbonate to flux the gangue material and the metallic oxides formed during the fusion, that are not fluxed by the litharge. It is well to aim at leaving little or no lead in the slag as silicate or as free litharge, as experience indicates that with high-grade ores the presence of an appreciable quantity of lead silicate and oxide in the slag causes low silver results, either by reason of the density of the slag or by dissolving the silver.

The charge must be made up according to the amount and nature of the pyritic material present. In many cases the following charge will be suitable:—

Ore . . . . . . 0·25 to 0·5 assay ton.
Red lead or litharge . . . 50 to 75 grammes
Sodium carbonate . . . . 10 ,, 15 ,,
Borax . . . . . . 10 ,, 15 ,,

When the ores are very pyritic it is preferable to effect the oxidation of a certain proportion of the sulphides by the addition of nitre as already stated.

(d) *Oxidation with Nitre.*—Nitre may be employed according to either of the following methods:—

I. To oxidise part of the metallic sulphides.
II. To oxidise the whole of the metallic sulphides.

Method I.—In this method the pyritic material present in excess of that required to act as a reducing agent is oxidised by the addition of nitre. When the proportion of nitre in the charge is insufficient to oxidise the whole of the pyritic material, and when litharge is also present, after the nitre has produced its action, the litharge acts in its turn on the sulphides that still remain unoxidised, producing metallic lead which carries down the silver. Therefore, by employing suitable proportions of nitre and litharge, any desired quantity of lead may be obtained. To ascertain how much nitre is required for a given ore it is necessary to first determine the reducing power of the ore, and from the result obtained to calculate the amount of nitre necessary to oxidise the excess of sulphides.

The approximate quantity of nitre required to oxidise various metallic sulphides is as follows:—

Approximate Oxidising Effect of Nitre.

	Part of nitre required to one part of metallic sulphide.
Iron pyrites	about 2 to 2½ parts.
Mispickel, copper pyrites, fahlerz, zinc blende .	,, 1½ to 2 ,,
Antimonite	,, 1½ parts.
Galena	,, ⅔ part.

Experience has shown that, as a general rule, from 0·25 to 0·30 parts of nitre is required to oxidise the quantity of pyritic material which would act as a reducing agent to produce one part of lead; or, in other words, the addition of 1 gramme of nitre will decrease the weight of the lead button by 4 to 5 grammes. Conversely, the weight of the lead button may be increased by from 4 to 5 grammes by diminishing the quantity of nitre by 1 gramme. The result of an interesting series of experiments on the oxidation of metallic sulphides with nitre is given in Beringer's well-known book of Assaying, p. 97 (6th edition, 1900). The experiments show that 1 gramme of nitre " kept up " on the average 4 grammes of lead, the range being from 3·2 with acid slags to 5·3 with very basic slags. These facts serve to explain some apparently irregular results obtained in practice.

The reducing power of an ore is determined by a preliminary assay, and to get the best results it is advisable to use a charge of the same composition as the final assay charge.

The following charge may be used:—

Preliminary Assay.

Pyritic ore . . . . 5 grammes.
Litharge (not red lead) . . . . 100 ,,
Sodium carbonate . . . . . 10 ,,

Cover of borax, 10 grammes.

If the ore contains little silica, from 1 to 5 grammes of silica should be added.

The lead resulting from the fusion is weighed, and from this weight is calculated how much lead would be obtained from the quantity of ore to be used for the final assay (say 10 grammes). From the weight thus found deduct 25 (the desired weight of lead to collect the values) and divide the difference by 4; the figure obtained gives approximately in grammes the nitre required for 10 grammes of ore.

For example. Suppose the preliminary assay of 5 grammes of ore gave 40 grammes of lead, then 10 grammes of ore would yield 80 grammes of lead. Subtracting 25 we get 80 − 25 = 55, and dividing this by 4 we get 13·8 grammes, the weight of nitre required.

This rule may be used as a general guide for adjusting the nitre, but in practice irregular results are frequently obtained, as the action of the nitre is considerably influenced by the conditions of the fusion.

The difficulty of adjusting the quantity of nitre is largely dependent on the following facts :—

(a) The common commercial varieties of nitre, which are those usually employed for assay purposes, frequently contain impurities, consequently, the oxidising power will vary with the different samples employed.

(b) Sulphur, and in some cases the metals, are capable of two degrees of oxidation, and whether the lower or the higher oxide is formed by the action of the nitre will depend very largely on the conditions of the fusion. In this connection the character of the slag resulting from the fusion is an important factor. When the slag is acid owing to the presence of much siliceous material in the charge, the sulphur will, in most cases, be eliminated as sulphurous oxide (SO_2), and the lower oxide of the metal be formed. On the other hand, if the slag contains much soda, the tendency will be for the sulphur to be oxidised to sulphate and the metal converted to the higher oxide.

(c) Rapidity or slowness of fusion will also make a great difference in the effect of the nitre ; in the former case, the pyritic material sinks before the nitre has time to complete its action, and frequently produces a regulus, and in the latter case this does not occur. The decomposition of potassium nitrate is greatly facilitated by the presence of silica and of silicates.

In making up the charge the litharge is usually added in slight excess of that necessary to provide the required weight of lead. Sodium carbonate equal to twice the weight of the ore, or less, is the chief flux employed.

Part of the sodium carbonate added to the charge will be utilised in the formation of sodium sulphate, which is one of the products of the reaction of the nitre upon the sulphides present, and due allowance must be made for this fact. The sodium sulphate, on account of its extreme fluidity, floats as a watery liquid on the surfaces of the slag, and when present in large quantities causes inconvenience in the pouring of the charge.

If the ore is deficient in silica, some should be added.

The following is a suitable charge for most sulphide ores :—

Ore	10–20	grammes.
Litharge or red lead . . .	40–80	,,
Sodium carbonate	10–20	,,
Borax	3–5	,,
Sand or powdered glass . . .	0–10	,,

Nitre sufficient to yield a button of about 25 grammes of lead.

For ores with more than 50 per cent. of pyritic material the weight of nitre required usually varies from an equal weight to twice the weight of the ore.

When fusion is made in a muffle it is not advisable to use more than 20 grammes of nitre in a charge, otherwise the chargo is very liable to boil over.

The fusion of charges containing nitre is effected in large crucibles at a comparatively low temperature, care being taken to avoid a sudden increase of temperature, as fusion takes place somewhat quickly, and the slags are in most cases very fluid. When effervescence has ceased, and the contents of the crucible have become thoroughly liquid and tranquil, it is advisable to lift the crucible out of the fire and subject it to a rotary motion in order to thoroughly mix the liquid contents. If this treatment causes effervescence, it indicates that the action of the nitre is not complete, and the crucible must be returned to the fire and the fusion continued.

The button of lead obtained should weigh between 25 and 35 grammes, but if below or above this weight, the charge must be modified to produce the required weight.

The weight of lead obtained will vary somewhat according to the quantity and nature of the fluxes used and the conditions of the fusion. When the charge contains much soda, the tendency will be, as before stated, for sulphates to form ; and as the nitre oxidises a smaller quantity of pyritic material when the higher oxides of the metals are formed, a larger proportion of the pyritic material is left unoxidised, and acts as a reducing agent, thus producing more lead, if the charge contains sufficient litharge. Until experience has been gained, the first assay of an ore of unknown composition may be more or less unsatisfactory, but the result obtained from the fusion will indicate what modification of the charge is necessary to give a satisfactory result.

With proper precautions the nitre may be so adjusted that a malleable lead button of the desired weight is obtained.

The necessity of making a preliminary assay to determine the reducing power of the ore constitutes one of the chief objections to the nitre method. The assayer at a modern smelting works has no time for preliminary assays. With experience, however, the preliminary assay may frequently be dispensed with.

A good practical assayer will, by a preliminary examination of the ore, or by panning, make an estimate of the amount of reducing agent (pyrites, etc.) present, and from this estimation he will generally, though not invariably, decide correctly the amount of nitre required.

Method II. (Balling's method).—This method consists in adding sufficient nitre to effect the complete oxidation of the pyritic material present, and when oxidation is complete to add a mixture of litharge and charcoal to supply lead for collecting the precious metals.

By this method a quantity of nitre equal to twice the weight of the pyritic material in the ore is usually sufficient to effect complete oxidation, but in exceptional cases rather more than this amount may be required. Borax and sodium carbonate are added, as in the previous method, to flux the metallic oxides formed, but the red lead is omitted. The fusion is conducted as before, and when the action of the nitre is complete, and the contents of the crucible are in tranquil fusion, it is withdrawn from the fire and allowed to cool until the slag begins to thicken. A mixture of 40 grammes of red lead, 2 grammes of charcoal powder, and 10 grammes of borax is then added, the crucible returned to the furnace, and the charge kept in a molten condition for from ten to fifteen minutes. When all action has ceased, the contents of the crucible are poured into a mould in the ordinary way. By this means the globules of metallic lead obtained by the reduction of the red lead sink through the molten slag and collect the gold and silver.

Granulated lead is sometimes substituted for a mixture of lead oxide and charcoal, but unless the granules of lead are very small there is a great tendency for the result to be low, owing to the rapidity with which the lead sinks through the molten mass. Some assayers prefer to divide the above mixture into two parts, and to add the second portion after a short interval, but from the author's experience there appears to be no advantage in this procedure. Balling's method of complete oxidation requires a little more time than the method of partial oxidation, but the button of lead obtained by the former method is always of the required weight, and is usually comparatively free from impurities, and may generally be cupelled direct. Both methods, however, give equally satisfactory results when due precautions are taken.

IV. *Complex Ores.*—Ores containing large quantities of copper, antimony, zinc, etc., require to be assayed by special methods which are described on page 221.

2. **Cupellation of the Lead Buttons.**—The precious metals, after concentration in lead by either scorification or crucible fusion, are separated by cupellation. The operation of cupellation has been described on page 157. Care should be taken to clean the lead buttons from all slag by brushing before cupelling. If the slag is left adhering to the button, it will interfere with the " uncovering " of the lead at the start of the cupellation. The temperature employed should be as low as is compatible with the satisfactory oxidation of the lead. Success in cupellation is also dependent on the quantity of air passing through the muffle, which must be carefully regulated. When the air-current is too rapid, the cupel is cooled, and crystals of litharge (feathers) formed, which may completely cover the metal and protect it from further oxidation, and cause " freezing." When the air-current is too feeble, the completion of the assay is unnecessarily delayed. The temperature should be raised towards the end of the operation to prevent solidification of the silver before the lead has been removed, but excessive heat should be avoided. A plan adopted by many assayers is to place the cupel in the front part of the muffle at the start and keep it there until cupellation is nearly complete, and then to transfer it to the back of the muffle until the " flash " takes place and the operation is finished. The cupel is then gradually moved towards the door in order to cool the silver slowly and prevent "spitting." When the buttons weigh less than 5 milligrammes, spitting does not take place readily. In the case of very small buttons of silver, or silver and gold, it is very necessary to watch the end of the cupellation carefully, and promptly to remove the cupels almost immediately after the button has brightened, as a heavy loss of silver will take place if the buttons are left in the furnace. The buttons are detached from the cupel with a pair of pliers (fig. 48, page 42), and freed from adhering bone-ash by squeezing and brushing and then weighed. Buttons that are too small to be brushed may be cleaned by placing on the palm of the hand and rubbing with the finger. It is advisable to examine the buttons with a lens before weighing.

3. **Determination of Silver absorbed by the Cupel.**—When great accuracy is required the amount of silver absorbed by the cupel is determined as described on page 180, and allowed for in calculating the assay value of the ore (see page 212).

4. **Determination of Silver in the Slag.**—The amount of silver retained in the slag is not inconsiderable in the case of rich silver ores, and it is frequently necessary to determine it. This is done by the methods given on page 155.

5. **Assay of Metallics or " Scales."**—The " metallics " may be assayed by cupellation direct, or by scorification and cupellation. Before the metallics are assayed they must be weighed and their weight, and also the weight of the ore from which they have been obtained, carefully recorded. For cupellation direct, the " metallics " should be small in quantity and free from siliceous material, etc.: they are rolled up in a convenient weighed quantity of sheet lead and then cupelled.

For scorification the metallics are mixed with about 15 to 20 grammes of lead, or more if necessary, with the addition of a little sodium carbonate, if silver chloride is present.   The scorification is conducted in the usual way and the resulting lead button cupelled.   The button of precious metal obtained is weighed, and the results allowed for in calculating the value of the ore (see page 212).

Metallics vary considerably in their precious-metal content.   Some may be very rich in metallic gold or silver, while others may be contaminated with native (metallic) copper, etc., and contain comparatively little precious metal.

As a general rule, if the metallics appear to be rich in native gold or silver, and the total weight is more than about 2 grammes, they should be divided into two or more smaller lots and each lot scorified and cupelled separately. The resulting precious-metal buttons are then weighed together.

6. **Weighing the Silver Buttons.**—The methods of weighing are described on page 60.   The balance employed should be sensitive to $\frac{1}{20}$ milligramme or less.   Silver buttons from ore assays are generally weighed by direct weighing, the button being placed in the left-hand pan and the weights in the right. Care should be taken to see that the balance is in the proper adjustment before commencing to weigh.   The rider should be used for all weights less than 10 milligrammes.   The difference between the weights of the silver buttons from duplicate assays of average silver ores should not, as a general rule, exceed 0·5 milligramme.   The weight of each button should be recorded as soon as it is ascertained.

7. **Examination of the Silver for Gold.**—Gold is not unfrequently present in sensible quantity in argentiferous ores from California, South America, and other localities, and should always be looked for.

For this purpose one or more of the buttons or "prills" are flattened out by hammering and heated with dilute nitric acid in a test-tube or porcelain crucible ; the solution is poured off and the gold, which is usually left in the form of a black or brown powder, is washed with distilled water, dried, heated to redness, and, when cold, weighed (see page 207).

CHAPTER XIV.

THE ASSAY OF GOLD ORES AND CALCULATION OF RESULTS (GOLD AND SILVER).

Introduction.—Minerals containing gold are assayed in precisely the same way as ores of silver, but as the former usually contain a very small proportion of the precious metal, it becomes necessary to operate on a larger quantity of material.

Gold ores rarely contain more than a few ounces, often only a few penny-weights of gold to the ton; consequently, the button of gold obtainable from such quantities of ore as may conveniently be employed in assaying is often so small as to require more than ordinary care in its manipulation.

For example. When assaying an ore containing 5 dwt. of gold per ton, if only 20 grammes were taken for the assay, as is usual with silver ores, the button of gold obtained would be so minute that it would not only demand considerable skill and care on the part of the assayer, in handling it, but the weight of it would be barely perceptible even on a sensitive balance. In order, therefore, to lessen the strain on the assayer's attention and to bring the button of gold within the scope of the balance, it is usual in the case of gold ores to take charges of 50 grammes, 100 grammes, or even 200 grammes, the larger charges being used for very poor material. A drawback to large charges is the large amount of fluxes required to flux them and the consequent employment of crucibles of large capacity which necessitate the use of some special form of fusion furnace. If the fusion of gold ores is to be made in a muffle, smaller charges of 20 grammes must be used, and, if necessary, the lead buttons resulting from two or more fusions scorified together so as to concentrate the gold into one button of lead, which is then cupelled. In the author's opinion it is very desirable to employ larger charges of 50 grammes for gold-ore assays, as they are more likely to give results nearer to the average gold content of the parcel of ore from which they are taken, than the results obtained with small charges. It must be remembered also that the errors incident to the assay operations are probably multiplied by taking small quantities for assay.

The actual weight of ore taken for assay varies according to circumstances. If gramme weights are used, a charge of 50 grammes is usually taken for most gold ores; but if the gold content exceeds 5 ounces per ton, less may be taken. When the "assay ton" weights are used a suitable quantity of ore is 1 A.T. if the value in gold is from 0·5 ounce to 10 ounces or more per ton, and the balance for the final weighings is sensitive to 0·02 milligramme or less.[1] With poor residues 2, 4, or even 10 A.T. may be taken, and with very rich ores ½ A.T. may suffice. If more than 3 A.T. of ore are taken, the charge is divided between two or more crucibles, and the resulting lead buttons scorified down and cupelled. The same remark applies to charges of more than 100 grammes. It

[1] T. K. Rose, *loc. cit.*, p. 441.

may be well here to state that for the purpose of assay the author has considered a gold ore to be rich when assaying over 3 ounces of gold per ton, and poor when yielding less than 10 dwt. per ton. All ores yielding quantities of gold between these limits are considered to be average ores.

As the same methods of assay may generally be applied to both silver and gold ores, it follows that much of what has been said concerning the assay of silver ores applies also to gold.

Silver is invariably present in gold ores and remains alloyed with the gold obtained from the assay, and the main distinction between the methods of assaying ores of the two metals is that in the case of gold ores there is a further operation involved for the separation of the gold from the silver.

The ore is prepared for assay in the manner previously described, and should there be any metallic residue, it must be carefully collected and assayed separately. The ore must in all cases be reduced to a state of fine powder so as to pass through a sieve of at least 80 mesh. Complex ores should be crushed much finer. The remarks on fine crushing of silver ores apply equally to the case of gold ores. The ore is well mixed, and the portion taken for assay weighed accurately on the pulp balance to 0·01 gramme, unless very poor, as in the case of "tailings," when less exact weighing is sufficient.

With regard to the relationship between the pulp balance and weights and the assay balance and its weights, Whitby [1] warns assayers not to place too great reliance on fluxing scales with steel-knife edges and planes, as he has found the sensibility to diminish rapidly from 10 milligrammes on a 1000-gramme load to 100 milligrammes on much smaller loads, even when the balance has a case. In regard to the weights, it is always necessary to check them from time to time on a chemical balance. He gives the following table (XXVI.) of limitations in reporting assays :—

Table XXVI.

Charge.	Sensibility of balance.	
	$\frac{1}{10}$ milligramme.	$\frac{1}{100}$ milligramme.
0·5 assay ton, report to .	2·0 dwt.	0·20 dwt.
1·0 ,, ,, . .	1·0 ,,	0·10 ,,
2·0 ,, ,, . .	0·5 ,,	0·05 ,,
2·5 ,, ,, . .	0·4 ,,	0·04 ,,
5·0 ,, ,, . .	0·2 ,,	0·02 ,,
10·0 ,, ,, . .	0·1 ,,	0·01 ,,

The assay of gold ores comprises the following operations :—

1. Concentration of the gold in a button of lead by fusion in a crucible, or, more rarely, by scorification.
2. Cupellation of the auriferous lead.
3. Weighing the gold-silver button.
4. "Inquartation" and parting to separate the gold.
5. Weighing the gold.
6. Assay of metallics.

1. **Concentration of the Gold in Lead.**—The fusion or crucible method of assay is almost invariably adopted for gold ores. The scorification method is employed occasionally for rich ores, but it is not generally applicable to ordinary gold ores owing to the large quantity of ore required for assay.

[1] *Journ. Chem. Min. and Met. Soc. of S. Africa*, 1906, p. 271.

The fusion of the ore is usually effected in a wind furnace, a suitable crucible for the purpose being a G Battersea round. The fusion is conducted as previously described. The button of lead should weigh from 25 to 30 grammes, and should be soft and malleable. If it is hard or brittle it is usually desirable to submit it to scorification before attempting to cupel.

The ore to be assayed may belong to one of the several distinct classes, such as—

    I. Ores with siliceous gangue.
    II. Ores with basic gangue.
    III. Pyritic ores.
    IV. Complex ores.

The following are typical charges for the several classes of ore :—

I. *Ores with Siliceous Gangue.*—Quartz, which constitutes the gangue of most gold ores, is chiefly fluxed with sodium carbonate. About $1\frac{1}{2}$ parts of the dry carbonate will yield a fluid slag with 1 part of quartz, but to make the slag more fluid it is the practice of many assayers to add a small amount of litharge in excess of that required to supply the necessary weight of lead, so as to form lead silicate, which is more fluid when fused than sodium silicate.

A typical charge for quartzose ores would be—

Ore	50	grammes.
Sodium carbonate	70	,,
Litharge or red lead	40–50	,,
Charcoal	1·5	,,

Borax should be substituted for a part of the sodium carbonate for ores containing small quantities of iron oxide or other basic material.

Iron should be added when fusing ores containing pyrites in small amount, but in this case the charge should not contain any excess of litharge, as the iron will reduce any litharge not reduced by the carbon or other reducing agent employed.

II. *Ores with Basic Gangue.*—The most important basic material to be fluxed in gold ores is iron oxide.

The matrix of surface ores is largely made up of this oxide, and when ores composed mainly of pyrites are roasted, the product will consist of ferric oxide. To produce a fusible slag it is necessary to reduce the iron to the lower oxide (ferrous oxide, FeO), as stated on page 142, and this is effected by the addition of a suitable quantity of reducing agent to the charge. Ferrous oxide forms a fusible slag with silica, and when the ore is deficient in silica some must be added.

Sodium carbonate and borax are added, as the addition of alkaline fluxes results in the formation of double silicates, which greatly increases the fluidity of the slag. When suitable proportions of the different fluxes are used no difficulty is experienced in obtaining good slags even when the ores contain large quantities of ferric oxide. If, however, the slag contains any considerable percentage of ferric oxide, owing to insufficient carbon to reduce it to ferrous oxide, it may retain an appreciable amount of gold, although the slag may be perfectly fluid.

To get good results with ores consisting mainly of ferric oxide it is well to add sufficient charcoal to give a larger button of lead than usual, say 40 grammes.

For ores ranging from moderately basic (with about 30 per cent. of ferric oxide) to very basic (practically all ferric oxide), the following proportions may be taken :—

	Moderately basic.	Very basic.
Ore	50 grammes.	50 grammes.
Litharge . . .	40 ,,	50 ,,
Charcoal powder . .	2·5 ,,	4 to 6 ,,
Borax	20 ,,	30 ,,
Sodium carbonate . .	30 ,,	10 to 15 ,,
Sand or powdered glass .	Nil.	10 to 15 ,,

For ores containing silicates of alumina, lime, and magnesia the proportion of alkaline flux should be increased, since these silicates are practically infusible. A mixture of 40 grammes of borax and 20 grammes of sodium carbonate usually gives satisfactory results, but the proportion of the two fluxes must be altered to suit requirements. Manganese oxide must be reduced to the lower state of oxidation as in the case of iron. It requires about twice the amount of charcoal for its reduction to the lower oxide as compared with iron oxide, and due allowance must be made for this fact when large proportions of manganese oxide are present. Should the ores contain small quantities of the oxides of copper, antimony, etc., the quantity of litharge should be slightly increased, the amount required being usually from 60 to 70 grammes for 50 grammes of ore.

Whitby [1] gives the following charge for an ore containing 50 per cent. calcite, siliceous matter, and a little pyrites :—

Ore	1 assay ton or 50 grammes.	
Litharge	1¼ ,,	62·5 ,,
Sodium carbonate . .	⅓ ,,	17·0 ,,
Charcoal	2 ,,	

This charge gives a lead button weighing about 20 grammes.

III. *Pyritic Ores.*—The sulphides chiefly associated with gold ores are iron pyrites and arsenical pyrites (mispickel).

Ores containing not more than about 25 to 30 per cent. of pyrites may be satisfactorily assayed by the method in which metallic iron is used as a desulphurising agent, the iron combining with the sulphur to form ferrous sulphide, which passes into the slag. When the percentage of pyrites is much above this, the lead button resulting from the fusion is very liable to be sulphury, and to be accompanied by a matte.

Ores containing a large proportion of metallic sulphides should be roasted. The portion of ore to be used for the assay is weighed, and then roasted in a large roasting dish or in a crucible, as described on page 150.

Many assayers hesitate to roast gold ores, on the assumption that precious metal is lost during the operation. Although this may be true with certain classes of ore, it may be remarked that the loss of gold sometimes experienced is, in many cases, probably mechanical, due to want of care in conducting the operation. A considerable loss may take place if the ore is stirred too vigorously during the operation. Another very possible source of loss is the retention of gold in the slag, due to the presence of ferric oxide which has not been reduced owing to insufficient reducing agent in the charge when fusing the roasted ore.

It is true that the oxidation of an ore by roasting takes up a considerable amount of time and requires careful attention to get satisfactory results, and the tendency of modern assay practice is to make use of other methods of oxidation as far as possible. But although the methods of oxidation described under "Silver Ores" are somewhat more expeditious than oxidation by roasting, their application is limited in the case of gold ores, owing to the large quantity of material that generally has to be taken for assay. As pointed out above, the

[1] A. Whitby, *Journ. Chem. Met. and Min. Soc. of S. Africa*, 1906, p. 269.

use of iron is limited to ores containing only moderate amounts of pyritic material, while the large quantities of nitre that would have to be used if this method of oxidation were employed, make its use impracticable in the case of gold ores, except in special cases mentioned later.

However, to save time, many assayers use the minimum quantity of ore permissible for the assay, and make the fusion with metallic iron or with nitre; but with gold ores the results are generally lower than those obtained by roasting, although the risk of precious metal being retained in the slag, owing to the presence of dissolved sulphides, is less with gold ores than with silver ores. Of the two methods, the use of iron is the more satisfactory; but if an ore gives an appreciable amount of matte, or speise, it is best to repeat the assay on a smaller quantity of ore, or to roast before fusion (consult also "General Remarks on Pyritic Ores," page 211). After roasting, the ore consists mainly of ferric oxide, and is fluxed accordingly on the lines indicated for basic ores, especial attention being given to the amount of reducing agent added to the charge for the fusion. An experienced assayer will frequently adjust the proportion of reducing agent correctly from the colour of the roasted ore.

2. **Cupellation of the Auriferous Lead.**—The cupellation of the lead buttons from gold-ore assays differs very little from that of lead obtained from silver assays. Gold is invariably accompanied by more or less silver, and when the proportion of silver exceeds that of the gold, and both metals have to be determined, the cupellation must be conducted exactly as in the silver assay, the temperature being moderated to lessen the cupellation loss and the button cooled slowly to avoid spitting. When the gold predominates the temperature should be higher at the finish than in the case of silver cupellation, as the melting-point of gold is 100° C. higher than that of silver. Very small gold buttons should be withdrawn from the muffle as soon as cupellation is finished. The characteristic glow which takes place with gold just after solidifying is a valuable help in finding the position of very minute buttons. When assaying material from which minute buttons may be expected, the cupellation should be conducted on a fine-faced cupel (page 47). Gold buttons are more globular than silver buttons of the same size, and adhere less to the cupel. The colour of the button left on the cupel affords a useful indication of the proportion of silver present. If the button is yellow, it contains less than half its weight of silver. With more than 50 per cent. of silver the button is white.

The button is removed from the cupel with a pair of pliers, squeezed to loosen the bone-ash adhering to its lower surface, and then cleaned by brushing ready for weighing.

As previously pointed out, the losses of gold and silver during cupellation are mainly due to absorption by the cupel. They amount to about 1 per cent. of the gold and 10 per cent. of the silver in gold ores of moderate value containing small quantities of silver, if the lead button is free from impurities, such as copper and tellurium, when charged into the cupel.[1]

3. **Weighing the Button of Gold and Silver.**—The button is weighed on a good assay balance and the weight recorded for the purpose of calculating the amount of silver contained in the ore. The silver in the button is subsequently removed by solution in nitric acid, and the weight of the residual gold taken, when the difference between the two weighings represents the silver. When it is not required to determine the amount of silver that accompanies the gold, the weighing of the button of gold and silver may be omitted.

4. **Inquartation and Parting.**—The button is treated with nitric acid to separate or part the gold from the silver it contains. Nitric acid will remove silver from alloys of gold and silver without attacking the gold when these

---

[1] T. K. Rose, *Journ. Chem. Met. and Min. Soc. of South Africa*, Jan. 1905.

metals are present in certain proportions.   If the silver is deficient more must be added, otherwise some of the silver will remain undissolved.

This operation is called "inquartation," the word coming from the older assayers, who held the view that the alloy for parting should consist of one-quarter gold to three-quarters silver.   As pointed out in *Bullion Assaying*, chapter xviii., the modern practice is to use the ratio of 2 or $2\frac{1}{2}$ of silver to 1 of gold when parting large buttons of gold such as are obtained in bullion assays, but in parting the small buttons from ore assays it is considered necessary to have at least four times as much silver as gold present. When the amount of gold is very small, it is convenient to use a large proportion of silver in parting.   T. K. Rose [1] recommends varying the proportion of silver according to the weight of the gold, and gives the following as the most suitable proportions of silver for different weights of gold :—

Proportion of Silver to Gold for parting small Gold Buttons.

Weight of gold.	Weight of silver.	Ratio of silver to gold.
Less than 0·1 milligramme.	2 to 3 milligrammes.	20 or 30 to 1
About 0·2 ,,	2 ,,	10 to 1
,, 1·0 ,,	6 ,,	6 to 1
,, 10·0 ,,	40 ,,	4 to 1
More than 200·0 ,,	450 ,,	$2\frac{1}{4}$ to 1

The ratio of silver to gold in the buttons obtained from ordinary gold ore assays is very variable, so that in practice the proportion of silver is not always under control.   If, however, it is less than 4 to 1, silver must be added to bring it up to this ratio at least.

The proportion of silver in the button left on the cupel is estimated from its colour.   A pale yellow button always contains more than 60 per cent. of gold and requires the addition of silver.   If the button appears white it is usually assumed that it already contains at least a sufficiency of silver, in the absence of any knowledge to the contrary.   But a perfectly white button may not part completely, and if on parting it does not lose at least two-thirds of its weight, a correct proportion of silver must be added and the button again parted.   Very frequently, however, the silver in the buttons will be in excess, and on parting the residual gold will tend to break up and demand very careful treatment to prevent loss.   With a little practice the proportion of silver in a gold button may be estimated very closely, and in judging of the amount no ordinary error will very seriously affect the result.   When the addition of silver is necessary, the button with the additional silver is wrapped in as small a piece of lead-foil as possible, and cupelled, and the resulting button parted.   Sometimes the addition of silver is effected in the case of small buttons by fusion on a piece of charcoal or on a cupel, by means of the blowpipe ; but this method, although expeditious, is not to be recommended, as it frequently occasions loss.

If the addition of silver is known by previous assays of the ore to be necessary, it may be added directly to the charge in the crucible or to the lead during the first cupellation.   This course cannot, however, be adopted when it is essential to determine the silver in the ore.

After "inquartation" the button of gold and silver is cleaned, flattened by a hammer on a clean anvil, and "parted" by dropping into nitric acid.

The parting is effected in a test-tube or a glazed porcelain crucible.   The

[1] *Journ. Chem. Met. and Min. Soc. of S. A.*, 1905, v. 165–168 ; and *Precious Metals*, p. 204.

strength of the nitric acid used for parting is of importance: authorities differ, however, as to the best strength of acid to employ for small gold buttons. Keller [1] recommends dilute acid consisting of 1 part of acid to 9 parts of water, and states that in this strength of acid the gold almost invariably remains in a coherent mass, even when the silver is 500 times as much as the gold. This strength of acid is recommended also by Fulton [2] for ordinary assay purposes. The buttons should be boiled in the acid for at least ten to fifteen minutes in order to ensure parting.

According to A. M'A. Johnston, [3] dilute nitric acid (1 of acid : 3 of water) is used on the Rand, South Africa, the flattened button being dropped in when the acid is slightly warm. Action starts at once, and unless the silver is in large excess the gold does not break up. The acid is then brought quickly to boiling-point.

Nitric acid of specific gravity 1·25 (4 of acid : 3 of water) is considered by T. K. Rose [4] as suitable for use with all buttons, and he expresses the opinion that there is no advantage in using more dilute acids for poorer alloys.

The acid employed should be free from chlorine, otherwise the dissolved silver will be precipitated and contaminate the gold.

The freedom of the nitric acid from chlorine is ensured by adding to it a little spent acid containing some silver nitrate.

In parting the buttons, the acid must in all cases be previously heated to boiling, as the gold does not in that case break up into such fine particles as when cold acid is used. The acid attacks the button instantly and violently, turning it black and giving off nitrous fumes. Little buttons containing only a small proportion of gold are dissolved in a few seconds, and decanting of the acid from the residual gold may be at once proceeded with. If the amount of gold is large, the boiling is continued for some minutes. Ten minutes boiling is enough for 10 milligrammes of gold and 40 milligrammes of silver. The acid is then poured off, the residue washed at least twice with hot distilled water by decantation, and if the button is very large, fresh acid of specific gravity 1·30 (2 of acid : 1 of water) is added. The boiling is then continued for five or ten minutes longer, when practically all the silver will have been dissolved.

With the buttons from almost all gold ores no second treatment with acid is required.

The gold usually remains as a single piece if it weighs less than 0·1 milligramme, even if the proportion is only 1 part of gold to 40 or 50 parts of silver. [5] If the ore is rich, the gold sometimes, though rarely, breaks up if hot acid is used, and invariably breaks up if the parting is begun with cold acid. In the event of the gold breaking up, the washings must be very carefully decanted, otherwise the finer particles may float and be lost. The method of dealing with the residual gold varies slightly according to whether a test-tube or a porcelain crucible has been used for parting.

Parting in Test-Tubes.—If a test-tube is used for the boiling, 3 or 4 c.c. of acid is usually quite enough. The test-tubes may be heated by supporting them on a piece of asbestos card suitably heated, or by merely resting them in a beaker of boiling water. When the parting is complete and the acid decanted, the parted gold is washed with hot distilled water as stated, and transferred to an unglazed porcelain crucible or "annealing" cup (fig. 140, page 271). To effect this the test-tube is completely filled with water and the annealing cup inverted over its mouth. The test-tube and cup are then quickly inverted together so that

[1] Keller, *Trans. Am. Inst. Min. Eng.*, 1908, xxxvii. p. 3.
[2] Fulton, *Fire Assaying*, 2nd edit., 1911, p. 108.
[3] *Rand Metallurgical Practice*, "Assaying," 1912, vol. i. p. 305.
[4] T. K. Rose, *Metallurgy of Gold*, 5th edit., 1906, p. 456.
[5] T. K. Rose, *loc. cit.*, p. 456.

14

the tube stands mouth downwards in the cup, and the gold falls steadily through the water into the cup.   When all the gold has settled, the test-tube is removed in such a way as not to disturb the precious metal.   This is best done by lowering the cup, with the test-tube still resting in it, into a porcelain basin filled with water.   As soon as the neck of the test-tube is immersed, the cup can safely be drawn away from under it and then lifted out of the water.   The test tube should not be taken away first, otherwise the gold may be washed out of the cup by the rapid flow of water which takes place when the test-tube empties.

The water in the cup is poured off and the cup dried at a gentle heat and then gradually raised to dull redness.   If the cup is dried too quickly the last drops of water may boil and cause loss of gold by ejection.

The final heating or annealing may be performed in a muffle or over a bunsen or spirit flame.

At a dull red heat the soft black gold assumes its ordinary yellow colour, shrinks, and hardens so that it can be transferred to the pan of a balance and weighed.   In some smelting works the parting is done in flasks similar to that shown in figs. 134A and B, page 262, but of small size.

Parting in Glazed Crucibles.—Parting may also be effected in small glazed porcelain crucibles or cups which are used for both boiling and annealing.   This method of working, which is largely adopted in mine-assay offices, has the advantage that there is no transference of the gold while in its soft state.   On the other hand, the boiling of the acid in a small porcelain crucible requires more care in adjusting the temperature, otherwise there may be loss by ejection. This danger is avoided by using watch-glass covers for the crucibles.   Small Berlin porcelain crucibles, 1 inch in diameter across the top and $\frac{3}{4}$ inch in depth, are very suitable for the purpose.

The crucible, half filled with dilute nitric acid, is heated on a piece of asbestos sheet, on a hot plate, until the acid is nearly boiling.   The flattened button is dropped into it and the heating continued with, at most, gentle boiling until parting is complete, which even with large buttons takes place in from five to ten minutes.   The crucible is then filled with distilled water, which cools it sufficiently for easy handling; and when the gold has settled, the liquor is poured off.   The gold is washed in three successive decantations, by means of a wash bottle; and when the last drops of water have been carefully drained off, the crucible is dried and then heated below redness to anneal the gold, precaution being taken that bright metallic gold is obtained.   Excessive heating must be avoided, as the glaze on the porcelain crucible softens at a bright red heat and the gold sticks to it.   When cold, the gold is transferred directly to the pan of the balance and weighed.

If the crucible shows a black stain after heating for the annealing of the gold it indicates insufficient washing.   The stain is due to metallic silver resulting from the reduction of silver nitrate by the heat.

When a large number of assays have to be dealt with, it is sometimes convenient to number the crucibles with rouge (iron oxide) made into a thin paste with water.

The amount of silver retained by the gold when parted in crucibles is stated by J. Phelps to be about 0·5 per cent., an error which is inappreciable in the assay of ores.

5. **Weighing the Gold.**—The fine gold from the parting should be weighed on a balance turning to $\frac{1}{100}$ milligramme at most, with a half milligramme rider for use on the beam.   Direct weighing is usually preferred, but the gold can also be weighed by the method of substitution, by which the bias of the balance is eliminated.   The principle of this method, as described on page 61, is to counterpoise the gold, then to remove it from the pan, and compensate for the

loss of weight by a rider.   Although requiring a little more time, this method is to be recommended when extreme accuracy is desired.

The method of "weighing by swings" is frequently employed to determine the weight of very minute buttons of gold (see page 62).

It is the practice of some assayers, when dealing with very minute buttons, resulting from the assay of a sample in duplicate, to first weigh approximately the two buttons separately to see that there is no undue discrepancy between the different weights, and then to weigh them together accurately.   This, by the law of averages, gives a result of greater accuracy.

General Remarks on the Assay of Pyritic Ores.—There is considerable diversity of opinion between assayers as to the best method of assaying pyritic gold and silver ores.

With a view to determining the best method to be adopted for ores of this class, experiments have been made from time to time, and the conclusions of some of the more recent investigators may be summarised as follows:—

Experiments were carried out by Keys and Riddell [1] upon a rich silver ore in order to determine the effects of each of the fluxes ordinarily used in the crucible assay upon the silver yield, and to show what type of charge gives the best results on an ore which is high in copper and sulphur and at the same time a typical roasting ore.   They concluded:

(1) That a dead roast is not so accurate as a run with a large excess of nitre.

(2) An excess of borax causes low value determinations, especially in the presence of sulphur, copper, and iron.

(3) An excessively acid flux fuses with great difficulty, and it is hard to handle the buttons, and when the ores contain matte-forming materials the value determinations are low.

(4) A very basic flux is open to the same objections.   In ores containing copper the loss is not so great, but in zinc ores the results are high.   Neither method is practical.

(5) Ores high in copper cannot be run by any crucible method without scorifying the lead buttons.

(6) Direct scorification of copper-bearing ores does not give as high values as a combination of the crucible and scorification methods.

(7) For ores containing much copper or zinc, a large excess of litharge in the charge greatly improves the buttons, decreasing the time and temperature of fusion with very close actual valuations for the silver content of the ore, and the method is quick, easy, and accurate.

Borrowman [2] gives details of the assay of an ore consisting of almost pure iron sulphide, using the following desulphurising agents:—(1) Nitre alone, (2) preliminary roasting, (3) iron nails, (4) nitre and iron nails.   The full charges are given in all cases.   It is concluded that with such an ore the maximum amount of nitre can be used with impunity, and that where great accuracy is desired the more troublesome nitre method should be used in preference to the convenient "nail" fusion.

For the assay of sulphides consisting mainly of iron pyrites, with small amounts of copper (up to 3·5 per cent.), zinc sulphide (4 to 8 per cent.), etc., and from 4 to 20 per cent. silica (insoluble), Fulton [3] observes that, "where gold only has to be determined in ores of this character, the roasting method is satisfactory."   Roasting, however, proves unreliable for silver, and in many cases (as at Leadville) the silver contents of these sulphides are the most important.

The best method, after many trials, was found to be the nitre fusion on comparatively small lots of ore, the details being as follows:—

[1] *Bulletin, Amer. Inst. Min. Eng.,* 1911, lv. 559–568.
[2] *J. Ind. Eng. Chem.,* 1910, ii. 251–252.   [3] *Fire Assaying,* 2nd edit., 1911, p. 148.

Four assays are made on 0·25 assay ton, each with 3 or 4 assay tons of the following stock flux, the amount depending on the reducing power of the ore, *i.e.* on the amount of sulphide present.

Stock flux.				Proportionate amount in 4 A.T. of flux.		
Litharge	.	. 8 parts.		Ore	. . . 0·25 assay ton.	
Nitre .	.	. 1·5 ,,		Litharge	. . 62 grammes.	
Sodium carbonate	3·0 ,,			Nitre .	. . 12 ,,	
Borax glass	. 1·5 ,,			Sodium carbonate	24 ,,	
Sand .	.	. 1·5 ,,		Borax glass	. . 11 ,,	
				Sand .	. . 11 ,,	

Cover of salt or soda.

The temperature of fusion is brought up gradually to a yellow heat. The buttons are usually clean, and separate well from the slag.

Another method which may be used on this type of ore is the nitre-iron method. This has the advantage that no preliminary assay is necessary to determine the amount of nitre for the proper size lead button, but that only sufficient nitre is added to oxidise partially the sulphide, the iron nails being relied upon to decompose the balance of the ore. On ores of the class described the following charge is successful :—

Ore	.	.	.	.	.	.	0·5 assay ton.
Sodium carbonate	.		.	.	.	.	25 grammes.
Litharge .	.	.	.	.	.	.	30 ,,
Nitre	.	.	.	.	.	.	15 ,,
Sand	.	.	.	.	.	.	8 ,,
Borax glass	.	.	.	.	.	.	8 ,,
Iron nails	.	.	.	.	.	.	2 or 3 ,,

Thin cover of borax glass.

The nitre and sand are decreased with a decrease in the reducing power of the ores.

Calculation of Results.—The amount of metal contained in ores of the common metals is usually reported in percentage or parts in a hundred, but with gold and silver ores the proportion of precious metal is so small that it is necessary to state the amount present in ounces or pennyweight per ton. Several methods of reporting the assay value of precious-metal ores are in use, and are described below. The calculation of the assay value of the ore from the weight of metal obtained from the assays is a matter of simple arithmetic, but since gold and silver are sold by troy weight, whilst the ton is avoirdupois, it is of importance to remember that the ounces in the two systems are not the same. There are 7000 grains in 1 lb. avoirdupois and 2240 lbs. in an English ton ; therefore there are 2240 × 7000 = 15,680,000 grains in a ton.

There are 480 grains in 1 ounce troy, consequently there are $\dfrac{15,680,000}{480}$

= 32666·6 troy ounces in 1 ton avoirdupois. With this data the full calculation can be made if necessary. Thus, suppose 20 grammes of ore gave a button of silver weighing 0·0795 grammes (after deducting silver in litharge, etc.), then :—

$$\text{Silver per ton (2240 lbs.)} \quad \frac{2240 \times 7000 \times 0.0795}{20} = \frac{1,246,560 \cdot 0000}{20} = 62,328 \text{ grains.}$$

$$\frac{62,328}{480} = 129 \cdot 85 \text{ ounces.}$$

In practice it would be too fatiguing to calculate fully each assay result, and it is usual to refer to a set of tables which give either directly, or by means of simple addition, the produce per ton.

Table XXVII. gives figures for computing from the percentage of gold or silver the troy weight of precious metal either per ton of 2240 lbs. or per ton of 2000 lbs.

Table XXVII.—Table for computing from the Percentage of Gold or Silver the Assay Value in Ounces and Decimals of an Ounce per Ton.

Per cent.	Per ton of 2240 lbs.		Per ton of 2000 lbs.	
	Ounces.	Equivalent in pennyweight.	Ounces.	Equivalent in pennyweight.
0·0001	0·033	0·65	0·029	0·58
·0002	·065	1·31	·058	1·17
·0003	·098	1·96	·087	1·75
·0004	·131	2·61	·117	2·33
·0005	·163	3·27	·146	2·92
·0006	·196	3·92	·175	3·50
·0007	·229	4·57	·204	4·08
·0008	·261	5·23	·233	4·67
·0009	·294	5·88	·263	5·25
0·001	·327	6·53	·292	5·83
·002	·653	13·07	·583	11·67
·003	·980	19·60	·875	17 50
·004	1·307	26·13	1·167	23·33
·005	1·633	...	1·458	29·17
·006	1·960	...	1·750	...
·007	2·287	..	2·042	...
·008	2·613	...	2·333	...
·009	2·940	...	2·625	...
0·01	3·267	...	2·917	...
·02	6·534	...	5·834	...
·03	9·800	...	8·750	...
·04	13·067	...	11·667	...
·05	16·335	...	14·584	...
·06	19·600	...	17·500	...
·07	22·870	...	20·417	...
·08	26·133	...	23·333	...
·09	29·400	...	26·250	...
0·1	32·667	...	29·167	...
·2	65·330	...	58·340	...
·3	98·000	...	87·500	...
·4	130·670	...	116·667	...
·5	163·330	...	145·835	...
·6	196·000	...	175·000	...
·7	228·700	...	204·165	...
·8	261·330	...	233·334	...
·9	294·000	...	262·500	...
1·0	326·667	...	291·667	...

The percentage of gold or silver in the ore is calculated from the weight of ore taken for assay and the weight of the resulting button of precious metal as follows :—

$$\frac{\text{Weight of gold or silver button}}{\text{Weight of ore taken for assay}} \times 100 = \text{percentage of metal in the ore.}$$

The method of using the table to ascertain the quantity of precious metal per ton of ore is best shown by an example.

Calculation of Silver-Ore Results.—Taking the simplest case of a silver ore assayed by the fusion method.

Suppose a button of silver weighing 0·2135 gramme was obtained from 20 grammes of ore, then,

$$\frac{0·2135}{20} \times 100 = 1·0675 \text{ per cent. silver.}$$

From the table select the values corresponding to each figure of the weight, thus :

1·0	per cent.	= 326·67	ounces per ton (2240 lbs.).	
·06	,,	= 19·60	,,	,,
·007	,,	= 2·29	,,	,,
·0005	,,	= ·16	,,	,,
Total, 1·0675		348·72		

Add these together, and the assay value of the ore is 348·72 ounces per ton (2240 lbs.). If the result is to be returned on the ton of 2000, the figures would of course be selected from the short-ton column accordingly.

The granulated lead and litharge used for assay purposes invariably contain more or less silver, and it is consequently requisite to determine beforehand (see page 145) the amount of silver present, and make a corresponding deduction from the weight of the button resulting from the assay of the ore.

Also, when the slag and cupels have been separately treated for the recovery of the precious metal they contain, the results of these assays must be allowed for in determining the assay value of the ore.

Thus, using the same example for illustration, we should get the following corrected result after making the above allowances.

"*Corrected*" *Assay Result.*—Ore taken for assay, 20 grammes.

Silver obtained by crucible assay . . .	0·2135	gramme.
Silver recovered from slag	0·0052	,,
Silver recovered from cupel . . .	0·0093	,,
Total . .	0·2280	,,
Deduct silver in red lead	0·0016	,,

Corrected result for silver in 20 grammes of ore 0·2264 = 1·1320 per cent.

From the table we find 1·1320 per cent. = 369·79 ounces per ton (2240 lbs.).

Calculation of Silver in Ore with Metallics.—When an ore contains metallics or scales (see page 111), as is frequently the case with precious-metal ores, the calculation becomes more complicated. The metallics have to be collected and assayed separately, and their precious-metal content calculated independently of the result obtained by the ordinary assay of the ore.

For the purpose of calculation it is necessary to take separately the weight of the metallics and the weight of the ore from which they are obtained. The following example will serve as an illustration :—

Total weight of sample	350·0	grammes.
The metallics weigh	1·25	,,
Hence the sifted portion weighs . .	348·75	,,

Twenty grammes of the sifted portion, when assayed, gave 0·1315 gramme of silver, equivalent to 0·6575 per cent.   The whole of the metallics, when scorified and cupelled, gave 0·831 gramme of silver, equivalent to 66·48 per cent.   The total silver content of the ore is calculated as follows :—

Calculate separately the content (best stated in per cents.) of precious metal in each portion of the sample. Multiply each content by the weight of the portion it represents. Add together the resulting products, and divide by the weight of the whole sample.

This may be stated thus :—

Let A = weight of the sifted portion.

B = weight of the metallics.

C = assay value of the sifted portion in per cents.

D = assay value of the metallics in per cents.

Then,

$$\frac{A \times C + B \times D}{A + B} = \text{total silver (or gold) content of the ore in per cents.}$$

Taking the above example for illustration we have,

$$\frac{348\text{·}75 \times 0\text{·}6575 + 1\text{·}25 \times 66\text{·}48}{348\text{·}75 + 1\text{·}25} = \frac{229\text{·}303125 + 83\text{·}10}{350}$$

$$= \frac{312\text{·}403125}{350} = 0\text{·}8923 \text{ per cent.}$$

Referring to the table we find

0·8	per cent.	=	261·33	ounces.
·09	,,	=	29·400	,,
·002	,,	=	·653	,,
·0003	,,	=	·098	,,
0·8923			291·481	

The total silver content of the ore is therefore 291·5 ounces per ton (2240 lbs.).

Calculation of Gold-Ore Results.—In calculating the assay value of a gold ore the gold content and the silver content are calculated separately. The weight of the fine gold from parting is subtracted from the weight of the gold-silver button from the cupellation, and the difference gives the weight of the silver.

Example.—Weight of ore taken, 50 grammes.

Weight of gold-silver button	0·0177	gramme.
Weight of gold from parting	0·0105	,,
Difference equals silver	0·0072	,,
Deduct silver in litharge	0·0015	,,
Silver in ore	0·0057	,,

Gold in 50 grammes ore 0·0177 = 0·0354 per cent. gold.
Silver in 50 ,, ,, 0·0057 = 0·0114 ,, silver.

From the table we ascertain that

0·0354 per cent. = 11·56 ounces gold per ton 2240 lbs.
0·0114 ,, = 3·73 ,, silver ,, ,,

In the case of a gold ore containing metallics separate assays of the ore and the metallics are made, and the resulting buttons of gold and silver parted in the usual way.

The total yield of fine gold and of fine silver is then calculated independently, as previously described.

Calculation for Assay-Ton System.—The calculation of the assay value of an ore is very much simplified when the quantity taken for assay has been weighed in assay tons (A.T.). As previously explained, the assay-ton system is so devised that if 1 assay-ton of ore yields a button of gold or silver weighing 1 milligramme, then 1 ton of ore yields 1 ounce troy of precious metal per ton of ore. When quantities either larger or smaller than 1 assay ton are taken for assay, the weight of the button of precious metal in milligrammes, divided by the assay-ton weight of the ore taken, will give the number of ounces per ton, thus :—

$$\frac{\text{Weight of button in milligrammes}}{\text{Weight of ore taken in assay ton}} = \text{number of ounces per ton.}$$

The following is an example :—
One-tenth (0·1) assay ton of silver ore gave by scorification 0·0413 gramme of silver. Then,

Silver from scorification assay	.	.	. 0·0413 gramme.
Silver from slag assay	.	.	. 0·0021 ,,
Silver from cupel assay	.	.	. 0·0020 ,,
			0·0454 ,,
Deduct silver in granulated lead used	.	.	0·0016 ,,
Corrected weight of silver in $\frac{1}{10}$ A.T. ore	.		0·0438 ,,

·0438 gramme = 43·8 milligrammes.

$$\frac{43\cdot8 \text{ mgms.}}{\frac{1}{10} \text{ A.T.}} = 43\cdot8 \times 10 = 438 \text{ ounces of silver per ton.}$$

The assay value of a gold ore is calculated in the same way, a separate calculation being made for the fine gold and for the fine silver.

For example :—Ore taken, 2 A.T.

Weight of gold-silver button	.	·0076 gramme = 7·6 milligrammes.
Weight of fine gold	. .	·0002 ,, = 4·2 ,,
Weight of fine silver	. .	·0034 ,, = 3·4 ,,

$$\text{Gold per ton } \frac{4\cdot2 \text{ mgms.}}{2 \text{ A.T.}} = 2\cdot1 \text{ ounces per ton.}$$

$$\text{Silver per ton } \frac{3\cdot4 \text{ mgms.}}{2 \text{ A.T.}} = 1\cdot7 \text{ ounces per ton.}$$

The corrections for silver in litharge, etc., have been omitted to simplify the calculations.

Calculation for Ores with Metallics (Assay-Ton System).—The pulverised ore and the metallics are separately assayed, and from the results obtained the total weight (in grammes) of silver (or gold) in the entire sample is calculated. The weight thus obtained, multiplied by 29166·66 and divided by the original weight of ore taken in grammes, gives the value in ounces per ton of 2000 lbs.

The following example will illustrate the method of calculation :—

Let us suppose that a sample of auriferous silver ore, on being pulverised and passed through a 90-mesh sieve, gave the following weights :—

Sifted ore	142·56 grammes.
Metallics or scales	1·14 ,,
	143·70 ,,

Each portion when separately assayed gave the following results :—

Sifted Ore, by crucible assay.

One-half assay ton (14·5833 grammes) yielded :—

Silver and gold together	0·1085 gramme.
Gold by parting	0·0021 ,,
Silver by difference	0·1064 ,,
Silver in the litharge	0·0006 ,,
Silver in the ore	0·1058 ,,

Metallics, by scorification assay.

The whole of the metallics (1·14 grammes) yielded :—

Silver and gold together	0·8326 gramme.
Gold by parting	0·0073 ,,
Silver by difference	0·8253 ,,
Silver in test lead	none.

From these results the value of the ore in silver and in gold is calculated separately.

Calculation of the Silver.

(a) *In Sifted Ore.*—If one-half assay ton yields 0·1058 gramme of silver, then the number of grammes of silver in 142·56 grammes is

$$142·56 \times \frac{0·1058}{14·5833} = 1·034 \text{ grammes.}$$

(b) *In the Metallics.*—If 1·14 gramme yields 0·8253 gramme, then (as the whole of the metallics were used) the number of grammes of silver will obviously be

$$1·14 \times \frac{0·8253}{1·14} = 0·8253 \text{ gramme.}$$

The total weight of silver in the entire sample is therefore :—

Weight of silver in sifted ore . . .	1·034 grammes.
Weight of silver in metallics . . .	·8253 ,,
Total weight of silver in entire sample .	1·8593 ,,

Then the number of ounces of silver per ton of ore is

$$29166·66 \times \frac{1·8593}{143·70} = 377·3805 \text{ ounces per ton.}$$

Calculation of the Gold.

The number of ounces of gold per ton is calculated in the same way.

(a) *In Sifted Ore*—One-half assay ton yielded 0·0021 gramme, therefore:—

$$142 \cdot 56 \times \frac{0 \cdot 0021}{14 \cdot 5833} = 0 \cdot 0205 \text{ gramme.}$$

(b) *In the Metallics.*—The whole of the metallics yielded 0·0073 gramme, therefore:—

$$1 \cdot 14 \times \frac{0 \cdot 0073}{1 \cdot 14} = 0 \cdot 0073 \text{ gramme.}$$

The total weight of gold in the entire sample is

Weight of gold in sifted ore . . .	0·0205	gramme.
Weight of gold in metallics . . .	0·0073	„
Total weight of gold in entire sample .	0·0278	„

Then the number of ounces of gold per ton of ore is

$$29166 \cdot 66 \times \frac{0 \cdot 0278}{143 \cdot 70} = 5 \cdot 647 \text{ ounces per ton.}$$

Agreement of Results.—The permissible difference between the results of duplicate assays varies according to circumstances. With low-grade ores there should be practically no difference between the results, while with high-grade ores small differences are allowable according to the richness and nature of the ore.

As a general rule, assays of average silver ores should agree very closely, and if properly conducted are of great accuracy. The results of duplicate assays should not differ from each other more than 2 or 3 ounces troy per ton. Should a greater difference be found, an additional assay should be made. (See also page 192.)

With regard to gold ores, the difference must be much smaller than in the case of silver ores. The results of duplicate assays on ores containing under 5 ounces of gold per ton should not differ more than 1 dwt. per ton or even 0·5 dwt. per ton for low-grade ores. With richer material assaying up to, say, 20 ounces per ton, a difference of 2 dwt. per ton is permissible.

In the purchase of gold and silver ores the difference to be allowed between the assay result of the buyers' representative and the result of the sellers' representative is mutually agreed upon before purchase.

Reporting the Results.—Several methods are in use for reporting the assay value of precious-metal ores. As previously stated, in England and Australia the quantity of silver and gold in ores is reported in ounces and decimals of an ounce troy per ton of 2240 lbs. avoirdupois, known as the English or long ton.

In Canada, the United States, and Africa it is reported either in ounces and decimals of an ounce troy per ton of 2000 lbs. avoirdupois, or in the number of dollars and cents worth of precious metal per ton of 2000 lbs. avoirdupois. The ton of 2000 lbs. is known as the American or short ton, and is a sort of compromise between the avoirdupois and metric system. When the quantity of precious metal is less than an ounce it is reported in pennyweight and decimals of a pennyweight.

The old method of reporting results in ounces, pennyweight, and grains troy per ton avoirdupois is now being discontinued. The following is an illustration of the general method of reporting the assay values of gold and silver ores and products. (Sample dried at 100° C.)

Fine gold . 3·24 ounces per ton of 2000 lbs.
Fine silver . 2·40   „   „   (or 2240 lbs. as required).

The values are usually reported to the second place of decimal.

The method of reporting "on the ton of 2000 lbs." instead of "on the ton of 2240 lbs." is now being very generally adopted.

In reporting the value of low-grade ores and products, the use of the term trace or traces is to be deprecated, unless when especially designating its application. "Thus in mine samples many mining men prefer that no intermediate stage be adopted between 0·5 dwt. of gold and nil, any distinction between these, except with special work, being generally guesswork on the part of the assayer. In reduction works (i.e. cyanide works) samples, reports should be limited to the use of 0·01 dwt. or nil; here again any traces found being · merely a speck, unweighable, and sometimes even indeterminable as gold." [1]

The method of reporting the gold and silver content in money value is chiefly confined to America. In calculating the money value of an ore, a troy ounce of fine gold is considered to be worth 20·67 dollars. Fine silver is valued according to the prevailing market price of the metal, which fluctuates considerably. The nominal value of a troy ounce of fine silver may be taken as 60 cents.

As an example, the money value of an ore having an assay value of 5·5 ounces of fine gold per ton and 35·75 ounces of fine silver per ton of 2000 lbs. would be reported thus :—

Gold, 5·5 ounces at 20·67 dollars per ounce   .   . 113·68 dollars.
Silver, 35·75 ounces at 0·60 dollar per ounce .   .   21·45   „

Total value of ore in gold and silver per ton of
    2000 lbs. .   .   .   . 135·13 dollars.

This would be termed a 135·13 dollar ore.

American "dollars on the ton of 2000 lbs." may be converted into "ounces in the ton of 2240 lbs." by dividing by 1·1544 in the case of silver, and by · 18·457 in the case of gold.

If the value of an ore is to be reported in English currency the value of 1 ounce of fine gold is taken as £4·247 (or £4, 4s. 11·45d.), while the value of fine silver varies but may be taken at 2d. or 2½d. above the market quotation for "standard" or coinage silver (which contains 925 parts of silver and 75 parts of copper in 1000 parts).

As pointed out by Beringer,[2] the practice of reporting assay results in money value is objectionable. "The prices of metals vary with the fluctuations of the market, and if the assayer fixed the price, the date of his report would be all important; if, on the other hand, he takes a fixed price which does not at all times agree with the market one, it leaves a path open for the deception of those unacquainted with the custom."

When it is necessary to state or estimate the money value of an ore, etc. (other than gold), it should be accompanied by the assay value, and the basis on which the former has been calculated from the latter should be stated.[3]

It is the practice of some assayers to report the value of gold ores in terms of "bullion," but this method of reporting is to be strongly deprecated unless the fineness (i.e. parts of fine gold in 1000 parts) of the bullion is also stated.

[1] A. M'A. Johnston, Rand Metallurgical Practice, vol. i., 1912, "Assaying," p. 320.
[2] Text-Book of Assaying, 11th edit., 1910, p. 9.
[3] Recommendation of the Inst. of Min. and Met., Trans. Inst. Min. and Met., vol. xx., 1910, p. 530.

The money value of bullion is dependent on the proportion of fine gold it contains ; thus the value of 1 ounce of bullion 942 fine would be about £4, while bullion 412 fine would only be worth about 35s., so that the statement that an ore contains so many ounces of "bullion" gives little indication of the money value of the ore.   Such reports are very misleading to the uninitiated, and are only favoured by a certain class of company promoters for reasons that had better not be stated.   If the assay value of gold and silver is to be reported as "bullion," it should be accompanied by a statement of the proportion of fine gold and fine silver.

Assay reports should state the exact condition of the sample as to dryness when assayed : it should be stated whether the sample was assayed in the condition in which it was received, or dried at —°C.   As a general rule, the assay value of gold and silver ores are reported on the dried ore for the reasons stated on page 125.

The amount of moisture is usually stated on the report.   Two methods are in use for reporting the amount of moisture present in ores, etc.   On the Continent the moisture is invariably reported in percentage, but in England it is still a very general practice to report the moisture in grains per pound, thus : if the material contains 1 per cent. of moisture the report will read "moisture $\frac{70}{7000}$"

meaning that there are 70 grains of moisture in 7000 grains or 1 lb. avoirdupois.[1]

The common limit for moisture in exported ores is from $\frac{20}{7000}$ to $\frac{50}{7000}$.

How long this old Cornish method of reporting moisture, which has survived in England to the present day, will continue to be used, in face of more modern and simpler Continental procedure, it is difficult to say.

The determination of moisture in consignments of ore at British ports or works is carried out by the joint samplers of buyer and seller, who may or may not have anything to do with the subsequent assay of the sample.

Examples of forms used by assayers for reporting the assay value of auriferous and argentiferous ores and products are given on page 14.

---

[1] H. L. Terry, *Min. Mag.*, vol. vi., 1912, p. 62.

CHAPTER XV.

THE ASSAY OF COMPLEX GOLD AND SILVER ORES.

THE term complex ores includes all auriferous and argentiferous ores containing excess of metals such as copper, antimony, tellurium, etc., which interfere more or less in the ordinary methods of assay and therefore require special treatment.

In many cases complex ores are satisfactorily assayed by combined wet and dry methods. The ore is treated with acid to remove the interfering metal or metals, and the precious metals collected into a button of lead by fusion of the insoluble residue after precipitating the silver in the solution.

Scorification may sometimes be applied to the assay of ores of this class when they are sufficiently rich to permit of small quantities being taken for assay.

The following complex ores are considered :—

1. Arsenical ores.
2. Antimonial ores.
3. Cupriferous ores.
4. Telluride ores.
5. Ores containing lead, zinc, nickel and cobalt, tin, bismuth, etc.

1. **Arsenical Ores.**—Arsenic is most frequently associated with gold ores as mispickel (arsenical pyrites) and occasionally as oxides.

As already mentioned, the presence of arsenic in an ore gives rise to the production of compounds which tend to diminish the accuracy of the assay. It is therefore necessary to eliminate the arsenic, and this is usually effected by subjecting the ore to an oxidising roasting or to the oxidising action of nitre. Much diversity of opinion exists as to which is the better of these two methods, many assayers hesitating to roast arsenical ores on the ground that, owing to the volatile character of arsenic, mechanical loss of the precious metals is caused. It has been pointed out that this may occur when too high a temperature is employed at the outset, thus causing a rapid disengagement of the volatile constituents of the ore, but with due precaution this may be avoided. The formation of arseniates is mostly to be feared, as they are liable to remain undecomposed and cause loss of gold and silver in the slag during fusion. The tendency to form arseniates is greater when nitre is employed as the oxidising agent, and on that account oxidation by roasting is preferable for the elimination of the arsenic. The ore is roasted, with all the precautions previously mentioned under pyritic ores, until completely oxidised; then from 5 to 10 per cent. of charcoal or anthracite powder is added, and the roasting continued until all the carbon is burnt out. Owing to the rapidity with which arsenic vaporises, great care is needed in regulating the temperature. In the early stages of the roasting it is necessary to employ a very low temperature and admit only a limited supply of air.

An attempt to eliminate arsenic by rapid oxidation would be attended with the danger of converting it into arseniates of the metals present in the ore, such as iron, copper, lead, etc. When once formed, arseniates are not readily decomposed, as they resist a high temperature and are only slowly converted into sulphates at a red heat by the sulphuric acid formed during the later stages of the roasting.

As before stated, with the exception of arseniate of copper, the decomposition of the arseniates of the metals most frequently associated with gold ores is effected by the addition of charcoal or anthracite powder, but in many cases prolonged heating is necessary; it is therefore desirable to avoid their formation as far as possible by taking the precautions already mentioned. The expulsion of arsenic, and also of antimony, is favoured by the presence of iron pyrites, as the sulphur distilled from the pyrites in the early stages of the roasting tends to eliminate them as sulphides.

The product obtained by roasting the ore is composed largely of ferric oxide, and is fluxed according to the directions given for the treatment of basic ores, to which class it belongs. By the addition of a small quantity of potassium cyanide to the charge for the fusion, the decomposition of any arseniates remaining in the roasted product is greatly facilitated. In exceptional cases the button of lead may be slightly brittle owing to the presence of arsenic. When an appreciable quantity of arsenic is present in the lead the effect on cupellation is somewhat similar to that produced by antimony. If present in large quantities, it may be removed by scorification. When the oxidation of the arsenical compounds is effected by means of nitre, the ore may be treated by the nitre methods previously described.

If arsenical ores are fused direct with the addition of metallic iron, the lead button will be accompanied by a layer of spiese which tends to retain values. Some assayers use the iron-reduction method and scorify the speise with the lead button after crushing it to a coarse powder.

2. **Antimonial Ores.**—Gold and silver ores containing antimony minerals such as stibnite, jamesonite, etc., are usually assayed by the nitre method. Oxidation of the antimonial compounds may be effected by roasting, but owing to the very fusible and volatile nature of stibnite, etc., it is very difficult to satisfactorily roast these ores without caking, especially when the percentage of stibnite is large. The addition of pure siliceous sand to the ore before roasting assists in preventing the formation of lumps.

C. O. Bannister [1] has shown that careful roasting, with the addition of a little charcoal powder towards the end of the operation, followed by a suitable fusion, gives good results. No loss of precious metals takes place in the roasting of antimonial ores when proper precautions are taken, and this method of oxidation is sometimes used by assayers. But in general practice few assayers have the time and patience needed to roast these ores, and the method of oxidation with nitre is most generally employed.

The amount of nitre required is determined if necessary by a preliminary assay, as previously described. The charge must contain excess of sodium carbonate to induce the formation of sodium antimonate. When the ore contains a large proportion of antimony, say over 50 per cent., it is usually more satisfactory to run two charges of 20–25 grammes each, and combine the resulting lead buttons, than to make one fusion on 50 grammes of ore.

The following charge is suitable for ores containing about 75 per cent. of stibnite with gangue mainly silica :—

Ore	20–25 grammes.
Red lead	80 ,,
Sodium carbonate	60 ,,
Borax	10–20 ,,
Nitre	14–20 ,,

[1] *Trans. Inst. Min. and Met.*, 1906, vol. xvi. p. 94.

The fusion should be conducted slowly and at a low temperature in order to keep the button of lead as free as possible from antimony. The button of lead should weigh from 45 to 50 grammes.

With due attention, the nitre may be so adjusted that a malleable button, comparatively free from antimony, will be obtained, and may be cupelled direct. If the button is brittle, comparatively hard, and white, it generally contains antimony, in which case it must be scorified with, if necessary, the addition of more lead before being cupelled.

According to Rivot,[1] antimony does not materially interfere with cupellation if less than 1 per cent. is present. If the lead contains much more than this amount, the antimony in volatilising takes gold with it, and it also forms antimonate of lead, giving a pale yellow scoria, and causing the cupel to crack at the edges.

This last effect is very characteristic of antimony, and is more likely to occur with bone-ash cupels than with magnesia cupels. The button of lead is scorified at a moderate red heat with free access of atmospheric air. By this means both the lead and the antimony are quickly oxidised, but the antimony in greater proportion than the lead, a fusible compound of antimony and lead oxide being formed, and the lead practically freed from antimony. The slags from comparatively rich ores should be cleaned in the manner already described.

It is necessary to clean the slags from antimonial silver ores and also in the case of gold ores, if an exact determination of the silver accompanying the gold is required.

Considerable corrosion of the crucible takes place with the highly basic charge employed, but this may be minimised by adding silica to the charge, or, better still, 10 to 20 grammes of fireclay or any common clay. The latter is recommended by Holloway,[2] who points out that it is similar in nature to the pot itself, and its fine state of division renders it more readily acted upon by the fluxes than is silica alone.

The presence of much silica in the charge tends to increase the amount of antimony in the lead button.

Combined Wet and Dry Method.—The following method, involving solution of the ore, may be conveniently employed for antimonial ores as a check upon other methods. It possesses the advantage of producing lead buttons practically free from antimony, and thus avoiding the necessity of scorification previous to cupellation. The weighed quantity of ore (from 25 to 50 grammes) is treated with concentrated hydrochloric acid, and heated until decomposition is complete; the solution is diluted with a small quantity of water in which a few crystals of tartaric acid have been dissolved. If sufficient tartaric acid has not been introduced, this dilution will cause the precipitation of some of the antimony as oxychloride. The insoluble residue is allowed to settle, and when the solution is clear as much of the liquid as possible is poured off through a filter without disturbing the residue. The residue, if small, is finally transferred to the filter, allowed to drain, and then dried. The filter paper is burnt, and the ash, with the insoluble residue, is scorified with 10 times its weight of granulated lead and a cover of borax, or it is fused with the following fluxes:—

Red lead	30 grammes
Charcoal powder	1·5 ,,
Sodium carbonate and borax, varying together up to .	30 ,,

The fusion is effected in the ordinary way, and the resulting button of lead cupelled direct. It may be remarked that the oxidised ores of antimony,

---

[1] Docimasie, *Traité d'analyse des substances minerales.*
[2] *Trans. Inst. Min. and Met.,* 1906, vol. xvi. p. 97.

cervantite (Sb_2O_4), and scnarmontite (Sb_2O_3) are only partially soluble in hydrochloric acid, the former being practically insoluble. They may be satisfactorily assayed by fusion direct. The oxides, however, are of somewhat rare occurrence, the ore most commonly met with in commerce being the sulphide (stibnite, Sb_2S_3), which is readily soluble in hydrochloric acid.

3. **Cupriferous Ores.**—Ores of this class are usually very troublesome to assay correctly by the fusion method, as they are generally poor in gold and silver, and frequently contain a large proportion of copper.

Owing to the readiness with which copper compounds are reduced during the fusion of the ore, some difficulty is experienced in passing the copper into the slag, and thus preventing the formation of cupriferous lead buttons. Copper sulphide is decomposed by lead oxide, the sulphur and a part of the copper being oxidised and retained in the slag ; the rest of the copper passes into the reduced lead, thus rendering it hard and difficult to cupel. One part by weight of copper requires about 16 parts of lead for cupellation, and though a smaller quantity of lead will pass the copper into the cupel, yet a sensible amount of gold and silver would in that case accompany the lead and copper, so that an incorrect result would be obtained (Percy). Under these circumstances it becomes necessary to add enough lead oxide to the charge not only to collect the gold and silver and oxidise the sulphur present, but to yield a button of lead sufficiently large to satisfactorily carry into the cupel all the copper present in the lead. With ores containing much copper this would give rise to the production of lead buttons inconveniently large for cupellation direct. In order to obtain a lead button sufficiently free from copper for cupellation, the proportion of copper in an ore should not, according to J. Loevy, exceed 6 per cent.[1] Copper ores (whether oxidised or containing sulphur) which do not contain more than 6 per cent. of copper may be readily reduced by fusion, and yield a lead button which is poor in copper and can be at once subjected to cupellation. With ores containing more than 6 per cent. of copper, Loevy reduces the proportion by the addition of silica. Thus to an ore with 12 per cent. of copper is added an equal weight of sea sand. If the ore contains 24 per cent., three times its own weight of sand is added. The mixture of ore and sand are then fluxed and fused in the usual way, the larger charges being divided into several smaller charges and the buttons resulting from the fusions concentrated by scorification into one lead button weighing 30 to 35 grammes.

In practice the fusion assay is seldom attempted for ores rich in copper, as they are more satisfactorily assayed by scorification or by a combination of wet and dry methods. For scorification from 2 to 3 grammes ($\frac{1}{10}$ of assay ton) of ore are scorified with from 15 to 20 times its weight of granulated lead, with the addition of 0·25 gramme of borax, and, if necessary, 0·5 to 1·0 gramme of silica.

If the resulting lead button is hard from the presence of copper, it should be re-scorified, with the addition of fresh lead if necessary. In cupelling the lead button the temperature should not be too low at the finish, otherwise the button of silver and gold will retain copper and give too high a result.

In the combination wet and dry method the ore is first treated with nitric acid to remove the copper by solution. The silver present is also dissolved, and is precipitated by the careful addition of sodium chloride. The insoluble residue containing the precious metals is then scorified or fused in a crucible with fluxes for the recovery of the gold and silver.

To facilitate the settling of the finely divided precipitate of silver chloride and of the suspended gold it is usual to add a solution of lead acetate, and precipitate the lead by the addition of sulphuric acid   The precipitate of lead

[1] *Chem. Zeit.*, 1911, xxxv. p. 278.

sulphate thus formed settles and carries down the gold and silver, leaving a clear solution.

The method of procedure differs in details with different assayers, but the following method is in very general use :—Stir 25 grammes (or 1 assay ton) of ore into a No. 5 beaker containing 100 c.c. water. Add 50 c.c. commercial nitric acid (sp. g. 1·42). Cover and allow to stand till action apparently ceases. Add another 50 c.c. nitric acid and boil gently until the greater part of the free acid has been expelled. Dilute to 500 c.c., and allow to settle. To precipitate the suspended gold and dissolved silver, add 2 to 4 c.c. normal salt solution, 5 c.c. sulphuric acid, and 10 c.c. lead acetate. Set aside to settle over-night; filter, wash, and dry. The filtered solution, which should be bright and clear, contains the copper, and may be neglected. The siliceous residue contains the gold and silver.

If the amount of residue is small, it may be scorified. The dried filter is carefully burnt on a scorifier placed in the mouth of the muffle at a low heat, then scorified with about 40 grammes of granulated lead and 1 gramme of borax. The resulting lead button is cupelled, and the gold and silver parted as usual.

In practice, by far the larger number of ores contain so much insoluble material that they are not adapted for scorification. If the ores are pure sulphides or carbonates, as is seldom the case, the residue can be scorified. Usually, however, they are composed chiefly of quartz or insoluble silicates, and consequently it is best to treat them by a crucible method.

The residue may be dried in a roasting dish, then heated at a gentle heat to burn the filter paper, and mixed with—

Lead oxide	.	.	.	.	40 grammes or 1·5 A.T.
Sodium carbonate		.	.	30 "	or 1·0 A.T.
Borax .	.	.	.	.	10 "
Argol .	.	.	.	.	1·5 "

With cover of borax, 10 grammes.

The charge may be mixed in the crucible with a spatula. Fuse for thirty minutes, then cupel, part, and weigh.

Some assayers put about half the lead oxide into the crucible, add the dry residue and filter paper, then burn the paper, and add the remainder of the lead oxide and other fluxes. The charge is then well mixed and fused as before.

The combination method requires careful manipulation to obtain accurate results.

The results for gold are usually a little low : the cause of this has not been fully investigated, but is probably due to the solvent action of copper salts. In precipitating the silver in solution excess of sodium chloride must be avoided, as silver chloride is appreciably soluble in a strong solution of salt. By the combination method, lead buttons free from copper are obtained, so the operation of cupellation is not liable to be interfered with.

4. **Telluride Ores.**—Gold and silver ores containing tellurium and selenium have received considerable attention from assayers within recent years owing to the difficulty experienced in the determination of the precious metals in these ores as compared with ordinary ores.

It may be well to remark that the most recent experiments on the assay of telluride ores have shown that the extraordinary assay losses which have been recorded from time to time by various assayers must be regarded as exceptional, and that even under ordinary working conditions the losses in assaying gold and silver tellurides are not excessive.

Research has shown conclusively that the loss due to volatilisation of the tellurides is not nearly so serious a factor as was generally supposed. There

15

226 THE SAMPLING AND ASSAY OF THE PRECIOUS METALS.

seems little doubt that many of the excessive losses recorded in assaying telluride ores are attributable to want of uniformity in the sample owing to insufficient grinding and mixing rather than to losses in the conduct of the assay operations. Although the association of gold and tellurium is usually considered to be exceptional, it has been pointed out by Kemp,[1] and confirmed by Sharwood,[2] that "tellurium is a more widely distributed and more common associate of gold than has been generally appreciated." As shown later, the chief telluride ores come from Colorado and West Australia.

Selenium has also been found associated with gold and silver ores, and as its behaviour during assay is very similar to that of tellurium, the precautions necessary for the assay of telluride ores apply in general also to selenium ores.

It must be remembered that, although the telluride mineral may be enormously rich in both tellurium and gold, the average telluride ore is one containing only pennyweight, or, at the most, a few ounces of gold per ton, the tellurium seldom much exceeding the gold in amount, and the telluride being often so finely disseminated through the gangue that no portions of it can be picked out.[3] A telluride ore yielding 1 ounce of gold is not likely, for instance, to contain more than 0·005 per cent. of tellurium, and the iron and other minerals which are commonly associated with it interfere seriously with the ordinary methods of detection.

The telluride can to a great extent be separated from the gangue by panning down the finely ground ore, but the concentrates, although far richer in tellurides, contain also more of the interfering minerals. It is, however, generally advisable to perform this concentration before applying tests, when sufficient ore is available.

Tests for Tellurium.—The following methods will be found of use in the testing of picked pieces :—

A piece of rich telluride, preferably a small picked piece of the telluride itself, or some of the concentrates, will, if heated with concentrated sulphuric acid, give a carmine coloration to the solution when the heat has become sufficient to cause evolution of white fumes of sulphur trioxide. The colour is intensified on adding a small scrap of tin-foil to the solution, bringing it to the boil, and then cooling somewhat. This addition of tin is essential in the case of oxidised tellurides. The colour gradually disappears on further cooling, and entirely fades in time, free tellurium being deposited. The colour, however, reappears on again heating. Dilution also precipitates free tellurium with destruction of the colour.

On heating the ore with strong nitric acid, diluting somewhat, filtering, evaporating the filtrate with sulphuric acid until heavy fumes of sulphur trioxide are evolved, and adding tin-foil, the same carmine colour is produced. This method is more delicate than that in which the sulphuric acid is used direct. In both these tests the presence of iron interferes with, and often completely masks, the colour due to tellurium.

In testing for tellurium in ores by the sulphuric-acid method there is a possible source of error, as it has been shown by Sharwood,[4] that manganese peroxide, if present in the ore, also gives a fine purple colour when treated with strong sulphuric acid and then allowed to stand for a few minutes. The colour is probably due to the formation of permanganic acid.

While this test is an excellent one for clean particles of unoxidised tellurium

[1] *Mineral Industry*, vol. vi., 1897, pp. 295–320.
[2] *Economic Geology*, vol. vi., 1911, p. 22.
[3] Holloway and Pearse, "The Assay of Telluride Ores," *Trans. Inst. Min. and Met.*, 1908, vol. xvii. p. 174.
[4] *Economic Geology*, 1911, vol. vi. p. 85.

minerals, it often fails in the presence of large amounts of sulphides such as pyrite, and is useless in the case of oxidised compounds unless previously reduced.

G. T. Holloway[1] has described a simple and generally applicable method of isolating tellurium from ores for testing purposes by concentrating it with the gold and silver in a lead button obtained by the ordinary fire assay, it being possible to treat as much as 50 or 60 grammes of ore in this way. A clean portion of the lead button is then dissolved in nitric acid, diluted considerably, and the tellurium precipitated by means of lead-foil, the deposit thus obtained being then dried and tested by concentrated sulphuric acid.

Too much stress cannot be laid on the necessity for careful sampling of telluride ores. Owing to the irregular distribution of the values in these ores very fine crushing is required to get a true sample. Experience shows that telluride ores should be crushed to pass a sieve of at least 120 mesh.

Tellurium has a great affinity for gold and silver, and also for lead, and when present in an ore in comparatively large quantity some of it will pass into the lead button with the precious metals in the fusion assay and exert its influence in the cupellation as described on page 169. With average ores the amount of tellurium left in the gold or silver button is usually insufficient to make any appreciable difference to the assay, provided proper precautions are taken to prevent it from passing into the lead button.

It is therefore very essential to remove as much tellurium from the gold and silver before cupellation as possible, and in the assay of telluride ores experience has shown that this is best accomplished by a crucible fusion with an oxidising charge containing lead oxide as the oxidising agent.

When ores containing telluride minerals are fused in the presence of a large excess of lead oxide, the tellurium is oxidised, forming probably lead tellurate and in the presence of soda, sodium tellurate. According to S. W. Smith,[2] tellurium reacts with litharge at moderate temperatures (700° to 900° C.) before the formation of silicates has taken place, in accordance with the equation :—

$$2PbO + Te = Pb_2O + TeO.$$

During the formation of silicates there is a partial reversal of this reaction, although the greater part of the tellurium which has once been oxidised dissolves to a red glass in the slag. When litharge has reacted sufficiently to form silicates there is no longer any oxidising effect on tellurium.

There appears to be general agreement among workers on telluride ores as to the precautions necessary for successful crucible fusion, although greater or less importance is attached to particular observances by individual writers. The best practice with regard both to silver and gold telluride ores appears to be to make a crucible fusion with a fairly large quantity of litharge or red lead, get a large lead button, and then either to cupel the lead button direct or to scorify the button only partially and then cupel.

The compositions of the charges recommended by different assayers have the same general character, a large excess of lead oxide (usually litharge) being their chief characteristic.

Holloway and Pearse[3] state that the passage of tellurium into the slag appears to be more complete in the case of a basic than an acid slag. Every worker on telluride ores has insisted on the necessity for careful fritting and slow fusion at moderate temperatures. The reasons for this are explained by the conclusions of S. W. Smith, quoted above.

It will be evident that, if the litharge does not exert its full oxidising effect before the reduction of lead takes place, the tellurium will combine with the

[1] Loc cit., p. 175.
[2] Trans. Inst. Min. and Met., 1908, vol. xvii. p. 473.   [3] Ibid., p. 189.

lead instead of passing into the slag. It is doubtful, however, whether under the most favourable conditions all the tellurium can be removed from gold and silver by such an oxidising fusion, owing to the fact that the metallic lead formed during the fusion acts as a reducing agent and reduces some of the tellurium which has been oxidised and passed into the slag.

Quick fusion invariably gives low results and tends to increase the proportion of tellurium passing into the lead. The following results obtained by W. H. Merrett [1] on a telluride ore containing some pyrites, in addition to an ordinary basic gangue, may be given as an example.

In the "ordinary" fusion the fritting and subsequent fusion were conducted slowly, the charge being poured after it had been an hour in the fire.

In the "quick" fusion the fritting and fusion were hastened so that the charge was ready to pour in about twenty minutes.

The results were as follows :—

	Gold. Ounces per ton.	Silver. Ounces per ton.
Ordinary fusion	1·50	1·10
Quick fusion	1·42	0·81

The gangue and the general nature of telluride ores vary so greatly that no charge suitable for general use can be recommended, but it is generally agreed that sufficient lead oxide should be present to yield a large button (not less than 30 grammes) of metallic lead, even when the ore is not rich in precious metal, and even when a very small charge of ore is used.

The charge should contain a comparatively large proportion of sodium carbonate, but borax should be sparingly used : as a general rule, the amount of borax should not exceed one-third the weight of sodium carbonate in the charge.

The following charges for various ores may be taken as examples of those used for telluride ores :—

Charges for Crucible Assay of Telluride Ores.   (*Holloway and Pearse.*) [2]

	Nature of ore.					
	Shaley ores.		Highly siliceous ores.		Basic ores.
	Rich.	Average (5 ozs. and less).		Rich.	Average.	
	grammes.	grammes.		grammes.	grammes.	
Ore.	10	20	50	10	20	50	See
Red lead.	120	100	120	120	100	120	note.
Flour	7	6	7	7	6	7	
Sodium carbonate	30	40–50	60	40	50	60	
Borax (glass)	10	15–20	20	5	5	10	

For Basic Ores.—The charge depends entirely on the nature of the ore, and is a matter for experiment. The charge should be such as will produce a fluid and homogeneous slag at a low temperature, and should be as basic as possible without danger to the crucible. In all cases, of course, ordinary borax may replace the borax glass, nearly twice as much being used, and litharge may replace red lead with slight reduction in the amount of flour.

---

[1] *Trans. Inst. Min. and Met.*, vol. xvii., 1908, p. 199.   [2] *Ibid.*, p. 188.

For ordinary siliceous telluride ores of only slight reducing power (*i.e.* containing little pyrites) the following charge is recommended by Fulton [1] :—

Ore	0·5 assay ton (15 grammes).
Litharge	100·0 grammes.
Charcoal	1·1 ,,
Sodium carbonate . . .	30·0 ,,
Borax glass	6·0 .,
Litharge to cover . . .	10·0 ,,

From 10 to 20 milligrammes of silver foil may be added for parting purposes and to reduce the cupellation loss.

Tellurium is chiefly met with in the ores of the Cripple Creek district, Colorado, and of Kalgoorlie and other districts in Western Australia. The ores from Cripple Creek and Kalgoorlie present appreciable differences in chemical composition, the prevailing rocks in the former district being mainly andesitic breccia, lying upon granite, whilst in the latter they are schistose, impregnated with disseminated pyrites.

The following remarks, by T. A. Rickard,[2] on the ores of these districts may prove useful for fluxing purposes.

The Cripple Creek ore contains a high proportion of alumina, ranging from 15 per cent. in the granitic ores to 25 per cent. in those occurring with phonolite and allied rocks; from 55 to 70 per cent. of silica, the amount being highest in the granitic ores; from 1·5 to 2 per cent. of sulphur; and a little lime and magnesia. The ores contain about 0·08 per cent. of tellurium, 0·05 per cent. gold, and 0·01 per cent. silver. The Kalgoorlie ore contains from 2 to 12 per cent. of alumina; from 5 to 15 per cent. of secondary carbonates of lime and magnesia; from 45 to 60 per cent. of silica; 1 to 2·5 per cent. of titanic acid; and from 4 to 8 per cent. of sulphur. In both ores the percentage of iron is a little higher than that of sulphur, but in about the same ratio.

The stock flux recommended quite generally by Cripple Creek assayers for the ordinary siliceous ores is given below.[3] With 0·5 assay ton of ore about 75 grammes of the stock flux is used, and the charge would contain the different constituents in the following amounts :—

	Stock flux.		Charge for 0·5 assay ton of ore.
Litharge	30·0 parts.	45·5 grammes.
Flour	1·0 ,,	1·5 ,,
Potassium carbonate . .	7·0 ,,	10·5 ,,
Sodium carbonate . .	6·0 ,,	9·0 ,, ·
Borax glass . . .	5·5 ,,	8·5 ,,

It is frequently recommended that the charge for the fusion of telluride ores should be covered with a layer of salt, but there seems little doubt that the use of salt, either in the charge or as a cover, is highly objectionable and invariably yields a lower result.

As previously stated, the fusion of telluride ores must be conducted slowly and at a low temperature, special attention being given to the fritting stage.

The lead button obtained should be large, in order that the proportion of tellurium and gold to the lead may be small for cupellation. This large excess of lead is necessary to assist the removal of the tellurium during cupellation. With sufficient lead the tellurium, although less easily oxidisable than lead, is expelled before the final stage of the operation, and does not exert its injurious

---

[1] Fulton, *Fire Assaying*, 2nd edit., 1911, p. 186.
[2] *Eng. and Min. Journ.*, 1900, lxx. (21), 611.　　　[3] Fulton, *loc. cit.*, p. 181.

effects. An excess of tellurium in the lead button, at the end of the cupellation, causes heavy losses of gold and silver, and, as previously stated, may result in the subdivision of the precious metal or even in complete loss by absorption if the amount of tellurium present is equal to one-tenth of the lead. Subdivision and complete absorption cannot, however, occur if sufficient lead is present. In the case of many telluride ores the amount of tellurium in the lead button is comparatively very small when the assay has been properly conducted and the amount left towards the end of the cupellation is too small to exert any injurious effect.

Opinions differ as to whether the lead buttons should be cupelled direct or submitted to a partial scorification before cupellation.

According to Holloway and Pearse,[1] cupellation after a moderate amount of scorification gives a better result than the direct cupellation of the lead button, but they point out that, if scorification be carried too far, or if the lead be allowed to remain long in contact with the slag, the subsequent loss on cupellation is increased. The cupellation losses are very large after excessive scorification, and are probably caused by a reabsorption of tellurium which is reduced by the lead from the litharge in the scorifier. Holloway and Pearse recommend that the lead button (if not less than 50 grammes) from the crucible assay should be scorified to reduce its weight by one-third, but not more than one-half, and then cupelled.

Since, however, the scorification of lead buttons tends to concentrate tellurium in the button, most assayers prefer to cupel the large lead button direct.

The relative elimination of tellurium during cupellation and scorification is shown in the following results of experiments by S. W. Smith.[2] A series of lead buttons weighing 20 grammes and containing 0·05 gramme of gold and 0·05 gramme of tellurium were submitted to cupellation and to scorification, and the operations interrupted at intervals for the determination of tellurium.

Table XXVIII.—Elimination of Tellurium during Cupellation and Scorification. (S. W. Smith.)

Cupellation.			Scorification.		
Cupelled until lead weighed grammes.	Per cent. of original lead remaining.	Per cent. of tellurium in the lead button.	Scorified until lead weighed grammes.	Per cent. of original lead remaining.	Per cent. of tellurium in the lead button.
20·00	100·00	2·48	20·00	100·00	2·48
11·86	59·00	3·25	6·145	30·7	5·94
4·25	21·00	2·49	2·085	10·4	12·90
1·635	8·15	1·80			

It will be noted that during cupellation the tellurium tends at first to concentrate in the lead and then begins to be eliminated. In the case of scorification there is a more pronounced concentration of the tellurium, particularly towards the end of the operation.

The tellurium may be partly removed from the lead button by "soaking" it under an excess of litharge at a temperature between 700° and 900° C., and such treatment of the lead button from the crucible fusion, prior to cupellation,

[1] Trans. Inst. Min. and Met., 1908, vol. xvii. p. 183.   [2] Ibid.

is recommended by S. W. Smith as a desirable substitute for scorification. A low temperature should be employed both for the scorification and cupellation of the lead.

As previously stated, the presence of silver reduces the loss of gold during the cupellation of buttons containing tellurium, the silver acting as a diluent. It is for this reason that many assayers make an addition of silver, either to the charge for the fusion of the ore, or to the lead button at the beginning of the cupellation. When the assays of telluride ores are carefully conducted, and proper precautions taken, the losses on average ores of, say, 10 ounces and less are very little above those experienced in the assay of average non-tellurous ores.

As, however, the tendency for precious metals, especially silver, to be retained in the slag is somewhat greater in the assay of telluride ores than with ordinary ores, it is very desirable to clean the slags from the fusion of all but low-grade ores. The slags, and also the cupels if necessary, may be cleaned by the methods previously described.

With regard to other methods of assay for telluride ores, experience has shown that the scorification method is unsatisfactory and gives low results. Owing to the smallness of the charge which can be taken, only rich ores can be assayed by scorification. The fluxing is less complete in scorification than in the crucible assay, and the tellurium is not so satisfactorily removed ; also there is a tendency at the scorification temperature for the lead to be oxidised at the expense of tellurium oxide, which is reduced and enters the lead button and thus becomes concentrated in the lead. Better results are sometimes obtained by the addition of various fluxing materials, but if considerable amounts of these have to be used, the scorification assay approaches the crucible assay in character with none of its advantages.

Combined wet and dry methods of assay have been recommended for the assay of telluride ores, but, as the result of a large number of trials on rich, average, and poor ores, Holloway and Pearse found that even when every precaution was taken the results were lower than by direct fusion, and that duplicate assays seldom agreed within reasonable limits.

The result of their work, which confirms that of other workers, indicates, however, that should a case arise in which preliminary treatment with acid is necessary, the following precautions, which are probably known to most assayers, should be adopted :—

Dilute sulphuric acid (1 to 4), where its employment is permissible, gives less loss than nitric acid. When nitric acid is used, as is generally the case, the weaker the acid the better. One volume of acid to three or four of water is a convenient strength. Care should be taken in adding hydrochloric acid, or salt, to precipitate any dissolved silver, to use the smallest possible quantity, lest gold be dissolved. Ample time should be allowed after such addition for the silver to precipitate, and it is even advisable to test the filtrate to be sure that no gold has passed into solution.

5. **Ores containing Lead.**—Galena, lead sulphide, the chief ore of lead, is found in most mining districts and is frequently associated with zinc blende and copper pyrites. As galena invariably contains more or less silver, this metal is always tested for and its amount determined. Galena is not usually rich in silver, but it is frequently found in association with rich silver minerals. In the case of lead ores, the silver is generally calculated in ounces per ton upon the lead in the ore as determined by dry assay, as well as upon the ore itself, so that it is usual first to determine the percentage of lead by the fusion method and then to cupel the resulting lead button to determine the silver.

Lead ores are easily concentrated by mechanical operations, so that the samples received for assay are generally comparatively pure. To determine the lead, weigh up 25 grammes of ore, mix with an equal weight of sodium carbonate and 2 grammes of argol; place in a crucible, then cover with a sprinkling of sodium carbonate or borax, and insert a piece of hoop iron bent into a ∩-shape to reduce the lead sulphide.   Place in a furnace heated to, but not above, redness, and cover the crucible.   In about twenty minutes the charge will be fused, and when bubbles of gas are no longer being evolved and tranquil fusion has taken place, the iron is withdrawn and any adhering buttons of lead washed off by dipping it a few times in the slag.   Cover the crucible, leave it in the furnace for a minute or two longer, and then pour. Detach the slag, when cold, by hammering, and clean by brushing or washing in hot water and weigh. The weight of the button multiplied by 4 gives the percentage of lead.   Duplicate assays for lead should agree within 0·5 per cent., and in practice they frequently agree within 0·2 per cent.   To determine the silver, cupel the lead and weigh the resulting silver button.   The silver per ton of lead is calculated as follows :—

Suppose 25 grammes of ore gave 15·6 grammes of lead (equal to 62·4 per cent.) which on cupellation yielded a silver button weighing 0·01 gramme. Then,

$$\frac{0\cdot01 \times 100}{15\cdot6} = 0\cdot0642 \text{ per cent. of silver.}$$

From the table given on page 213, 0·0642 per cent. is equal to 20·97 ounces of silver per ton (2240 lbs.) of lead.

For the assay of oxidised lead ores, argol or charcoal is employed as the reducing agent instead of iron.   Ores comparatively poor in lead may be assayed for silver by the fusion method, but more or less litharge or red lead must be added to the charge to obtain a lead button of suitable weight.   The lead button is cupelled, and the slag cleaned in the usual way if the ore is rich.   The usual corrections for silver in the litharge used, etc., are made, and the silver calculated in ounces to the ton of ore.

The silver in argentiferous lead ore may also be determined by scorification. In this case take from 5 to 10 grammes of ore, 30 grammes of lead, and 0·5 gramme of borax.   Scorify, cupel, and calculate to ounces to the ton.   Unless the ore is rich, scorification is liable to give slightly lower results.

Zinciferous Ores.—Zinc is most frequently met with in ores as the sulphide, zinc blende.   As a general rule, zinc blende contains very little silver, but occasionally it is comparatively rich in silver.   Ores containing a large percentage of zinc are somewhat difficult to assay owing to the fact that zinc oxide is not readily fluxed, and if a good fluid slag is not obtained the results are likely to be low.   Hall and Popper[1] have worked out the best conditions for the crucible assay of zinciferous gold and silver ores by direct fusion.

As the result of a large number of experiments, they concluded "that the amount of litharge used should be just sufficient to give a lead button large enough to collect the gold and silver, since lead oxide in the slag seems to interfere with the complete decomposition of the ore, and prevents the formation of a slag which is free from lumps, and which can be readily poured."

The amount of sodium carbonate should be from four to five times that of the ore.   Borax glass should be added in amount sufficient to prevent the charge from being entirely basic, and to assist in fluxing the gangue minerals not

[1] *School of Mines Quart.*, 1904, pp. 355–358.

acted on by sodium carbonate alone. If necessary, the amount of argol sufficient to reduce the whole of the litharge should be added; whilst, if the ore contains more than 15 per cent. of pyrites, the addition of a couple of iron nails will prevent the formation of a brittle button. The following charge was found to be the most suitable for a 20-gramme crucible, the quantities being given in assay tons and in grammes :—

Ore	.	.	.	.	.	¼ A.T. or 10 grammes.
Sodium carbonate	.	.	.	.	1½ A.T. or 40 ,,
Borax glass	.	.	.	.	⅜ A.T. or 15 ,,
Litharge	.	.	.	.	⅝ A.T. or 25 ,,

The most suitable temperature, according to Hall and Popper, is 750°–775° C., and the time required for the fusion is thirty to thirty-five minutes.

The method gives good results with zinc ores containing up to 7·5 per cent. of copper. As zinc does not alloy to an appreciable extent with lead, the lead buttons from the fusion assay of zinciferous materials do not become contaminated with zinc.

A crucible fusion with nitre may be satisfactorily employed for zinciferous ores, the quantity of nitre being varied according to the amount of zinc blende present. From 1·5 to 2 parts of nitre are required to oxidise 1 part of zinc blende.

A suitable charge for ores containing about 75 per cent. of zinc blende is the following :—

Ore	.	.	.	.	.	20 grammes.
Litharge	.	.	.	.	.	50 ,,
Sodium carbonate	.	.	.	.	20 ,,
Borax	.	.	.	.	.	15 ,,
Nitre	.	.	.	.	.	20 ,,
Sand	.	.	.	.	.	5 ,,

The sand is reduced in quantity or entirely omitted, according to the amount of silica present in the ore.

Fusion with nitre is recommended by Lay[1] for the assay of concentrates consisting of zinc blende and galena such as are obtained at Broken Hill, Australia.

The silver-lead ore at Broken Hill varies widely in its nature, but an average analysis of the crude ore at present being sent to the concentration mill shows it to contain 36 per cent. of silica, 14·4 per cent. of lead, and 19·4 per cent. of zinc, the whole of the silver present being credited to the latter item.[2] For Broken Hill concentrate, containing 60 per cent. of lead, 11 per cent. of zinc, and 30 ounces of silver per ton, Newman[3] states that fusion with excess of lead oxide gives satisfactory results. The charge recommended consists of a quarter assay-ton of concentrate, 30 grammes of red lead, borax, sand, and soda, the latter in slight excess. The resultant lead button weighs about 30 grammes, and separates well from the slag. It is scorified to 10 or 12 grammes, and cupelled at a low heat. Ores containing zinc blende may be roasted previous to fusion, but it is difficult to roast these ores "sweet," as it requires a full red heat to decompose any sulphate that forms.

When blende is heated in a current of air decomposition begins at a temperature of about 450° C, with the formation of sulphate and oxide, but the proportion of sulphate is not considerable; its absolute amount increases with

[1] *Mineral Industry*, vol. xiii., 1905, p. 287.
[2] *Bulletin of Amer. Inst. Min. Eng.*, 1909, pp. 763–793.
[3] J. M. Newman, *Trans. Inst. Min. and Met.*, vol. xx., 1910–11, p. 412.

the temperature, though but slightly (to 6–8 per cent.), but the ratio of sulphate to oxide is smaller at high than at low temperatures.[1]

Zinc blende being practically infusible has no tendency to clot, and thus permits of a somewhat high temperature being employed in the earliest stage of the roasting. Zinc blende is not, however, infrequently accompanied by galena, and if this sulphide is present even in small quantity it is very liable to cause clotting in roasting. There is no loss of gold caused by roasting low-grade gold ores containing zinc blende, but a loss generally occurs with the increase in the amount of gold. This loss is largely mechanical, and in part due to the volatilisation of zinc oxide which causes loss of gold and silver. When the roasted product contains a large proportion of zinc oxide, difficulty may be experienced in satisfactorily melting the charge. Zinc oxide combines with silica in definite proportions with the production of silicates difficult to fuse, but with suitable porportions of borax and alkaline carbonates more fusible compounds may be obtained. With borax, zinc oxide yields a very fluid slag. With litharge it requires at least eight times its weight to form a fusible compound. The following charge is suitable for fluxing an ore, mainly zinc blende, after roasting :—

Ore (before roasting)	.	.	.	.	.	25 grammes.
Litharge	.	.	.	.	.	40 ,,
Charcoal	.	.	.	.	.	1·5 ,,
Borax	.	.	.	.	.	30 ,,
Sodium carbonate	.	.	.	.	.	10 ,,

The fusion is conducted at a comparatively high temperature and the resulting lead cupelled.

Scorification is frequently employed for zinciferous ores, but if much zinc is present the results are not generally satisfactory, as zinc oxide, as stated above, is only soluble in litharge with difficulty. There is a great tendency for insoluble scoria and crusts to form on the scorifier.

For the scorification of ore consisting mainly of zinc blende the ore is usually mixed with about twenty times its weight of granulated lead and its own weight of borax. Silica equal to half the weight of ore taken is sometimes added to ores consisting largely of zinc blende.

Some assayers employ a combined wet and dry method for zinciferous ores. The ore is first decomposed by treatment with hydrochloric acid and the zinc removed by filtration after diluting with water. The residue is then dried and scorified or fused in a crucible as directed for copper ores on page 225. Accurate results may be obtained by this method if care is exercised in manipulation.

Ores containing Nickel and Cobalt.—Arsenical, nickel, and cobalt silver ores are now being mined in the Cobalt district, Ontario. These ores are very rich in silver, often yielding several thousand ounces to the ton. For example, the shipments from one mine average 2500 ounces of silver per ton.[2] The high nickel, cobalt, and arsenic content of the silver ores of the Cobalt district, together with the presence of metallic silver, renders the sampling and assaying especially difficult. D. K. Bullens[3] has described the methods of sampling in present use on two of the mines, the main features being the careful reduction in size of the big sample, the use of mechanical samplers, and the separate assaying of the fine ore and "metallics." The comparative efficiency of scorification and crucible assays is discussed, and it is stated that, with the exception of very rich ores high in nickel, cobalt, and arsenic, the crucible assay gives the

---

[1] Krutwig, *Congrès de Chim. Pharm.*, Liège, 1906, Sec. XI., p. 419.
[2] *Trans. Inst. Min. and Met.*, 1909, vol. xviii. p. 167.
[3] D. K. Bullens, *Eng. and Min. Journ.*, 1910, vol. xc. pp. 809–810.

best results. Ores from the oxidised zone present no special difficulty, as the nickel and cobalt are readily prevented from entering the lead button by an excess of litharge. If, however, more than 0·5 per cent. of nickel oxide enters the button, the latter becomes covered with a scum of the oxide during cupellation and finally freezes. For sulphide ores a strongly basic slag, high in soda and approaching to a sub-silicate, is employed, as arsenic, in addition to nickel and cobalt sulphides, may be slagged off by an excess of an alkaline flux at a low temperature. The fusion must, however, be conducted rapidly in order to prevent the formation of a speise, so that only small amounts of those ores which have a high nickel, cobalt, and arsenic content can be used, as little as 0·1 or even 0·05 of an assay ton being occasionally employed. For low-grade silver ore rich in nickel and cobalt a flux containing an excess of litharge has proved satisfactory. The sulphide assay using metallic iron may be employed for comparatively high arsenic-nickel-cobalt ores if a large excess of the alkaline flux is used.

For the scorification of the ores high in silver the following charge is recommended :—

Ore	0·05 to 0·10 assay ton.
Granulated lead	65 to 75 grammes.
Borax glass	3 to 5 ,,
Sand	1 to 3 ,,

Slags and cupels should be treated for the recovery of the silver, and corrections made. It is found to be desirable to check results for silver from time to time by wet analysis. The methods of sampling and assaying silver ores containing the arsenides of cobalt, nickel, and silver, and also metallic silver, have also been described by Handy.[1] Details of the sampling are given on page 107.

The samples are assayed in a gas-fired muffle furnace by the crucible fire-assay method, using the following fluxes :—(1) "Silver" flux containing 40 per cent. each of sodium and potassium carbonates and 20 per cent. of borax glass; (2) litharge; (3) borax glass for cover. The charge for the crucible contains 0·2 assay ton of pulp, 30 grammes of "silver" flux, 60 grammes of litharge, and 1 gramme of flour. These are mixed by rolling in a rubber cloth, transferred to the crucible, covered with borax glass, and heated to a medium red heat for twenty minutes. The door of the muffle is then shut and the crucible heated to a bright yellow for thirty minutes, the contents poured into a conical iron mould, and when cold hammered free from slag. The metal is then cupelled at as low a temperature as possible, using a freely oxidising atmosphere without excessive draught; the temperature and draught control are important factors. The following table gives examples of results obtained by independent assayers :—

Sample.	Silver, oz. per ton.	
	A.	B.
Number		
37396	142·46	142·64
37691	57·07	53·79
37252	2521·38	2518·95

[1] "Sampling and Assaying of Silver Ores containing Cobalt, Nickel, and Arsenic," J. O. Handy, *Eighth Int. Cong. Appl. Chem.*, 1912, Sect. IIIA. (abstract, *Journ. Soc. Chem. Ind.*, 1912, vol. xxxi. p. 880).

Auriferous Tin Ore.[1]—Native gold is not infrequently found in tin ores and in the alluvial deposits of tinstone (cassiterite, tin oxide), the chief ore of tin. The samples obtained by panning alluvial deposits may be very rich in gold, and contain 25 ounces or more of gold to the ton. The crucible assay gives excellent results, both for silver and gold present ; and as 25 grammes can easily be worked off, it is the most suitable method for most tin ores containing precious metal.

Soda is the chief flux used, as it forms with tin oxide sodium stannate, which is readily fusible. The ore should be crushed to pass a sieve of at least 90 holes to the linear inch. The following charge will give a very good slag with almost all tin ores :—

Ore	25 grammes
Litharge or red lead	60 ,,
Charcoal	1·5 ,,
Sodium carbonate	40 ,,
Borax	10 ,,

It will be noted that a large amount of sodium carbonate is used in the charge as compared with the amount of ore taken, and that only sufficient charcoal is added to reduce half the lead oxide in the charge. By this means no tin is reduced during the fusion, and the resulting lead button, weighing about 30 grammes, is free from tin and can be cupelled direct without a preliminary scorification. The slags are quite satisfactory, being readily fusible, very fluid, and quite uniform in appearance.

Experience has shown that it is necessary to clean the slags from all auriferous tin ores carrying over 1 ounce of gold to the ton. For this purpose the slag is collected and fused with—

Litharge or red lead	30 grammes
Charcoal	1·5 ,,
Sodium carbonate	10 ,,

The resulting lead button is cupelled and parted separately to obtain the values left in the slag. In assaying poor ores containing only a few dwt. of gold per ton it is advisable to work on 50 grammes of ore, and in this case more satisfactory results are obtained by making two separate fusions, each of 25 grammes, scorifying the resulting lead buttons together and then cupelling, than by making one fusion only on 50 grammes of ore.

The scorification method invariably gives low results for auriferous tin ores, and this is due to the unsatisfactory nature of the slag obtainable. Even when sodium carbonate, equal to the weight of the ore taken, is added to assist in the formation of a fusible slag of sodium stannate, the slags are never quite satisfactory. In the case of rich ores the results may be improved by taking small quantities of ore, such as 2·5 to 3 grammes, and using 5 grammes of sodium carbonate ; but in this case any error occurring in the assay is largely magnified in calculating the result, and thus it is not advisable to use this method.

The amalgamation method of assay described on page 239 is used by some assayers for the determination of gold in ordinary alluvial tin ore, because of the necessity for operating on large quantities of such " patchy " ores, first, in order to obtain a fair sample, and second, on account of the extremely low gold values which they commonly carry.

Auriferous Bismuth Ores.—Although not of frequent occurrence in nature, bismuth accompanies various ores of silver, and is also found occasionally associated with gold ores.

[1] C. O. Bannister, "Assay of Auriferous Tinstone," *Trans. Inst. Min. and Met.*, 1906, vol. xv. p. 513.

The extent to which gold is found in combination with bismuth ore is referred to on page 343.[1]

Bismuth is almost invariably found in nature in the metallic state; but occasionally it is met with as sulphide in bismuthine, and as carbonate in bismutite. It is also found in combination with tellurium associated with gold. Assays of auriferous bismuth ores are conducted like those of oxidised gold ores, but on account of the volatility of bismuth it is of importance that a readily fusible slag should be obtained, and for this purpose comparatively large quantities of sodium carbonate or of borax should be employed. When much bismuth is present the quantity of red lead should be increased in order to effect its oxidation; a little nitre may also be added. Ores containing bismuth should not be roasted, as clotting invariably takes place owing to the low melting point of bismuth.

It is not essential to pass the whole of the bismuth into the slag, but it is desirable to do so, otherwise it alloys with the lead, and although bismuth cupels as satisfactorily as lead, the amount of gold and silver absorbed by the cupel and volatilised is greater when bismuth is present.[2] The time occupied by the cupellation is also lengthened, as bismuth is not quite so readily oxidised as lead.

The fusion of the ore is conducted at a comparatively low temperature. After pouring the charge the mould should be left undisturbed until the slag is quite cold, as the effect of bismuth when combined with lead is to form more readily fusible alloys which remain in a molten condition for a longer period than lead. If the proportion of bismuth does not exceed that of the lead, the alloys are malleable, but the malleability is diminished with an increase of bismuth. The lead buttons obtained from the fusion, if inconveniently large, may be reduced in weight by scorification, but the temperature employed for this purpose should be low on account of the volatility of bismuth.

The effect of the presence of bismuth during cupellation is to reduce the surface tension of the molten alloy as in the case of tellurium, although the effect is less with bismuth. If the lead button contains a large excess of bismuth and a very small quantity of silver, the alloy of bismuth and silver may "wet" the cupel in the last stage of the cupellation and be partially absorbed. There is less tendency for this to take place with large silver buttons.

Bibliography (Tellurium in Assaying).[3]

1883. DIVERS and SHIMOSE, *Jour. Chem. Soc.*, vol. xliii. p. 319.

1895. F. and C. HEBERLEIN, *Berg. und Hütt. Ztg.*, 1895, p. 41 ; and *Mineral Industries*, vol. iv. p. 480.

1896. F. C. SMITH, *Trans. Amer. Inst. Min. Engineers*, vol. xxvi. p. 485.

1896. H. VAN F. FURMAN, *Amer. Inst. Min. and Metallurgy*, vol. xxvi. p. 1103 ; *Manual of Practical Assaying*, 1899, p. 399.

1897. J. F. KEMP, "Geological Occurrence and Associations of Tellurides," *Mineral Industries*, vol. vi. p. 295.

1897. RICHARD PEARCE, *Colo. Sci. Soc.*, April 1, 1895 ; and *Eng. and Min. Jour.*, Jan. 30, 1897, and April 17, 1897 ; and *Trans. Amer. Inst. Min. Engineers*, vol. xviii. p. 449 ; and *Jour. Chem. Soc.*, 1896, A, ii. 612.

1898. C. H. FULTON, "Assay of Telluride Ores," *Jour. American Chem. Soc.*, vol. xx. p. 586 ; also *Mineral Industries*, vol. vii. p. 451 ; also *New York School of Mines Quarterly*, vol. xix. p. 419 ; also *Jour. Chem. Society*, 1899, A, ii. 63.

---

[1] Consult also *Journal of Society of Chemical Industry*, vol. xii. p. 816, 1893.
[2] E. A. Smith, *Journal of Chemical Society*, vol. lxv. p. 624, 1894.
[3] Compiled by Sydney W. Smith, *Trans. Inst. Min. and Met.*, vol. xvii., 1907–1908, p. 474, with additions by the author.

1899. R. W. LODGE, *T.Q.*, 1899, xii., 171 ; and *Mineral Industries*, vol. viii. p. 397 ; and *Notes on Assaying*, 1905.

1900. T. A. RICKARD, "Telluride Ores," *Eng. and Min. Jour.*, Nov. 17 and 24, 1900, p. 611.

1901. E. A. SMITH, "Assay of Complex Gold Ores," *Trans. Inst. Min. and Met.*, vol. ix., p. 343.

1902. C. VINCENT, "Tellurium in Silver Ingots," *Bull. de la Soc. Chimique* (3), 27, 23, 1902 ; also *Chemical News*, June 20, 1902.

1902. HALL and LENHER, *Jour. American Chem. Soc.*, vol. xxiv., 1902 ; and *Chemical News*, Dec. 26, 1902.

1905. W. F. HILLEBRAND and E. T. ALLEN, "Comparison of Wet and Crucible Fire Methods for the Assay of Gold Telluride Ores," *U.S. Geol. Survey*, Series E 44, Bulletin No. 253 ; also *Jour. Chem. Industry*, April 15, 1905 ; also *Chemical News*, March 25, 1906.

1905. E. C. WOODWARD, "Cupel Losses in Telluride Ores," *Western Chemist and Metallurgist*, vol. i., No. 4 ; also *Mining Magazine*, Sept. 1905.

1905. J. C. BAILAR, "Cupel Absorption with Telluride Ores," *Western Chemist and Metallurgist*, vol. i., No. 4 ; also *Mining Magazine*, Sept. 1905.

1907. G. T. HOLLOWAY and L. E. B. PEARSE, "The Assay of Telluride Ores," *Trans. Inst. Min. and Met.*, vol. xvii., 1907-1908, pp. 171-209.

1908. S. W. SMITH, "The Behaviour of Tellurium in Assaying," *Trans. Inst. Min. and Met.*, vol. xvii., 1907-1908, pp. 463-476.

1908. T. K. ROSE, "The Alloys of Gold and Tellurium," *Trans. Inst. Min. and Met.*, vol. xvii. p. 207 ; Contributed Remarks on the paper by Holloway and Pearse, *Trans. Inst. Min. and Met.*, vol. xvii. p. 171 ; "Roasting Gold Ores containing Tellurium," *Brit. Assoc. Report*, 1897, p. 623.

1908. G. BORROWMAN, "Some Observations on the Assay of Telluride Ores," *Jour. American Chem. Soc.*, vol. xxx., 1908, pp. 1023-1027 ; also *Chemical News*, vol. xcviii., 1908, pp. 180-181.

1909. H. PÉLABON, "Fusibility of Mixtures of Gold and Tellurium," *Compt. rend.*, 1909, vol. cxlviii., pp. 1176-1177.

1911. W. J. SHARWOOD, "Notes on Tellurium-bearing Gold Ores," *Economic Geology*, vol. vi., 1911, pp. 22-36.

CHAPTER XVI.

SPECIAL METHODS OF ORE ASSAY.

1. **Amalgamation Assay.**—This method is used to determine the amount of "free" gold present in an ore, *i.e.* the gold capable of being extracted by mercury. It must therefore be regarded as a test to ascertain the amount of precious metals that can be recovered from the ore by amalgamation rather than as a method of assay capable of giving the total gold content of the ore.

From 500 to 1000 grammes of the sample of ore are pulverised sufficiently fine to pass through an 80- or 100- mesh sieve, and any gold particles that remain on the sieve are mixed in with the powdered ore. The ore is placed in a porcelain mortar, made into a stiff paste with a little water, from 15 to 20 grammes of mercury added, and the whole ground thoroughly with the pestle for an hour. The mercury used must be free from gold and silver. As the commercial mercury may contain an appreciable amount of gold and silver it is desirable to run a blank assay on 50 c.c. to determine the amount present. The consistency of the mass after adding water must be such that the globules of mercury do not sink in it, but are broken up during the operation into very small particles, which are uniformly distributed through it. When amalgamation is judged to be complete the mass is reduced to a thin pulp by the addition of water, and then stirred for a few minutes to facilitate the collection of the mercury at the bottom. The ore is then separated from the mercury by washing it out with a stream of water from a piece of hose attached to the water-supply. The over-flow of tailings and slime is carefully collected in a large gold washing pan. If the mercury is very finely divided or floured a little sodium amalgam should be added to assist in collecting it. When collected the mercury is transferred to a porcelain basin, washed with water until quite free from sand, dried with filter paper, and then weighed to check any serious loss.

Instead of separating the mercury as described, the entire contents of the mortar may be washed out into a gold washing pan and the mercury carefully separated from the ore by panning, the washings being poured off into another pan and kept. Even when the mercury is separated with a water-jet as described above it is advisable to pan the tailings to make sure that no mercury has been washed over. The tailings from the panning are allowed to settle, the surplus water poured off, and the pan placed on a hot plate to dry. When dry the ore is thoroughly mixed and several portions of one assay ton or more taken and assayed by crucible fusion in order to determine what amount of gold has escaped extraction by mercury.

The treatment of the mercury varies according to circumstances. If the amount of silver as well as gold in the ore recoverable by amalgamation is to be determined the mercury is submitted to a crucible fusion. The mercury is transferred to a crucible containing a charge of 30 grammes of litharge, 0·5 gramme of argol, and 10 grammes of sodium carbonate. The crucible is placed in the furnace at a very low heat and the heat raised very gradually until the charge is fused. There should be a good draught to prevent the escape of mercury fumes into the room. The lead from the fusion is cupelled and the resulting button of gold and silver weighed. The weight in milligrammes, divided by the weight of ore taken in grammes and multiplied by 29·166, gives the ounces of gold per ton (2000 lbs.) present as "free" gold that can be recovered from the ore by means of amalgamation. This figure, added to the result obtained from the assay of the tailings, gives the total contents of the ore in ounces per ton. The size of the crushed ore will influence the results from the amalgamation test; but with ores generally, crushed to about 80 mesh, the yield of gold by amalgamation may be from 80 to 90 per cent. of the actual gold present in the ore. The yield is generally low with ores containing "rusty" gold, i.e. gold which does not readily amalgamate because the particles are coated with iron oxide.

If it is required to make the return in fine gold and fine silver, the button of gold and silver from the cupellation, after weighing, must be inquarted, and parted and the gold weighed. The fine gold and fine silver per ton of ore is then calculated in the usual manner.

When the yield of fine gold only is required it is usual to add enough silver to part the gold to the charge for the fusion. Another method of treating the mercury to determine the fine gold is to dissolve it in dilute nitric acid, filter, transfer the paper to a scorifier, add 30 grammes of test lead and sufficient silver foil to ensure parting, and cover with 2 grammes of borax glass. The scorifier is charged into the muffle, the paper burned, and scorification allowed to proceed until the lead is reduced to about 20 grammes. The charge is then poured, the lead cupelled, the gold-silver button parted, and the gold weighed.

Some assayers, after filtering, dry and burn the paper, then wrap the residue in a small piece of lead foil and cupel direct. When a large number of amalgamation tests have to be made, it is usual to make the tests in jars with tight screw covers. For instance, at the Homestake Mine, South Dakota,[1] the ore is placed in a bottle of 250 c.c. capacity with spring clamps and rubber washers. The charge consists of 100 grammes of ore, 150 c.c. of water, and 2 c.c. of pure mercury from a burette. The stopper is clamped, the bottle wrapped in a piece of cloth and placed in a mechanical shaker with a horizontal reciprocating motion, and shaken for two hours. The machine holds twenty-four bottles. When amalgamation is complete the mercury is collected and treated by scorification in a manner similar to that described above.

2. **Blowpipe Assay.**—This method of assay may be described as a fusion assay in miniature, as it comprises all the operations of an ordinary assay on a very small scale. Though less exact than the furnace methods, it is possible by its means not only to detect the gold and silver in an ore, but also to determine its amount quantitatively with a fair degree of accuracy. Since a modern prospector must be prepared to make an assay even if only approximate, this method is of importance, as it would obviously be impossible, in prospecting expeditions, to carry the cumbrous apparatus required to make an ordinary assay.

For a full description of the methods of assaying by means of the blowpipe the student is referred to Plattner's well-known work on Blowpipe Analysis.[2]

Before making the assay the selected portions of ore are freed as far as

[1] A full description of this method, by W. T. Sharwood, is given in Fulton's *Fire Assaying*, 2nd edit. 1911, pp. 154-156.
[2] Translated by Prof. Cornwall, 8th edit., New York, 1902.

possible from all earthy matter and then finely pulverised. A weighed quantity, usually 100 milligrammes of the powdered ore, is then mixed with an equal weight of borax and about 1 gramme of granulated lead. The whole is wrapped in tissue paper, placed in a cavity made in a piece of charcoal, and heated in the reducing flame of a blowpipe until fusion is complete. If pyritic material is present, the oxidising flame may have to be used for part of the time. In some blowpipe outfits small fireclay crucibles and a carbon furnace are provided for the fusion. The lead is then separated from the slag, placed on a small bone-ash cupel, and heated in the oxidising flame until it is all converted into litharge. The resulting button of silver or gold, which is too minute to permit of weighing, is carefully measured on an ivory scale, which at once gives the percentage amount of precious metal in the ore.

The scale for this purpose consists of a strip of polished ivory on which are drawn two very fine and distinct lines emanating from the zero point at one end and diverging towards the other end. The lines are divided by cross lines into 100 equal parts, numbered from zero upwards, with numbers which correspond to weights of the buttons in milligrammes.

Thus, if a small button of silver is placed in the space between the two lines, using a magnifying glass to assist the eye in moving it up or down until the diameter of the button is exactly contained within the lines themselves, the diameter of the button can be determined, and from this measurement its weight is found by reference to the figures on the scale. In the case of buttons from gold ores it is usual to separate the gold from the silver by parting in nitric acid.

3. **Chlorination Assay.**—This method is used in determining the amount of gold present in an ore capable of being extracted by chlorine.

It is applicable to earthy or oxidised ore direct and to ores containing pyrites after roasting "dead." From 200 to 300 grammes of ore (through 60 mesh) are slightly moistened with water, and placed in a glass bell jar about one litre capacity, with an opening at the bottom for the introduction of a current of chlorine gas (see fig. 114). A layer of broken

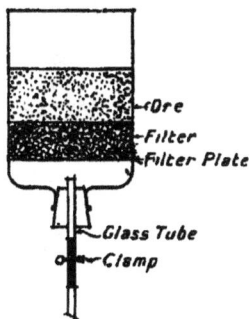

Fig. 114.

quartz or glass, coarse at the bottom and fine at the top, is first put into the jar to act as a filter, and on this is placed the moistened ore, which should not fill the jar more than two-thirds full. Chlorine is then passed into the jar, and when the odour of the gas is noticed above the ore, a cover is put on, the stream of chlorine stopped, and the whole left for one or two hours; after which the reaction is complete if chlorine is still in excess. Hot water is now run through the ore to dissolve the gold chloride formed, and the solution boiled to expel the free chlorine.

A solution of ferrous sulphate and a little hydrochloric acid is added, which precipitates the gold in the metallic state.

The precipitate is collected on a filter, then washed, dried, wrapped in lead-foil, cupelled and weighed. The result obtained is compared with the assay value of the ore before treatment with chlorine. This method fails if more than about 20 per cent. of silver is present, as the silver chloride formed encrusts the gold and protects it from the action of the chlorine.[1] The chlorine may be generated by gently heating a mixture of manganese dioxide and hydrochloric acid or, instead of employing hydrochloric acid, the materials from which the acid is

[1] T. K. Rose, *Metallurgy of Gold*, 5th edit., p. 465.

16

prepared, namely, sodium chloride and sulphuric acid, may be used. Suitable quantities for one test are as follows :—

Manganese dioxide	.	.	.	.	3 parts or 24 grammes.
Sodium chloride	.	.	.	.	4　　,,　　32　　,,
Sulphuric acid (commercial)	.	.	10·5　　,,　　84 c.c.
Water	.	.	.	.	.	7　　,,　　56 ,,

By adding the water and acid gradually, the gas-supply will run about 1¼ hours.[1] If necessary the tailings left in the jar may be dried and assayed, but care must be taken not to include any of the filter bed when removing the ore.

4. **Chloridised Silver Ore Assay.**—This test is used to determine the amount of silver present as silver chloride in silver ores that have been chloridised or roasted with salt to convert the silver sulphide present into silver chloride preparatory to the metallurgical treatment of the ore.

Two assays are made, one on the sample of roasted ore as received, and the other on another portion of the same sample after treatment to extract the silver chloride. The difference between the two results gives the amount of silver converted into chloride by roasting the ore. The assay value of the roasted ore is determined by scorification on 3 grammes of ore in the usual manner.

Another 3 grammes are placed in a beaker and digested with a warm 2 per cent. solution of sodium thiosulphate (hyposulphite), which readily dissolves the silver chloride from the ore. The treatment is continued until a small portion of the filtered liquid contained in a test-tube darkens but slightly and does not lose its transparency upon the addition of a few drops of sodium sulphide solution.

When free from silver chloride, the ore is filtered and washed with warm water, then dried, the filter paper burned, and the residue scorified with 30 grammes of lead and 1 gramme of borax. The lead is cupelled and the silver button weighed. As a confirmatory test the silver in the clear filtrate may be precipitated as sulphide by the addition of sodium sulphide or by acidulating the solution with hydrochloric acid and warming. The precipitate is filtered off, and dried, then gently heated to expel the sulphur and burn the filter paper, and the residue wrapped in a little lead-foil and cupelled. Any silver sulphate contained in the ore will also be dissolved by the hyposulphite solution and thus be included as chloride. If, as is sometimes the case, it is desired to know the amount of silver sulphate present, a third quantity of 3 grammes of the ore is treated with warm water, which will dissolve the silver sulphate only. The residue is then dried and scorified as before. The amount of silver present in the roasted ore as chloride is ascertained as shown in the following example :—

(1) Assay value of roasted ore　.　.　.　.　195·5 ounces per ton.
(2) Assay value after treatment with hyposulphite .　12·3　　,,　　,,

Difference equals silver present as silver chloride .　183·2　　,,　　,,

Therefore, of the total silver in the chloridised ore, the percentage of silver present as chloride will be :—

$\dfrac{183\cdot2}{195\cdot5} \times 100 = 93\cdot71$ percentage of silver present as chloride in total silver content of ore.

The amount of silver present as sulphate is determined by subtracting the result of the third assay from the result of the first assay and then calculating the percentage. If the ore contains gold the amount present must be determined

[1] Lodge, *Notes on Assaying*, p. 225.

by parting the silver buttons and deducting the weight of the fine gold from the weight of the buttons from the cupellation.

The amount of silver present as chloride may also be determined by digesting 5 grammes of the roasted ore in a beaker with dilute ammonium hydrate which dissolves the silver chloride. The solution is then filtered, the residue washed with water, and the silver precipitated from the clear solution by acidulating with nitric acid. The precipitate of silver chloride is washed, dried, wrapped in a piece of lead-foil, and cupelled. The resulting button of silver is weighed and the percentage calculated in the usual manner. This method is more expeditious than that first described, and when carefully performed the results are quite satisfactory.

5. **Assay by Panning.**—This operation consists in washing auriferous gravel or crushed rock by hand in a pan in order to obtain the particles of gold. The

FIG. 115.   FIG. 116.

pan is largely used by prospectors for rapidly arriving at an approximate estimate of the value and distribution of gold in a deposit or vein.

The pan (fig. 115) is usually made of stiff sheet-iron, with a flat bottom about 12 inches in diameter, and has sides from 5 to 6 inches in height, sloping outwards at an angle of about 30° to the bottom. Some pans are made with a riffle formed by the thickening or bulging inwards of the side, situated about half-way up the latter and running about half-way round the pan. A pan with a riffle is shown in fig. 116. ·

The method of using the pan is as follows :—

From 2 to 4 lbs. of the crushed ore or gravel is shovelled into the pan and submerged in water until the contents are well soaked, in the meantime being stirred with the hand. The pan is then held by the two sides, inclined away from the operator, and raised until the gravel is only just covered with water. Then a half-circular motion is given to the pan, which will cause the finer particles of ore to float off, leaving the coarser material exposed to view. After a second stirring, the coarser material and any pieces of rock are re-

FIG. 117.

moved, care being taken to wash them properly. This process is repeated until all the gravel has been swept away by the water, and until gold and a certain amount of black sand (magnetic iron oxide) alone remains. From this point onwards great care must be exercised as, the black sand being heavy, the gold is liable to float away with it and so be lost. When the panning has been successfully performed the "colours," *i.e.* yellow specks of gold, are seen distinct from the black sand. The actual motions given to the pan vary with different operators ; some use a circular movement until the black sand is reached, whilst others let the water flow on and off the pan without any circular motion whatever. To be of any value a panning must be made with care, that is, with due regard to the laws governing hydraulic separation.

Panning is very frequently performed in a *batea* shown in fig. 117, which differs from the gold-washing pan in not having a flat bottom. It is made of wood turned in a lathe, about 20 inches in diameter, conical, or more rarely basin-shaped, and about 3 inches deep in the centre so that the angle at the apex is about 160°. The gold collects at the lowest point and clings to the wooden surface under conditions such that it would slide over iron. The batea consequently is more rapid and effective in obtaining a "colour" than the pan, especially when the gold is fine. The best material for the batea is mahogany cut with the direction of the grain vertical to the surface of the implement.[1]

The batea is used chiefly in South America, while the pan is in more general use in the United States, Australia, and South Africa.

During panning a loss may result in float-gold, *i.e.* very minute particles of gold, so light that they are often floated away and lost. "The 'colour' or showing of gold obtained on panning an ore carrying iron pyrites (and nearly all unoxidised gold ores carry more or less iron pyrites) is decidedly smaller than that obtained from an oxidised ore of the same assay value. Two reasons probably contribute to this, one being the fact that when mixed with iron pyrites the gold does not separate so well, both on account of the particles being finer, and also because the heavy pyrites tends to cover it up. The other reason is the fact that the pyrites in most cases is, in itself, much richer than the rest of the ore, and thus locks up and conceals an appreciable proportion of the gold."[2]

The amount of gold recovered by panning is determined in several ways. Those who have had considerable experience in panning very frequently judge the value from the apparent amount of the gold particles or "colour" left on the pan. Such guesswork is, however, of little value, as it is almost impossible to value the material obtained by panning by inspection. In some cases the pannings are determined by comparison with weighed standards of gold dust, contained in small glass tubes, but since different samples of gold dust vary considerably in character and physical condition, some being coarse and some very fine, the standards employed should be especially prepared from gold dust obtained from the district in which the panning tests are being made.[3]

The gold may be weighed when sufficient in amount, after drying and separating the black sand by blowing. The residue in the pan is dried and placed in a "blower," consisting of a shallow metal or horn scoop open at one end, similar to that shown in fig. 77, page 65. This is held with the open end away from the operator and the sand removed by gentle blowing, leaving the particles of gold behind. Unless the blowing is done with extreme care gold is very liable to be lost. The gold when free from sand is weighed.

A more general method is to add a little mercury to amalgamate with the gold, and after collecting the liquid amalgam carefully, to volatilise the mercury and then accurately weigh the gold bullion on a portable assay balance. From this weight the value in free gold per ton is calculated. The fineness of the gold can be easily determined by inquartation with the blowpipe and parting in nitric acid. The fineness of the bullion is of importance, as it makes a material difference in the value of the gold-pannings. An objection to the use of mercury is that the gold from many alluvial deposits does not amalgamate readily. The gold is sometimes recovered from the mercury by treating the amalgam with nitric acid. The method yields unsatisfactory results, because the nitric acid dissolves out a portion but not all of the silver present in the bullion, especially in the case of coarse gold, which becomes coated and held by the mercury, but is not dissolved in it, so that these particles of gold retain a large and varying proportion of their silver after treatment with the nitric acid.[4]

[1] T. K. Rose, *Metallurgy of Gold*, 5th edit., 1906, p. 47.
[2] W. Frecheville, *Trans. Inst. Min. and Met.*, vol. xviii., 1909, p. 494.
[3] Consult S. J. Lett, "The Use of Standards in reading Gold Pannings," *Trans. Inst. Min. and Met.*, vol. xviii., 1909, pp. 482-494.
[4] G. T. Holloway, *Trans. Inst. Min. and Met.*, vol. xii., 1903, p. 462.

A very satisfactory method of recovering the gold is to transfer it carefully from the pan to a small piece of lead-foil about 4 inches square which is then wrapped up and cupelled. The bullion button is then inquarted and parted so that the final result is given in fine gold.

The results of the pannings of crushed ores such as auriferous quartz are best returned in ounces or pennyweight of fine gold per ton of 2000 lbs.

Where 2 lbs. of quartz are taken, the following are the weights of gold dust corresponding to the attached values [1] :—

Weight of gold dust.		Equivalent in ounces and pennyweight per short ton.
Grains.	Grammes.	
0·024	0·00156	1 dwt. per ton of 2000 lbs.
0·048	0·00311	2 ,, ,, ,,
0·096	0·00622	4 ,, ,, ,,
0·144	0·00933	6 ,, ,, ,,
0·192	0·01244	8 ,, ,, ,,
0·240	0·01556	10 ,, ,, ,,
0·480	0·03110	1 ounce ,, ,,
0·960	0·06221	2 ,, ,, ,,
1·920	0·12442	4 ,, ,, ,,

Data
{ 1 lb. avoirdupois = 7000 grains troy.
1 ton (2000 lbs.) = 907,185 grammes.
1 gramme = 15·43285 grains.
1 grain = 0·064798 grammes (0·0648 used).

The assay values of alluvials should be reported in grains and decimals of a grain of "fine" gold, or in pence (at 2d. per grain of "fine" gold) or cents per cubic yard.

It is recommended by the Institution of Mining and Metallurgy [2] that, in the absence of specific information as to boulders, etc., 1 cubic yard of ordinary alluvial be taken as equivalent to 3000 lbs. (1½ short tons). Samples of alluvial deposits are frequently very poor in gold, yielding only 1 or 2 grains of gold per cubic yard, it is therefore advisable to use 6 lbs. of material for the panning.

Although the results obtained by panning cannot be taken as a correct index of the gold contents of an ore, the pan, as previously stated, is undoubtedly, in experienced hands, an efficient method of arriving at an approximate estimate of the value and distribution of gold in a deposit or vein, provided the showing in the pan is checked by some assays of typical samples. When the character of an ore is fairly well known and carries most of its value in "free gold," the relative amount of gold that different samples contain can be ascertained without any great difficulty. Panning can only be learned by experience. Students may practise the use of the pan by weighing out 2 grammes of dressed tin ore, mixing it with 2 lbs. of powdered quartz, and then panning as described. The success of the operation is judged by drying and weighing the tin oxide left on the pan.

[1] Lett, loc. cit., p. 485.   [2] Trans. Inst. Min. and Met., 1911, vol. xx. p. 530.

CHAPTER XVII.

VALUATION AND SAMPLING OF BULLION.

Introduction.—The term bullion is conveniently restricted to "the precious metals, refined or unrefined, in bars, ingots, or any other uncoined or unmanufactured condition, whether contaminated by admixtures with base metal or not."

As shown later, bullion varies greatly in composition, and gold or silver may be present in any proportion up to nearly 100 per cent. The base metals most frequently present in bullion are copper, iron, lead, bismuth, antimony, zinc, and tin. Platinum and its allied metals and tellurium and selenium are also present at times.

Although the bullion produced by the various methods of ore treatment is often subjected to rough refining operations at the mine or mill when it is melted and cast to bring it to a marketable form, the bars are always impure and frequently brittle and require careful handling.

When the bullion is brought to London, New York, Philadelphia, Melbourne, St Petersburg, and other great centres to be sold, it is in the first instance melted and assayed in order that its value may be ascertained. After the value has been agreed upon, the bullion is at once refined, the base metals being removed, the gold and silver separated, and fine bars of each obtained. These bars of refined gold and silver being of a high degree of purity, are in a condition to be used for minting, or for the various industrial purposes to which the precious metals are applied.

The degree of purity or fineness of bullion is always expressed in parts per thousand, the pure metal being described as 1000 fine. Thus a bar of bullion containing 91·3 per cent. of gold is said to be 913 fine. The weight, fineness, and name or mark of the assayers are usually stamped upon each bar of bullion. Bullion may be classified as follows :—

I. **Gold Bullion.**
    (a) *Unrefined gold bullion.*
    (b) *Cyanide gold.*
    (c) *Refined gold bullion.*

II. **Silver Bullion.**
    (a) *Unrefined silver bullion.*
    (b) *Doré silver.*
    (c) *Refined silver bullion.*

III. **Base Bullion.**
    (a) *Lead containing gold and silver.*
    (b) *Copper containing gold and silver.*

I. **Gold Bullion.**—Bars of gold bullion usually weigh 200 or 400 ounces each.

(a) *Unrefined Bullion* often contains from 2 to 3 per cent. or more of base metals. Silver is always present, in amounts varying from about 1 per cent.

up to more than half the total weight. According to M'Arthur Johnston,[1] the fineness of the "mill" bullion resulting from the amalgamation, on the Rand, seldom exceeds 925 fine, and is practically never below 860 fine. The analysis of typical West Rand mill bullion is given below. The bars are stated to be homogeneous.

(b) *Cyanide Bullion.*—This is produced by means of the cyanide process, which always yields impure bullion, the impurities being chiefly lead, zinc, and silver. "Cyanide gold" is sometimes called "base bullion," but it is convenient to restrict this name to lead and copper bullion described later.

The presence of 20 or 30 per cent. of lead, zinc, and iron in gold bullion gives it a worthless appearance, bearing some resemblance to cast iron, but with an earthy fracture. Cyanide gold may vary from 999 fine down to 500 or 600 fine. It is very seldom, however, that cyanide gold is below 800 fine, as the penalty now enforced by the buyer, should this figure not be obtained, is sufficient inducement for the smelter to guarantee that degree of fineness (see page 249).

The following are analyses of typical mill bullion produced by amalgamation, and typical cyanide gold, resulting in both cases from the treatment of the ores on the Rand, South Africa.[2]

Analysis of Typical Rand Bullion (1912).

	Mill gold.	Cyanide gold.
Gold	89·48 per cent.	83·46 per cent.
Silver	9·44 ,,	8·57 ,,
Lead	...	7·10 ,,
Iron	0·17 ,,	Trace.
Zinc	...	0·50 per cent.
Copper	0·33 ,,	Trace.
Nickel	0·62 ,,	0·25 per cent.
	100·09 per cent.	99·88 per cent.

Most bullion, even if it has the outward appearance of gold, is brittle before it is refined.

(c) *Fine or Refined Gold Bullion* is practically free from silver, and contains from 990 to 1000 parts of gold per 1000.

The purity of the gold produced by the refining processes in general use is as follows :—By sulphuric acid parting, 996 to 998 fine ; by chlorine gas, 996 fine ; and by electrolytic processes, 999·8 to 999·9 fine.

II. **Silver Bullion.**—Silver bullion is usually cast into bars which weigh from 1000 to 1200 ounces each.

(a) *Unrefined Silver* differs in content of silver from 500 to 995 per 1000, and may or may not be free from gold. It contains base metals, chiefly consisting of copper and lead, but bismuth, antimony, zinc, tellurium, selenium, etc., are occasionally present also.

(b) *Doré Bullion.*—The term "Doré" silver or "Doré" bullion is applied to such silver bars as contain gold. The proportion of gold seldom exceeds one-third of the weight of the bar and is usually much less. The silver content of "Doré" bars is generally above 500 parts per 1000, the remainder being gold

[1] *Rand Met. Practice*, 1912, p. 331.   [2] A. M'A. Johnston, *ibid.*, p. 333.

and base metal. Bars of mixed bullion containing both gold and silver are seldom cast of a greater weight than 600 ounces in mills.

(c) *Refined Silver* bars invariably contain above 990 parts of silver per 1000. The refining processes now in general use yield silver bars of the following fineness :—By sulphuric acid parting, 996 to 998 fine ; by chlorine gas, 990 fine ; and by electrolytic processes, 999·8 fine.

Cupelled silver may be as fine as 997 or 998 or even higher.

Silver is sold in the state in which it arrives in this country, usually as bars of various sizes, unless of low quality, when it is generally remelted here by one of the Bank of England melters. In exceptional cases it is necessary partially to refine it to raise it to marketable fineness.

III. **Base Bullion.**—This name is applied to the bars or pigs of argentiferous and auriferous lead and copper resulting from the smelting of gold and silver ores with ore containing lead or copper.

(a) *Base Lead Bullion.*—The pig lead contains from about 0·1 per cent. of silver upwards, 95 per cent. or more of lead, and some copper, antimony, etc. The proportion of silver in the lead is generally kept below 2 per cent. in order to avoid undue losses in the slag during smelting. The lead frequently contains more or less gold.

(b) *Base Copper Bullion.*—Owing to the great affinity of gold and silver for metallic copper, this metal is also used for the concentration of the precious metals in smelting processes. The copper containing the gold and silver is cast into bars or pigs constituting base copper bullion.

Base bullion is more fully discussed on page 312.

Sale of Bullion.[1]—An assayer should be familiar with the manner and terms of selling gold and silver bullion. The following notes are therefore given.

Gold is, as is well known, the *standard of value*, its price being fixed at the British Mint at 77s. 10·5d. per standard ounce troy of 916·6 fineness ; and at the United States Mint at 20·6718 dollars per ounce troy. These mints receive unlimited quantities at those prices. However, there is nothing to prevent a private banker from bidding higher if he chooses. Of course, no one will pay much above the coinage rate, because gold can always be obtained from the mints on the basis of 77s. 10·5d. and 20·6718 dollars per ounce respectively, but gold coin loses something by abrasion and is subject to further loss in shipment, therefore there is a margin for bidding on gold bullion. Shipments of bullion to London by the gold-mining companies are received by their banks. On the day of arrival, bids are asked from the Bank of England, and from foreign banks which may want the gold for export. The highest bidder gets it. The market for gold is intimately associated with the banking transactions whereby international balances are settled. In the purchase of ores, and for all ordinary purposes, gold has only the fixed coinage value.

"It may be noted that gold may be sent in to the London Mint for coinage by anyone who has some in his possession, the only restrictions being that the average fineness must not be less than 916·6 per 1000, the standard fineness of the gold coinage, and that the gold must be suitable for conversion into coin without further refining. The gold is converted into coin at the rate of £3, 17s. 10½d. per standard ounce troy without any charge or deduction whatever. The price of standard gold in the London market varies from £3, 17s. 9d. to £3, 18s. 0d., but it is never profitable for ordinary owners of gold bullion to send it to the Mint, because the conversion into coin usually takes some weeks, and the interest on the value of the gold amounts to over 2d. per ounce per month. The result is that the Bank of England alone sends gold into the Mint. The metal, as previously stated, is usually in the form of refined ingots, each weighing about 400

[1] Consult *Eng. and Min. Journ.*, vol. lxxxiii., 1907, pp. 437–438.

ounces and containing over 99 per cent. of gold."[1] The ingots, after weighing, are sampled and assayed and a notification of the estimated value of each consignment sent to the bank. Unless an objection is made, the ingot is then passed for melting and conversion into coin.

At the Australian Mints and at the Mints and Government assay offices in the United States, unrefined gold is received for treatment. It is assayed and valued, and a charge for treatment made to cover the cost of refining and coining.

The Silver Market is also intimately connected with banking transactions, large exports being made from London to the Far East in settlement of balances. The silver market of the world, more than any other metal market, is determined by London. The silver brokers of London meet daily at 2 p.m. and fix a price for that day. They are bound to sell at the price thus fixed any quantity required, but are not bound to take any quantity that may be offered. If the quantity offered is more than can be disposed of, it necessarily goes over to the day following, and then has an effect in fixing the price for that day. The London quotation is for standard silver, *i.e.* 925·0 fine.

The New York price for silver fluctuates generally in correspondence with the London price, being lower than the London price by an amount representing commissions, expressage, insurance, interest, and other charges between the two markets, and also is affected by fluctuations in the rate of exchange.

Silver is commonly sold in ingots a little over 1000 ounces in weight, 999·0 fine, for cash, immediate delivery.

The quotation above referred to is frequently called the "official price," and is commonly adopted in business and by the press. Occasionally there are quotations for "commercial bars," which are a little higher than the "official price." These represent the retail business that is done with jewellers, manufacturers of silverware, etc., a business into which the large refiners do not enter. The large transactions in silver are rather complicated, and, as stated above, are closely involved with banking.

"Silver bullion is no longer received for free coinage at any Mint. The usual practice at Mints at the present day, with currency on a gold basis, is to purchase silver bullion with Government funds and to coin it into pieces required for circulation at rates giving a large percentage of profit."[2]

Prices and Terms of Sale.—The following prices and terms of sale in London, adopted in 1906 and now in force, are given by Mr Arthur Claudet.[3]

Gold Bullion.—All gold is melted, and bars weighing over 1000 ounces troy are melted into two bars.

Melting charges, ¼d. per ounce on all ordinary gold bullion.
" " ⅜d. " on gold dust and amalgam.
Refining " 3d. " on weight after melting.

Deduction on Cyanide Gold.—If the gold report is below 700 per 1000 parts, and the gold and silver together do not equal 900—deduction 4 millièmes on gold. (1 millième = 0·5 milligramme, see page 60.)

If the gold report is 700 to 800 parts and the gold and silver together do not equal 900—deduction 3 millièmes. If gold report is above 800—no deduction.

The full gold and silver contents are paid for less the above deductions, and $\frac{1}{16}$ of 1 per cent. commission on the net out-turn is allowed the seller.

Silver is paid for at the price of fine silver.

Silver Bullion from January 1st, 1906, will be dealt with in millièmes (parts per 1000), instead of, as hitherto, on the trade report (see note over page).

Bars will be weighed to the ¼ ounce, instead of, as heretofore, to the ½ ounce.

[1] T. K. Rose, *Precious Metals*, p. 222. [2] *Ibid.*, p. 224.
[3] *Inst. Min. and Met. Bulletin*, No. 16, Jan. 11, 1906.

Parting silver will be refined on the following terms.

All gold and silver contents will be accounted for without deduction.

Refining charge on
gross weight.

1. Silver containing no gold and up to 100 millièmes   ¾d. per ounce.
2.   ,,   ,,   100   ,,   225   ,,   1d.   ,,
3.   ,,   ,,   226   ,,   350   ,,   1½d.   ,,
4.   ,,   ,,   351   ,,   500   ,,   2d.   ,,

No gold reported under 0·3 millième will be paid for. No bullion under headings 3 and 4 will be purchased except on "dip" assays. All bullion containing 500 millièmes of gold will be treated according to the conditions under gold bullion.

Note.—Formerly the silver was reported as better or worse in half-pennyweight per lb. than standard silver (925 per 1000), and the gold in grains per lb. of gross weight of the bar. The abolition of this system now brings English reports into line with foreign ones and is practically a decimal one, the same as is the case with gold bullion.

Melting, Refining, and Casting of Bullion.—The assayer is frequently called upon to melt and cast bullion. It may be well, therefore, to describe briefly the melting and refining of small lots of crude bullion, and the casting and sampling of the bars for assay purposes. For full details of these operations the excellent description given by Dr T. K. Rose in *Metallurgy of Gold* in the same series should be consulted.[1]

Bullion is generally melted in graphite crucibles in a wind furnace such as that described on page 18. A fire-clay crucible (fluxing pot) is sometimes used for melting very small lots of bullion.

All crucibles must be thoroughly annealed before being used, as they are very liable to crack when placed in a hot fire unless this precaution is taken. The crucible should be stored in a warm place near the furnaces for several days before it is wanted. When required for use it is first placed on the top of the furnace or in the ashpit for a few hours, and then held over the open furnace by means of the crucible tongs for a few minutes until it is thoroughly warm throughout. The crucible is then lowered, rim downwards, upon the burning fuel. As soon as the rim becomes red hot, all risk of cracking is past, and the crucible may be turned over and placed in position for the reception of the bullion. Salamander crucibles are well annealed before being sold and do not therefore require such careful annealing before use. The crucible stands on a fire-brick about 3 inches thick, placed on the fire-bars. The fuel is built up round the pot until it reaches to its rim, and the fire urged until the pot is uniformly heated to a full red heat. Borax is then charged into the crucible by means of a scoop to act as a flux for the base metal oxides and so assist the bullion to melt. As soon as the borax has melted, the introduction of the bullion is commenced. Before commencing to melt, the ashpit of the furnace should be cleared in order to avoid unnecessary trouble in case of accident.

When the quantity of bullion is small it may be introduced by means of a scoop or wrapped in paper. With large quantities it is safer to use the shoot shown in fig. 118, which is held with the left hand so that the tube is inside the crucible, while the bullion is transferred to it in a scoop by the right hand. Large pieces of metal are added by the crucible tongs. Bars of metal to be remelted should be placed on the top of the furnace to warm before being introduced. When the first supply has melted down more is added in the same manner, and the operation repeated until the crucible contains sufficient metal to form a bar. The crucible is not allowed to become more than two-thirds full

---

[1] The account given here is mainly based on Dr Rose's description.

at any time. During melting a cover is kept on the crucible as much as possible.

When the metal is thoroughly molten it is ready for casting, unless the bullion is impure, in which case it is usually subjected to partial refining before casting. If the bullion is of a high degree of purity, containing little dirt or base metals, very little flux is added, a spoonful or two of sodium carbonate and nitre being enough. In this case the slag is not skimmed off but poured with the metal. If the bullion is very base, however, it is usual to refine it partially by adding nitre and borax, a little at a time, and skimming off the slag when all action has ceased. The nitre gives up oxygen and converts the base metals into oxides, which are dissolved by the borax. The slag formed is very corrosive and rapidly destroys the pot. To prevent this corrosion, bone-ash is sprinkled on the molten metal, the surface cleared a little in the centre, and the nitre and borax added. The slag formed is absorbed by the bone-ash, which protects the crucible from attack. If, after the slag is skimmed off, the metal shows no sign of oxidation on exposure to the air, it is ready for pouring, but if the refining is not complete the operation is repeated as often as necessary.

When silver bullion has been treated for a long time by nitre and raised to a high degree of fineness, it is affected by a peculiar bubbling which is probably due to the evolution of oxygen previously absorbed from the nitre. If this continues it is necessary to stir continuously with a graphite rod, keeping the surface covered with charcoal powder, until the bubbling ceases. Iron and zinc can be removed from bullion by the use of nitre; but the oxidation of lead is more tedious, and bismuth, tellurium, and copper are very troublesome. When lead is present, alternate additions of sal-ammoniac and nitre are recommended. Iron is used to remove arsenic, antimony, sulphur, etc. The molten metal is stirred briskly with an iron rod for a few minutes. Antimonides, arsenides, or sulphides are formed and separate from the metal.

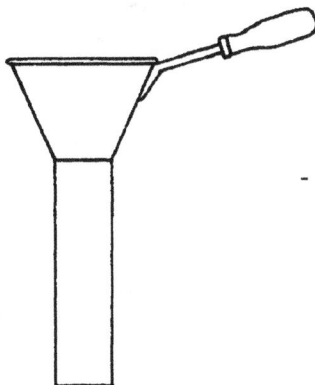

Fig. 118.

The ingot mould for receiving the metal after refining is cleaned thoroughly inside by rubbing with emery paper and oil, or with a dry piece of pumice stone. It is then wiped with an oily rag, or blackleaded inside, and placed on the top of the furnace until it is too hot to touch with the hand but not sufficiently hot to ignite the oil. When the bullion is ready to pour, the mould is placed on a level surface and a little oil poured into it. Any cheap, non-volatile oil is suitable for the purpose, and sufficient should be used to cover the bottom of the mould to a depth of from $\frac{1}{8}$ to $\frac{1}{4}$ of an inch according to the size of the mould. The charge must be stirred thoroughly before pouring, so as to obtain a bar as homogeneous as possible to ensure a correct assay. The stirring is usually done with a graphite rod made expressly for the purpose. It is annealed and raised to a full red heat before being used, and is held firmly by a pair of tongs. With very small meltings it is sufficient to lift the crucible out of the furnace with the tongs and to give it a rotary motion just previous to pouring. Since segregation may occur on cooling, a sample for assay is often dipped out immediately after stirring as described later.

To cast the metal, the crucible is lifted from the furnace usually with basket tongs, and the contents poured rapidly but steadily into the mould, the crucible being held for a little time in an inverted position to allow the last portion of the metal to flow from it. The oil is ignited and burns on the top of the cast metal, thus keeping it from tarnishing.

In small castings the slag is allowed to flow out and remain on the top of the metal in the mould; in large castings the slag is usually skimmed off with a ladle before pouring. The bar, while still too hot to be handled, is turned out by inverting the mould, and the slag is separated by a few light taps with a hammer. The bar is then momentarily dipped into water to assist in the complete removal of the last fragments of the slag, or it may be dipped first into dilute sulphuric acid and then into clean water, the bar retaining sufficient heat after removal to expel all moisture. This treatment gives a bar free from tarnish, which is then stamped. In casting the bar, a mould of suitable size for the weight of metal to be cast should be selected. It may be noted that 1 cubic inch of fine gold weighs 10·168 troy ounces, and 1 cubic inch of fine silver weighs 5·568 troy ounces. Table XXIX., of sizes and capacities of bullion moulds of the shape shown in fig. 119, will be found useful.[1]

Table XXIX.—Sizes and Capacities of Ingot Moulds.

Inside measure.			Capacity in ounces.		Weight.
Length.	Width.	Depth.	Gold.	Silver.	
inches	inches.	inches.			lbs.
1	$\frac{5}{8}$	$\frac{1}{2}$	4	2	1
$1\frac{1}{2}$	1	$\frac{3}{4}$	10	5	1
$2\frac{1}{4}$	$1\frac{1}{16}$	1	25	12	1
$3\frac{1}{4}$	$1\frac{1}{4}$	$1\frac{1}{4}$	50	25	3
$3\frac{1}{2}$	2	2	95	50	6
4	2	$1\frac{3}{4}$	100	56	7
$4\frac{1}{4}$	$2\frac{1}{4}$	2	136	76	9
$4\frac{1}{2}$	$2\frac{1}{4}$	$2\frac{1}{4}$	180	100	10
5	$2\frac{1}{2}$	$2\frac{1}{4}$	244	134	10
$5\frac{1}{4}$	$2\frac{3}{4}$	$2\frac{1}{2}$	250	140	10
$5\frac{1}{2}$	$2\frac{3}{4}$	$2\frac{3}{4}$	295	166	11
$5\frac{1}{2}$	3	$2\frac{3}{4}$	365	200	12
$5\frac{3}{4}$	3	$2\frac{3}{4}$	375	208	13
$6\frac{1}{4}$	$3\frac{1}{4}$	$3\frac{1}{4}$	550	300	15
$6\frac{3}{4}$	$3\frac{1}{4}$	$3\frac{1}{4}$	620	340	19
$7\frac{1}{2}$	$3\frac{1}{2}$	$3\frac{1}{2}$	780	400	28
8	$3\frac{3}{4}$	$3\frac{1}{2}$	910	500	35
9	$3\frac{3}{4}$	$3\frac{1}{2}$	1015	600	36
$9\frac{1}{2}$	4	$3\frac{1}{2}$	1285	700	40
$9\frac{1}{2}$	$4\frac{1}{2}$	$3\frac{1}{2}$	1448	800	41
10	4	4	1470	800	42
$10\frac{1}{2}$	4	4	1650	900	55
11	$4\frac{1}{2}$	4	1830	1000	65
11	$4\frac{1}{2}$	$4\frac{1}{2}$	2200	1200	72
$11\frac{1}{2}$	5	5	2750	1500	76

Liquation and Segregation in Bullion.—It is well known that when two or more metals are melted together and allowed to cool, it seldom happens that the resulting alloy solidifies or freezes as a whole and at a definite temperature. Usually one portion freezes first, rejecting another portion of different composi-

[1] These moulds are made by Messrs Fraser & Chalmers.

tion, which then solidifies at a lower temperature. This property is known as *liquation*.[1]

The researches of Peligot, Roberts-Austen, Rose, and others have shown that practically no segregation occurs in alloys of gold with either silver or copper or with both when rich in gold and free from all impurities; but when base metals other than copper are present, segregation occurs and the solidified metal is not uniform in composition. The result is that the exact value of the bullion cannot be determined, and the buyer must make some allowance to guard against loss. In the case of bullion containing a large proportion of base metals such as cyanide gold, a considerable amount of liquation generally takes place, the precious metals being driven outward or inward in the cooling mass according to the nature and amount of impurities present. The question of segregation in connection with the alloys of gold, silver, lead, and zinc produced in cyanide mills has been investigated by Matthey.[2] He found that one such ingot weighing about 120 ounces contained 662 parts per 1000 of gold at the bottom corner, and only 439 parts at the top. In another case, when 16·4 per cent. of lead and 9·5 per cent. of zinc were present, the standard fineness of an ingot weighing 400 ounces as shown by actually separating the whole of the precious metals, was, gold 614·0, silver 75·8, and its true value £1028, while the value as deducted from the average of fourteen assays made on it (gold 576·0, silver 90)

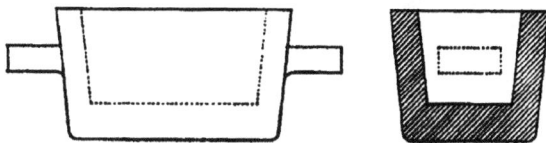

Fig. 119.

would have been only £965. Seven dip assays made on this ingot varied from 562 to 622 fine. Other cases of irregular distribution were even more remarkable. Some results obtained by Stockhausen with cyanide bars are given on page 256.

In the case of silver bullion the bars are rarely perfectly homogeneous even when over 950 fine, but the amount of segregation is in many cases too small to have any serious effect.

More or less segregation always occurs in silver bars containing copper, the silver being concentrated towards the centre, or to the outer surface of the cooling mass, according to whether silver or copper is in excess. The presence of other base metals in silver produces similar effects. For example, Gowland and Koga[3] have shown that when silver bullion containing 984·4 parts of silver and 15 parts of bismuth is melted and cast into an open ingot mould so as to give an ingot weighing about 1000 ounces, the portions of the ingot which remain longest fluid are richer in silver than the others.

Silver and gold present in bars of lead and of copper are also subject to the same irregularity of distribution, the segregation in some cases being considerable and causing great difficulty in determining the value of the bars.

The liquation of base bullion is discussed on page 312.

Sampling Bullion.—Since segregation may occur in cooling either gold or silver bullion, the method of taking pieces of metal for assay purposes is of considerable importance. When the bar is clean, small portions must be taken from different parts for assay; and if no liquation has taken place, the difference

[1] *Alloys*, E. F. Law, p. 35.
[2] *Proc. Roy. Soc.*, vol. lx. (1896), p. 21, quoted by Rose; *Metallurgy of Gold*, 5th edit., p. 20.
[3] *Journ. Chem. Soc.*, vol. li. (1887), p. 410.

between the assay results of the several portions should be very small. Before sampling a bar, it should be stamped with a distinguishing number, and the same number placed with each sample taken from the bar. This number should represent the bar through every stage of the assay by which its value is determined. Some assayers stamp the initials of their name on the cut faces so that no portion can be removed after it leaves their hands without detection.

When the bar has been pickled by dipping in acid after casting, the surfaces must be well scraped before sampling, as the outside is usually finer than the interior, in consequence of the pickling. For this reason some assayers always reject the first portions of metal removed in taking the sample.

Fig. 120.

Four methods are in use for sampling gold and silver bars, viz. (1) cutting, (2) drilling, (3) dipping, (4) granulation. The sampling of base bullion is described on page 312.

1. **Sampling by Cutting.** — This method is in very general use, and consists in cutting pieces from a top corner and a diagonally opposite corner of the bar, as shown in fig. 120. Experience proves that with bullion of high grade these samples are usually representative of the whole bar.

In the case of fine gold ingots weighing 400 ounces which are received from the Bank of England for coinage, a single sample is cut from the middle of one of the lower edges of the ingot.[1]

Fine silver bars are usually sampled in the same way. Cutting samples are taken with a narrow steel chisel, which must be used with care, otherwise the chip when cut off is liable to "fly away" and be lost. From 1 to 1·5 gramme is taken for each cut. A sampling machine is usually employed if a large number of bars have to be dealt with. Sampling by taking cuts is not always satisfactory if the bar is brittle, as a much larger piece may break off with the chip than is required. The sample is prepared for assay by flattening on a bright anvil with a hammer or by rolling.

2. **Sampling by Drilling or Boring.**—In this method the bar is drilled, top and bottom, with a small drill. This may be done in a lathe or with a ratchet drill. An ½- or a ³⁄₁₆-inch bit is generally employed for the purpose. A twist-drill as shown in fig. 121 is frequently used. The bar to be sampled should be placed in a clean metal tray in order to retain the drillings so that no loss may occur. The surface borings resulting from the first few revolutions of the drill should be rejected. Those that follow, to the extent of a little more than 1 gramme, are collected and reserved for assay, each lot being kept separate.

Fig. 121.

One boring is usually taken from a top corner of the bar and one from an opposite bottom corner, as in the case of a cut sample. Some assayers mix the two lots of drillings to form one sample. This is the general practice in South African gold mills. When it is known that the bar is not uniform in composition, four borings are sometimes taken, two from the top and two from the bottom, as shown in fig. 122. The four lots of drillings may be assayed separately, or the two lots from the top may be mixed to form one sample, and the two lots from the bottom mixed for a second sample according to circumstances.

[1] T. K. Rose, *Metallurgy of Gold*, 5th edit., p. 471.

The character of the drillings varies according to the nature of the bullion. With tough bars the drillings are soft, long and spiral in form, and can be rolled and cut up for mixing. When there is a tendency for the bars to be brittle the drillings are hard and short. In this case they are well mixed and the portion for assay selected at random.

The mixing of drillings from different parts of the bar is to be deprecated, unless it is known from the source of the bullion that the bar is practically uniform in composition.

3. **Sampling by Dipping.**—The sample is taken when the metal is in a molten condition and just prior to pouring. The metal is melted in a graphite crucible with a covering of borax, thoroughly stirred to ensure uniformity, and one or two dip samples taken by means of a graphite stirrer, with a hollow cavity at one end. The sample may also be taken by a small graphite crucible fastened by iron wire to an iron rod or by means of an iron ladle, but the introduction of iron into the melted metal is not considered to be desirable.

The sample is dipped out with a borax cover, and poured into an iron mould or allowed to cool in the dipper. About 2 grammes of metal are usually taken. The borax is detached, and when quite clear the metal is flattened and rolled for assay.

In some cases one sample is dipped from the top of the molten metal and another dipped from the bottom.

This method of sampling is always used in cases where the solidified bar is not uniform in composition. It is very generally employed for sampling cyanide gold bars. These bars are also sampled by drilling, but the results are not always trustworthy.

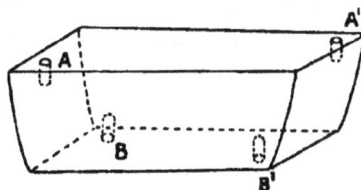

FIG. 122.

4. **Sampling by Granulation.**—The metal is granulated by pouring into water. When the bullion has been melted, and if necessary refined, it is well stirred, and two samples are taken with a scoop or small ladle, one from the bottom of the molten mass, and the other from the top. Each ladleful is poured, slowly and carefully, and in a narrow stream, into a clean copper bowl or wooden bucket containing warm water, which is slowly rotated with a paddle. The granulated metal resulting from each ladleful is kept separate. The water is drained off and the granules carefully dried. After drying, the granules should be hammered flat or rolled, in order to remove any enclosed water. This method of sampling is sometimes adopted for bullion of high grade, but it is not satisfactory for impure bullion, as the pouring into water may result in partial oxidation of some of the base metals present and thus alter the composition of the sample so that it is not representative of the mass.

The importance of employing a suitable method of sampling, especially in the case of bullion that is subject to liquation when cast into bars, is shown by the irregularity of the assay results obtained by Stockhausen,[1] when drill and cutting samples are taken from cyanide bars. It is pointed out that the variation is sometimes due to imperfect mixing and sometimes to liquation, and experience shows that a dip assay (from molten metal) alone is to be relied upon with this material. In the case of two bars exhaustively examined by means of carefully checked assays of the dipped sample, and of drillings and cuttings taken from different parts of the bars, the assays gave the following results :—

[1] " Liquation in Cyanide Gold Bars," *Chem. Met. and Min. Soc. S.A.*, 1897.

Assays of Cyanide Gold Bars sampled by Different Methods.

	Average assay.	Extremes.	Maximum difference.
	Bar 1.		
Dip sample	809·0	808·9–809·6	0·7
Drillings	810·3	810·0–811·0	1·0
Cuttings	807·2	804·0–810·9	6·9
	Bar 2.		
Dip sample	679·5	679·1–680·3	1·2
Drillings	680·0	679·5–680·2	0·7
Cuttings	677·5	676·3–678·9	2·6

From an extensive series of tests made by Mr F. P. Dewey [1] in 1909 at the San Francisco Mint, it was found that, as a rule, in the cyanide bars from several California plants the chip samples taken from the outside of the bars would be about 2·5 per 1000 fine less in gold than the borings when taken away from the edges of the bar, and that the borings gave satisfactory samples of the bars. Mr Dewey also gives the results obtained by sampling, in three different ways, three cyanide bars high in gold (about 835 fine) and very low in silver (8·0 fine) and each weighing about 1500 ounces. "The assays show a wide variation on the chip samples. While the drill-sample assays are fairly concordant for this class of material, the dip-sample assays agree much better, and are to be preferred." These conclusions confirm those of Stockhausen, quoted above.

In many instances the composition of the bullion is such that a preliminary refining becomes necessary before the gold can be determined with accuracy. **Preparation of the Sample for Assay.**—The metal selected for the assay piece by any of the methods of sampling described is prepared for assay by flattening on an anvil with a polished hammer, weighing about

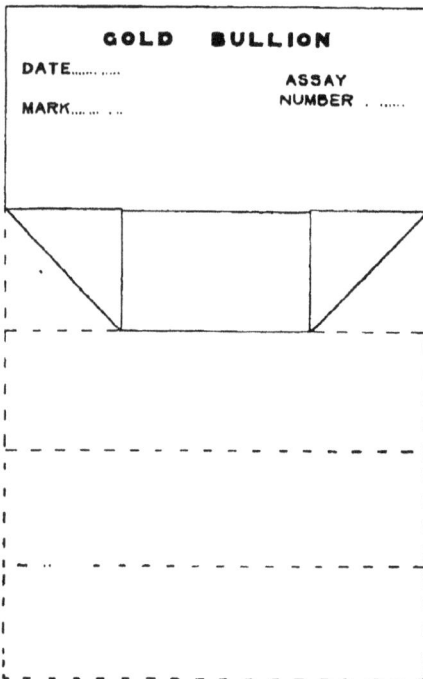

GOLD BULLION

DATE........ .....

MARK....... ...

ASSAY NUMBER . ...

FIG. 123.

[1] F. P. Dewey, "The Assay and Valuation of Gold Bullion," Spokane meeting of *Am. Inst. Min. Eng.*, Sept. 1909.

11 lbs., until thin enough for cutting conveniently with shears. The anvil and hammer should be kept perfectly clean and only used for this purpose.

In the case of samples of bullion that is more or less brittle, a heavy iron ring about 8 inches in diameter and 1¼ inches high is placed on the anvil to enclose the assay piece and retain any metal that tends to fly away. Some assayers only use the anvil to flatten the assay piece to a convenient thickness for rolling, and then pass the metal through the rolls until of the desired thickness.

Bullion samples are usually wrapped in "assay papers" or "dockets" about 6½ inches by 4 inches, on which the number and any other details relating to the sample are recorded. In many cases the papers bear the name of the assayer. The general method of folding the assay paper so that the sample shall not fall out is shown in the accompanying illustration, which gives the details of the folding and is self-explanatory (see fig. 123). Tough paper of good quality should be used, otherwise the metal is liable to cut through and occasion loss.

Owing to the high value of the precious metals, the unused portion of the sample is returned with the gold cornet from the assay, when the assay value of the bullion is reported, or the value of the metal taken for sample is allowed for in the settlement for the purchase of the bullion.

CHAPTER XVIII.

THE ASSAY OF GOLD BULLION.

Apparatus used in Bullion Assaying.

THE following apparatus is used in connection with bullion assaying.

Hammers and Anvil.—These are used for flattening bullion "assay pieces" and gold and silver buttons. The face of both the hammer and the anvil should be kept quite bright and clean, and they should be used for the above purpose only. Suitable anvils and hammers are illustrated and described on page 41.

Rolling Mill.—The ordinary jewellers' rolling mill (fig. 124) is the form usually adopted for assay purposes; but considerable care should be devoted to accuracy of manufacture, as it is of great importance to be able to roll large numbers of slips of metal of the same thickness. In one form of mill the rolls are accurately adjusted by one motion, as shown in the illustration. The size of the rolls used is dependent on the nature of the work to be done. A pair of 2-inch rolls is often used, but 3-inch rolls will be found a more generally convenient size. The rolls should be fixed on an iron or wooden stand bolted firmly to the floor. The handle should have a radius of not less than 15 inches, and time may be saved by introducing an inclined plane of sheet brass between the two uprights of the mill, by means of which the strips, after passing through, are immediately returned to the front.

A section of an inclined plane or shoot is shown in fig. 125. The rolls should be kept quite clean and bright and should be used for bullion work only. In assay offices where a large amount of general assaying is done it is usual to have an additional and cheaper pair of rolls for rolling out metals such as lead for cupellation, etc., and for "breaking down" small flat ingots of fine gold or silver.

Steel Forceps.—The ivory-tipped forceps used for the manipulation of weights are not convenient for handling bullion, etc., when weighing out the quantity for assay, steel forceps are therefore generally employed. The forceps illustrated in fig. 126 are very suitable for the purpose: they are about 5 inches long.

Shears.—Metal shears with blades about $1\frac{3}{4}$ inches in length are employed in cutting off the portion of metal for assay. The shears should be strong, of good quality, and have a keen cutting edge. The cutting pliers illustrated in fig. 50 and fig. 51, page 42, are also useful for cutting bullion samples.

File.—In some assay offices the adjustment of the weight of the portion of metal for assay is completed by means of a fine file, about $\frac{3}{4}$ inch wide, which is in many cases fixed on a shallow wood tray in front of the assayer's balance. The file should be flat and smooth; if too rough, the adjustment of the weight is rendered more difficult.

Sample Rack.—This consists of a block of hard wood, such as oak or walnut, provided with a number of slits $\frac{3}{16}$ inch wide cut at an angle of about 80°, as shown in fig. 127. It is used to stack the assay papers (fig. 123) after the sample of bullion has been weighed, and is placed either to the left or right of the balance when weighing out the samples. The rack is about 2½ inches wide, 1⅛ inches thick, and the length is varied according to the number of slits required, usually twenty or less. Other forms of sample racks are in use, but that described will be found very convenient.

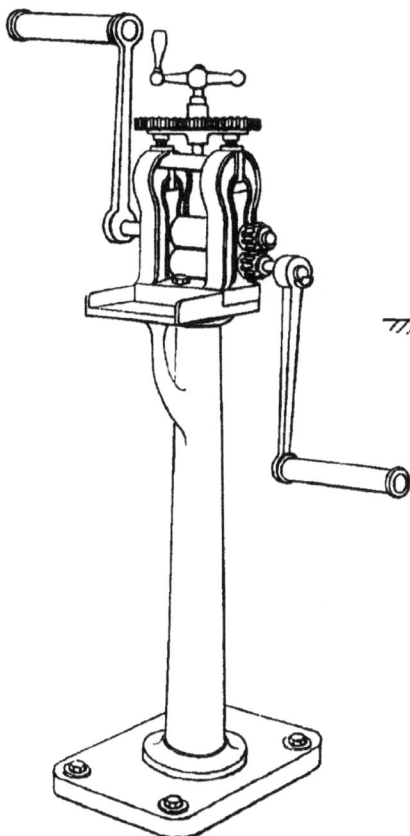

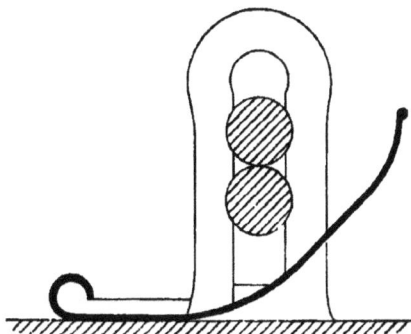

Fig. 125.

Fig. 126.

Fig. 124.   Fig. 127.

Trays.—Several forms of trays are employed to keep the assay pieces and the buttons produced in cupellation, etc., in order through the various operations. The tray shown in fig. 128 is of iron, and those in fig. 129 and fig. 130 are of wood. A piece of sheet zinc is inserted in the bottom of the latter to protect the wood from burning through the tongs becoming heated when charging in a tray full of lead buttons for cupellation. The compartments are 1¼ inches square and about ½ inch deep. The sloping sides of the compartment enable larger numbers to be pasted on than is possible in the case of the tray in fig. 129, so that the number of any particular assay is more readily seen.

Lead-Foil (test lead).—This should be free from silver, and must be tested for silver before use (see page 145).

Lead-foil in strips about 1½ to 2 inches wide, and rolled to a thickness of

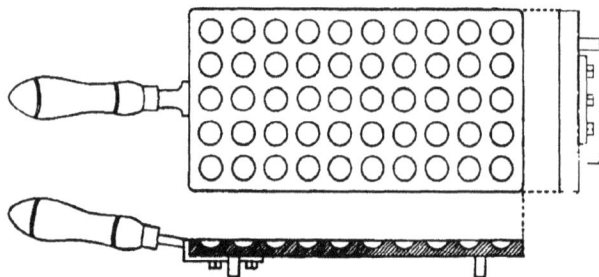

FIG. 128.

from No. 2 to No. 5 Birmingham metal gauge (0·009 inch to 0·014 inch), is usually employed. For bullion assays the required quantity is cut off, weighed,

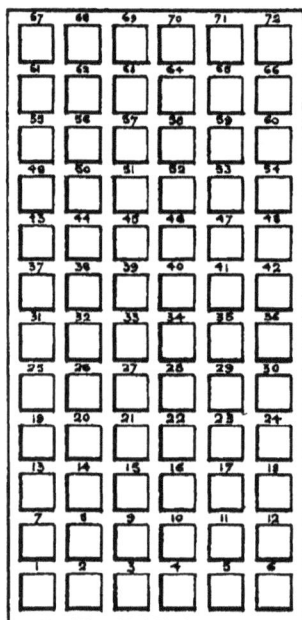

FIG. 129.

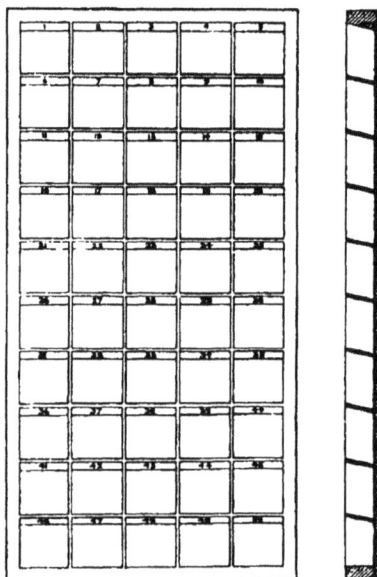

FIG. 130.

and folded into a packet for the reception of the portion of bullion weighed for assay.

It is important that the packet should be securely folded after the bullion has been introduced, otherwise part of the sample may be lost by cutting through the lead.

A convenient and very usual method of folding the lead is shown in the accompanying illustrations, fig. 131. A shows the lead ready for folding. B, the lead with left-hand corner folded over to the right, and then the right-hand corner folded over to the left, thus forming a packet for receiving the weighed sample. C shows the packet, after inserting sample, pressed flat and folded under on the line a–b. D the packet with the corner c_1 folded on the top, to make it convenient for rolling. E the packet rolled between the fingers as indicated by the arrow.

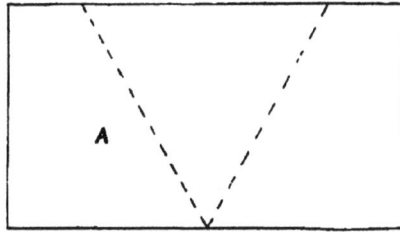

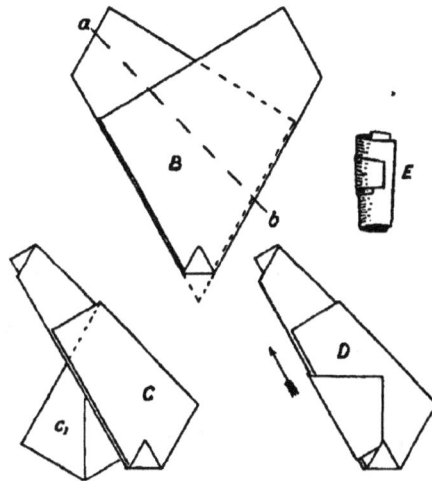

By folding in this way the sample is always enclosed in several thicknesses of lead and does not therefore tend to cut through.

When a large number of assays of the same kind have to be made the lead-foil is conveniently purchased cut ready to size and weight.

If varying weights of lead are required, much time is saved by measuring off a strip of foil of uniform width instead of weighing.

A convenient device[1] for this purpose is shown in fig. 132. It consists of a sheet-iron plate with an oblong hole cut in the middle, and graduated into a number of divisions corresponding to different weights of lead in grammes. The graduations are made by noting the positions occupied by different weights of lead-foil of uniform width.

FIG. 131.

To use the instrument, the end of the

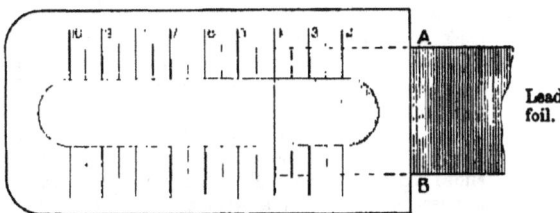

FIG. 132.

strip of lead-foil is placed under it in position against the lines corresponding to the weight of lead desired, the plate held down firmly, and the lead then torn

[1] The author is indebted to Mr W. F. Lowe, A.R.S.M., of the Chester Assay Office, for first suggesting this device. The one described is a modification of that used by Mr Lowe.

off along the line A-B. The illustration shows the position of the strip for obtaining 4 grammes of lead.

A plate is very readily made for use with any desired thickness and width of lead. The lead packets should always be prepared by the assayer himself, and a quantity kept on hand ready for use.

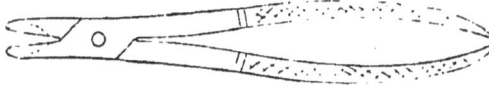

FIG. 133.

Balling Pliers or Nippers.—These are used for squeezing the lead-foil packets into the form of a bullet for convenience of manipulation. They consist of a pair of strong steel pliers with smooth concave chops as shown in fig. 133. The diameter of the cavity in the chops corresponds to the size of lead bullet desired: from ¾ inch to ½ inch is a very usual size for squeezing bullion assays.

The lead to be "balled," after being folded as described above, is placed between the chops fully opened, and then squeezed repeatedly by alternately opening and closing the pliers held in the right hand. The chops of the pliers are lightly supported between the thumb and second or third finger of the left hand and the lead is gently rotated each time the pressure is released, by a slight upward movement of the first finger, thus gradually causing it to be converted into a ball. With a little practice the pliers can be readily manipulated and a batch of forty or fifty assays "balled up" in a very few minutes.

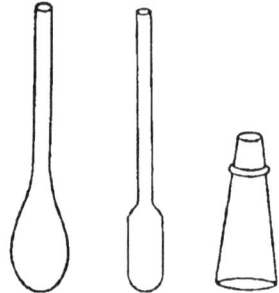

FIG. 134A.   FIG. 134B.   FIG. 134C.

When the lead is balled in this way the assays are much more convenient for charging into the furnace than when the lead packets are irregular in shape.

Parting Apparatus.—The parting operation, to be subsequently described, in which the "cornets" are treated with boiling nitric acid, may be conducted either in glass flasks or in platinum cups.

Flasks.—Several forms of glass flasks are employed as shown in figs. 134A, B, and C. When flasks are used, a series is so arranged over a number of bunsen rose gas burners, that several can be heated simultaneously. A section of a burner with a flask in position on an enlarged scale is shown in fig. 135. The burners are suitably arranged in a fume cupboard with flues connected with a chimney for exhausting the fumes. An arrangement for heating ten parting flasks is shown in front elevation in fig. 136A and in section in fig. 136B. The cupboard is glazed at the sides, and has a window in front which slides up and down. The flasks are kept in position on the burners by means of a wire support as shown in fig 135, or more generally by means of a perforated strip of sheet iron as shown in fig. 136. The top of the burner is made slightly concave (see fig. 135) so as to keep the bottom of the flask in the centre of the gas jets.

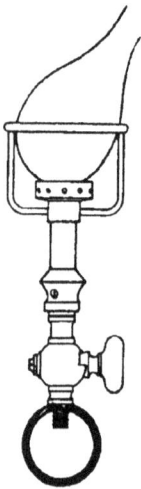

FIG. 135.

Another form of flask sometimes used for parting is illustrated in fig. 134c. In this case the flask is heated on a hot plate covered with asbestos board.

Platinum Cups.—When it is desired to treat several cornets simultaneously, short thimble-like platinum cups are used. A cup is shown at A, fig. 137; at

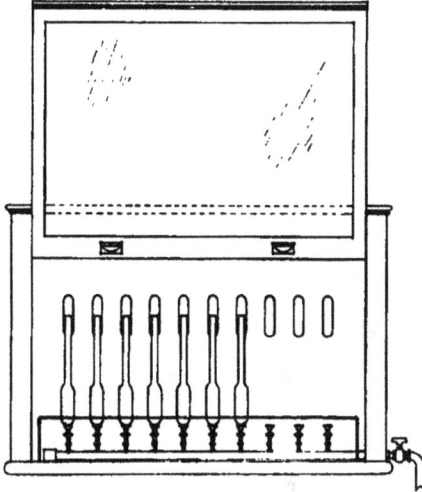

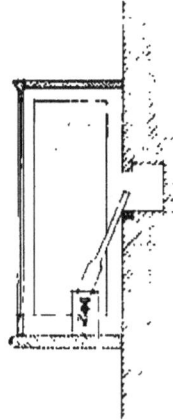

FIG. 136A.   FIG. 136B.

the bottom there are five slits, which are fine enough to prevent any gold from escaping, and large enough to allow of the free passage of nitric acid or water.

A modern design of frame made of stout platinum wire for holding twenty cups is illustrated in fig. 137. It is supported on four legs, and is provided with a handle. The cups are slightly enlarged at the top, as shown, to support them in the tray. One of the cups is provided with a thin wire handle and is placed in the left-hand corner at the back of the frame so as to identify No. 1 in the series. A platinum hook fitted in a wooden handle is used for lifting the frame in and out of the hot acid (fig. 137, B). The frame, with its cups, is immersed

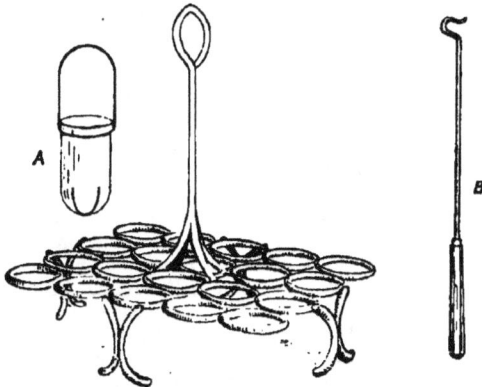

FIG. 137.

in boiling nitric acid in a large glass beaker or porcelain boiler. Platinum boilers were formerly used, but are now being discontinued.

Platinum parting frames to hold any desired number of cups are obtainable from Messrs Johnson, Matthey & Co., Hatton Garden, London, the well-known platinum refiners.   Frames to hold fifty or sixty cups are in general use in bullion assay offices.

Proof Gold and Silver.—Gold and silver of a very high degree of purity is used for "check" or "proof" assays.   The preparation of proof gold is described on page 70, and of proof silver on page 75.   ·The metals are usually rolled to sheet conveniently thick for cutting up with the shears.   Proof silver used for inquartation is sometimes stamped into discs of convenient size and weight, or drawn into flat wire and pieces of the required weight cut off with gauged pliers. Some assayers consider it more convenient for weighing out to granulate the silver, and use only the granules that will pass through a sieve of about 13 mesh.   Others use the precipitate of metallic silver obtained by the reduction of silver chloride with iron, after drying and without melting.

The Assay of Gold Bullion.

Introduction.—Gold is always determined by the "parting assay," which consists in cupelling the bullion with from two to three parts of silver, followed by the removal of silver by boiling in nitric acid, and the weighing of the gold residue.   Cupellation removes the greater part of the base metals present in the bullion as oxides dissolved in litharge, and an alloy of gold and silver is left on the cupel.   This is "parted" by nitric acid, which dissolves the silver and leaves the gold unattacked.   In the assay of refined gold bullion the gold alone is determined—silver, if any, being disregarded.   Unrefined gold bullion contains some silver and variable quantities of base metals, and it is usually necessary to determine the proportion of both gold and silver.

Determination of Gold only.—The method of conducting the "parting assay" varies slightly as to details of manipulation in different assay offices, but the following method of procedure may be regarded as descriptive of the general practice.   The parting assay of gold in bullion comprises the following five distinct operations :—

1. Preparation of the assay piece for cupellation.
2. Cupellation of the assay piece.
3. Preparation of the assay piece for parting.
4. Parting and annealing the cornets.
5. Weighing the cornets and reporting.

1. **Preparation of the Assay Piece for Cupellation.**—The sample or assay piece is selected and prepared for assay by flattening or rolling in the manner previously described.   Each assay piece is placed in an assay paper, and its number and any other details relating to it duly recorded on the paper.   The assay piece is then prepared for cupellation by weighing about 0·5 gramme and wrapping it in lead-foil with the silver necessary for parting.

Weighing—In the assay of gold bullion the "millième" system of assay weights is used, a millième being 0·5 milligramme.   In this system, fully described on page 60, the half-gramme weight is stamped 1000, and the decimal subsidiary weights stamped 900, 800, etc., down to 0·5.   These stamped numbers denote the number of half-milligrammes (*i.e.* millièmes) contained in the weight.   Ordinary weights in the gramme system may of course be used, each milligramme corresponding to 2 millièmes in the assay system.   The special millième set of "gold assay weights" is, however, very generally used in all bullion assay offices and in many mines where bullion assays are being made regularly.   The report finally made gives the number of parts (in millièmes and tenths) of pure gold in 1000 parts of the bullion.   In the Colonies it is usual for assayers on gold mines to use for weighing their parted gold the set of weights in which the weight marked 1000 weighs 1 gramme, and therefore the weight marked 1 equals 1 milligramme.[1] This set of assay weights is described on page 60, and is generally used for the assay of silver bullion only.

The assay piece is in many cases first rough weighed (i.e. adjusted approximately) by an assistant upon a "preparing balance" capable of turning to 1 or 2 millièmes About 1001 millièmes are weighed from each piece, in duplicate or in triplicate, and the weighed pieces folded in a small square of paper and tucked inside the paper containing the rest of the sample or placed in wooden partitioned trays until they can be checked by the assayer. The method of weighing is conducted as follows:—The assayer seats himself directly before the balance, having the assay piece in a convenient position in front of him. · The 1000-millième weight is placed on the right-hand pan of the balance and portions of the assay piece added to the left-hand pan until in slight excess. The portions are adjusted approximately to the exact weight by cutting with shears and filing. In using the shears the metal to be cut must be held firmly between the forefinger and thumb of the left hand, and care taken to keep the plane of the assay piece perpendicular to the cutting faces of the shears. The adjustment of the assay piece is completed by carefully filing on a very fine-grained file. With practice, weighing becomes a very quick operation, a skilful assayer being able to perform forty to fifty weighings in an hour. The "fine" weighing or final adjustment of the assay pieces is effected on a "fine" balance. "When the assay is reported to $\frac{1}{10,000}$ part, it is evident that the fine balance used must clearly indicate a difference in weight of 0·1 per 1000 or 0·05 milligramme. With such a very sensitive balance it is extremely tedious to adjust the assay piece to exactly half a gramme, and in practice it is usual to adjust it until almost correct, and then to note the slight excess or deficiency in weight and record it on the assay paper as a plus or minus 'weighing in' correction. Thus an excess weight of three-tenths of a millième would be recorded on the corner of the assay paper as +3. This correction is allowed for in weighing the final gold cornet. It is convenient to have the balance adjusted so that one subdivision of the ivory scale traversed by the pointer corresponds to 0·05 milligramme (i.e. one-tenth of a millième)."[2] The weight of the assay piece should not differ from 1000 by more than four subdivisions on the scale (i.e. four-tenths), and if the deviation is greater than this the assay piece is adjusted until it comes within this limit. In some bullion assay offices "substitution weighing," described on page 61, is adopted.

The weighed assay piece is wrapped in lead-foil (folded as previously described) together with the fine or proof silver necessary for parting, and some copper unless this metal is already present in the bullion being assayed. As each lead packet is prepared it is placed in order in a numbered compartment of a wooden tray and its position duly recorded on the assay paper. The lead packets are next squeezed in the balling nippers to form "bullets," and the assays are then ready to be charged into the furnace.

The amount of silver, copper, and lead used in the assay varies according to circumstances.

Experience has shown that the alloy for parting should not contain less than two nor more than three parts of silver to one part of gold. If the silver is in large excess, the residual gold, instead of holding together in a form easy to manipulate, falls to a powder which requires great care in its treatment to avoid loss. If the silver is insufficient the gold protects it from the action of the acid and some will remain undissolved, so that the alloy will not be properly "parted."

[1] H. T. Durant, Mining Mag., vol. v., 1911, p. 435.
[2] T. K. Rose, "Practice at the Royal Mint," vide Metallurgy of Gold.

The modern practice is to aim at getting an alloy with $2\frac{1}{2}$ parts of silver and 1 part of gold. After a long series of experiments at the Royal Mint, however, it has been found that the most exact results are obtained by using a ratio of silver to gold of about $2\frac{1}{8}$ to 1.[1]   In the American Mints a ratio of 2 of silver to 1 of gold is used.

If silver is contained in the gold bullion, it is allowed for in calculating the quantity of silver to be added.   In order to form an alloy of gold and silver in the correct proportions for parting it is obvious that the amount of silver already present in the bullion must be known, and if this is not previously known with a fair degree of accuracy a preliminary assay must be made.

Preliminary Assay.—Experienced assayers can frequently judge the fineness of the gold bullion with sufficient accuracy by the colour and hardness of the metal, and add the correct amount of silver to ensure parting.   In some cases the source of the bullion is sufficient indication of the amount of silver required. With cyanide gold it is generally necessary to make a preliminary assay.   When the composition of the bullion is quite unknown, it may be determined by means of the touchstone described on page 280, or more satisfactorily by a simple ʹ cupellation and parting.   For this purpose weigh up two lots of 100 milligrammes (0·1 gramme) of the bullion, add 300 milligrammes of silver to one lot, and then wrap each lot in 3 grammes of lead.   Cupel both side by side.   Flatten the button containing the added silver and boil in 15 c.c. of dilute nitric acid (1 to 1) for fifteen minutes; the resulting gold is washed, dried, heated to redness, and weighed.

The weight in milligrammes gives directly the percentage of gold.   The weight (in milligrammes) of the other button gives the percentage of gold and silver together; the difference between the two gives the percentage of silver. The remainder will be base metal.   Thus :—

Weight of gold and silver = 98·4 milligrammes (or per cent.).
Weight of gold   = 94·0   ,,   ,,

Difference equals silver   4·4   ,,   ,,

and 100–98·4 equals 1·6 base metal.

If the base metal consists of copper, its presence will be indicated by the appearance of the cupel.   In cases where it is known that the alloying metal is copper only, or that silver is not likely to be present in large quantity, a simple cupellation with lead is alone necessary to determine the fineness of the bullion (see page 329).   When the composition of the alloy has been ascertained, the amount of silver to be added for parting is readily determined.   For example, suppose the bullion has the composition given above and that 0·5 gramme of it is to be taken for an assay.   Then the 0·5 gramme will contain :—

	Composition per cent.		Proportion in 0·5 gramme.
Gold	94·0	Gold	0·470 gramme.
Silver	4·4	Silver	0·022 ,,
Base	1·6	Base	0·008 ,,
	100·0		0·500 gramme.

The total silver required to part the gold is $0·47 \text{ gramme} \times 2·5 = 1·175$ grammes.   Allowing for the 0·022 gramme of silver already present, the amount of silver to be added is $1·175 - 0·022 = 1·153$ grammes.   As previously stated, the silver is incorporated with the gold, and at the same time the base metals, such as copper, eliminated by wrapping in sheet lead and cupelling.

The quantity of lead to be used is dependent on the weight of bullion taken for assay and on the proportion and nature of the base metals present. Table XXX. shows the proportions recommended for ordinary gold bullion of different degrees of fineness when using 0·5 gramme for assay.[1]

Table XXX.—Weight of Lead to be taken for Cupellation of 0·5 gramme Gold Bullion.

Fineness of bullion (gold and silver together).	Weight of lead (for 0·5 gramme bullion).
950 to 1000 fine.	2 grammes.
900 to 950 ,,	4 ,,
800 to 900 ,,	6 ,,
700 to 800 ,,	8 ,,
600 to 700 ,,	10 ,,
500 to 600 ,,	12 ,,
100 to 500 ,,	12 ,,

When more than about 6 grammes of lead are required, some assayers wrap the assay piece in a part of the lead and then in cupelling charge in the excess lead either before or after introducing the assay piece containing the assay piece.

M'Arthur Johnston states[2] that when the bullion is below 700 fine it is best to scorify 0·5 gramme with from 20 to 24 grammes of lead before cupelling and to use checks and submit them to exactly the same treatment.

If the base metal alloyed with the gold consists mainly of copper as in the case of industrial gold alloys, other quantities of lead than the above are used, as described in Chapter XXI. If the bullion is known to be free from copper it is usual to add a piece of copper foil or wire weighing from 10 to 20 milligrammes. The copper is added as the small quantity that remains in the button tends to prevent "spitting" or vegetating after cupellation and also to increase the malleability of the button for rolling.

2. Cupellation of the Assay Piece.—The cupellation may be conducted in any of the muffle furnaces described in Chapter III., but gas-muffle furnaces are preferable for bullion assays when a good supply of coal gas is available, and are now very generally used. Opinions differ as to the number of assays that should be cupelled at the same time when a large amount of bullion work has to be dealt with. Some assayers cupel only twenty assays at the same time, in order to avoid extreme variations of temperature,[3] while at the Royal Mint seventy-two assays constitute a "fire," the cupels used being in sets of four (see page 48). In most assay offices a "fire" will consist of twenty assays or less. The cupels should be arranged carefully in rows of four abreast in the central part of the muffle and surrounded by blank cupels, so that the temperature may be kept as uniform as possible for those cupels in which the assays are placed. The well-known French cupels (Deleuil) are excellent for bullion assays. The cupels are baked in the furnace for at least half an hour before use, and the muffle brought to a good orange-red heat before charging the assay into the

[1] Recommended by the Committee appointed by the Chemical, Metallurgical, and Mining Soc. of S. Africa to investigate the causes of the differences in gold bullion assays, see Proc. of the Society, vol. ii., 1897-99, Appendix I., p. 791.
[2] Rand Metallurgical Practice, 1912, p. 306.
[3] Vide A. C. Claudet, Trans. Inst. Min. and Met., vol. xvi. (1906-07), p. 137.

cupels. When the cupels are ready the assays are charged into them singly with tongs or by means of some charging device such as that described on page 183. The charging in is performed carefully, but as rapidly as possible, so as not to cool the muffle unduly, and the muffle door is then partly closed. The exact temperature employed for cupellation, provided it is not too low, is of less importance than uniformity of temperature, as in bullion assaying check assays (see page 276) are invariably used to correct uniform errors due to high temperatures. At the same time excessive heating should be avoided. The conditions for gold cupellation are discussed in Chapter XII.

The time occupied by the cupellation will necessarily vary with the amount of lead used; with 2 grammes of lead it is complete in from eight to twelve minutes. The completion of cupellation takes place in the front rows of cupels and proceeds regularly backwards. The cupels are sometimes removed from the furnace whilst the assay is still molten, and if any detached globules of metal are observed, they are washed down very carefully by a circular movement of the cupel when removed from the furnace, but this requires considerable experience in manipulation. The method of removing the assays while still molten is adopted at the Royal Mint and other offices, where all the cupels are removed at the same time on a slab, as described on page 182. With single cupels, each cupel as it is withdrawn is placed in its proper position in the cupel tray. The characteristic " flashing " of gold ensues in a few seconds, and is most marked in the purer buttons, in which little copper or lead remains. Slight effervescence may occur in these cases, but if the bullion being assayed contains more than 50 parts of copper per 1000, the buttons from the cupellation are never sufficiently freed from base metals (notably copper) for "spitting" to take place (T. K. Rose). To avoid losses by "spitting," etc., most assayers allow the assays to remain in the furnace until the buttons have solidified. To effect this the muffle-door is closed and the temperature lowered. The buttons when "set" (solidified) should show a depression on the top, and no "vegetation" (fig. 138, a). The detection and the effect of the presence of metals of the platinum group during cupellation and the subsequent parting are discussed in Chapter XXV.

3. Preparing the Assay Piece for Parting.—The buttons are removed from the cupels by means of a pair of sharp-nosed pliers and squeezed to free them from adherent bone-ash. They are then cleaned on the under side with a stiff brush and placed in the compartments corresponding to their cupels in a metal or wooden tray. The complete removal of the bone-ash is not of extreme importance, since bone-ash is readily dissolved by nitric acid on parting. The buttons are flattened on an anvil with a polished hammer weighing about 7 lbs. (figs. 44 and 45), both perfectly clean and bright and used for this purpose only. A heavy blow is first delivered on the centre of the button, the diameter of which is thereby increased to about half an inch. A blow is then given on the edge, so directed as to elongate the metal in one direction, and a similar blow is next given on the opposite edge, reducing it to the form shown at b, fig. 138. The flattened buttons are then annealed at a low red heat on a clean "fireclay" tile in the muffle or by means of a blowpipe. In some assay offices the buttons are annealed on a piece of iron gauze which is heated by means of a "grid" burner. At the Royal Mint the annealing is effected in an iron tray especially made for the purpose, which is placed in the muffle.[1] The object of the annealing is to soften the buttons to facilitate rolling, the hammering having hardened them. After annealing, the buttons are passed in succession through the "flatting" rolls to form elongated "fillets" all of the same length (about 2¼ inches) and thickness, the rolls being adjusted so that one passage through them reduces the buttons to the required thickness, which is about 0·01 inch (No. 3 Birm.

[1] Consult Rose, *Metallurgy of Gold*, and Percy, *Metallurgy of Silver*.

metal gauge), or about the thickness of an ordinary visiting-card (see *c*, fig. 138). The edges of the "fillets" should be quite smooth and not "ragged," otherwise loss may take place during the boiling.

After being rolled they are again annealed to soften them, and then separately rolled up between the finger and thumb, or round a glass rod, into a "cornet" or spiral (*d*, fig. 138). It is the practice in some of the American assay offices to stamp a distinguishing letter or number at one end of the fillet before it is rolled into a "cornet," and then to commence rolling at the unstamped end so that the number remains in sight when the cornet is completely rolled. All the cornets should have the same number of convolutions, and care is taken to leave that which was formerly the lower side of the button, outside, for the reason stated on page 428. This face is always dull and is easily recognised.

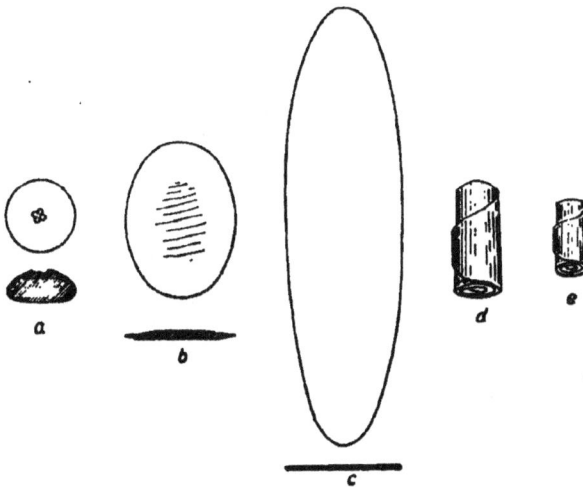

FIG. 138.

a, button (plan and sectional elevation); *b*, button after flattening (plan and sectional elevation); *c*, fillet (plan and sectional elevation); *d*, cornet before annealing; *e*, cornet after annealing.

4. Parting and Annealing the Cornets.—The "cornets" prepared as described are now ready for "parting" the silver from the gold by treatment with acid. This is effected either in platinum cups contained in a frame or in glass "parting flasks" such as are described on page 262. Parting frames and cups made of fused silica are now coming into use as a substitute for platinum, and appear to give satisfaction.

(*a*) *Parting in Platinum Cups.*—These are now used whenever possible, as they effect a considerable saving of time. The cornets are placed in the frame in their respective cups, and the frame immersed in boiling dilute nitric acid and boiled for ten to fifteen minutes, or until no more nitrous fumes are observed. The frame is taken out and drained from acid liquor, then washed by dipping in and out of a vessel of hot distilled water, drained again, and immersed in a second lot of stronger boiling nitric acid. The cornets are boiled in this for from fifteen to twenty minutes, when they are drained and washed as before, and are then ready for annealing.

It is usual to keep a stock of the acid suitably diluted to the two strengths required for the parting. These are known as the "parting acids" The first or No. 1 parting acid is the weaker and is used in the first attack on the cornets. After treatment in this acid nearly all the silver has gone into solution as silver nitrate. The second or No. 2 parting acid is stronger, and is used to remove the small quantity of silver still remaining in the cornet. The proper strength of parting acid is of importance. When the ratio of gold to silver in the cornet is 1 to 2½, the specific gravity generally recommended for No. 1 acid is about 1·2 (equal parts of acid and water), and for No. 2 acid a specific gravity of about 1·3 (2 parts of acid to 1 of water). In practice, the No. 1 acid is usually prepared by diluting No. 2 acid after use, the small proportion of silver nitrate this contains not being harmful for this purpose, but an advantage, since it serves to remove any chlorine that may be present. The nitric acid used should be free from hydrochloric acid or chlorine in any form, and a small quantity of silver is kept in solution, in the stock of acid, so that chlorine, if present, would be instantly detected. The nitric acid should also be free from sulphuric acid. The washing of the cornets after treatment in No. 1 acid is frequently omitted, but experience proves that the "final" cornets retain less silver when they are well washed with distilled water at this stage. Platinum boilers are

FIG. 139.

still in use in many offices, but, owing to the high price of platinum, glass and porcelain boilers are being introduced. The platinum boilers are fitted with a hood of platinum, the lower edge of which drops into a channel extending round the upper edge of the boiler and kept filled with water to form a "water joint." The top of the hood is connected by means of platinum tubes with any suitable arrangement for removing the fumes and condensing nitric acid.[1]

With frames for parting twenty cornets or less, the boiling is usually effected in glass beakers (preferably of Jena glass) with a clock-glass cover, a separate beaker being used for each parting acid. The beakers are placed on a gauze-covered quadripod stand and heated by means of two bunsen burners with "rose" tops.

Bumping is very liable to occur during the boiling, especially with the stronger acid, and to prevent this the addition of a small ball of well-burnt clay of about the size of a pea is generally recommended, as it lessens the tendency to irregular and dangerous boiling. Bumping is also very effectively prevented by the use of the simple device shown in fig. 139, which the author has used for many years with entire satisfaction.[2]

It consists of a piece of glass rod, bent to form a semicircle, and to which are fused three short legs, about 1 inch long, of glass tubing of about ⅛-inch bore, so that the top end is sealed while the lower end is left open. The apparatus is simply placed in the beaker containing the cold acid, so that the open ends of the tubes rest on the bottom of the beaker, and is allowed to remain there during use. The diameter of the apparatus is such that it fits close to the sides of the beaker and allows ample room in the middle for introducing the platinum frame.

---

[1] Platinum boilers such as are used at the Royal Mint are fully described and illustrated in Percy's *Metallurgy of Silver*, p. 263.

[2] This device was suggested to the author by a paper on "Prevention of Bumping during Boiling," by Hayward Scudder, *Journ. Amer. Chem. Soc.*, 1903, vol. cxxv. pp. 163–165.   *Theory.*
—In a general way the theory, of the action of such a tube is that, when heated, the air in the capillary expands and passes through the liquid in bubbles. The vapour of the liquid gradually replaces the air, and the stream of bubbles is continuous as long as the temperature around the capillary is at the boiling-point of the liquid. This constant bubbling prevents superheating and consequent explosive boiling.

After the final washing of the cornets, the frame is placed on a pad of felting or a folded duster to absorb the last drops of water. It is then introduced into the muffle at a bright red heat, kept especially for the purpose, and allowed to remain there until the cornets are thoroughly red-hot. The cornets, which before annealing are of a dull brown colour and friable, now assume a bright gold-yellow colour, diminish considerably in bulk, and harden. The annealing must not be conducted at a temperature sufficiently high (orange-red) to cause adhesion between the platinum and the gold. The cornets when cold are ready for weighing.

(b) *Parting in Flasks.*—The flasks most frequently used are either bulbs with long necks (fig. 134A) heated on rose burners of special construction, as described on page 262, or they are small, flat-bottomed conical flasks which are conveniently heated on a hot plate (fig. 134c). The bulb flask is generally used in England, while in America conical flasks are more frequently employed.

About 50 c.c. of the first parting acid are placed in each flask and raised to boiling point. The flask is then withdrawn and tilted a little to one side, whilst the cornet is cautiously dropped into it. After introducing the cornet, the hand should be withdrawn promptly, to avoid scalding by the sudden issue of hot acid vapours. The flask is replaced on the burners or hot plate, and the acid kept boiling for ten or fifteen minutes (*i.e.* for about five minutes after nitrous fumes cease to be given off). The flask is then withdrawn with a piece of stout paper folded into a strip, or with a pair of wooden tongs, and the acid liquor is carefully decanted off into a clean beaker, which is best placed upon a white glazed tile, so that any detached particles of gold accidentally carried over during decanting are readily detected. The decanted silver nitrate solution is transferred to a bottle marked " Waste Silver," and the silver subsequently recovered by precipitation as chloride.

The residual gold in the flask is washed by adding distilled water and again decanting. About 40 c.c. of the second parting

FIG. 140.

acid, previously heated, are then poured into the flask, which is now replaced on the burner or hot plate and the boiling continued for fifteen or twenty minutes. As it is at this stage that bumping is most liable to take place, the heating of the flasks should be carefully regulated. The addition of clay peas or a piece of capillary glass tube lessens the tendency to bump, and in the case of flat-bottomed conical flasks it may be entirely prevented by adopting the device previously described (see conical flask illustrated in fig. 165, page 419).

After boiling with the second acid, the flask is removed and the acid carefully decanted off. The liquor is preserved in a bottle marked " Acid Waste," and is used for making up the first parting acid, as previously indicated. The flask is then washed twice with hot distilled water by decantation. If any small particles of gold have become detached from the cornet, time must be allowed for them to settle before each decantation. After the last decantation the flask is filled to the brim with water, a small unglazed porous crucible (annealing cup, fig. 140) inverted over the mouth, and then the flask and crucible are quickly but carefully inverted together. The pure gold, which is of a dark brown colour and exceedingly fragile, falls steadily through the water and rests in the crucible.

When time has been allowed for even the finest of any gold particles to settle into the crucible, the flask is gently raised until the mouth is on a level with the edge of the crucible, when with a quick side-motion the flask is removed, the water from it of course running to waste. If the flask is raised too quickly, the inrush of air will cause a rapid displacement of the water, with consequent risk of the gold being washed out of the crucible. The satisfactory manipulation of the flask requires experience, and beginners will do well to use a basin, as described on page 209, as a safeguard. The water in the crucible is poured off, and the crucible and cornet carefully dried. The cornet is then annealed at a red heat over a gas flame or in the muffle, when the gold assumes its characteristic bright yellow colour and hardens, so that it is in a state fit to be transferred to the balance and weighed.

Comparative Advantage of Platinum Cups and Parting Flasks.—The use of platinum cups effects a great saving of time in decanting and washing, the manipulation is less tedious than with parting flasks, and the treatment to which the cornets are subjected is more uniform, so that the correction afforded by the use of check assays is more trustworthy. There is also a considerable saving in acid when a large number of partings have to be made. Platinum cups can, however, only be employed where the amount of silver present bears such a relation to the gold that no cornet is in danger of breaking up—as it will be obvious that, if one cornet in a series breaks up, fragments may adhere to a number of others. An objection to the use of platinum cups is the impossibility of detecting in any individual cornet the presence of such metals as platinum

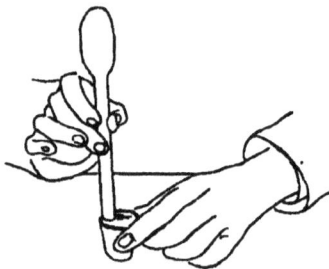

FIG. 141.   FIG. 142.

and palladium, which impart a straw-yellow or orange colour respectively to the acid, so that less information on the samples of bullion being assayed is obtained.

If the approximate proportion of gold in the bullion or gold alloy is unknown, so that it is not certain that the cornet will remain entire in the acid, it must be boiled separately, and the use of flasks becomes necessary. When the parting is effected in flasks, the results always differ more or less from those obtained in the platinum apparatus, especially with high-grade bullion, even when check assays are used, as it is impossible to ensure that all the cornets are subjected to exactly the same treatment, however carefully the heating of the flasks may be effected. As these irregularities are eliminated when all the cornets are parted under exactly the same conditions, it is very desirable that platinum cups or the recently introduced fused quartz cups should be employed in all cases where a large number of assays of bullion or industrial gold alloys of approximately known composition have to be made. Flasks are used in all metallurgical works where a large number of bullion samples of unknown and irregular composition have to be assayed, and are of course used for all preliminary assays of bullion. In some gold mills the parting is done in test tubes with slits in the bottom, the glass tubes being used in place of the platinum cups.

5. **Weighing the Cornets.**—The platinum frame is taken to the balance room, and the cornets, when cold, are weighed on the balance and in the same pan in which the assay pieces were weighed out originally. The balance must readily indicate differences of 0·05 millième per 1000, or $\frac{1}{40}$ milligramme, when

loaded with a weight of ½ gramme in each pan. In order to ensure this high degree of sensitiveness, it is very necessary that assay balances should be used with the greatest care, and periodically examined by some experienced person. The weights and riders used should also be compared with a standard set from time to time. Further remarks on balances and weights and methods of weighing will be found in Chapter V.

In weighing the cornets, the "check" or "proof" assay (see page 276) is weighed first, and the excess or deficiency in weight (usually excess) is applied as a correction to all the cornets from assays worked with it. The correction or "surcharge" to be applied varies according to the nature of the bullion being assayed, and is fully discussed subsequently. An excess or plus surcharge is subtracted from the weight of each cornet, and the minus surcharge added. The surcharge correction is conveniently applied by means of a light rider. The "weighing in" correction is also allowed for. Thus, if the original weight taken was, say, 1000·4 (recorded as +4), it is sufficient to deduct 0·4 from the final weighing. It must be pointed out that in this correction 0·4 of bullion is reckoned as fine gold, but the error is inappreciable when bullion differing but little from pure gold is under examination, and it will remain inappreciable if the difference from the correct weight is kept at less than 0·5. In the case of bullion 900 fine, nine-tenths of the weighing-in correction must be added to or deducted from the weight of the cornet. If the bullion is 500 fine, one-half of the weighing-in correction must be applied ; and in general, if the weighing-in correction is x and the weight of the cornet is y, the correction to be applied in weighing out is [1]

$$\frac{xy}{1000}.$$

With the millième system of weights the weight of the cornet (after correction) at once indicates, without further calculation, the fineness of the bullion expressed in parts (in millièmes and tenths) of pure gold in 1000 parts of the bullion.

A. O. Watkins [2] has described two new methods (in use at the Perth Mint, Western Australia) of weighing the cornets, by which the balance readings give the fineness directly, i.e. the weights of the cornets corrected for surcharge.

Auxiliary Balance for weighing Cornets (fig. 143).—In order to lessen the time of weighing when dealing with a large number of cornets of widely differing weights, an auxiliary balance of special design has been invented by Mr Robert Law,[3] of the Melbourne Mint, where it has been in use for some years.

It is so arranged that it gives the approximate weights of the cornets without the use of any weights. The cornet is placed in a pan attached to one end of the beam, and its weight is recorded by means of a long pointer which is attached to the other end of the beam and moves over a curved scale. The position of the pointer when it comes to rest denotes the percentage of gold present. The cornet is then transferred to the pan of an ordinary balance, and the necessary weights, already determined by the auxiliary balance within narrow limits, are placed on the pan.

This balance not only effects a saving of time, but also saves the wear on the very sensitive assay balance and on the weights.

Reporting.—The degree of accuracy now attained in most assay offices reduces the probable error in the report of a gold bullion assay to about 0·1 per 1000 ; but, to prevent the error from rising above this amount, all weighings

[1] T. K. Rose, *loc. cit.*, p. 488.
[2] Paper before Nat. Hist. and Science Soc. of Western Australia, 1912 ; also *Chem. News*, vol. cvi , 1912, p. 248.
[3] *Chem. Soc. Journ.*, May 1896, p. 526.

must be correct to 0·05 per 1000, which is not always the case in ordinary bullion assays.[1]

Mr A. C. Claudet,[2] Bank of England assayer, states that with fine gold bullion (*i.e.* over 990·0 fine) it is difficult, in spite of all precautions, for the reports to be nearer, on individual bars, than 0·1 to 0·2 per millième, and he considers this to be the limit of accuracy that can be obtained, outside research work.

Gold bullion purchased by the United States Mints is reported to the nearest quarter millième.

With mine gold bullion the results are very generally reported to the nearest ½ millième for the gold (*i.e.* 0·5 per 1000); thus, 899·2 is reported 899·0 fine,

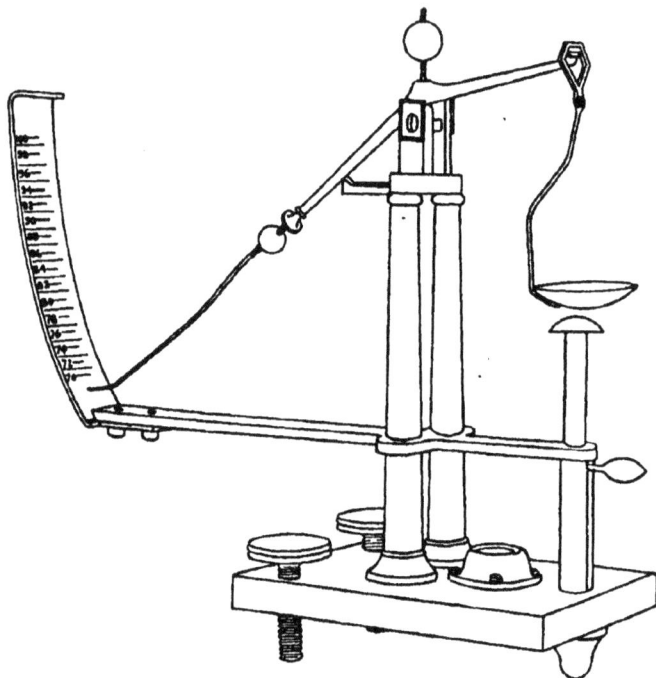

FIG. 143.

899·3 as 899·5 fine, 899·7 as 899·5 fine, and 899·8 as 900·0 fine. It is, however, perhaps more satisfactory for the assayer to report the exact result obtained for the assay.

Remarks on the Parting Assay : Surcharge.—The accuracy of the result of the parting assay is liable to be diminished by certain errors incidental to the process; the weight of gold, therefore, as indicated by the balance, does not represent the amount originally present in the assay piece.

The errors are both positive and negative, and are due to certain losses and gains, viz. :—

(*a*) A loss of gold on the cupel, and to a less extent by solution in the acid.

(*b*) An apparent gain of gold due to the retention of silver in the parted gold

[1] T. K. Rose, *Metallurgy of Gold*, p. 471.
[2] *Trans. Inst. Min. and Met.*, vol. xvi., 1906–7, p. 139.

" Both errors are small, and as they are of an opposite character they tend to neutralise each other. Hence they are altogether without effect on the accuracy of the assays of ores when the total gold is reckoned in milligrams. And even with the larger amounts present in bullion assays their influence is so small that an uncorrected result is still fairly accurate; the resultant error would not be more than one part in two or three thousand " (Beringer).

The errors may be summarised as follows:—

(a) *Losses of Gold.*—As fully discussed in Chapter XII., gold is lost during cupellation, chiefly by being carried into the cupel. This constitutes by far the largest proportion of the loss in the gold assay. The amount varies with the conditions of the cupellation, and it is considerably affected by the presence of base metals, especially copper. An increase in the percentage of copper in the assay piece is accompanied by an increase in the loss by absorption. The amount varies with the proportion of copper, being about 0·3 or 0·4 per 1000 when the copper in the gold does not exceed 10 per cent., and more if larger quantities of copper are present.

The *loss of gold by solution in nitric acid* during parting is very small, but experiments have proved that gold is actually dissolved. Beringer states that on a 500-milligramme assay piece of bullion it may amount to from 0·05 to 0·15 milligramme, *i.e.* from 0·01 to 0·03 per cent.[1] According to Rose,[2] 8 per cent. of the loss in the parting assay is due to the gold going into solution in the nitric acid. F. P. Dewey[3] has shown that, on boiling about 25 grammes of gold in concentrated nitric acid for two hours, the gold passed into solution to the extent of 660 milligrammes per litre, and he observes that the solution of the gold is influenced as much by the temperature (120° C.) required to boil the concentrated acid as by the strength of the acid.

(b) *Silver retained by the Gold after Parting.*—The gold obtained by parting always retains more or less silver, the amount varying in a marked degree with the proportion of gold to silver in the alloy before parting. A series of experiments made by Dr Rose[4] at the Royal Mint shows that the amount of silver retained by the cornet is least when the ratio of 2 parts of silver to 1 part of gold is used, and a larger or smaller ratio than this results in a less pure gold after parting. If the ratio of silver to gold is lower than 1·75 to 1, the cornets are not properly parted by boiling in two acids in the ordinary way. In the assay of bullion, the parted gold resulting from treatment with the first acid retains on an average 0·25 per cent. of silver.

After treatment in the second acid, the gold cornet retains from 0·06 to 0·09 per cent. of silver, according to the length of time the boiling is continued. The amount can be reduced to about 0·05 per cent. by a third boiling. Mr W. F. Lowe, working on half an ounce of gold cornets obtained in assaying gold bullion in the ordinary way, found it to contain 0·123 per cent. of silver.[5]

Occluded Gases.—Cornets retain about twice their volume of gases (mainly carbon monoxide) in occlusion after annealing.[6] This amounts to 0·02 per cent. by weight, which is reckoned as silver retained. The quantity of gas retained appears to vary slightly with the temperature at which the annealing of the cornets is conducted. If they are not heated sufficiently strongly, they remain porous and retain gas.

Surcharge.—*Net Losses and Gains.*—The net sum of the above losses and gains incurred in the various operations is called the "surcharge," since, in the assay of high-grade bullion, the silver retained is usually in excess of the gold lost, so that

[1] *Text-Book of Assaying*, 10th edit., p. 154.   [2] *Metallurgy of Gold*, p. 484.
[3] *Journ. Am. Chem. Soc.*, vol. xxxii., 1910, p. 318.
[4] *Loc. cit.*, p. 486.   [5] *Journ. Soc. Chem. Ind.*, vol. viii., 1889, p. 687.
[6] Graham, *Phil. Trans. Roy. Soc.*, 1886, p. 433.

the cornet will weigh more than the gold originally present in the assay piece; if the reverse is the case, the result is considered by some assayers to be less accurate. When there is a gain in weight it is referred to as a "plus surcharge," while a loss in weight is designated a "minus surcharge." With material rich in gold, the silver retained more than compensates for the loss, and, as just stated, the surcharge is positive; but with a decrease in the proportion of gold the loss is greater and the surcharge is negative. Thus the surcharge will usually amount to about 0 (nil) for bullion of about 700 to 800 fine; above that there will be a "plus surcharge," and below that a "minus surcharge." The surcharge is reported in parts per 1000 in the same way as the fineness of bullion. Thus a surcharge of +0·25 means that the gold as weighed was 0·25 part per 1000 more than the gold actually present, while a surcharge of −0·25 means that on the whole there was a loss of 0·25 part per 1000 in the assay. As previously stated, a "plus surcharge" is subtracted from the weight of the cornet, and a "minus surcharge" added.

The actual amount of the surcharge not only varies with the fineness of the bullion, but is also influenced by the conditions under which the assay is conducted, such as the temperature of the furnace, strength and purity of acid, etc.

The loss of gold by absorption during cupellation is increased by employing too high a temperature, while by prolonged boiling in acid, and especially by annealing the fillets at a high temperature, the amount of gold dissolved in the acids is increased.[1]

With fine gold bars (say over 990 fine) assayed under normal conditions there is always a "plus surcharge" varying from about +0·7 to +1·0 part per 1000. Claudet[2] states that during the operations the assay approximately loses 0·3 millième of gold and retains 1 to 1·3 millième of silver, not dissolved out in the parting process; the difference between these is the surcharge.

With bullion from about 800 to 900 fine, cupelled in batches of thirty-six assays, Gowland[3] found that the surcharge varied from 0·5 to 0·7 per 1000, the average being about 0·55.

The surcharge on mill and cyanide bullion is stated by M'Arthur Johnston[4] to vary from 0 to 0·5 part per 1000.

In assaying English standard gold (916·6 fine), Dr Rose[5] has found the surcharge, under normal conditions of working, to be 0·4 to 0·8 part per 1000. With this surcharge the total loss of gold is about 0·4 per 1000, of which about 82 per cent. is absorbed by the cupel, 8 per cent. is dissolved in the acid, and the remaining 10 per cent., which is unaccounted for, is probably volatilised. These ratios, however, vary considerably, according to the conditions of working.

The surcharges obtained in the assay of industrial gold alloys containing gold, silver, and copper are given in Table XLII. page 330.

Assaying with Checks or Proofs.—From the statements in the preceding paragraphs it will be seen that the errors in the parting assay, although comparatively small, are of sufficient magnitude to invalidate more or less the accuracy of gold assays; and in mints and places where bullion assays must be made with the highest attainable accuracy, the surcharge is always determined by experiment, and the proper correction is made in the reports on the bullion. Since the losses and gains are dependent on so many conditions, the surcharge is best determined by the use of "checks" or "proofs" consisting of pieces of pure gold which are subjected to the various operations side by side and under

---

[1] T. K. Rose, loc. cit., p. 484.
[2] Trans. Inst. Min. and Met., vol. xvi., 1906–7, p. 138.
[3] Trans. Inst. Min. and Met., vol. xvi., 1906–7, p. 142.
[4] Rand Metallurgical Practice, vol. i., 1912, p. 306.
[5] Journ. Chem. Soc., vol. lxiii., 1893, p. 710.

identical conditions with the assay pieces. The composition of the checks is the same as the composition of the bullion being assayed, in order to make the assays absolutely comparable. Being, therefore, of the same composition and weight, and undergoing exactly the same treatment, the checks may reasonably be expected to have the same surcharge as the assays which they imitate. The checks are made up from the data obtained by a preliminary assay.

Suppose the preliminary assay of a gold alloy gave the following composition :—

Gold,	0·4600	gramme, equal to	920·0	parts per 1000.	
Silver,	0·0174	,,	,,	34·8	,,	,,
Copper,	0·0226	,,	,,	45·2	,,	,,
	0·5000			1000·0	

For the ordinary assay using 0·5 gramme (1000 millièmes) the total weight of silver required would be 0·460 gramme × 2½ = 1·150 grammes. Deducting the silver already in the alloy, we get 1·150 − 0·0174 = 1·1326 grammes, the actual weight of silver to be added to the assay piece.

The check assay is then made up from the following weights of pure metals, viz. :—

Pure gold	. . . . .	0·4600 gramme.
Pure silver	. . . . .	1·1500 ,,
Pure copper	. . . . .	0·0226 ,,

These quantities are wrapped in the same weight of lead as is being used in the other assays, and then the check and ordinary assays are cupelled, parted, etc., as nearly as possible under the same conditions. Suppose the parted gold from the check weighed 0·4604 gramme, the gain in weight or plus surcharge is 0·0004 gramme, and this weight would be deducted from the results of all the assays worked with it. Since the amount of surcharge varies with bullion of different fineness, a surcharge correction is never applied except to bullion of the same fineness as that represented by the "check assay" it was calculated from.

It will be evident that, if the gold used in making up the checks is not of the highest degree of purity, an error will be introduced, and the amount of gold found in the bullion, after correction, will be in excess of the truth. The preparation of the fine or proof gold and silver for check purposes is described on pages 70 and 75 respectively.

If a sample of proof gold is not pure an allowance is made. Thus, if it is only 999·7 fine, a deduction of 0·3 is made from all results of assays checked by it. The fineness of the gold, when carefully prepared by the method given on page 70, is usually taken as 1000 fine, but for very accurate work it may be regarded as 999·9 fine. In this case a deduction of 0·1 must be made, and this has proved to be a very close approximation to the correct one.[1]

It is the practice of some assayers to use "cornets" for "pure" gold in making up checks; but unless the gold is first melted into a bar and carefully assayed to determine its purity, the practice is to be deprecated, since, as previously stated, parted gold invariably contains more or less silver.

The number of checks to be used for a given number of assays varies according to the nature of the bullion to be assayed. Thus, for the assay of bullion over 990·0 fine, one check is usually used with each nineteen assays, and its relative position is always the same—the eleventh cupel from the front—the cupels being arranged in five rows of four.[2]

[1] T. K. Rose, loc. cit., p. 487.

[2] A. C. Claudet, "The Assay of Gold Bars," Trans. Inst. Min. and Met., vol. xvi., 1906–7, p. 137.

For bullion of lower fineness, such as cyanide gold bars, a larger number of checks are used. Assayers on the Rand, South Africa, use one check for every four bullion assays, that is, three checks in a batch of twenty cupellations.[1] At the Royal Mint, London, a batch of seventy-two assays of standard gold (916·6 fine) includes six checks.[2]

In order to get correct results in the parting assay of gold and to utilise the checks to the greatest advantage, it is very necessary that all the operations should be carried on exactly under the same conditions. A very systematic and uniform method of working must be adopted, and although the details of practice may differ from those described, equally accurate results may be obtained so long as all the assay pieces and checks are treated alike.

Limits of Accuracy of the Gold Bullion Assay.—The degree of accuracy of the gold bullion assay as conducted in most assay offices is probably about 0·1 per 1000, but in 1893 T. K. Rose[3] showed that by the use of suitable precautions the probable errors could be reduced to ±0·02 per 1000 in the case of high grade refined gold.

The probable errors in the results of gold assays made in the course of routine work in the assay office of the Royal Mint, London, and by Bank of England assayers, have been more recently investigated by J. Phelps, of the Royal Mint.[4]

The degree of accuracy obtained in gold bullion assays in the United States mints and assay offices, and at the Mint Bureau at Washington, has been discussed by F. P. Dewey,[5] of the Mint Bureau, Washington. In Table XXXI. an attempt has been made by Mr Phelps to compare the probable errors of a single gold bullion assay made in these different assay offices, as deduced from the figures given in the papers quoted.

The figures assigned to the Bank of England assayer's results and to those of the United States Mint are deduced from the comparison of the results with subsequent assays of the same material at the Royal Mint; the others are calculated by means of the usual formulæ from the differences of individual results from their mean.

Table XXXI.—*Comparison of Results of Gold Bullion Assays made in Different Assay Offices.*

Authority or assayer.		Probable error per 1000.
Royal Mint {	Ordinary assays—Single assay . .	±0·043
	Trial plate comparisons—Single assay .	±0·023
Trade assayers {	Single assay	±0·081
(London) {	Triple assay	±0·057
United States Mint—Ten ingots by two assays . .		±0·062
United States Mints (F. P. Dewey)—Single assay .		±0·150

Stamping the Gold Bars.—In mine and other assay offices where the gold bars after being assayed are sold, they are weighed and then stamped with steel punches (see page 42) giving the following data: Name or mark of the assayer; number of the bar; total weight of the bar in troy ounces and decimals; fineness of the gold; fineness of the silver; and the date. In America, the total money value of the bar in dollars is also added.

[1] *Rand. Met. Prac.*, 1912, p. 306.   [2] T. K. Rose, *Precious Metals*, p. 210.
[3] *Journ. Chem. Soc.*, vol. lxiii. pp. 700–714.
[4] *Ibid.*, vol. xcvii., 1910, pp. 1272–1277.   [5] *Trans. Inst. Min. Eng.*, 1909.

Determination of Silver in Gold Bars.—In the case of mine gold bullion, the amount of silver present in association with the gold is usually required. The silver is generally determined by difference. Two or more assays for gold are made by parting in the usual way, and duplicate assay pieces of 0·5 gramme are at the same time cupelled without the addition of silver, but with sufficient lead to remove, as far as practicable, all the base metals present. The difference in weight between the gold-silver button so obtained and the amount of gold determined by the ordinary parting assays gives the amount of silver present in the bullion. The weight of lead to be used is generally guessed at from experience: if the button left on the cupel is not quite bright, it is re-cupelled with the addition of a very small quantity of lead and weighed again. In practice, this assay for the determination of silver is utilised as a preliminary assay to ascertain the approximate fineness of the bar. The colour of the gold-silver button indicates, to an experienced assayer, the right proportion of silver to be added for the gold assay proper.

Another method of determining the silver in gold bars is described on page 280.

Parting Assay of Gold by means of Cadmium (Balling's method).[1]—Cadmium may be substituted for silver in the operation of parting; it is very fusible (melting-point 322° C.), readily unites with the gold, and renders it possible to effect the union of the metals in a short time in a porcelain crucible over a lamp.

The resulting gold-cadmium alloy is brittle, and cannot be flattened, but some assayers crush it to powder to facilitate parting. This method can be used for the determination of both the gold and the silver in bullion.

(a) *Determination of the Gold.*—A weighed quantity of the bullion, usually 0·5 gramme, is melted in a porcelain crucible with cadmium in the proportion of 2½ to 1 of gold, and sufficient potassium cyanide to completely cover the metals during the fusion to protect them from oxidation. If an appreciable amount of silver is present in the bullion, less cadmium is used, so that the combined weight of cadmium and silver will be equal to 2½ times that of the gold. The fusion should be complete in about five minutes, and, after cooling, the cyanide is dissolved out with hot water and the clean button of alloy placed in a flask with 20 c.c. of water. The water is boiled and 40 c.c. of nitric acid added in four successive additions of 10 c.c., and the boiling continued for an hour. The acid solution is then carefully decanted off, and the gold residue washed by boiling for a few minutes in distilled water. The gold is then dried, annealed, and weighed.

When the proportion of cadmium is 2½ times the gold, the gold after parting is obtained in a coherent form. In the case of gold bullion practically free from silver an increase in the proportion of cadmium is recommended by some assayers. After parting the gold retains its original form, but shows upon its surface numerous cracks from which the cadmium has been dissolved. Parting is sometimes effected in two acids, the button being boiled once with dilute nitric acid of 1·2 specific gravity, and twice with acid of 1·3 specific gravity. The boiling in the first acid is continued for not less than half an hour, or longer, according to the amount of gold present. Each boiling in the second acid is continued for ten to fifteen minutes. By this method the losses of gold and silver incidental to cupellation are entirely avoided. A similar method, employing zinc in place of cadmium, had been recommended previously by von Juptner.[2]

A "check" assay is run at the same time, a preliminary assay being made to ascertain the approximate composition of the bullion.

[1] *Oestr. Zeitsch. für Berg. und Hüttenwesen*, 1879, p. 597; and 1880, p. 182.
[2] *Zeitsch. für anal. Chem.*, 1879, p. 104, quoted by Rose, *Met. of Gold.*

(b) *Determination of the Silver.*—Balling's method of parting is recommended by some assayers for high-grade bullion in which it is desired to determine the silver, as difficulties are encountered when attempts are made to determine the very small amounts of silver present in this class of bullion by the usual methods previously described.

The content of silver in the bullion can be determined in the parting solution either by precipitation as chloride or volumetrically after it has been carefully poured off from the gold.

The following method for the determination of the silver is in use at the Great Boulder Perseverance Mine, Western Australia.[1] The acid solution from the parting, including the washings, is diluted to 150 c.c., 10 c.c. of ferric alum indicator added, and the silver titrated with a standard solution of ammonium thiocyanate containing 1·6 grammes of the pure salt to the litre (see Volhard's method, page 303). The thiocyanate solution is standardised against pure silver, each c.c. being equivalent to about 4·483 parts of silver per 1000. A "check" is run at the same time, the composition of the bullion having been ascertained by a preliminary assay as previously stated.

Whitehead[2] recommends the following method of procedure. The 0·5 gramme of bullion is heated in a small crucible with 10 grammes of potassium cyanide, and when fused 1 gramme of cadmium is added. The molten alloy is poured on to a slab, and when cool it is washed to free it from cyanide. The alloy is powdered, transferred to a flask along with 1·004 grammes of pure silver foil, and then treated with nitric acid of specific gravity 1·278. When solution is complete the total silver is determined either as chloride or by means of standard thiocyanate. The result is compared with a check assay made with 1·004 grammes of pure silver, dissolved in the same amount of nitric acid. The difference between the two results, of course, gives the amount of silver in the bullion.

Determination of Gold in Doré Bullion.—Small quantities of gold in silver bullion are determined by parting in nitric acid and subsequent collection and weighing of the finely divided gold. From 0·5 to 1 gramme of the silver bullion is weighed out on an assay balance, wrapped in 10 grammes or more of lead according to the quality of the bullion, and then cupelled. The resulting button is weighed (if it is desired to determine the silver), and then flattened and parted in dilute nitric acid of 1·2 specific gravity. The gold is frequently left in the form of powder: it is carefully collected, washed, dried, ignited, and weighed. When the amount of gold present is comparatively large, some assayers add a weighed quantity of fine gold so as to make the proportion of gold to silver 1 to 2½, and then part in the ordinary way as for a gold bullion assay. The amount of gold added is of course allowed for in calculating the results.

If the silver is to be dissolved by boiling the assay but once with nitric acid, at least eight parts of silver to one part of gold must be present. It is frequently necessary to make a preliminary assay, and it is desirable to have the regular assay accompanied by a check.

The amount of gold in silver bars is frequently reported in grains per pound when the amount is small.

Assay by the Touchstone.—This old method of determining the fineness of a gold alloy consists in rubbing the alloy to be tested on a small block of hard, smooth, dark stone (usually lydianstone), called a *touchstone*, and comparing the appearance and colour of the streak thus produced with those made by a

[1] E. H. Taylor, *Journ. Chamber of Mines of West Australia*, July 31, 1908; also *Eng. and Min. Journ.*, vol. lxxxvii., 1909, p. 543. Consult also J. E. Clennell, *Eng. and Min. Journ.*, vol. lxxxiii., 1907, p. 1909.

[2] *Journ Franklin Inst.*, 1891, vol. cxxxii. pp. 365–369.

series of small bars of carefully prepared alloys of definite composition, known as "touch-needles" (figs. 144 and 145). The effect of the action of a drop of nitric acid and of dilute aqua-regia on these streaks is also noted; the streak from the less pure alloy will be more readily acted upon, with the production of a more or less green colour, according to the proportion of copper present. Several series of touch-needles are usually employed, consisting of alloys of gold and copper, gold and silver, and gold, silver, and copper, the alloys being made either to correspond to legal standards or in series in which the proportion of gold gradually decreases, say, from 1000 fine to 500 fine with a difference of 20 parts between each needle. For the sake of convenience the touch-needles are frequently soldered to the points of a star-shaped piece of metal, as shown in fig. 145. Each needle is stamped with a number denoting the fineness.

The valuation of an alloy is made by determining to which of the touch-needles the streak it produces most nearly corresponds. In order to get correctly the streak of the alloy to be tested, the surface of the metal should first be slightly filed away, as this may have been made somewhat richer than the bulk of the alloy by boiling with acid to remove the base or inferior metal from the surface, as in the "colouring" process used by goldsmiths.

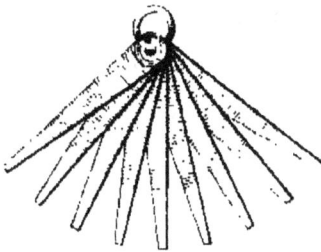

FIG. 144. FIG. 145.

The touchstone is generally used for the approximate assay of small articles of jewellery which it would be necessary to destroy in order to obtain samples for a correct assay; it is also of use to the assayer in determining the approximate fineness of gold bullion and of industrial gold alloys; but it cannot be relied upon for very accurate assay, although it yields sufficiently good results for a preliminary test, and for some purposes is sufficiently exact. It requires, however, a sharp and very practised eye. "The trial is more sensitive for alloys below 750 fine than for higher standards. The amount of gold in alloys between 700 and 800 fine can be determined correct to five parts per 1000." [1]

In some cases where the same class of bullion is frequently being assayed the touch needles are made from pieces cut from the gold bars. Thus, in assaying large quantities of gold dust from Korea, Gowland [2] put aside pieces cut from the ingots of the gold dust, which varied from about 790 to 990 per 1000 of gold, and made touch needles from them. By the use of these he was able to determine the proportion of gold present to within 5 or 10 millièmes (i.e. five to ten parts per 1000), and thus to decide on the amount of silver to be added for parting.

Wet Methods of Assay of Gold Bullion and Alloys.—The assay or complete analysis of gold bullion can be made by the ordinary chemical methods,

[1] T. K. Rose, *Metallurgy of Gold*, p. 497.
[2] *Trans. Inst. Min. and Met.*, vol. xvi., 1906–07, p. 142.

but difficulty is sometimes experienced in getting the metal into solution, especially in the presence of silver in certain proportions.

The most rapid solvent for gold is hot aqua-regia (nitro-hydrochloric acid; a mixture of three parts of hydrochloric acid with one part of nitric acid).

Alloys of gold and silver containing over 50 per cent. of gold are difficult to dissolve in acid; nitric acid has little effect on them, and nitro-hydrochloric acid (which dissolves the gold) converts the silver into insoluble chloride, which protects the alloy from further attack. The alternate action of nitro-hydrochloric acid and ammonium hydrate eventually results in the dissolution of all the gold, but the method is a tedious one.

When the gold contains less than about 15 per cent. of silver, nitro-hydrochloric acid dissolves all the gold, while the whole of the silver is left as chloride; for this purpose, however, the metal must be in a very thinly laminated state.

If it be desired to dissolve out the silver and base metals in gold that is not attacked by nitro-hydrochloric acid, and to determine them in the wet way, the gold is best alloyed with a sufficiency of some other metal to render it amenable to the attack by acid. Cadmium is the metal generally recommended for this purpose, and the alloy is made by melting the gold with five times its weight of cadmium, as described on page 279.

Bullion containing at least 60 per cent. of silver is readily acted upon by boiling nitric acid, the silver being dissolved and the gold left behind in a brown, porous state. As previously pointed out, the gold obstinately retains about 0·1 per cent. of its weight in silver, and to separate this it is necessary to dissolve it in nitro-hydrochloric acid. Copper or other base metal can replace silver without materially affecting the results of the action of nitric acid. After gold is dissolved in a mixture of hydrochloric and nitric acids for the purpose of analysis, care must be taken that none of the latter remains undecomposed after the solution is effected. The nitric acid must always be expelled from the solution by warming with successive additions of hydrochloric acid. Since gold chloride (and some base metal chlorides such as copper) are volatile, the acid solution must not be heated too strongly, or loss of gold will occur.

In the analysis of a solution containing gold, as well as other metals precipitable by sulphuretted hydrogen, it is usual first to remove the gold in the metallic state, by means of oxalic acid. If platinum is also present, it is removed, after filtering off the precipitated gold, by evaporating the solution with ammonium chloride, which causes the precipitation of the platinum as ammonium platinum chloride. The platinic chloride is not reduced by oxalic acid (see Chapter XXIV.).

The quantity of bullion or alloy to be taken for analysis varies according to circumstances. From 1 to 5 grammes of bullion are usually enough, but a much larger amount is necessary when examining high-grade bullion for small quantities of impurities. The metals and quantities of each generally found in bullion are shown in the examples of analysis of native gold on page 77, and the analysis of bullion on page 247. The presence and determination of metals of the platinum group that may be found in bullion are discussed in Chapter XXV.

When the solution of the bullion or alloy has been effected, the various metals present in solution are determined by the usual methods of separation. The various precipitants for gold are discussed on page 68, but the following remarks by Dr Rose [1] may be of value in the choice of a precipitant for the gold in any particular case.

"Ferrous sulphate and sulphurous acid act well in strongly acid (HCl) solutions; oxalic acid, sulphuretted hydrogen, and ammonium sulphide act best in

[1] *Metallurgy of Gold*, 5th edit., p. 496.

presence of small quantities of hydrochloric acid. The solution should be dilute (say 1 part of gold in 300 of water), so that other metals may not be carried down by the gold. Ferrous sulphate gives a very finely divided precipitate which is difficult to wash by decantation without loss; precipitation is slow in cold solutions. Oxalic acid causes plates and scales to form which are readily washed and are very pure; it acts best in boiling liquids, but a temperature of 80° for forty-eight hours suffices; in the cold or in the presence of much hydrochloric acid or alkaline chlorides the action is very slow and partial; a large excess of the precipitant must be present. Oxalic acid is used for solutions containing metals of the platinum group, which are not precipitated by it. Alkaline oxalates act better than the free acid.

"Sulphurous acid is an excellent precipitant for most solutions. It acts rapidly and completely in the cold, and does not readily precipitate other metals, except tellurium. Sulphuretted hydrogen is used in the absence of all other metals whose sulphides are insoluble in hydrochloric acid.

"In all cases careful consideration must be given to the nature of the base metals present, and the precipitant which will not render any of them insoluble must be selected."

The following method of separation of gold, silver, and copper in an alloy may be given as an example:—

Analysis of Gold, Silver, Copper Alloy.—The solution resulting from the treatment of the alloy with nitro-hydrochloric acid is evaporated nearly to dryness to expel the greater part of the free acid, water is added, the solution filtered, and the insoluble residue of silver chloride is washed, dried, and weighed. The solution is heated, and oxalic acid added to precipitate the gold in the metallic state. It is washed, dried, transferred to a porcelain crucible, the filter completely burnt, and the gold ignited and weighed. From the filtrate copper may be precipitated as sulphide by means of sulphuretted hydrogen, and weighed as such, or it may be converted into oxide by re-solution and precipitation with sodium hydrate (NaOH).

Should the alloy contain zinc, this metal will be left in the solution. It may be determined by evaporating the solution down to expel sulphuretted hydrogen and the zinc precipitated by means of sodium carbonate. The precipitate is then dried, ignited, and the metal is weighed as zinc oxide (ZnO). If much zinc is present, small quantities are very liable to be carried down as sulphide, with the precipitated copper sulphide; it is necessary, therefore, to dissolve the latter and reprecipitate the copper.

Analysis of Gold Alloys by means of Ether.—It has been found by F. Mylius[1] that the known solubility of gold chloride in ether may be utilised for the accurate analysis of gold alloys, especially coinage, in which the gold may be determined with a loss of not more than 0·01 per cent. The solution of the alloy in dilute aqua-regia, after filtering off any silver chloride, is repeatedly extracted with ether (the presence of the free acid being necessary), and the ethereal extract is dealt with as below; the aqueous layer may then be used for the detection or determination of other metals. For the extraction, the solution should contain 5 to 10 per cent. of metal and 5 to 10 per cent. of total hydrochloric acid: the shaking is repeated four or five times, a total volume of 50 c.c. of ether being used for 20 c.c. of solution containing 1 gramme of metal: 10 c.c. of water are added to the ethereal extract, the ether is distilled off, and the residual solution is warmed for some minutes with sulphurous acid, after which the precipitated gold is filtered off, washed, ignited, and weighed. Gold may thus be separated from all other metals, and the method furnishes a simple

[1] *Z. anorg. Chem.*, 1911, lxx., 203–231, abstract *Journ. Soc. Chem. Ind.*, vol. xxx., 1911, p. 547.

means for the preparation of very pure gold, containing less than 0·001 per cent. of impurity. For very accurate determinations, the principal portion of the gold is removed from the solution by two extractions with ether; the aqueous layer is then concentrated, the chlorides are converted into sulphates, and the residual gold precipitated by sulphurous acid or hydrazine; the precipitate is then dissolved in a small volume of acid and the solution is extracted several times with ether. As an example, a detailed analysis of gold coinage and suitable apparatus therefore are described, the detection or determination of small quantities of silver, lead, iron, arsenic, antimony, tellurium, nickel, platinum, palladium, and iridium being discussed, and analyses of the gold coinage of various countries, as effected by this method, being given; all the coins examined contained, besides copper and silver (0·2 to 0·4 per cent.), very small percentages of lead, iron, platinum, and palladium, and a trace of arsenic. An analysis of certain brittle coins pointed to the presence of a slight excess of lead as the possible cause of brittleness. In the course of the investigation it was found that hydrochloric acid is able to dissolve gold in the presence of cupric chloride; the compound $CuAu_2Cl_8,6H_2O$ was also prepared, in the form of very soluble lustrous olive-green prisms, by evaporating a solution containing cupric chloride (1 mol.), and chlorauric (2 mols.).

Volumetric Determination of Gold.—The gold in solutions of auric chloride may be determined volumetrically by the following methods :—

Oxalic Acid Method.—Franceschi[1] precipitates the gold with a measured excess of potassium oxalate and then determines the amount of free oxalate by titration with potassium permanganate. When a gold chloride solution free from nitric acid is boiled for some minutes with potassium oxalate in excess, the gold is all reduced and separates according to the equation,

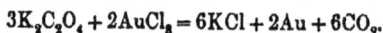

$$3K_2C_2O_4 + 2AuCl_3 = 6KCl + 2Au + 6CO_2.$$

Thus 8·3 parts of potassium oxalate precipitates 6·533 parts of gold.

The reagents required are :—

(a) Solution of potassium oxalate containing 8·3 grammes of the dry salt per litre.

(b) Solution of potassium permanganate which is exactly reduced by an equal volume of potassium oxalate.

Oxalate is added in excess to the gold solution and the mixture boiled for a few minutes. The solution is cooled, the gold filtered off and washed. The filtrate, with the washings, is rendered acid with sulphuric acid and the excess of oxalate solution titrated back with decinormal permanganate. Then, if

n = number of c.c. of oxalate solution
n_1 =   ,,   ,,   permanganate solution

the gold reduced is represented by $(n - n_1)$ 0·0065, the reactions being

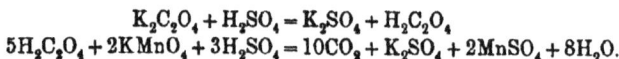

$$K_2C_2O_4 + H_2SO_4 = K_2SO_4 + H_2C_2O_4$$
$$5H_2C_2O_4 + 2KMnO_4 + 3H_2SO_4 = 10CO_2 + K_2SO_4 + 2MnSO_4 + 8H_2O.$$

Ferrous Sulphate Method.—Juptner[2] proposes a somewhat similar method, but precipitates the gold with ferrous sulphate. In this case the gold chloride solution is mixed with a measured excess of ferrous sulphate solution, which reduces the chloride to metallic gold; the non-oxidised portion of the ferrous oxide is then determined by titration with permanganate.

[1] *Bull. Chim. Farmac,* 1894, xxxiii. p. 35.
[2] *Oestr. Zeitsch. f. Berg. und Hüttenwesen,* 1880, p. 182, quoted by Kerl, *Assaying,* 1889, p. 182.

The reactions are expressed by the following equations :—

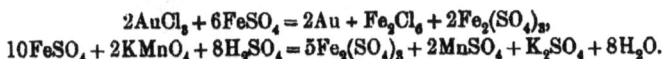

$$2AuCl_3 + 6FeSO_4 = 2Au + Fe_2Cl_6 + 2Fe_2(SO_4)_3,$$
$$10FeSO_4 + 2KMnO_4 + 8H_2SO_4 = 5Fe_2(SO_4)_3 + 2MnSO_4 + K_2SO_4 + 8H_2O.$$

Hence, 2 of gold correspond to 6 of iron, or 1 of gold to 3 of iron; designating the portion of iron oxidised in titrating with permanganate by m and the content of gold sought by x, we have the proportion

$$3Fe : Au = 168 : 196 \cdot 7 = m : x$$
$$x = \frac{196 \cdot 7}{168} m = 1 \cdot 172 \ m.$$

The solutions are prepared as follows :—

Dissolve 6 grammes of ferrous ammonium sulphate in one litre of water acidulated with sulphuric acid. These 6 grammes contain $6 \times 0 \cdot 14285 = 0 \cdot 857$ gramme of iron, and 1 c.c. of this solution precipitates exactly 1 milligramme of gold.

The permanganate solution is prepared so that 1 c.c. will exactly. correspond to 1 c.c. of the ferrous sulphate solution. It is then only necessary to deduct the number of c.c. of permanganate used in titrating from the amount of ferrous ammonium sulphate solution added to the gold solution, in order to determine the amount of gold, in milligrammes, in the solution of the alloy.

Solutions of ferrous ammonium sulphate rapidly precipitate definite quantities of gold from solutions of auric chloride with a marked degree of concordance, provided that no free nitric acid or nitrates are present which would affect the re-solution of the finely divided precipitate. The strength of the ferrous solution is important: if it is too weak the precipitate will remain partially suspended, and if too strong the burette readings are less exact, as each drop will obviously throw down a larger amount of gold. The strength of solution usually employed in practice is such that 1 c.c. will precipitate 25 milligrammes of gold (see below).

Potassium Iodide Method.—Peterson [1] recommends treating the solution of auric chloride with excess of potassium iodide solution, which precipitates the gold as aurous iodide, AuI, and liberates free iodine, which is determined by titration with standard sodium thiosulphate. The reactions are as follows :—

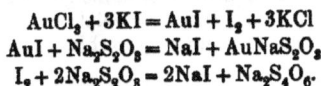

$$AuCl_3 + 3KI = AuI + I_2 + 3KCl$$
$$AuI + Na_2S_2O_3 = NaI + AuNaS_2O_3$$
$$I_2 + 2Na_2S_2O_3 = 2NaI + Na_2S_4O_6.$$

Thus 3 molecules of sodium thiosulphate are required for 1 molecule of auric chloride.

According to Peterson, the amount of thiosulphate required is $1\frac{1}{2}$ times greater than that necessary for the conversion into tetrathionate in the usual way. Gooch and Morley [2] have proved experimentally that this statement is incorrect, and that the thiosulphate reacts in the usual manner upon the iodine set free by the action of the potassium iodide upon gold chloride. These authors have found the method to be very accurate for the determination of very small quantities of gold. The gold should not exceed 0·01 gramme, and if in the metallic state it is dissolved in chlorine water, and the excess of chlorine is then removed by adding ammonium hydrate, boiling, and acidifying with hydrochloric acid : this treatment should be repeated. After diluting to 200 c.c., 10 c.c. of the liquid are mixed with 0·02 gramme of potassium iodide, and the liberated

[1] *Zeits. anorg. Chem.*, 1899, vol. xix. p. 59.
[2] *Amer Journ. Science*, 1899, vol. viii. pp. 261–266.

iodine determined by titrating with N/1000 solution of sodium thiosulphate, using starch as indicator; it is advisable to add a very slight excess and titrate back with N/1000 solution of iodine until a faint pink tint is detected.

The volumetric determination of gold is used in mints and refineries where bullion is refined by electrolytic processes, for determining the gold content of the electrolytes of auric chloride, and of wash waters.

The methods in use at the San Francisco Mint refinery have been described by H. French,[1] who states that precipitation with oxalic acid is unsatisfactory, since the finely divided gold requires at least twenty-four hours in which to settle, and the ferrous sulphate method is employed.

As stated above, the most satisfactory results are obtained with a ferrous sulphate solution of such strength that 1 c.c. will precipitate 25 milligrammes of gold. It is prepared by dissolving 153·5 grammes of pure ferrous ammonium sulphate in warm distilled water and diluting to one litre. About 1 c.c. of concentrated sulphuric acid is added to maintain the solution in the ferrous state. Such a solution will remain unaltered for months.

A permanganate solution is prepared by dissolving 12·3 grammes of potassium permanganate and diluting to one litre. This solution has the power of oxidising an equal volume of the standard ferrous solution to the ferric state.

Usually, after weighing and dissolving the permanganate, 5 c.c. will oxidise from 4·9 to 5·1 c.c. of the ferrous sulphate solution. The permanganate solution and the ferrous sulphate solution are brought into a perfect balance of chemical-conversion power by the addition of from 10 to 20 c.c. of water to the stronger solution. For all practical purposes these two solutions give satisfactory results, so long as they are within one-fiftieth of their theoretical proportion of strength. Occasionally it may be desirable to be more exact, and then the amount of ferrous iron in solution is determined by titration with a standard solution of potassium permanganate in the manner adopted for iron assays, care being taken to prevent any oxidation of the ferrous sulphate solution before titrating. As a rule the crystals of ferrous ammonium sulphate average between 14·2 and 14·3 per cent. of iron.

The actual precipitating power of the ferrous ammonium sulphate solution is usually checked by the gravimetric method, a carefully measured quantity, say 5 to 10 c.c., of the ferrous sulphate solution being added to a solution of gold chloride, and the resulting precipitate of metallic gold weighed. After precipitation the gold is allowed to settle, the clear solution decanted through a filter, then the residue of gold boiled in dilute hydrochloric acid, and washed on to the filter with distilled water.

Particles or films of gold adhering to the sides of the beaker are removed by means of a small piece of filter paper. After carefully washing the gold until all trace of iron is removed, it is transferred to a porcelain crucible and dried, then heated to redness, cooled, and weighed. The results are a little higher than when the dried precipitate is cupelled. To be exact the gold should be carefully wrapped in a small piece of lead-foil and cupelled. In this way a careful check is made upon the titrations.

The usual procedure of the volumetric analysis of gold solutions, as practised in the laboratory of the San Francisco Mint Refinery, is as follows:—Unless the approximate percentage of gold in solution is known, a preliminary test is made to determine the probable amount of ferrous sulphate solution required.

Having determined this, a measured quantity of the gold solution, usually 5 c.c., is run from a pipette into a beaker, 10 c.c. of hot water added, and about one-fifth c.c. of sulphuric acid shaken in from a "dropping" bottle. Then,

[1] *Mining and Engineering World*, vol. xxxvii., 1912, pp. 853–855. See also "Volumetric Determination of Gold," V. Lenher, *Min. and Scientific Press*, S. Francisco, June 21, 1913.

according to the judgment of the operator, the standard ferrous sulphate solution is run in rapidly from a burette and in sufficient quantity to turn the faint yellow tint of the gold chloride solution to a pale green in which a turbid mass of finely divided brownish-black gold is suspended. The contents of the beaker are vigorously stirred with a glass rod and allowed to settle.

If the orange-yellow colour prevails over the green, another c.c. of the precipitant is added until an excess is obvious by the permanent pale-green hue of the solution in the beaker. After allowing the precipitate to settle, water is added to the solution, increasing its volume to nearly 100 c.c. Then the permanganate solution is run in carefully from a burette until all of the excess of ferrous sulphate solution has been oxidised, as indicated by the last drop of the permanganate turning the solution a permanent pink. To calculate the result:—If 5 c.c. of the gold solution were taken and 18 c.c. of the standard ferrous sulphate solution had been run in, effecting a complete precipitation as above described, plus a slight excess, and the latter had been determined by adding 3 c.c. of permanganate, then 15 c.c. was the exact volume of the ferrous sulphate solution required to precipitate the gold contained in 5 c.c. of the sampled solution. Therefore 15×0.025 gramme $= 0.375$ gramme of gold in 5 c.c. (i.e. $\frac{1}{100}$ part of a litre) of the gold solution tested, and one litre of this solution would contain (0.375×200) 75 grammes of gold.

For the determination of gold in solutions containing less than 1 gramme per litre, the iodide method described above is more exact. Usually 10 c.c. of the gold solution are taken. The gold is first precipitated by potassium iodide as aurous iodide, and redissolved in an excess of the reagent. A clear solution of starch is then added, and the blue colour produced is destroyed by thiosulphate. Then standard iodine solution is run in until a faint pink tint is detected. From the number of c.c. of the iodine solution of known strength added, the number of milligrammes of gold precipitated is calculated. As in the case of the ferrous sulphate solution, the weight of gold it is capable of precipitating is previously ascertained by careful standardising.

Electrolytic Assay of Gold.—Gold may be accurately deposited from cyanide solution. Potassium cyanide is added to the gold solution in the proportion of 2·5 grammes per 100 c.c. of solution, and a current is passed through the solution between platinum electrodes. According to Classen,[1] the current density should be about 0·3 ampère per square decimetre, the fall of potential 2·7 to 4 volts, and the temperature 50° to 60° C. The gold is deposited in about 1½ hours: it is washed with water and alcohol, dried, and weighed together with the cathode. The increase in weight of the cathode equals the amount of pure gold in the quantity of alloy taken.

S. P. Miller[2] states that from 150 c.c. of a solution containing 1 gramme of potassium cyanide and 0·1291 gramme of gold, the latter is deposited completely in 2½ hours with a current of 1·8 volts and N.D.$_{100}$ $= 0.2 - 0.04$ ampère during the first 1½ hours and then increased so as to maintain a voltage of 1·8. The temperature employed is 55° to 65° C.

Messrs Perkin & Prebble[3] recommend the use of ammonium thiocyanate in place of the potassium cyanide which is generally recommended. With a current density of 0·4 to 0·5 ampère per square decimetre the deposition of 0·05 to 0·08 gramme of gold is complete in 1½ to 2 hours at ordinary temperatures.

The solutions employed contained 5 grammes of ammonium thiocyanate in 120 to 150 c.c., the gold chloride solution which was used for the tests being

[1] *Quantitative Analysis by Electrolysis.*
[2] *J. Amer. Chem. Soc.*, 1904, vol. xxvi. pp. 1255-1269.
[3] F. M. Perkin and W. C. Prebble, "Electrolytic Analysis of Gold" (paper before the Faraday Society), *Elec. Chem. and Met.*, 1904, vol. iii. pp. 490-494.

added in measured volume. The gold is deposited on a platinum electrode in the usual way. The completion of deposition is judged by testing a sample of the electrolyte with stannous chloride after the thiocyanate has been decomposed with sulphuric acid.

Removal of Deposited Gold.—"The best way of removing deposited gold from platinum electrodes is by means of a solution of potassium cyanide containing an oxidising agent such as hydrogen peroxide, sodium peroxide, or an alkali persulphate."[1]

The electrolytic method of assay is used in some mills for the determination of the gold in cyanide solutions resulting from the cyanide process. In this case the cathode consists of a cylinder of lead-foil notched at the base to facilitate circulation of the liquid. The solutions, which vary in gold content, are electrolysed for four hours with a current varying from 0·06 to 1·2 ampere.

The gold separates as a bright yellow deposit. When precipitation is complete, the cathode is washed in water, dried, rolled up, then scorified with test lead, and cupelled. Details of this method, as recommended by Crichton, are given on page 376.

Electrolysis of Platiniferous Gold.—Balling [2] has proposed the separation of gold from platinum and metals of the platinum group by electrolysis. A weighed quantity of the platiniferous gold, in the form of thin sheet, is used as the anode, while an accurately weighed sheet of pure gold or of platinum is used as the cathode. Both sheets are placed in a solution of gold chloride and subjected to the action of an electric current. The gold anode is dissolved and is deposited on the cathode, while the platinum metals are liberated and fall to the bottom as a grey-black powder, which is carefully collected and assayed for platinum, etc.

[1] T. K. Rose, *Metallurgy of Gold*, 5th edit., p. 469.
[2] Balling, *Fortschritte im Probirwesen*, Berlin, 1887, p. 116.

CHAPTER XIX.

THE ASSAY OF SILVER BULLION.

THE silver in silver bullion and alloys is determined by one of the four following methods, namely :—

1. By cupellation.
2. By titration with common salt (Gay-Lussac method).
3. By titration with thiocyanate (Volhard method).
4. By conversion into chloride, which is then weighed (Indian Mint method).

Cupellation Assay of Silver.

This method of assay has been largely displaced by wet methods, but is still in wide use. It is accurate to about 0·5 part per 1000 when nearly pure silver is being assayed. As the operation of cupellation has been already described in Chapter XII., it will only be necessary here to point out the precautions which must be observed in conducting the operation.

The cupellation assay of silver may be divided into three distinct operations, namely :—

1. The preparation of the assay piece.
2. Elimination of impurities by cupellation.
3. Weighing the buttons and reporting the results.

1. **The Preparation of the Assay Piece.**—The sample is selected by any of the methods described in Chapter XVII., but additional precautions are necessary to ensure that the assay piece represents the mass from which it is taken; for, as previously stated, alloys of silver and copper are peculiarly liable to undergo liquation. The assay piece is flattened and a portion adjusted to an exact weight, usually 0·5 gramme, in the manner already described under the assay of gold (page 264). The assay piece is enclosed in a packet of sheet lead, the weight of which bears a definite ratio to that of the copper present. The lead should be at least five times the weight of the assay piece, and should be increased in proportion to the amount of copper present: the exact proportions usually recommended are given below. The corners of the lead packets are squeezed down so as to fit the cupels by pliers or are "balled" in the balling nippers (fig. 133) especially designed for the purpose.

The assay pieces thus prepared are placed in their proper order in a wooden tray with numbered compartments, from which they are transferred direct to the muffle. The amount of lead to be employed in the cupellation depends on the composition of the alloy under examination; it is obviously necessary, therefore, that this should previously be approximately determined.

An experienced assayer will frequently be able to decide this with sufficient accuracy from the colour and hardness of the metal, or the composition may be roughly ascertained by means of the touchstone, see page 306, or by noting the amount of oxidation produced on a clean piece of the bullion when gently heated as described on page 307.

Preliminary Assay.—When necessary a preliminary assay may be made by cupelling 0·5 gramme of the bullion with 7 grammes of sheet lead, and working the assay in the usual way. The silver button is cleaned and weighed. The loss will give the amount of base metal present : this is usually copper, and its presence may be detected by the colour of the cupel.

The composition of the bullion having been approximately determined, the following Table XXXII. gives the proportion of lead to be used with one part of bullion of different fineness.

Table XXXII.—Proportion of Lead required for Cupellation of Silver Bullion.

Fineness of bullion. Silver in 1000 parts.	Parts of lead for one part of bullion.	Ratio of lead to base metal.
1000 to 950	5	100 to 1
950 to 900	7	70 to 1
900 to 800	8	40 to 1
800 to 700	10	33·3 to 1
700 to 600	12	30 to 1
600 to 500	14	28 to 1
500 to 250	16 to 18	33·3 or 24 to 1
250 to 0	18 to 20	18 or 20 to 1

For the cupellation of English standard silver (925 per 1000), six times its weight of lead is usually employed.

2. **Elimination of Impurities by Cupellation.**—The assay pieces are introduced into the cupels, arranged in rows corresponding to those on the tray, the muffle being at a bright red heat. The cupellation is conducted in the manner previously described on page 159. The proper adjustment of the heat and gradual cooling of the assays are of the first importance. It is advisable to keep the temperature as low as possible, so as to reduce the volatilisation of the silver and its absorption by the cupel to a minimum ; but, on the other hand, care must be taken to avoid all risk of the buttons setting before the oxidisable metals have been removed. The heat should be so adjusted that only slight fumes are visible over the cupels, and the fused metal should be very luminous and clearly distinguishable from the cupel. With too low a temperature crystals of litharge collect round the button, while with an excessive heat the fumes rise rapidly to the crown of the muffle. With "patent" magnesia cupels the conditions of cupellation differ somewhat from those just described for bone-ash cupels (see page 166).

In cooling down the muffle prior to withdrawing the cupels, special care must be taken to prevent "spitting." The methods generally adopted for preventing this are discussed on pages 160 and 184.

When the cupellation has been satisfactorily performed the silver buttons are bright, crystalline, and slightly depressed in the centre, and dull white on the under surface. The presence even of a minute quantity of copper renders a silver button less globular. After all the buttons are set, the cupels are withdrawn and placed in their proper positions in the cupel tray. The buttons are then removed by means of a pair of pliers, squeezed to loosen the adhering bone-ash, and cleaned with a stiff brush. It is the practice in some assay offices to remove the buttons from the cupels, and give each button a blow on its side with a small hammer to loosen the adherent bone-ash before it is brushed.

The buttons are now ready for the final operation of weighing.

3. **Weighing the Buttons and Reporting.**—In ascertaining the absolute amount of pure silver present in the assay piece, the method of weighing and the system of "weighing in" correction, described on page 273, is employed.

Checks for Silver Assays.—The results of silver bullion assays are controlled by check assays of pure silver or of silver with the addition of copper or other base metal to make them of the same composition as the assay pieces. The checks are made up in the same manner as gold checks, previously described. The losses during silver cupellation are discussed on page 170.

The absorption of silver by the cupel is large, amounting to as much as 10 parts per 1000, and varies considerably, being increased by a higher temperature of the furnace and by the presence of a larger amount of copper or of lead. It is therefore necessary to use several checks in different parts of the muffle; and, since the loss of silver is much greater than in the case of gold, and varies considerably throughout the muffle, it is necessary that the checks should be more numerous. It is a common practice to place one check in each row of assays in the muffle, and to vary the position of each in the rows. Thus in a batch of forty assays of fine silver bullion including eight checks, arranged in the muffle in eight rows of five abreast, the positions of the checks would be 4, 7, 15, 18, 21, 29, 32, 40, cupel No. 1 being at the extreme left of the back row.

The correction to be applied to an individual assay piece is deduced from the checks in its immediate neighbourhood. Considerable skill and knowledge are necessary in applying such corrections to the weights obtained, as well as a thorough knowledge of the manner in which the temperature varies in the muffle. The variations of temperature of a muffle are determined in the first instance by working off a complete set of fine silver cupellations. This must be done each time a new muffle is fitted.

The amount of the correction due to cupellation loss need not exceed 10 parts per 1000. When carefully conducted the cupellation assay is, as before stated, accurate to about 0·5 per 1000 when nearly pure silver is being assayed.

Reporting the Results.—The results of silver bullion assays are reported in parts per 1000 to the first place of decimal, to the nearest 0·5 per 1000. Thus a result 995·6 is reported 996 fine and 994·4 as 944 fine. The results of assay pieces from the top and bottom of the silver bars should agree closely; in the case of fine silver bars, within 0·5 per 1000. In making the report, either the average of the results or the lowest result is given: it is customary to give the lowest result of assays of fine silver. If the cupellation has been carefully performed and proper checks used, there is no difference between the results with silver bars over 980 fine. Bars between 980 and 725 fine show a difference of 0·5 to 3 parts per 1000; from 720 to 710 fine, no difference, or only an infinitely small one.[1] This is due to the retention of small quantities of base metals, which counteract the loss of the silver absorbed by the cupel. The greatest differences occur in low-grade bullion below 400 fine. Very considerable differences may occur between the assay results if the bars have not been properly melted and mixed before casting.

Wet Assay of Silver.

Apparatus employed.—The following apparatus is used in connection with the wet methods of silver assaying :—

Weights.—The special set of "silver assay weights," described on page 60,

[1] Kerl, *Assaying*, 1889, p. 146.

are very frequently used in the assay of silver bullion. In this system the numbers stamped on the weights denote their weight in milligrammes. Thus

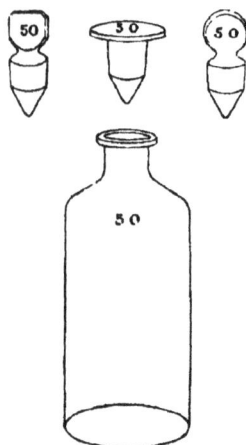

FIG. 146.

the 1-gramme (*i.e.* 1000 milligrammes) weight is stamped 1000, the 0·5-gramme (500 milligrammes) weight 500, etc. It will be noted that in this system the value of the weights is double that in the "millième" system used in gold assaying, in which, as already stated, the 0·5-gramme (500 milligrammes) weight is stamped 1000. It is very important in offices where the two systems are in use to keep very carefully the weights distinct, as the confusing of the weights may lead to serious error in reporting the value of the bullion.

Bottles.—Round, narrow, stoppered bottles, capable of holding 8 to 10 ounces (225 c.c. to 300 c.c.), both bottles and stoppers being numbered, are usually employed. The bottles should be of good white glass and capable of being heated in a water bath without cracking. The usual shape of bottle is shown in fig. 146 with three kinds of stoppers in common use. Some unnumbered bottles, provided with a "sand-blasted" patch, should be kept in stock to replace any numbered bottles in the series that are accidentally broken. The required number is best painted on the ground patch with black enamel paint. The number on the bottle into which each assay piece is introduced is noted on the corner of the assay paper in which the sample is wrapped.

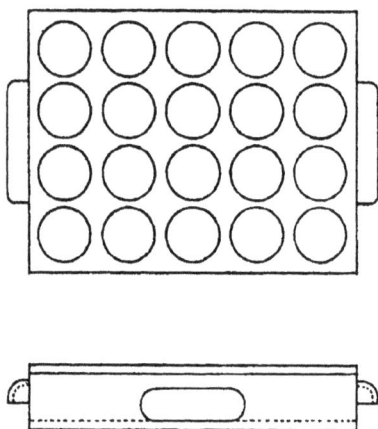

Scale

FIG. 147.

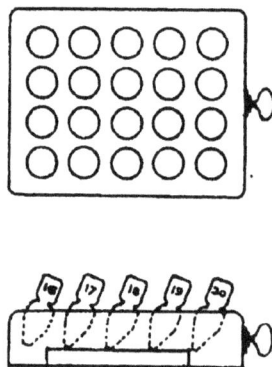

FIG. 148.

Water Bath.—A water bath for heating the bottles is best made of copper with a loose diaphragm for keeping the bottles in position. A water bath to hold twenty bottles is similar in size and shape to the wooden tray shown in fig. 147.

The bath is supported on an iron quadripod stand and heated by three or four tube air-gas burners.

Bottle Trays.—When a large number of assays have to be dealt with it is convenient to work them in batches of twenty, and to facilitate the handling of the bottles they are placed in wooden trays (see fig. 147). The tray has a thin wooden top, with holes to admit the bottles, so that the bottles are always kept in position: the top is made of three-ply wood to prevent warping. A hole is cut in each side of the tray, as shown, to permit of a cloth being inserted to wipe the bottom of the tray in the . event of a bottle breaking. The trays should be well made and be provided with wooden handles. They are distinguished by numbered enamelled plates.

Stopper Trays. — The numbered stoppers to the bottles are kept in their proper order in the tray shown in fig. 148. This is made of a block of hard wood with holes drilled in a diagonal direction so that when the stoppers are placed in position the numbers are readily seen. The tray is placed in front of the bottle tray to the right of the operator when titrating the assays, as described later. The trays are distinguished by numbers corresponding to those on the trays containing the bottles to which the stoppers belong.

Bottle Cabinet.—In offices where several hundred assays are made daily, some convenient method of storing the assay bottles becomes a necessity. A very convenient form of cabinet for the purpose is shown in fig. 149. It holds 16 trays, each containing 20 bottles—a total of 320 bottles—and their stoppers, which are placed on a series of shelves between

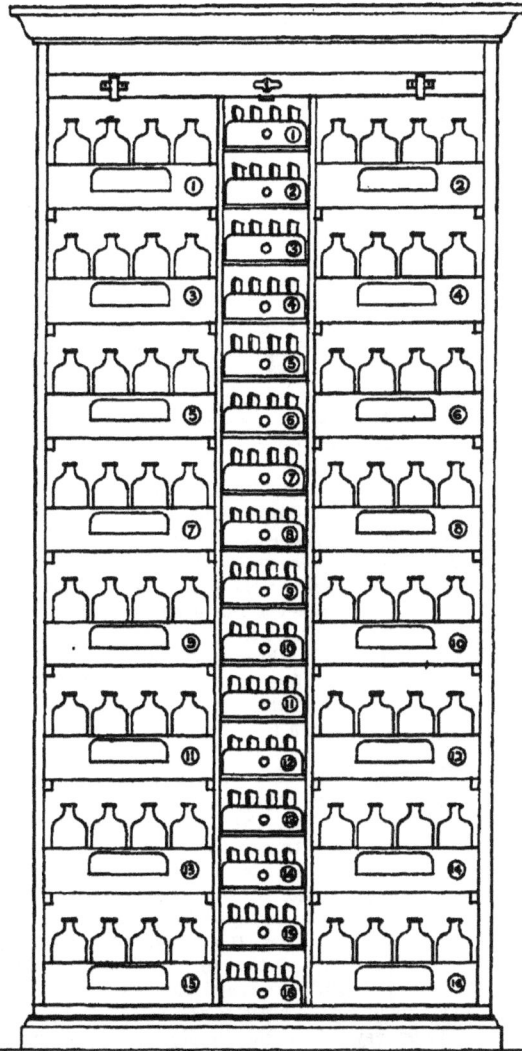

FIG. 149.

the bottle trays, as shown.   The bottle trays are slipped into position on runners. The cabinet is provided with a roll front to keep out the dust.

Burettes.—A burette of large bore and with large opening for quick delivery is used for adding the requisite quantity of acid to the assays.   A 200-c.c. burette, 24 inches long and $1\frac{1}{4}$-inch bore, graduated into 10 c.c., is very suitable for adding the acid to a batch of twenty assays.   The burette is held in the left hand and passed over the mouth of each bottle in succession, the time occupied in adding acid to twenty assays being about forty seconds.   A hand pipette, similar to that shown in fig. 153, is frequently used for the purpose, but is not so suitable.

Pipettes.—A large pipette is required for adding 100 c.c. of a standard solution to each assay.   The form of pipette, devised by Stas, shown in fig. 150, is used for the purpose.   The solution is introduced from below by means of an india-rubber tube, and the pipette when full is closed at the top by the finger with the left hand.   The tube is then removed with the right hand, the bottom of the pipette gently "wiped" between the thumb and first finger, and the solution allowed to run into the bottle placed ready to receive it.   The bottom end of the pipette is sometimes sand-blasted for a short distance to prevent the slipping of the rubber tube.   The pipette is fitted with a cup, and also provided with a glass hood or cap to catch any overflow when filling.   The hood is connected by rubber tubing to a receptacle to receive the overflow of solution, which can be used again.

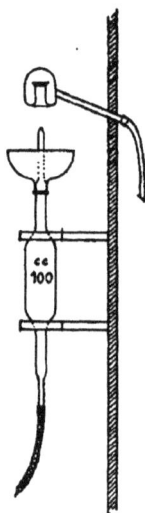

The Stas pipette is one of the most important pieces of apparatus used in silver assaying, and it must be made with great care.   It is not so important that the pipette shall deliver exactly 100 c.c. as that in two consecutive deliveries the volume shall not differ by more than 0·005 c.c.   In well-made pipettes the error in the deliveries is often not more than 0·002 c.c.

Where large numbers of assays are frequently made, a set of four pipettes is mounted in a suitable case side by side. Experience proves that there is no advantage in having more than four pipettes in a series.   A convenient form of case is shown in front elevation in fig. 151, and side elevation in

Fig. 150.

fig. 152.   The doors are glazed, and hinged so that they can be swung right back to be out of the way : one door partly open is shown in fig. 151.   The overflow from the pipettes is caught in a glazed earthenware trough from which it passes into a jar beneath the bench on which the case stands.   When the assay bottles are placed on the stand they fit into a "guide," so that they are at once in the correct position under the pipettes.   Holes are bored in the stand, in direct line with the pipettes, so that any "drops" from the pipettes fall into the earthenware trough.   The tube for supplying solution to the pipettes passes through a clip fixed to the floor and operated by means of the foot, so that both hands are left quite free for manipulating the pipettes and bottles.   A standard solution of common salt may be conveyed from the supply tank to the pipettes through a tin-lined lead pipe, but a glass tube should be used for thiocyanate solutions.

In working with a series of pipettes it is of the utmost importance that each pipette should deliver the same volume of solution.   All pipettes should be accurately tested before being taken into use.   For accurate assay work it is well to have the pipettes standarised at the National Physical Laboratory. Before working off a batch of assays, the pipettes should be filled and emptied

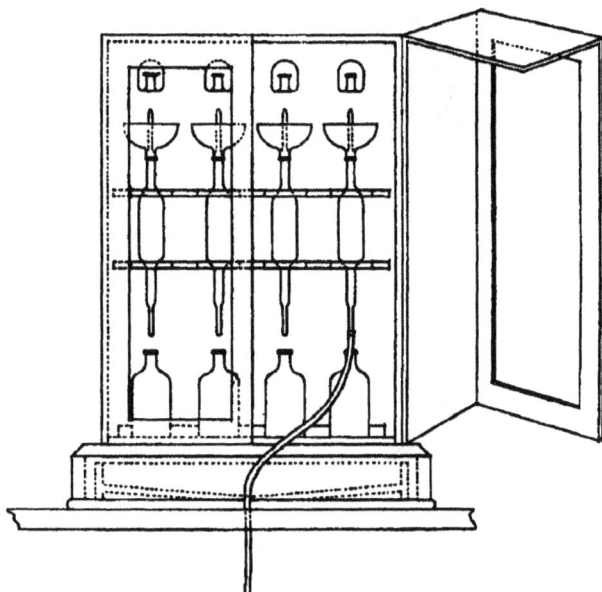

Fig. 151

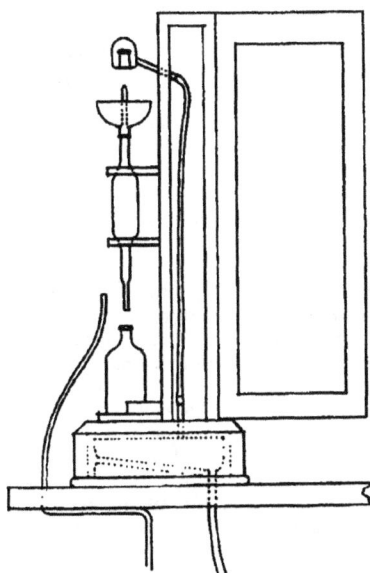

Fig. 152.

several times until the volume of solution delivered is constant.   The pipettes should be cleaned periodically with a warm solution of caustic soda.

Small Pipettes.—For the addition of dilute solutions small hand pipettes are sometimes used (fig. 153).   They have a capacity of 10 c.c., graduated in $\frac{1}{2}$ c.cs. Most assayers use ordinary 50-c.c. burettes for the addition of dilute solutions.

Automatic Pipette.—For diluting a large number of assays with the same volume of water or dilute indicator solution, as in the Volhard method described later, the automatic pipette shown in fig. 154 is convenient.   It is fitted with a two-way cock, and has a large outlet tube to give quick delivery.   The pipette is connected with a large supply of solution by means of a rubber tube, and is filled from below in the same way as the Stas pipette.   The overflow solution is caught in the hood and delivered into a separate vessel placed to receive it.

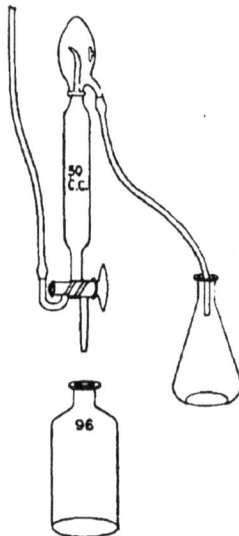

FIG. 153.   FIG. 154.

Testing Shelf.—The end of the reaction in the Gay Lussac method, described subsequently, is best seen against a black surface.   A shelf painted dead-black should be fixed in front of a window, which may be glazed with yellow glass, to receive the bottles during the addition of the decimal solution; it is provided with a blackened upright back 3·5 inches high, extending along its entire length, and the bottles are so arranged on it that the light passes through the upper part of the liquid they contain.

"When artificial light has to be used, it has been found that by placing screens so that the light from an electric arc-lamp passes through the upper portion of the silver solution, while the eyes of the observer are protected from the direct rays, accurate readings of the 'clouds' or fine layers of floating silver chloride can be made without any undue straining of the sight."   The facilities thus secured are especially valuable during the fogs which are more or less frequent in large cities in winter.[1]

Shaking Machine.—Shaking is always resorted to in the wet method of silver assay to induce the clearing of the solution.   It may be conducted by hand, two bottles being shaken at a time; but if many assays have to be dealt with, it is more convenient to employ some form of shaking machine or agitator.

One of the shaking machines in use in large assay offices is shown in fig. 155. It is made of copper, but may be made of wood, and holds twenty bottles.   The general arrangement of the appliance is shown by two elevations.   The shaker has a reciprocating movement with a 2-inch throw, the speed of the fly-wheel being 300 revolutions per minute.   The power is supplied by a $\frac{1}{4}$ H.P. continuous-current motor.   The box in which the bottles are placed is fitted with a diaphragm to keep the bottles in position, and springs are fitted in the lid, as shown in the enlarged section, to prevent the stoppers from coming out during the shaking.   The lid, which is in two halves, is secured by a clamp.   Springs are fitted to the side of the box to break any sudden fall of the lid, which might

[1] *Eighteenth Mint Report*, 1887, p. 53.

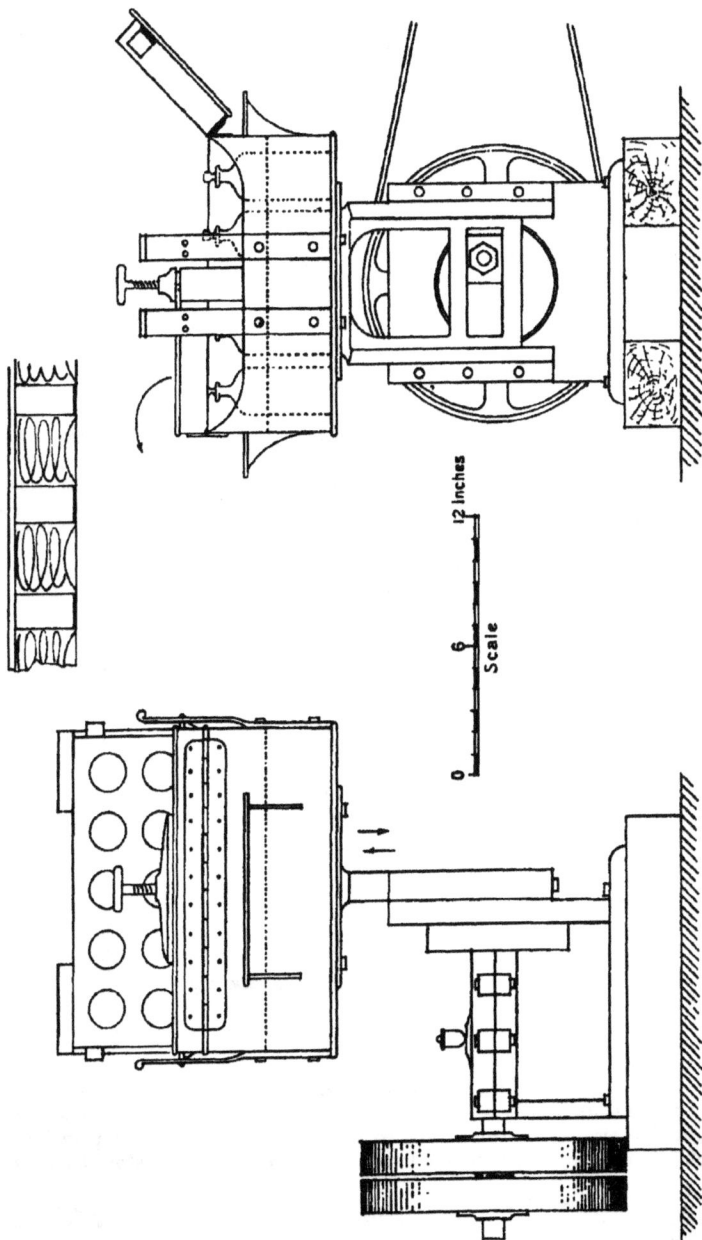

Fig. 165.

result in breakage of the bottles. When the assays are worked off in batches of twenty in the trays described (fig. 147), a considerable saving of time is effected by having the shaker so arranged that the tray is placed directly into it without the necessity of transferring the bottles separately. A shaker of this description is in use at the Sheffield Assay Office.

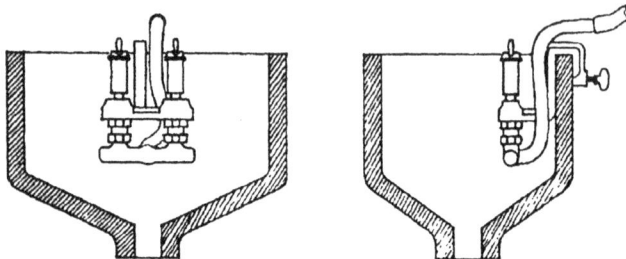

FIG. 156.

Bottle-washing Apparatus.—An important consideration in assay offices dealing with a large number of wet silver assays is the washing of the bottles. This may be accomplished very rapidly by means of the patent [1] automatic washing jets shown in fig. 156. These jets, which are connected up with an ordinary water-supply by means of a hose pipe, are so constructed that when a

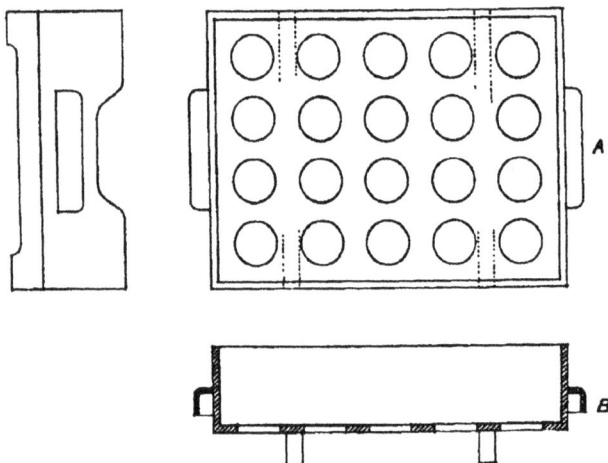

FIG. 157.

bottle is inverted over them and gently pressed down, a jet of water rises in the bottle and washes it. It is convenient to use two jets, clamped to the side of a sink, so that two bottles can be washed simultaneously. A jar is placed below the sink to receive the precipitate from the bottles.[2] The operator places a tray

[1] Proprietors of patent, Messrs D. G. Binnington, Kingston Machine Works, Hull.

[2] The silver chloride from the assays is allowed to settle, the clear liquor syphoned off, a little hydrochloric acid added, and the chloride then reduced by plates of wrought iron, as described on page 75.

of bottles on one side of the sink, then takes a bottle in each hand, empties the contents, holds the bottles over the jets for a few seconds, then returns the bottles to their places. With this apparatus a batch of twenty bottles and stoppers is readily washed in 1½ minutes. After being washed, the "drainer" shown in plan in fig. 157,A, and elevation in fig. 157,B, is placed over the tray of bottles, and the drainer and tray inverted together, and the bottles allowed to drain until quite dry (see fig. 158). The bottles are then transferred to the bottle cabinet ready for use.

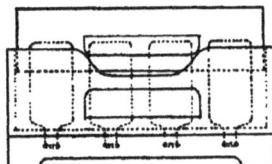

Fig. 158.

Gay Lussac Method (sodium chloride).— This method was invented by Gay Lussac in 1832, and has displaced the cupellation process in almost all mints and in many private assay offices. It is the most accurate method of silver-bullion assaying at present known. It is a volumetric method, in which the volume of a standard solution of sodium chloride required for the precipitation of a known amount of silver in solution as silver nitrate is measured. The end of the operation is judged from the appearance of an exceedingly faint cloud produced by sodium chloride in a solution from which almost all the silver has been precipitated.

Three solutions are required, namely :—

1. Standard or normal solution of sodium chloride (or of sodium bromide).
2. Decimal solution of sodium chloride.
3. Decimal solution of silver nitrate.

When large numbers of assays are made, it is usual to prepare periodically a considerable bulk of these solutions, which should be preserved in well-closed vessels of stoneware or glass.

1. "*Normal Salt*" *Solution*.—This is made up of such a strength that 100 c.c. will precipitate about 1 gramme of silver: this requires 5·416 grammes of sodium chloride per litre. This is usually termed the "normal" solution, although not correctly of "normal" strength. The strength of the solution is standardised by means of check assays of pure silver.

2. "*Decimal Salt*" *Solution*.—This, as its name implies, is one-tenth the strength of the normal solution: it is frequently called the decinormal solution. It is prepared by taking 1 part by volume of the normal solution and adding to it 9 parts of distilled water. A convenient method is to introduce the contents of the 100-c.c. Stas pipette into a litre flask and fill up with water. As 1 c.c. of this solution is equal to one-tenth c.c. of the normal solution, it will obviously precipitate 0·001 gramme of silver.

3. "*Decimal Silver*" *Solution*.—One gramme of pure silver is dissolved in a small quantity of pure dilute nitric acid, and the solution diluted to one litre. One c.c. of this solution will contain 0·001 gramme of silver: it follows therefore that 1 c.c. of the decinormal salt solution will just precipitate the silver in 1 c.c. of the decimal silver solution.

Standardising the Solutions.—A weighed quantity, in duplicate, of proof silver of exactly 1·003 grammes is transferred to an 8-ounce (205-c.c.) glass bottle and dissolved in 15 c.c. of dilute nitric acid (sp. gr. 1·26) and heated until the nitrous fumes are expelled. When cold, 100 c.c. of the normal solution is run into the bottle from the Stas pipette (see fig. 150), and the contents are shaken for three or four minutes either by hand or a mechanical shaker, until the silver chloride has curdled and will settle to the bottom, leaving the supernatant liquid clear. If the normal solution is made up correctly, it will have precipitated just 1·0 gramme (1000 milligrammes) of silver, leaving ·003 gramme (3 milligrammes) unprecipitated.   When the precipitate has settled, the solution is tested by the addition from a pipette or burette of 1 c.c. of the decimal salt solution, which is allowed to run down the side of the bottle so that it forms a layer resting on the assay solution.   If any silver remains in solution, a cloudy layer will be formed near the surface of the liquid.   This is best observed against a black background, and the shelf described on page 294 is used for the purpose.   If a cloudiness is seen, the bottle is again shaken to clear the liquid, and then another c.c. of decimal salt added.   If this second addition of decimal salt solution gives a precipitate, the shaking and settling are repeated and further additions of decimal salt made of 1 c.c. at a time, shaking between each addition, until no further cloud appears.

An experienced assayer soon learns to judge by the density of the cloud whether only part of the c.c. has been used up, and readily decides when all the silver has been precipitated.   The end reaction is marked by the gradual formation in the course of about ten minutes of a very faint cloud, the value of which to the fourth of a c.c. (or 0·25 milligramme) is judged by the assayer.   In very exact work experienced assayers judge the clouds even more closely.   In this way the exact strength of the normal solution is determined in duplicate.   As an example, suppose the total additions of decimal salt were 4 c.c. (equivalent to 0·004 gramme of silver), and that the cloud produced by the last addition was declared to be equivalent to only one-half of a c.c. (equal to 0·0005 gramme), then $4 - 0·5 = 3·5$ c.c., the amount of solution actually required to complete the precipitation of the silver.

Since 100·35 c.c. of normal solution was required to precipitate 1·003 grammes of silver, the strength of the solution is calculated in terms of 100 c.c. as follows.   Thus :

$$100·35 : 1·003 :: 100 : x$$
$$x = 0·9995 \text{ gramme, or } 999·5 \text{ milligrammes.}$$

The strength of the solution is finally recorded on the bottle as follows :—

$$100 \text{ c.c.} = 999·5 \text{ mgs. Ag.}$$

Some assayers prefer to weigh out exactly 1 gramme of silver for the check assays, and in this case the strength of the "normal" solution is adjusted so that after the addition of 100 c.c. to the assay solution 1 or 2 milligrammes of silver will still be left in solution and thus ensure the production of a cloud when decinormal salt solution is added.

If the strength of the solution is incorrect to the extent of more than 2 points (i.e. 2 milligrammes), it should be corrected by the addition of either water or salt, and then restandardised.   A solution that is too strong is dealt with as follows :—

If the first addition of decimal salt solution causes no cloudiness, the normal solution contains an excess of salt, and 5 c.c. of decimal silver solution (equivalent to 5 milligrammes of silver) are added to give an excess of silver in solution, and then the additions of decinormal salt are proceeded with in the usual way. The 5 milligrammes of silver must be allowed for in calculating the strength of the solution (see note on page 302).

The Assay.—It is evident from the preceding paragraph that the method requires a practically constant quantity of silver, that is, one which varies by a few milligrammes only in each determination, and thus avoids undue additions of the decinormal solutions.   If many additions have to be made, the operation not only becomes tedious, but the solution also ceases to clear after shaking, so that it becomes impossible to determine the finishing point.

Under these conditions the composition of the bullion must be known approximately before the weight to be taken for assay can be determined. If this is not known, a preliminary assay must be made by cupellation or by the Volhard method, described later. If the composition is, say, approximately 900 fine, then the weight to be taken is

$$\frac{1 \cdot 003}{900} \text{ grammes, or } 1 \cdot 115 \text{ grammes (1115 milligrammes)}.$$

This amount of bullion is weighed out in duplicate, placed in bottles, dissolved in acid, and then titrated as described above.

The fineness of the bullion is calculated as follows :—

Suppose the strength of the normal solution is 100 c.c. = 1001 milligrammes of silver, and that 100·2 c.c. of normal solution was used in the titration (*i.e.* 100 c.c. normal salt + 2 c.c. of decimal salt), then :

$$100 : 1001 :: 100 \cdot 2 : x$$
$$x = 1003 \text{ milligrammes,}$$

which is the amount of silver in the 1·115 grammes of bullion taken. The fineness of the bullion is calculated thus :—

$$1115 : 1003 :: 1000 : y$$
$$y = 898 \cdot 65 \text{ parts per 1000,}$$

which is the fineness of the bullion.

If the assay contains less than 1 gramme of silver in solution, the first addition of the decinormal salt solution of course produces no precipitate. Five milligrammes of silver in solution (*i.e.* 5 c.c. of decimal silver solution) is then added, and the assay proceeded with in the usual way, 5 milligrammes of silver being deducted from the amount found. Thus, employing the sample given above, suppose the total additions of normal and decimal salt solution required to precipitate all the silver in solution (including the 5 milligrammes added) amounted to 100·2 c.c., then the quantity of normal solution used to precipitate the silver actually present in the bullion will be 100·2 − 0·5 = 99·7 c.c., and

$$100 : 1001 :: 99 \cdot 7 : x$$
$$x = 997 \cdot 99 \text{ milligrammes,}$$

the amount of silver in the bullion taken, and

$$1115 : 997 \cdot 99 :: 1000 : y$$
$$y = 895 \cdot 05,$$

which is the fineness of the bullion. Report the result as stated on page 291.

Since the volume of normal solution measured varies with the temperature, a difference of 5° C. making a difference of 0·1 c.c. in measuring the 100 c.c., it is always desirable to run a check assay with each batch of assays.

In the case of low-grade bullion where it is necessary to take a large quantity in order to obtain 1 gramme of silver in solution, it is usually more satisfactory to weigh 1 gramme of the bullion and add a carefully weighed quantity of fine silver to make up the total silver to about 1 gramme. By adopting this method, excess of copper in the solution is avoided and the end point decided without difficulty, which is not always possible when the solution is intensely coloured with copper.

Limit of Accuracy of the Gay Lussac Method.—As ordinarily carried out by experienced assayers the method is accurate to 0·1 part per 1000, but with proper precautions and the expenditure of sufficient time it is capable of giving results accurate to 0·01 per 1000.[1]

Interfering Metals.—Mercury interferes with the method, as it is precipitated as mercurous chloride (Hg_2Cl_2) with the silver, but its presence is detected by its prevention of the darkening of the silver chloride when exposed to light. The addition of 20 c.c. of sodium acetate and a little free acetic acid to the assay before titration will prevent the precipitation of the mercury. Mercury is sometimes present in mill bullion which has been retorted at too low a temperature.[2]

Tin also interferes, as it is converted into metastannic acid by the nitric acid, and remains as a white amorphous powder suspended in the solution, thus producing a cloudiness which interferes with the finish of the assay. The presence of as little as 0·05 per cent. of tin was found by Salas[3] to be sufficient to cause the obscuration of the end point, while 0·5 per cent. rendered the assay impossible. In order to prevent the formation of a precipitate of metastannic acid, the following procedure is recommended by Salas.

A quantity of bullion containing about 1 gramme of silver is accurately weighed and placed in the assay bottle. Two grammes of tartaric acid crystals and 3 or 4 c.c. of water are added, and the bottle is heated till the acid is dissolved. After cooling, 10 c.c. of dilute nitric acid (1 : 1) are added, and when the silver has dissolved, the solution is titrated as usual. Under these conditions the liquid remains clear, but it must be kept cool during and after the solution of the silver. If the bullion contains more than 5 per cent. of tin, a smaller quantity than 1 gramme should be taken for the assay and some pure silver added.

Silver bullion from Mexico and Bolivia frequently contains tin, and it is also met with (sometimes in large quantities) in silver bars resulting from the treatment of jewellers' sweeps. Tin is sometimes present to the extent of over 5 per cent. in English standard silver (925 fine) as a substitute for copper. Silver bullion containing tin may be satisfactorily assayed by cupellation or by the Volhard method.

Note on the End Point in Gay Lussac's Method.—Mulder[4] has shown that when the most exact chemical proportions of silver and salt are made to react on each other, the addition of either salt or silver to the clear supernatant liquid will produce a precipitate, indicating the presence of free silver nitrate and sodium chloride in a state of equilibrium, either of which is thrown down on the addition of the other. This is known as Mulder's "neutral point." No trouble is experienced in determining the end point in the ordinary assay if care be taken that rather more than 1 gramme of silver is present. When less silver is present, so that the solution contains excess of sodium chloride after the addition of 100 c.c. of normal salt solution, an excess of silver should at once be added, rather than an attempt made to finish the assay by the addition of decinormal silver. The existence of Mulder's "neutral point" is explained by Hoitsema[5] by the solubility of silver chloride. As this compound is also present in the solid state, the product of the concentrations of its ions must remain constant, and hence the additions of either silver or chlorine ions must cause a precipitation of the salt.

A similar neutral point is not observed in titrating with a bromide or iodide, because the solubility is less than the amount necessary to produce the apparent precipitate. It is for this reason that some assayers use a normal solution of sodium bromide instead of the chloride.

[1] T. K. Rose, *Precious Metals*, p. 212.   [2] Fulton, *Fire Assaying*, p. 180.
[3] L. E. Salas, *Trans. Amer. Inst. Min. Eng.*, March 1912, 267–278.
[4] *De Essayeer-methode vanhet zilver scheikundig onderzocht*, Utrecht, 1857.   Quoted by Percy, *Metallurgy of Silver*, p. 291.
[5] *Zeit. physikal. Chem.*, 1896, xx. 272–282.

Volhard Method (thiocyanate).—This method was proposed by J. Volhard in 1878, and is based on the fact that when ammonium thiocyanate (or other soluble thiocyanate) is added to silver nitrate, a white precipitate is produced consisting of silver thiocyanate, insoluble in nitric acid.

$$AgNO_3 + (NH_4)CNS = AgCNS + NH_4NO_3.$$

This reaction takes place even in the presence of a ferric salt, and it is not until the whole of the silver has been precipitated and the thiocyanate is in slight excess that the characteristic blood-red coloration of ferric thiocyanate makes its appearance.

A ferric salt, therefore (but obviously not the *chloride*), constitutes an extremely delicate indicator for this reaction, the end of which is very clearly defined. In conducting the assay, a ferric salt is added to the silver solution and a standard solution of thiocyanate then run in. The white silver thiocyanate settles readily on shaking, leaving the liquid clear; and a persistent red-brown coloration in the liquid when the thiocyanate is in excess indicates the finish. The silver solution must be cold in conducting the assay, and all the water used must be free from chloride, otherwise silver will be precipitated and errors introduced.

The following solutions are prepared :—

Standard Thiocyanate.—This is prepared either from ammonium thiocyanate or potassium thiocyanate, both of which are too deliquescent to allow of very exact weighing.

From 3·5 to 4 grammes of crystallised ammonium thiocyanate (or 4·5 to 5 grammes of potassium thiocyanate) are dissolved in a litre of distilled water. 100 c.c. of this solution are equivalent to 0·5 gramme of silver.

Standard Silver Nitrate.—Dissolve 5 grammes of fine silver in 50 c.c. of dilute nitric acid (1 : 1), boil to expel nitrous fumes, and dilute to 1 litre. 100 c.c. of this solution will contain 0·5 gramme of silver.

Indicator.—Several ferric salts are used as indicators. That most frequently employed is a saturated solution of iron alum, 2 c.c. being used for each assay. Another convenient ferric salt is the sulphate, a solution of which may be prepared by dissolving a few crystals of ferrous sulphate in water, adding about half the volume of strong nitric acid in order to oxidise the iron, and then briskly boiling the mixture for a few minutes to expel all the nitrous fumes.[1] This solution is slightly diluted with water, and 3 or 4 c.c. used for each titration.

A saturated solution of iron nitrate is also very suitable for the purpose. It is prepared by dissolving iron wire in dilute nitric acid (1 : 1). About 2 c.c. are used.

When a number of titrations are to be made, the ferric solution should be made up of some definite strength, and the same volume of it used for each assay. The addition of a short piece of thin iron wire to each assay bottle has been suggested; this dissolves at the same time as the silver, and supplies the ferric indicator.

Standardising the Solution.—The thiocyanate solution is standardised by transferring 50 c.c. of the silver nitrate solution to a flask, adding 10 c.c. of dilute nitric acid (1 : 1), 2 c.c. of ferric indicator, and diluting with 100 c.c. of water.

The solution of thiocyanate is then run in from a burette. As each drop falls into the flask, the red colour of ferric thiocyanate appears for a moment, but at once disappears on gently shaking the flask. The solution is cautiously

[1] Newth, *Chemical Analysis*, p. 365.

added until the first indications of a permanent red coloration are seen. The colour thus developed should remain permanent after shaking continuously. After shaking, the precipitated silver settles readily, enabling the colour of the clear supernatent liquid to be distinctly seen. From the volume used, the exact strength of the thiocyanate solution is ascertained, and the amount of dilution, or the addition of crystals, which it requires in order to bring it to the desired strength, is readily calculated. After the solution has been thus adjusted, it should be once more titrated.

To obtain reliable results, the nitrous fumes must be expelled, and the silver solution must be cold. The effect of variations of temperature may be seen from the following experiments by Beringer,[1] in which 20 c.c. standard silver nitrate were used.

Effect of Varying Temperature.

Temperature	10° C.	30° C.	70° C.	100° C.
Thiocyanate required	19·6 c.c.	19·3 c.c.	19·0 c.c.	18·6 c.c.

The effect of varying amounts of silver in the assay solution is as follows :—

Silver solution added	1 c.c.	10 c.c.	20 c.c.	50 c.c.	100 c.c.
Thiocyanate required	1·0 c.c.	9·7 c.c.	19·6 c.c.	49·4 c.c.	99·0 c.c.

From these results it will be seen that the error in the determination of the silver increases with an increase in the amount of silver present. This error is eliminated in the ordinary assay by the use of checks of the same composition as the silver being assayed.

The Volhard method, worked as described, is valuable for determining silver in bullion and alloys of unknown composition, where no more than an ordinary degree of accuracy is demanded. It is quick, easy in manipulation, and applicable under most of the usual conditions, and on that account is very frequently employed in determining the approximate composition of bullion which is to be more accurately assayed by the Guy Lussac method. The presence of other metals, as lead, copper, nickel, zinc, cadmium, exerts no disturbing effect so long as the solution contains free nitric acid, and the metals are not present in large excess. The effect of tin and large quantities of copper or nickel are discussed below.

The assay is conducted exactly as in the titration for standardising the thiocyanate solution, the silver being dissolved in 15 c.c. of dilute nitric acid (1 : 1), and diluted to a convenient volume. The greatest disadvantage of the method when worked as described is the brown coloration produced by the thiocyanate when the assay is nearly, but not quite, finished, and the slowness with which this is removed on shaking up with the precipitate. This is worse with large quantities of precipitate, and if about 1 gramme of silver is present, it gives an indefiniteness to the finish, which lowers the precision of the method to about 2 parts per 1000, a degree of accuracy which is useless for the assays of bullion.[2]

In determining silver in silver bullion by Volhard's method, the general practice is to work on the principle of the Gay Lussac method, by adding a measured quantity of normal solution to precipitate the bulk of the silver, and then to make successive additions of decinormal solution until the whole of the silver is precipitated. Checks of similar composition to the bullion are run with all assays to eliminate errors. Working in this way upon samples of pure silver, Van Riemsdigk[3] found that by the successive additions of decinormal ammonium

[1] *Text-Book of Assaying*, p. 122.   [2] Beringer, *loc. cit.*, p. 123.
[3] *Algemeen Verslag. Munt-Collegie*, 1878, Appendix 38.

thiocyanate solution, each sufficient to precipitate 0·25 milligramme of silver, results are obtained which are accurate to 0·25 part per 1000.

According to Van Riemsdigk, this marks the limit of accuracy of Volhard's method when the above method of working is adopted, as he points out that an appreciable colour cannot be obtained without an excess of about this quantity of the thiocyanate solution.

For some years it has been the practice at several assay offices to slightly modify the ordinary method of finishing the assay by using a "normal" solution sufficiently strong to give a permanent red colour to the check assay, and using this colour as a standard of comparison. By this means the finishing point of the assay is perceived more sharply and with greater certainty than is possible when the assay is finished in the usual way by cautiously adding the decinormal solution until the first indications of a permanent red coloration are seen.

In conducting the assay, a quantity of bullion is taken such that about 1 gramme of silver is present. It is dissolved in 10 c.c. of dilute nitric acid (sp. gr. 1·2) in a bottle of about 250 c.c. capacity, and heated in a water bath until all traces of nitrous acid are expelled. The solution is then diluted with 50 c.c. of water containing 2 c.c. of a saturated solution of iron alum decolorised by nitric acid, and 100 c.c. of "normal" ammonium thiocyanate is added from a Stas pipette. This solution contains about 7·04 grammes of ammonium thiocyanate per litre, and 100 c.c. is sufficient to precipitate about 1 gramme of silver and leave a very small amount of thiocyanate in excess, which in the presence of the ferric salt gives a pale red colour to the check assays. For the assay of fine silver bars a convenient strength for the normal solution is such that 100 c.c. is equivalent to about 1·0003 to 1·0005 gramme of silver.

The bottle is then well shaken for about two minutes, when the white precipitate of silver thiocyanate settles to the bottom and leaves the liquid clear. If some silver is still in solution, the liquid is colourless or retains the pale green colour due to the dissolved copper originally contained in the bullion. Decinormal thiocyanate is now run in from a burette until the colour *approaches* that of the check; the liquid is then well shaken and allowed to clear; after which the decinormal solution is cautiously run in, drop by drop, until the colour corresponds exactly to that of the check. The addition of 1 drop (0·05 c.c., equivalent to 0·05 milligrammes of silver) of decinormal thiocyanate makes a perceptible difference to the colour, which is readily recognised.[1] The assay should be well shaken before the final comparison is made.

Experience has shown that equally accurate results are obtained whether the amount of silver left in solution after the addition of normal solution is 1 milligramme or 20 milligrammes, or more.

The comparison of the assays with the check is best made by placing the bottles so that they catch the reflected light of a white surface. A shelf with a back to which a sheet of white opal glass is secured, and arranged so that it faces the direct light from a window, is very suitable for the purpose. A side light is liable to give rise to error of judgment in comparing the tints. If after adding 100 c.c. of normal solution the ammonium thiocyanate is in excess, the liquid is coloured red from the action on it of the ferric salt, the intensity of the colour varying with the amount of thiocyanate in excess.

In this case the assay may be finished off either by adding decinormal silver solution until the colour of the assay solution is brought into agreement with that of the check, or an excess of silver may be added and the assay completed

[1] J. W. Leather has drawn attention to the extreme delicacy of this reaction in an interesting paper on "The Determination of Small Quantities of Iron" (see *Journ. Soc. of Chem. Ind.*, vol. xxiv., 1905, p. 385.

in the ordinary way, the added silver being of course allowed for in calculating the results. The latter method is to be preferred. The method of calculating the results is the same as that described for the Gay Lussac method on page 301.

By finishing the assay as described, the exact composition of the bullion can be readily ascertained correctly to 0·2 per 1000, or even more closely.[1] When the bullion contains copper or nickel, the blue colour of the solution somewhat masks the colour of the ferric thiocyanate, and to obtain a satisfactory tint for comparison it is desirable to use a slightly stronger "normal" solution.

For example, the strength of the solution for assays of standard silver (925 silver and 75 copper) should be such that 100 c.c. will precipitate 1·001 gramme of silver. It is very important in assaying silver-copper alloys to use checks of pure silver, with the addition of copper approximately equal to that in the bullion. With silver-copper alloys below 900 fine more satisfactory results are obtained by taking about 0·5 gramme only for assay, and adding fine silver to bring up the total silver to about 1 gramme. When tin is present, the metastannic acid formed tends to remain in suspension and clouds the solution, but does not seriously interfere in comparing the final tints; to get accurate results, however, an equal amount of tin must be added to the check assays. By allowing the solutions to stand for five or ten minutes before making the final comparison, the bulk of the tin will settle with the silver precipitate, leaving the solution almost clear.

The author has obtained very satisfactory results with this method with bullion containing as much as 10 per cent. of tin.

The Volhard method is particularly suitable for the assay of silver plate at the hall-marking offices. The modification of the method described was proposed by Mr Arthur Westwood, the assay master of the Birmingham Assay Office, and the method has been used there ever since 1892, for doing over 1000 silver assays a day. Later it was introduced at the Sheffield Assay Office and elsewhere.

Combined Gay Lussac and Volhard Method.—To obviate the tediousness of determining the residual silver by adding successive small quantities of salt, in the last stage of the Gay Lussac method, Beringer[2] has proposed a combination of the Gay Lussac and Volhard methods which he states answers admirably when applied to the assay of silver bullion.

The combined method is conducted as follows:—

Prepare the assay as described for the Gay Lussac method and precipitate the bulk of the silver by adding 100 c.c. normal salt solution: shake till the liquor clears, and filter into a flask, washing with a little distilled water. Filtration from silver chloride is necessary, and the wash water must be free from chlorine. Add 2 c.c. of "ferric indicator" to the filtrate and titrate with a dilute thiocyanate solution made by diluting the ordinary standard solution to such an extent that 100 c.c. of the diluted solution shall be equivalent to 0·1 gramme of silver or 1 c.c. = 1 milligramme. To prepare this, multiply the *standard* by 1000 and dilute 100 c.c. of the standard solution to the resulting number of c.c. Thus, with a solution of a standard 0·495, dilute 100 c.c. to 495 c.c., using, of course, distilled water.

The titration is continued until the characteristic colour of the ferric thiocyanate is permanent. The absence of the silver precipitate from the solution enables the titration to be completed much more readily. An advantage of this

---

[1] "Assay of Silver Bullion by Volhard's Method," E. A. Smith, *Trans. Inst. Min. and Met.*, vol. xvi. (1907). p. 154.

[2] *Text-Book of Assaying*, 10th edit., p. 124. Consult also A. E. Knorr, *Journ. Amer. Chem. Soc.*, 1897, vol. xix. pp. 814–816.

modification is that an excess of 15 milligrammes of silver may be determined as easily and exactly as 5 milligrammes.

The solution is standardised by means of fine silver, about 1 gramme of which is accurately weighed and treated as described above. The standard of the salt solution is then found by deducting the number of c.c. of the thiocyanate used for the check from the number of milligrammes of the fine silver initially weighed out. Thus: Suppose 1015·0 milligrammes of silver required the addition of 13·5 c.c. of thiocyanate solution, equivalent to 13·5 milligrammes of silver, then the silver precipitated by the salt is 1015·0 – 13·5 = 1001·5 milligrammes (1·0015 gramme), which is the standard. The strength of the salt solution may therefore be expressed as 100 c.c. = 1001·5 milligrammes Ag.

The total amount of silver in the bullion taken for assay is calculated by adding to the standard of the salt solution the milligrammes of silver equivalent to the number of c.c. of thiocyanate used.

The fineness of the bullion is calculated in the ordinary way from the silver in the assay piece and the weight of bullion taken.

According to Knorr,[1] the limit of accuracy of this combined method of assay is 0·2 to 0·3 per thousand.

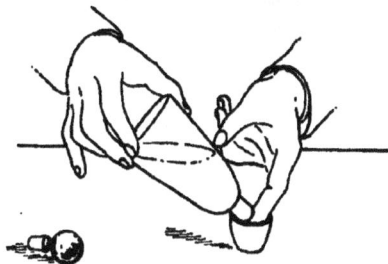 

FIG. 159A.   FIG. 159B.

Attempts have been made to increase the accuracy of the Volhard method as ordinarily worked by filtering off the precipitate obtained on the addition of 100 c.c. of normal thiocyanate and then finishing the assay by adding deci-normal thiocyanate to the clear filtrate.

By finishing the titration in a clear solution, free from precipitate, the end of the reaction is much more readily distinguished, but experiments by Hoitsema[2] have proved that the filtration of the solution impairs the accuracy of the results. He concluded, as the result of a series of experiments, that the filtration method is not reliable when more than about 1 milligramme of silver remains in the clear solution, the error increasing with the increase in the amount of silver remaining in solution. Thus in six tests with 1000, 1001, 1002, 1003, 1004, and 1005 milligrammes respectively of pure silver, after adding 100 c.c. of standard thiocyanate (100 c.c. = 1000 milligrammes of silver) and filtering, the amounts of dilute thiocyanate required to produce a pale red colour in the filtrate were 0, 1, 3, 7, 14, and 20 drops instead of there being a regular increase of 20 drops (equal to 1 milligramme of silver) in each case. In other tests also variable results were obtained, which are attributed by Hoitsema to absorption by the silver thiocyanate precipitate of soluble silver salts, and also of the soluble thiocyanate.

[1] *Journ. Amer. Chem. Soc.*, 1897, vol. xix. pp. 814–816.
[2] *Zeit. andew. Chem.*, 1904, vol. xvii. pp. 647–650.

Tests made by Beringer[1] and also by the author confirm the conclusions of Hoitsema that the precipitate must be allowed to remain in the solution when finishing the assays by Volhard's method.

Gravimetric Assay of Silver (Indian Mint Method).[2] —

This method was introduced by Dr Dodd, of the Calcutta Mint, in 1851, and is of a special character. It consists in adding an excess of hydrochloric acid to a solution of silver nitrate and weighing the silver chloride produced. The weight of silver bullion taken is 18·817 grains, which is equivalent to 25 grains of silver chloride if the bullion consists of pure silver. After precipitation, the bottle is well shaken, and the chloride is allowed to settle and is washed by decantation. The silver chloride is transferred to a Wedgwood cup by placing the latter over the mouth of the bottle and inverting it (see fig. 159). Afterwards the excess water is poured off, the chloride broken up by a glass rod, and dried. The drying is finished on a hot-plate, and the cake of chloride is weighed while still warm. A set of weights is used in which the weight marked "1000" weighs 25 grains, and the smaller weights are in proportion, so that the composition of the alloy is indicated without calculation.

In this method any gold present in the bullion remains unchanged and is weighed with the silver chloride.

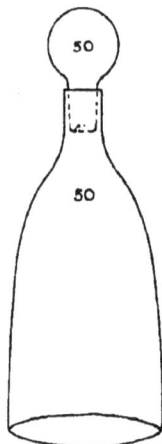

FIG. 160A.

The special form of bottle employed in this method is shown in fig. 160A and the Wedgwood crucible in fig. 160B. The method is practically confined to the Indian Mints, where the conditions of the country with regard to climate and cheap native labour render it suitable.

Other Methods of Silver Assaying.

FIG. 160B.

Touchstone Test for Silver.—The quality of silver is sometimes tested by means of the touchstone, as described for gold on page 280. If the streak is silver, it will dissolve immediately on wetting with a drop of dilute nitric acid; to confirm the presence of silver, a drop of dilute hydrochloric acid may be added, when a white precipitate of silver chloride will be produced. If no white precipitate appears after adding the hydrochloric acid, then the metal is probably German silver, a white alloy of nickel, zinc, and copper.

If a white residue remains after wetting the streak with nitric acid and before the hydrochloric acid is applied to it, then the metal is tin or an alloy containing tin. The following reagent is recommended by A. Steinmann[3] for testing the metallic streaks which are obtained when silver and its alloys are rubbed across the surface of the touchstone.

It consists of a mixture of nitric acid (sp. gr. 1·27), 40 c.c.; glacial acetic acid, 50 c.c.; and water, 50 c.c. According to Steinmann, the use of this reagent enables one to distinguish between pure silver and alloys of silver containing small quantities of copper, etc. The amount of silver in an alloy may be determined approximately by comparing the behaviour of the alloy towards the reagent with that of alloys of known composition. (See also "Test Acid," test-page 338.)

[1] *Text-Book of Assaying*, 10th edit., p. 123.
[2] Rose, *Precious Metals*, p. 217. Consult also F. T. C. Hughes, *Trans. Inst. Min. and Met.*, vol. xvii., 1907, p. 384.
[3] *Ann. Chim. Analyt.*, 1911, vol. xvi. pp. 165-167.

Oxidation Test for Silver-Copper Alloys.—When heated in a muffle or other oxidising atmosphere, the alloys of silver and copper are superficially oxidised, with the production of variously coloured films on the surface, according to which an approximate estimate of the amount of copper in the alloys may be determined. Thus the appearance of the surface of silver-copper alloys of different composition when heated in the muffle is as follows[1] :—

Composition of the Alloy.	Oxidation Colour.
Pure silver	Dull white.
950 fine (50 parts copper)	Dull greyish white.
900 ,, (100 ,, ,,)	Dull greyish white with black fringes along the edges.
880 ,, (120 ,, ,,).	Grey.
860 ,, (140 ,, ,,).	Almost black.
840 ,, (160 ,, ,,).	Quite black.

Silver does not oxidise on the application of heat; but on heating the highly cupriferous alloys of silver both metals are oxidised, the copper, however, to a much greater extent than the silver.

With a little experience this method can be employed as a preliminary test for approximately estimating the proportion of copper in silver alloys, but it ceases to be applicable in the case of alloys below about 850 fine. It is a valuable test for detecting the soldered seams in silver wares (see page 338).

Electrolytic Assay of Silver.—A method for the quantitative determination of silver by electrolysis was described by Luckow[2] as early as 1865; it consisted in passing a current through a silver nitrate solution containing not more than 8 to 10 per cent. of free nitric acid, but the silver is deposited in a very voluminous spongy condition. A thorough investigation of the conditions of silver deposition from solutions of the nitrate was carried out by F. W. Kuster[3] and H. von Steinivehr in 1898. As the results of a large number of experiments, they found that silver up to 2 grammes in weight could be deposited in compact adherent form, suitable for quantitative determination, from solutions of the nitrate containing from 1 or 2 c.c. of free nitric acid (1·40 sp. gr.) and 5 c.c. of alcohol, using an E.M.F. of between 1·35 and 1·38 volts. The temperature of the electrolyte should be 55°-60° C. With an E.M.F. above or below the limits named, unsatisfactory results will be obtained.

According to R. C. Benner and W. H. Ross,[4] silver may be accurately deposited from cyanide solutions by using gauze cathodes and stationary anodes. If the quantity of silver to be deposited does not exceed 0·15 gramme, high currents do not cause sponginess, and as much as 5 ampères may be used, reducing the time required to one-half of that required when 1 ampère is used.

Analysis of Silver Bullion (and Silver Alloys).—The chief impurities in silver bullion are gold, copper, and lead, but antimony, bismuth, zinc, tellurium, and selenium are sometimes present in small quantity. Complete analyses of silver bullion and of silver alloys are made by the ordinary chemical methods. The solution of the silver bullion is effected with dilute nitric acid (1 : 1), and the percentage of silver determined either volumetrically by the Gay Lussac method or gravimetrically as silver chloride, as previously described.

The impurities present may be determined as follows :—

Gold.—Small quantities of gold in silver bullion are usually determined by parting in nitric acid, and subsequent collection and weighing of the finely divided

[1] *L'Art de l'Essayeur*, par M. Ohaudet. Quoted by Percy, *Metallurgy of Silver*, p. 157.
[2] *Dingler's Journal*, Bd. clxxviii., 1865, p. 43.
[3] *Zeits. f. Elektrochem.*, vol. iv., 1898, pp. 451-455.
[4] *Journ. Amer. Chem. Soc.*, 1911, vol. xxxiii. pp. 493, 1106.

gold, as described on page 207.   When the quantity of gold is very small, from 5 to 10 grammes of bullion may have to be taken.

Since gold, when in a very finely divided condition, is soluble in nitric acid, Dewey [1] recommends the following method for collecting very small quantities of gold from solutions containing much silver.

A small quantity of strong sulphuretted hydrogen water is added to the silver nitrate solution freed from excess of nitric acid.   By this means any gold that may be in suspension or in solution is precipitated as gold sulphide conjointly with a small amount of silver sulphide.   The solution is well stirred, allowed to stand for some time (two or three hours), and the sulphides filtered off.   The precipitate is dried, carefully burnt, and the metallic residue wrapped in a small piece of sheet lead, cupelled, and the gold parted in the usual way.   The collection of the gold in a small quantity of silver chloride precipitate, instead of silver sulphide precipitate as described, was found to be less satisfactory.

Copper.—It is sometimes desirable to determine the small amounts of copper in silver bullion, and this is very frequently done by the colorimetric method. In this method advantage is taken of the deep blue colour imparted to copper solutions by the addition of ammonium hydrate, the depth of colour depending on the quantity of copper present.   The method is based upon the principle that equal volumes of solutions of an equally intense colour contain also equal quantities of copper.   By comparing solutions of an equal intensity of colour, one of which contains a known weight of copper, and taking the volume into consideration, a conclusion is formed as to the percentage of the copper contained in the one to be determined.

According to Beringer,[2] the quantity of copper in 100 c.c. of the assay solution should not be more than 15 milligrammes, or less than 0·5 milligramme.

The following method of working with a standard copper solution described by Beringer will be found to be very satisfactory :—

Standard Copper Solution.—Weigh up 0·5 gramme of electrotype copper, dissolve in 10 c.c. of nitric acid, boil off nitrous fumes, and dilute to 1 litre. 1 c.c. = 0·5 milligramme of copper.   For the determination of the copper in silver bullion the solution from the Gay Lussac assay may be used after filtering off the precipitate of silver chloride.

The whole or a measured portion of the clear solution (generally two-thirds, one-half, or one-fourth of the whole according to the amount of copper present) is placed in a white glass cylinder standing on a sheet of white paper and 10 c.c. of dilute ammonium hydrate added.   Into an exactly similar cylinder is placed 10 c.c. of dilute ammonium hydrate, and then water is added until the solutions in the two cylinders are of nearly equal bulk (say 200 c.c.).   The standard copper solution is then run into the ammoniacal water from a burette, the liquid being stirred after each addition until the colour approaches that of the assay.   The bulk of the two solutions is equalised by adding water.   Then more standard solution is added until the tints are very nearly alike.   Next, the amount added is read off from the burette, still more copper solution is run in until the colour is slightly darker than that of the assay, and the burette read off again.   The mean of the readings is taken, and gives the quantity of copper added.   It equals the amount of copper in the portion of assay solution taken.

If this portion was one-half of the whole, multiply by two ; if one-third, multiply by three, and so on.   When the quantity of copper is very small, it may be necessary to separate it with sulphuretted hydrogen from a solution of about 5 grammes of the bullion after removing the silver as chloride.

The filter paper containing the copper sulphide is dried and burnt.   The

[1] *Journ. Amer. Chem. Soc.*, vol. xxxi., 1910, p. 323.
[2] *Text-Book of Assaying*, 11th edit., 1910, pp. 44 and 203.

ashes are dissolved in 5 c.c. of dilute nitric acid, 10 c.c. of dilute ammonium hydrate added, and the solution filtered through a coarse filter into the glass cylinder, washing the paper with a little dilute ammonium hydrate. The solutions to be compared must, of course, contain free ammonium hydrate ; and about 5 c.c. of dilute ammonium hydrate in a volume of 50 c.c. of the solutions is usually recommended.

Another method of procedure frequently employed is to prepare a series of standard copper solutions as follows [1] :—

Weigh out 30 milligrammes of copper, dissolve in dilute nitric acid, and make up to 180 c.c. 6 c.c. = 1·0 milligramme copper.

Transfer five portions of 12 c.c., 18 c.c., 24 c.c., 30 c.c., and 36 c.c. respectively of the standard copper solution to tall white glass bottles, add to each 10 c.c. of dilute ammonium hydrate, and dilute up to 150 c.c.   These standards will be equal to 0·2, 0·3, 0·4, 0·5, and 0·6 per cent. of copper respectively, if 1·0 gramme of the bullion is taken for assay.

The ammoniacal copper solution from the assay is compared with these standards in succession, and the standard of the same depth of colour as the assay solution indicates the percentage of copper in the bullion.   This method is quicker than titration with standard copper solution, as described above, and is in very general use.

In the case of silver alloys containing a large percentage of copper, this metal may be determined volumetrically by means of potassium iodide and sodium thiosulphate solution or by gravimetric methods.

For the determination of the small quantities of other metals such as lead, zinc, etc., sometimes present in silver bullion, the standard works on chemical analysis must be consulted.

Industrial silver alloys may contain comparatively large amounts of zinc and cadmium ; these may be separated as described on page 339, and determined gravimetrically by the usual methods.

[1] J. E. Lathe, *Eng. and Min. Journ.*, 1912, vol. xciii. p. 1072.

CHAPTER XX.

THE ASSAY OF BASE BULLION (LEAD AND COPPER).

As previously stated, the term base bullion is conveniently restricted to argentiferous and auriferous lead and copper bars.

I. Base Lead Bullion.

Liquation in Lead Bullion.—Silver and lead mix together in all proportions while they are molten, but on solidification they separate so that the alloys are not homogeneous. Consequently, the argentiferous lead produced in smelting operations, and cast into pig-moulds of cast-iron, undergoes segregation, and the pigs or bars are not uniform in composition. A result of this is that the valuation of argentiferous and auriferous lead is a matter of difficulty. Moreover, in many smelting establishments the surface of the melted bullion is skimmed, and the clear lead ladled into the mould, until the latter is filled to within an inch of the top, and when it has solidified, the mould is filled completely. This results in a bar of good appearance, but composed of good lead above and below, and more or less dross in the middle, which has separated from the lead first cast, while practically no dross separates from the thin layer of clean metal cast on the top. Bars that are cast in this way are very difficult to sample accurately.

Large quantities of lead carrying gold and silver are sold to refiners in bars weighing about 80 to 110 lbs. each. The bars are approximately 24 inches by 5 inches at the top, 20 inches by 3 inches at the bottom, and $3\frac{1}{2}$ to 4 inches in depth. Some bars are cleaner from dross than others. Lead bars are received from South Australia, Queensland, Mexico, Greece, South America, and other places. It has been found that different parts of the same bar of lead carrying silver or gold vary very much in their contents of these metals, and this circumstance affects the sampling of argentiferous and auriferous lead bullion on a large scale. The top of a bar is, as a rule, poorer in silver and gold than the bottom, but when the gold contents become high, the reverse appears to be the case.[1]

Sampling Lead Bullion.[2]—The ordinary methods of sampling are as follows:—

1. Cutting chips out of the top and bottom of the bars.
2. Punching half-way through the bars.
3. Sawing, by a circular saw, through the bars.
4. Dip sampling after melting the bars in mass into clean lead and dross.

1. *Chip Sample.*—Chips are cut by a gouge or chisel out of the top and bottom of every bar, or alternately the top of one bar and the bottom of the next, the chips being taken from opposite sides and ends of the bars.

[1] Consult Percy's *Metallurgy of Silver*, p. 172; J. and H. S. Pattinson, *Journ. Soc. Chem. Ind.*, vol. xi., 1892, p. 321; G. M. Roberts, *Trans. Amer. Inst. Min. Eng.*, vol. xxviii., 1898, p. 418; A. C. Claudet, *Trans. Inst. of Min. and Met.*, vol. vi., 1898, p. 29.
[2] A. C. Claudet, *loc. cit.*

2. *Punching Sample.*—A punch, resembling a leather punch, such as those shown in fig. 161A and 161B, is very frequently used. It is driven about half-way through the bars by the use of a sledge hammer, and gives cores about 2 inches long and about ¼ inch in diameter. When sampling a large lot of bullion it is usual to sample five bars at a time. In this case it is customary to place the five bars side by side on the sampling platform, and to drive in the punch at one end of the first bar and at the opposite end of the last bar, and on the others in intermediate positions in such a manner that all the holes will be along a diagonal of the rectangle enclosing the bars. The bars are then turned over and similar portions punched out in the bottoms of the bars and along the other diagonal (see fig. 162A). Or one set of five may be sampled along the top and the next set along the bottom of the bars.[1] The chips or cores are then carefully melted into one bar, or those from the tops and bottoms melted separately into two bars.

FIG. 161A.   FIG. 161B.

The molten metal is well stirred and then poured into a mould to give a thin flat bar about 10 to 18 inches long, about 3 inches wide, and ½ to ¾ inch thick. This is sampled by cutting out portions from the ends and sides, as shown in fig. 162B, or cutting strips right through the bar, or by punching out pieces at regular intervals, diagonally across the bar. The four samples thus obtained are, as a general rule, each approximately of the weight required for an assay.

On assaying the bars, the results are compensated for enrichment on melting, if necessary. This method should only be employed with lead containing 10 to 20 ounces of silver per ton.

3. *Sawing.*—The bars are sawn transversely, either half-way through or right through, and usually alternately in the middle of one bar and at one end for the next bar, and so on. The sawings are either melted into small bars and assayed, or they are assayed without melting; in the latter case, the result must be corrected for impoverishment, due to the oil which has to be used as a lubricator when sawing the bars. Sawing is done by a circular saw, operated by steam power, and the teeth should be rather small, to prevent too frequent breakages, and to produce small-sized sawings.

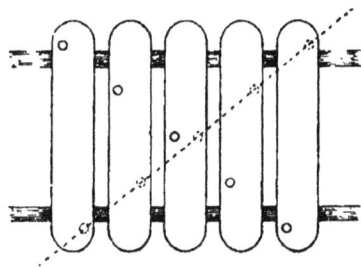

FIG. 162A.

FIG. 162B.

The "sawings," as the sample is termed obtained in this way, which may amount to two or three hundredweight, are thoroughly mixed and reduced in bulk by quartering in the usual way or by means of a mechanical sampler, such as Clarkson's. The final sample, weighing about 2 to 3 lbs., is sent to the assayer. It is contaminated with oil, and requires special treatment, as described later. This method may be satisfactorily used with lead containing silver up to 100 or 200 ounces per ton.

[1] Beringer, *Assaying*, p. 160.

314 THE SAMPLING AND ASSAY OF THE PRECIOUS METALS.

4. *Dip Sample.*—The lead is melted in large iron pots, holding from 10 to 30 tons, the charge is heated and skimmed until the molten lead is quite clean, the temperature is then raised as high as possible, the lead well stirred up, and a "dip" sample taken. The dross, amounting to from 1 to 5 per cent., according to the impurities in the lead, is broken down when cold and a sample taken. The two samples are then assayed, and the results corrected and calculated with regard to the original lead charge. This method takes the longest time, but is stated to yield the most reliable results; it should be employed in all cases where the silver is above 200 ounces per ton, and where the gold is above 1 or 2 ounces per ton.

A comparison between the different methods of sampling lead bullion is made in the following tables, XXXIII., XXXIV., and XXXV., which give the results of assays obtained by A. Claudet [1] for gold and silver in commercial lead bars from various sources.

Table XXXIII.—Comparison between Chip and Sawing Sampling. Argentiferous Lead.

Lot.	Total weight.	Silver per ton.		
		Chip sample (London).	Chip sample (foreign).	Sawing sample.
	tons cwts.	ounces.	ounces.	ounces.
1	13 5	114·20	120·20	116 20
2	12 4	104·80	110·65	107·75
3	12 14	132·90	140·65	136·25
4	12 17	135·85 [2]	137·75	135·50

Table XXXIV.—Comparison between Dip and Sawing Sampling. Argentiferous Lead.

Lot.	Total weight.	Silver per ton.	
		Dip sample.	Sawing sample.
	tons cwts.	ounces.	ounces.
1	19 4	119·80	120·20
2	22 14	131·10	132·50
3	25 10	125·40	127·40
4	18 16	126·10	127·55
5	68 7	70·80	72·20

Table XXXV.—Comparison between Dip and Sawing Sampling and the Precious Metals (Gold and Silver) actually recovered by refining the Lead.

Lot.	Total weight.		Sawings sample.	Dip sample (Colonial assay).	Recovered by refining.
	tons cwts.		ounces per ton.	ounces per ton.	ounces per ton.
1	2 13	gold .	1224·0	1305·0	1261·0
		silver .	305·0	296·0	296·0
2	6 0	gold .	...	342·0	340·0
		silver .	...	102·0	99·0

[1] *Loc. cit.*
[2] Average result. Top chips, 130·5 ounces. Bottom chips, 140·8 ounces.

In lot No. 2 two quantities of 1 ton each were sawn separately, the sawings melted and assayed. Dip samples were also taken and assayed. The results for the gold were as follows :—

	Gold per ton.	
	Sawing sample. ounces.	Dip sample. ounces.
Lot A, about 1 ton	303	312
Lot B, „	332	329

Assay of Lead Bullion.

Preparation of the Sample for Assay.—The sample of chips, punchings, or sawings sent to the assayer usually weighs about 2 or 3 lbs. For assay purposes it is carefully melted, and cast into an ingot mould to obtain a small " sample bar" or "assay bar" from which portions are cut for assay.

Great care has to be exercised during the melting to prevent oxidation, as this causes an enrichment of the precious metals and consequent error in the assay value of the lead unless the results are compensated for enrichment, as described below.

The error due to enrichment may be considerable when the lead contains impurities, such as antimony, copper, sulphur, etc., which produce much dross or scum, i.e. a mixture of oxides, sulphides, etc., contaminated with metallic lead.

In the case of samples taken by means of a saw, the "sawings" are always contaminated with the oil used for lubricating the saw. The sawings are first freed from oil by washing with ether, benzene, or carbon bisulphide and carefully dried. The removal of the oil may have decreased the weight of the sample by about 0·2 to 0·5 per cent.[1] The sawings thus cleaned are thoroughly mixed, and from 100 to 200 grammes carefully weighed and melted under 10 or 20 grammes of potassium cyanide to prevent the formation of dross.

A sample consisting of chips or cores is melted in the same way.

The molten metal is well stirred and then poured into a mould to give a thin, flat bar which necessarily varies in size according to the weight of the sample, but is usually about 10 to 15 inches long, about 3 inches wide, and ⅛ to ¾ inch thick. The mould should be warm, and in casting the metal the crucible should be moved backwards and forwards along the length of the mould so as to distribute the lead and ensure a uniform bar. Any slag adhering to the lead is readily removed by washing in hot water. When the bar is quite clean the portions required for assay are cut from the ends and sides, as shown in fig. 162B, or by punching out pieces at regular intervals, diagonally across the bar, or cutting strips right through the bar.

The four portions thus obtained are, as a general rule, each approximately of the weight required for an assay, viz. 15 to 20 grammes.

The addition of potassium cyanide during melting is very necessary, as with some classes of lead the loss of weight, due to oxidation of the impurities when the sample is melted alone, is very considerable. The percentage loss on melting is reduced to less than one-half by the use of potassium cyanide.

With samples of lead of average quality the loss when melted under potassium cyanide does not, as a general rule, exceed 0·25 per cent., and is frequently about half this amount.

This loss may at first sight seem to be of little importance, but, actually, the correction of the assay corresponding to this small loss makes a difference on lead containing, say, 500 ounces of silver per ton, of about 1 ounce per ton ;

[1] J. and H. S. Pattinson, loc. cit.

and this amount, when deducted from the assays of the sample, gives the true contents of silver in the original lead. With richer leads the correction is of course proportionately greater. With auriferous lead the correction for loss on melting is of considerable importance, owing to the greater value of gold compared with silver.

It is obvious that, if the sample has not been freed from oil before melting, the melting loss will include also the loss due to the oil, and an error would be made which would lead to an *excessive deduction* amounting probably to more than 2 ounces per ton in the case of lead containing 500 ounces of silver per ton.

Determination of Silver in Lead Bullion.—The four or more portions selected for assay are adjusted to exactly 15 or 20 grammes (or to one-half assay ton), according to the percentage of silver in the lead, and then carefully cupelled at a low temperature and the resulting buttons of silver weighed. If the lead is very impure, it should first be scorified with about 1 gramme of borax, and if necessary with the addition of about 30 to 50 grammes of sheet lead. The resultant buttons, which should weigh about 15 grammes, are then cupelled. If the lead is very poor in silver, from 40 to 50 grammes may be scorified down to about 15 grammes and then cupelled. In the case of lead rich in silver (say over 500 ounces per ton) the cupels are fused, as described on page 180, and the correction for cupel absorption made. Or the cupellation loss may be determined by wrapping each silver button obtained from the cupellation separately in 20 grammes of sheet lead, and recupelling side by side with two or more fresh lots of 20 grammes each of the sample of lead bar. Then weigh the resulting silver buttons, calculate the loss incurred, and add this to the weight of the silver buttons from the two fresh portions cupelled (see page 181).

It is usually preferable, however, in the case of rich argentiferous lead, to run check assays after first determining the approximate composition of the sample by a preliminary assay.

Determination of Gold in Lead Bullion.—To determine the gold the silver buttons, after carefully weighing, are flattened and parted in a small parting flask by heating with 15 to 20 c.c. of dilute nitric acid of about 1·2 sp. gr. until the silver is dissolved and all nitrous fumes expelled. The residual gold is washed, and heated for a few minutes with stronger nitric acid of about 1·3 sp. gr. and then washed, dried, annealed, and weighed in the usual manner.

Calculating and reporting the Results.—The silver and gold in lead bullion is always reported in ounces per ton, in the same way as for ores. The amount of silver may vary from 10 ounces to about 1000 ounces per ton, while gold may be present to the extent of 500 ounces per ton or even more, but is usually much less. As previously stated, the smelters generally keep the proportion of silver in the lead below 2 per cent. (650 ounces per ton), in order to avoid undue losses in the slag.

The results of the several assays should agree very closely, say within half an ounce per ton, on average bullion assaying from 200 to 250 ounces of silver per ton. With rich argentiferous lead assaying over, say, 500 ounces, the variation between the results will be greater than this, and it is advisable in assaying rich material to make a larger number of assays, from five to ten, and take the average of all the results. There should be practically no difference in the weights of the gold results, and if they are not almost identical the assays should be repeated. The assay value of the lead both in silver and gold is reported in ounces and decimals of an ounce to the second place of decimal on the ton of either 2240 lbs. or 2000 lbs. The former is used in England, but the latter is becoming more universal.

In calculating the assay value of the lead bars it is necessary, as previously stated, to make a correction for the loss during melting, otherwise the value reported will be too high. The following is an example :—

Weight of sample before melting . 200·0 grammes.
Weight of sample bar after melting 199·4 ,,
Loss on melting . . 0·6 ,, = 0·3 per cent.

Therefore 100 grammes of the original sample before melting is represented by 99·7 grammes after melting. If 20 grammes of the lead after melting gave 0·1671 gramme of silver, equal to 0·8355 per cent., then the percentage of silver in the original lead will be :—

$$\frac{0.8355}{100} \times 99.7 = 0.8329 \text{ per cent. of silver.}$$

And 0·8329 per cent. equals 272·08 ounces of silver per ton of lead of 2240 lbs. If gold is also present it is calculated in the same way.

With lead bars which yield a large amount of dross on melting, the dross is removed and sampled separately. It is then assayed, and the results obtained for the silver and gold present allowed for in calculating the total assay value of the lead bars, as shown below.

Assay of "Dross" from Lead Bullion.—The dross skimmed off from the surface of the molten lead contains, as a rule, from 80 to 95 per cent. of lead present as metal, lead oxide, sulphate, etc. The skimmings vary considerably in content of silver according to the richness of the bullion melted. They are assayed by a crucible fusion, the following being a suitable charge :—

Dross . . . . 20 to 30 grammes.
Sodium carbonate . . . . . 50 ,,
Borax . . . . 10 ,,
Argol . . . . 1·5 ,,
Litharge . . . . 20 ,,
A piece of hoop-iron or nails.

The charge is fused in the ordinary way as for silver-ore assays, the resulting lead button cupelled, and the silver button cleaned and weighed. If gold is present the buttons are flattened and parted. The slags from the fusion assay of dross from rich bullion should be collected and assayed to recover the silver, as directed on page 155.

Calculation of Assay Value of Lead Bars yielding Dross.—The value in silver and gold of impure lead bullion with dross is calculated from the following data :—

Melting.		Assays of bar.	Assays of dross.
Bar weighs . . . 4·23 lbs.		Silver . 374·22 ozs.	Silver . 776·15 ozs.
Dross . . . . 1·02 ,,		Gold . . 1·14 ,,	Gold . . 2·95 ,,
Total (after melting) . 5·25 lbs.		Per ton of 2000 lbs.	Per ton of 2000 lbs.

Then :—
The Silver in the Bar is :

$$\frac{4.23}{2000} \times 374.22 = 0.791475 \text{ ounces.}$$

The Silver in the Dross is :

$$\frac{1\cdot02}{2000} \times 776\cdot15 = 0\cdot395836 \text{ ounces.}$$

The total silver in the bar and dross is $0\cdot791475 + 0\cdot395836 = 1\cdot187311$ ounces.

Hence the assay value of the silver in the sample is :

$$1\cdot187311 \times \frac{2000}{5\cdot25} = 452\cdot308 \text{ ounces per ton of 2000 lbs.}$$

The gold is calculated in the same way.
The Gold in the Bar is :

$$\frac{4\cdot23}{2000} \times 1\cdot14 = 0\cdot002411 \text{ ounces.}$$

The Gold in the Dross is :

$$\frac{1\cdot02}{2000} \times 2\cdot95 = 0\cdot001505 \text{ ounces.}$$

The total gold in the bar and dross is $0\cdot002411 + 0\cdot001505 = 0\cdot003916$ ounces. Hence the assay value of the gold in the sample is :

$$0\cdot003916 \times \frac{2000}{5\cdot25} = 1\cdot49 \text{ ounces per ton of 2000 lbs.}$$

The assay value of the sample of impure lead bullion will therefore be :

Silver   .   .   452·31 ounces per ton of 2000 lbs.
Gold   .   .   1·49   ,,   ,,   ,,

Assay of Lead Bullion from "Cyanide Process" (see page 383).

II. Base Copper Bullion.

Liquation in Copper Bullion.—During the solidification of argentiferous and auriferous copper bars, a rearrangement of the constituents takes place so that the silver and gold present in bars of copper are subject to the same irregularity of distribution as in lead.   The evidence adduced from various assays made by Mr A. R. Ledoux [1] shows that all gold- and silver-bearing copper varies in richness at different points of the bars, and that while there is no absolute rule, yet, highly refined material free from arsenic generally has a concentration of the precious metals along the centre line, the maximum being at the middle point of the bar. The distribution of silver in copper bars is shown in Table XXXVI., which gives the results of assays of samples taken from different parts of the bars.

Table XXXVI.—Assays of Argentiferous Copper Bars (Hixon). [2]

Assay No.	Part of bar sampled.	Bar No. 1. Silver in ounces per ton.	Bar No. 2. Silver in ounces per ton.
1	End	358·1	231·2
2	End	403·4	247·4
3	Top	608	451·7
4	Bottom	366·2	203·6
5	Side	393·9	252·9
6	Side	372·5	235·9
7		423·6	351·4

[1] *Journ. Canadian Min. Inst.*, 1899, ii. 108–118.   [2] Hixon, *Lead and Copper Smelting*, p. 83.

Sampling Copper Bullion.—The sampling of gold- and silver-bearing copper bars is guided by the same principles as the sampling of lead bars.

Argentiferous copper bars are usually sampled by boring, by taking dip samples, or by sawing. Ledoux[1] recommends sampling by sawing, the sample being taken from five bars laid side by side. A diagonal line is drawn across the bars, and each bar sawn through at the point where the diagonal intersects the centre line of the bar, as illustrated for lead bars in fig. 162A.

Buyers and sellers are inclined to accept the "dip" method of sampling if done at their own works. In view of the want of uniformity in these bars, Ledoux urges that both buyer and seller should agree in their contracts as to the method of sampling and assay to be employed.

A comparison between three methods of sampling copper cast into the form of anodes at the Washoe Smelter, Anaconda, has been made by Wraith[2] with special reference to the silver content. The smelter's method of sampling consists in "shotting" a small portion of the molten stream of copper, during casting, by "batting" the stream into water with a wooden paddle and selecting the portion passing a 4-mesh and remaining on a 10-mesh screen. The refinery method consists in drilling a ½-inch hole through each fourth anode, a template the size of the anode being used having ninety-nine holes arranged equidistantly over the surface, each hole being used in turn. The total drillings are mixed and ground to pass a 16-mesh screen. A ladleful of copper was also taken from the molten stream and "shotted." It was found that the smelter and refinery methods of sampling give the true silver content of the copper and check within practical limits, but ladle samples tend to give high results on account of the silver segregating towards the portion that solidifies last.

Assay of Copper Bullion.

Determination of Silver in Copper.—Silver is found in most samples of commercial copper. The bars resulting from the smelting of argentiferous copper ores contain very variable proportions of silver, containing from 100 ounces to 1000 ounces or even more per ton. Low-grade bullion contains from 10 to 50 ounces of silver per ton.

To determine the silver, duplicate samples of the borings of about 25 to 30 grammes (or one assay ton) are weighed, transferred to a beaker, and 100 c.c. of water and 50 c.c. strong nitric acid (sp. gr. 1·42) added. When the violent action has ceased, another 50 c.c. of nitric acid is added, and the solution heated, stirring from time to time, until everything soluble is dissolved, and until the oxides of nitrogen have been expelled. The removal of these is facilitated by blowing a gentle current of air through the solution. A dilute solution of sodium chloride is then added to precipitate the silver as chloride. The usual strength of salt solution used contains 0·542 gramme of sodium chloride per 1000 c.c., and 1 c.c. of this solution will precipitate 1 milligramme of silver. An excess of about 5 c.c. above that required to precipitate the silver present should be added. If the amount of silver in the copper is small (say under 50 ounces), 10 c.c. of a concentrated solution of lead acetate (or nitrate) and 2 c.c. of strong sulphuric acid are added in order to obtain a precipitate of lead sulphate to aid the settling of the finely divided silver chloride. The precipitate is allowed to settle for twelve hours, the solution is then filtered through a double Swedish paper, and the precipitate washed into the point of the filter paper. The filter is carefully dried, and when dry transferred to a scorifier containing 2 grammes of granu-

---

[1] Loc. cit.

[2] W. Wraith, *Trans. Amer. Inst. Min. Eng.*, 1910, 209–214 (abstract from *J.S.C.I.*, vol. xxix., 1910, p. 493).

lated lead. This is placed in the muffle, heated just to dull redness, and the filter papers burnt off, but only until the flame disappears, and not to a white ash. Then add another 8 grammes of lead and 0·5 gramme of borax glass and raise the muffle temperature sufficiently to render the charge molten, and then pour, as no scorification is necessary. The lead button is cleaned and cupelled at a low temperature, and the resulting silver button weighed. Some assayers carefully burn the filter paper over a small piece of lead-foil and allow the ash to fall on to the foil, which is then wrapped up and cupelled. The results should agree within about 0·25 ounce of silver per ton.

With a view to avoiding loss of gold by solution, sulphuric acid is sometimes recommended in place of nitric acid for dissolving the copper, but the action in this case is much slower, and a longer time is required to effect solution. The sulphuric acid method is stated to give results equal to those obtained by the "all fire" method described below.

Argentiferous copper is also assayed by scorification or "all fire" method, as it is sometimes termed, but this method is not so suitable, as small quantities only can be taken for assay. When this method is used, from 5 to 10 separate quantities, each of about 3 grammes (or one-tenth assay ton), are separately scorified with 15 to 20 times its weight of granulated lead and a cover of 1 gramme each of borax and sand. The lead buttons obtained are cupelled separately, and the beads of precious metal are weighed, first separately, and then together according to the richness of the bullion. With high-grade material the cupels are ground up and fused in several lots, and the silver obtained from the cupellation of the resulting lead buttons is added to that of the first assay.

Determination of Gold in Copper.—When gold is present in the bullion it is left as an insoluble residue after treatment with nitric acid, as described for the determination of silver. Since, however, copper salts in the presence of nitric acid appear to exert a solvent action on finely divided gold, it is advisable not to heat the solution too strongly when dissolving the copper. The assay is conducted as previously described for silver, and if enough silver is present to allow the button from the cupellation to be parted, it is flattened and parted in dilute nitric acid in the manner described for gold ores. When the bullion contains gold and silver, both metals are usually determined in the same quantity taken for assay. If there is insufficient silver, a small quantity of silver nitrate may be added to the solution or a little silver-foil added during the cupellation. The scorification method is used as a check upon other methods, especially for gold. The results for gold should agree very closely.

Reporting the Results.—The gold and silver content of copper bars are generally reported in ounces and decimals of an ounce per ton either of 2240 lbs. or 2000 lbs., as in the case with lead bullion.

In the case of gold, the return when below an ounce is made in pennyweight and decimals of a pennyweight per ton.

Owing to the small amount of precious metals present in copper bullion it is very necessary, in order to obtain reliable results, to use not less than one assay ton of the copper borings and to make all the final weighings on a good assay balance.

Bibliography (arranged chronologically).

The Assay of Copper Bullion, etc.

1886.   J. W. WESTMORELAND, "The Determinations and Valuations of Copper in Ores and Products for Commercial Purposes, with some remarks on the Assay of Gold in Bar Copper," *Journ. Soc. Chem. Ind.*, vol. v., 1886, pp. 48, 63, and 277.

1892.   C. WHITEHEAD, "Estimating Small Proportions of Silver and Gold in Base Metals, Mattes, etc." *Journ. Anal. and Appl. Chem.*, vol. vi., 1892, pp. 262–266.

1894.   A. R. LEDOUX, " A Uniform Method for the Assay of Copper Materials for Gold
        and Silver," *Trans. Am. Inst. Min. Eng.*, vol. xxxiii., 1904.
1898.   " Assay of Copper Bullion," *Eng. and Min. Journ.*, vol. lxv., 1898, p. 223.
1899.   T. ULKE, "Commercial Methods of Copper Assaying and Analysis," *Eng. and
        Min. Journ.*, vol. lxviii., 1899, pp. 727–729.
1899.   A. R. LEDOUX, "The Sampling of Argentiferous and Auriferous Copper,"
        Journ. of Canadian Min. Inst., 1899.
1900.   R. W. VAN LIEW, " Causes of Losses in the Determination of Gold in Copper
        Bars, and a Method for overcoming them,' *Eng. and Min. Journ.*, vol. lxix.,
        1900, pp. 469 and 498.
1902.   T. B. SWIFT, "Determination of Gold in Copper Bullion," *Eng. and Min.
        Journ.*, vol. lxxiv., 1902, p. 650.
1904.   "All-fire Method for the Assay of Gold and Silver in Blister Copper,"
        Trans. Amer. Inst. Min. Eng., vol. xxxiii., 1904, p. 670.
1905.   C. C. SAMPLE, "Determination of Silver in Blister Copper," *Eng. and Min.
        Journ.*, vol. lxxx., 1905, p. 732.
1909.   F. F. HUNT, "Determination of Gold in Copper Bullion," *Eng. and Min.
        Journ.*, vol. lxxxvii., 1909, p. 465.
1910.   W. WRAITH, " Sampling Anode Copper, with special reference to Silver Content,"
        Trans. Amer. Inst. Min. Eng., 1910, pp. 209–214.
1910.   H. NISSENSON, "Determination of Gold and Silver in Black or Unrefined
        Copper," *Chem. Zeit.*, 1910, vol. xxxiv. p. 539 (abstract, see *Journ. Soc. Chem.
        Ind.*, 1910, vol. xxix. p. 761).
1911.   E. F. KERN and A. A. HEIMROD, "Determination of Gold and Silver in Copper,"
        Met. and Chem. Eng., 1911, vol. ix. pp. 496–499.

CHAPTER XXI.

THE ASSAY OF INDUSTRIAL GOLD AND SILVER ALLOYS.

I. Industrial Gold Alloys.

For the purposes of this chapter the term "industrial alloys" includes the gold alloys of definite composition especially prepared for industrial purposes as distinct from those usually included in the term "bullion," which, as previously stated, is conveniently restricted to "the precious metals, refined or unrefined, in bars, ingots, or any other uncoined condition, whether contaminated by admixtures with base metal or not."

The alloys of more or less indefinite composition resulting from the treatment of scrap, lemel, sweep, etc.—obtained during the working of gold and silver—are also included.

Composition.—The gold alloys most frequently used for industrial purposes are those made to conform to definite standards. The standards recognised by law in England are those containing 22, 18, 15, 12, and 9 carats or parts of gold in 24 parts; or, expressed decimally, 916·6, 750, 625, 500, and 375 parts of gold in 1000 parts. The alloy 916·6 fine is used for the gold coinage, and is known as "standard gold." In Ireland 20 carat or 833·3 per 1000 is a legal standard, but is seldom used. Alloys of standards, other than the legal standards, are also in use by jewellers for special purposes. The alloys usually consist of gold-copper, gold-silver, and gold-silver-copper. Alloys of the lower standards frequently contain zinc, but the proportion is generally small, and seldom exceeds 10 per cent. Platinum is also added in some few special cases; alloys containing this metal are considered in Chapter XXV. The alloys in general use for goldsmiths' work and jewellery vary considerably with regard to the ratio of copper to silver alloyed with the gold. Examples of the actual composition of some of these are given in Table XXXVII,[1] page 323.

More examples of 9-carat gold have been given than of any other standard, as this is the one most frequently used, more than 50 per cent. of the gold wares submitted for hall-marking being of this standard.

It may be pointed out that the gold alloys supplied for the manufacture of wares to be subsequently hall-marked must obviously assay up to the legal standard, and to ensure this it is the usual practice of the makers of these alloys to add a small quantity of fine gold, usually about 2 grains per ounce of standard alloy, in excess of that required by law. The makers also usually guarantee that the quality of the metal is equal to the standard stated on the invoice, and undertake, in the event of the articles made from the metal failing to pass the official test at the Assay Office, to make an allowance to the purchaser.

Alloys of hall-marking quality are sold as such to distinguish them from gold

---

[1] Further examples of compositions are to be found in Wigley, *Art of the Goldsmith*, and Gee, *Goldsmiths' Handbook*.

used for the manufacture of articles exempted from hall-marking, which very frequently assay below the legal standard. The standard gold alloys used in other countries for coinage and for plate are given on pages 339 and 340.

With regard to gold alloys used for soldering purposes, some of those used for soldering 9-carat and 12-carat gold contain cadmium, which is added to lower the melting-point.

These alloys were introduced as the result of the new regulation relating to the use of gold solder issued by the Assay Offices in 1909, which requires that "Gold wares of the two lower standards, namely, 9 carat and 12 carat, needing solder in the making shall, as a whole, as well as in every part thereof, assay at not less than the standard declared by the sender."

Table XXXVII.—Actual Composition of Industrial Gold Alloys of Legal Standard.

	22 carat or standard of 916·6.			18 carat or standard of 750.		
Gold . . .	916·8	917·8	917·0	750·6	753·4	752·0
Silver . . .	40·7	7·2	10·3	113·4	126·6	124·4
Copper . .	42·5	75·0	72·7	136·0	120·0	123·6
	1000·0	1000·0	1000·0	1000·0	1000·0	1000·0
	15 carat or standard of 625.			12 carat or standard of 500.		
Gold . . .	626·0	631·9	703·5	500·0	500·2	501·0
Silver . . .	107·0	96·6	76·5	165·5	125·3	234·2
Copper . .	267·0	271·5	220·0	334·5	374·5	264·8
	1000·0	1000·0	1000·0	1000·0	1000·0	1000·0
	9 carat or standard of 375.					
Gold . . .	378·0	377·0	378·5	378·3	375·9	376·3
Silver . . .	110·5	200·7	262·5	217·7	104·7	310·7
Copper . .	511·5	422·3	359·0	404·0	519·4	313·0
	1000·0	1000·0	1000·0	1000·0	1000·0	1000·0

Sampling.—The method of sampling gold alloys varies according to their condition. When in the form of ingots, slabs, or bars they are sampled in the ordinary way by cutting a single sample from a corner or the middle of one side. Experience proves that in the case of alloys of high standard these samples are usually representative of the whole mass. In dealing, however, with bars of low-standard alloys, it is very advisable to take two samples by cutting diagonally opposite corners from the upper and lower sides, as alloys containing a large proportion of copper are always subject to more or less segregation during solidification, even when every precaution has been taken in melting, and they are consequently not perfectly uniform in composition. As previously stated, the researches of Peligot, Roberts-Austen, Rose, and others have shown that practically no segregation occurs in the gold-silver and gold-copper alloys rich in gold and free from all impurities; but in the case of many of the ordinary trade bars of gold alloys, more especially those of low standard, the discrepancies frequently met with between the assays of samples taken from different parts of the bars prove that they are seldom uniform in composition, although in many cases the differences in the assay results are comparatively small.

The errors due to the non-homogeneous character of the alloys are, of course, to some extent modified by rolling the bars.   Still, the general experience is that assays of samples from the middle of fillets, especially of low-standard alloys, invariably differ slightly from samples taken at the edges.   These differences in the gold-content may be considerable when proper care has not been exercised in the preparation of the alloy.   The question of segregation in ternary alloys of gold-silver-copper is of considerable interest and importance, and one upon which more experimental work remains to be done.

The bars resulting from the treatment of scrap and lemel vary considerably in fineness, and are very seldom uniform in composition, owing to the presence of impurities, and also in some cases to want of care in melting.   To ensure getting a representative sample the bar should be remelted and well stirred before casting : the resulting ingot should be sampled by drilling in at least two or three different places.

When the alloys are in the form of sheet or wire, or of manufactured articles, considerable care has to be exercised in sampling, as the surface is invariably richer in gold than the interior owing to the "pickling" or "blanching" to which the metal is subjected to remove the coating of oxide formed during the operations of annealing, soldering, etc.   The removal of this enriched or "coloured" surface before sampling is very necessary, as it is obvious that if it is included in the sample the result obtained for the gold assay will be too high and will not represent the true fineness of the alloy.

In the case of sheet metal and wire, which are readily sampled by cutting off a piece with the shears or pliers, the errors introduced in the gold assay, when the coloured surface is included in the sample, are largely dependent on the thickness of the sheet [1] or wire and the standard of the alloy.   As a general rule, the errors are smaller with the alloys of higher standard, as there is less base metal to be dissolved out.   Assays made by the author [2] show that the errors on sheet metal may vary from 0·5 to 2 or 3 parts per 1000 or more above the correct standard of the alloy.

In the case of gold wares where it is usually necessary to scrape the surface in order to obtain a sample, the removal of the "colour" becomes a matter of extreme importance : if it is not removed before sampling, the errors on the gold assay will be much greater than those given above for cutting samples of sheet metal, but the extent of the errors will necessarily be dependent on the depths to which the scraping is carried, and also the area covered in taking the sample. In sampling gold-plate and articles of jewellery the assayer is confronted with difficulties which do not arise in connection with the sampling of sheet metal and gold bullion, as the area from which a sample can be taken in the former case is frequently very limited.   When a number of small articles of the same kind have to be tested one (or more) is selected as representative of the lot and is cut up for assay.   The errors arising from the inclusion of the coloured surface in the case of scraping samples may vary from a few parts per 1000 to anything up to 100 parts per 1000, or even more in exceptional cases.

At the assay offices for hall-marking it is the usual practice to remove the "colour" from gold wares by a preliminary scraping, or by "buffing" before scraping to obtain the sample proper.

Methods of Assay.—The methods employed for the assay of gold alloys are similar in principle to those universally adopted for the assay of ordinary gold bullion, and involve cupellation to remove the oxidisable metals and inquartation with subsequent parting in nitric acid to separate the gold.

The proportion of copper, however, in industrial gold alloys is greater than

---

[1] The minimum thickness allowed by the assay offices for gold wares is 0·0076 inch (No. 36 Imperial Standard Wire Gauge).
[2] *Chemical News*, vol. xciii., 1906, p. 225.

in ordinary gold bullion, as already shown by the analyses in Table XXXVII., page 323, and on this account more lead has to be employed for cupellation. The amount of lead required is dependent on the following facts:—Copper has a greater affinity for gold than for silver, and it is necessary to use a larger proportion of lead to ensure its oxidation when combined with gold than when in combination with silver. This proportion varies according to the composition of the alloy and the temperature used for cupellation, but experience has shown that for any given standard the amount of lead necessary for gold-copper alloys is about twice that required for silver-copper alloys of the same standard.

As the result of repeated experiment it has been found that the maximum quantity of lead required to eliminate copper in the case of gold-copper alloys is about thirty-two times the weight of alloy taken, while for silver-copper alloys about sixteen times will suffice. It must, however, be pointed out that although the greater part of the copper present in a gold-copper alloy will be eliminated during cupellation by the addition of a suitable quantity of lead, the resulting gold bullion in all cases obstinately retains a small quantity of copper which cannot be entirely eliminated even by a second cupellation with a fresh quantity of lead.

The separation of copper from gold is also greatly facilitated by the presence of silver, and as this metal is usually present in industrial alloys or is added for parting purposes, it has to be taken into account when making up the cupellation charge.

From these remarks it will be evident that the quantity of cupellation lead is largely regulated by the amount of copper and of silver alloyed with the gold. If insufficient lead is used the copper will not be completely removed, and if too much lead is employed the cupellation losses will be excessive.

It is the coexistence of these two facts that make it somewhat difficult to decide upon the best quantity of lead to use, especially when the approximate composition of the alloy is unknown, and it is not surprising to find a want of agreement on this point amongst different authorities, as Table XXXVIII. shows.

For alloys in which the alloying metal is copper only, the proportions recommended by D'Arcet, Kandelhardt, and Cumenge and Fuchs respectively, are as follows [1] :—

Table XXXVIII.—*Proportions of Lead for Cupellation of Gold-Copper Alloys.*

Gold in 1000 parts.	Amount of lead employed for one part of alloy.		
	D'Arcet.[2]	Cumenge and Fuchs.[3]	Kandelhardt.[4]
1000	1	1	8
900	10	14	16
800	16	20	20
700	22	24	24
600	24	28	24
500	26	32–34	28
400	34	32–34	28
300	34	32–34	32
200	34	32–34	32
100	34	30	32
50	34	28	32
0	...	11	32

[1] Quoted from *Metallurgy of Gold*, T. K. Rose.
[2] Pelouze and Frémy, *Traité de Chimie Générale.*
[3] *Encyclopædie Chimique*, vol. iii , L'Or, p. 154.　　[4] *Gold-Probirnerfahren*, Berlin, p. 3.

Kandelhardt's table is modified to make it uniform with the others.

As Dr Rose has remarked, "It is difficult to understand the small quantity of lead assigned by Cumenge and Fuchs to the alloys containing less than 200 parts of gold per 1000."

From the compositions given in Table XXXVII. it will be seen that the majority of the industrial gold alloys are ternary alloys of gold-silver-copper, and for the cupellation of these less lead than the quantities just given is required.

As the result of experiments on the cupellation of industrial gold alloys of different compositions, Riche suggested that the most suitable proportions of lead would be the means of the quantities recommended by D'Arcet for the cupellation of gold-copper and of silver-copper alloys respectively. The quantities obtained in this way are given in Table·XXXIX., which is compiled from D'Arcet's well-known tables.

Table XXXIX.—Lead for Cupellation of Gold-Silver-Copper Alloys.
(Suggested by Riche.)

Standard of alloy.	Lead employed for 1 part of alloy.		
Gold in 1000 parts.	D'Arcet's quantities for gold-copper alloys.	D'Arcet's quantities for silver-copper alloys.	Mean quantities for gold-silver-copper alloys.
1000	1	0·3	0·65
950	...	3	3
900	10	7	8·5
800	16	10	13
700	22	12	17
600	24	14	19
500	26	16–17	21·5
400	34	16–17	25·5
300	34	16–17	25·5
200	34	16–17	25·5
100	34	16–17	25·5

These quantities, recommended by Riche, agree fairly closely with those now in general use, as will be seen from Table XL., which gives the quantities of lead for the cupellation of gold alloys of the legal standards recommended by different assayers experienced in assaying these alloys. The table is compiled from notes kindly communicated to the author at various times. Riche's quantities are inserted for comparison, and are modified to make them uniform with the others.

Table XL.—Quantities of Lead used by different Assayers for the Cupellation of Legal Standard Gold Alloys.

Standard.	Amount of lead employed for 1 part of alloy.							
Gold in parts per 1000.	Carats.	Riche.	Assayer A.	Assayer B.	Assayer C.	Assayer D.	Assayer E.	Assayer F.	Assayer G.
916·6	22	8	8·3	14·4	13·0	12	8	6	16
750·0	18	15	16·6	14·4	15·6	16	15	8	16
625·0	15	18	16·6	16·8	22·4	22	18	22	24
500·0	12	21	16·6	16·8	22·4	22	20	28	30
375·0	9	25	25·0	16·8	24·0	24	24	38	30

The author prefers the quantities given in column E, which are very similar to those recommended by Riche.

As an interesting exception to the above quantities, it may be mentioned that Mr W. F. Lowe, of the Chester Assay Office, uses the same weight of lead for alloys of all standards, viz. about twenty times the weight of the sample taken.

The method of adding the lead varies with individual assayers, but when using the larger quantities necessary for low-standard alloys, it is not usual to add all the lead in one charge. The general practice appears to be to wrap up the assay piece and the necessary parting silver in a piece of lead-foil of convenient size, and to "ball up" separately the remainder of the lead required.

In charging the assays into the furnace, the lead containing the assay piece is charged either before or after the ball of extra lead, the second lot being charged as soon as the first lot has melted.

When a furnace load consists of assays of alloys of different standards, the charging in of the different quantities of lead should be so arranged that all the cupellations will finish at the same time.

It is with this object in view that Mr Lowe uses the same weight of lead for all standards.

Well-formed, clean, and bright buttons may be obtained with less lead when the latter is added in two portions. In this case the assay piece and parting silver is usually wrapped in two-thirds of the total lead and charged in first, and, after the cupellation is finished, the remainder is added in the form of a bullet. This practice is adopted for the three lower standards by the assayer B, hence the smaller quantities of lead given in the table for these alloys.

The practice is also adopted at the Royal Mint for the assay of industrial gold alloys of low fineness. In this connection Dr Rose [1] has pointed out that, in making an assay of a gold alloy for the determination of the gold only, the elimination of the whole of the copper is not necessary. The buttons for parting must be malleable, and should be clean and bright in order to make sure that no insoluble foreign material is left adhering to them. It is, therefore, necessary to remove most of the copper by cupellation, as otherwise the buttons blacken on cooling. The exact proportion of copper which can be left in the button without disadvantage has not been determined, and is not of practical importance, as, if the button is clean, the amount of copper it contains is immaterial.

The removal of copper entails an increase in the loss of gold by cupel absorption, and therefore in making an assay for gold only it is advantageous to eliminate as little as possible of the copper.

In considering the removal of copper it is easy to exaggerate the importance of the precise amount of lead to be used. In practice any quantity less than about eight times the weight of the gold alloy is inconvenient, because of the risk of small globules being "hung up" on the cupel and separated from the remainder of the assay piece. The upper limit of the quantity of lead is also largely a matter of individual choice and convenience, influenced by the length of time in working off, the size of the cupel, etc. An increase in the amount of lead, within moderate limits, has very little effect on the loss of gold ; thus Dr Rose compared the effects of 4 parts, 8 parts, and 12 parts of lead on 22-carat gold, and found the results after parting to be identical.

According to Dr Rose, there is little advantage in increasing the lead beyond about twenty times the weight of the alloy (Mr Lowe's choice), because in a diluted mixture of lead and copper very little copper is oxidised. In fact, most of the copper, as previously pointed out on page 168, is eliminated during the last stage of cupellation, just before the lead is worked off.

[1] *Journ. Inst. of Metals*, vol. iii., 1910, p. 130.

It was due to these considerations that the method of using successive doses of lead has been developed at the Royal Mint. The quantities of lead used there are as follows :—

Standard.	Amount of lead used for 1 part of alloy.		
	Lead for 1st charge.	Lead for 2nd charge.	Total lead.
916·6	8·0	...	8·0
750·0	8·0	...	8·0
625·0	8·0	4·0	12·0
500·0	8·0	8·0	16·0
375·0	8·0	8·0	16·0

The approximate surcharges by this method are those given in Table XLII.

The quantities of lead given are said to be convenient, and to give concordant results.

The absorption of gold by the cupel is greater by this method, but it is not excessive when precautions are taken to avoid too hot a fire : the buttons are always bright, well formed, and clean.

The exact method of procedure adopted in assaying industrial gold alloys varies according to requirements. For many commercial purposes an exact determination of the gold is all that is required ; in other cases a determination of the silver is also required, or the complete composition of the alloy may be desired.

Assay for Gold only.—For the determination of gold only, from 0·25 to 0·50 gramme (500 to 1000 millièmes gold assay weights) of the alloy is weighed and cupelled, with the addition of silver equal to two and a half times the weight of the gold assumed to be present, and sufficient lead to remove the copper and zinc if present.

The resulting gold-silver button is flattened and parted, and the gold cornet weighed in the ordinary way.

For the cupellation the quantities of lead given in Table XL. are used.

As described on page 274, certain losses and gains (the sum of which is called the surcharge) take place during the cupellation and parting of gold alloys, the losses being due to (1) absorption by the cupel, (2) volatilisation, and (3) solution in the acid, and the gains to the retention of silver by the gold cornet. The amount of the surcharge varies with the conditions of working and the composition of the alloy, the general experience being that the losses under all three heads are greater as the percentage of copper in the alloy increases.

With regard to the losses in cupellation, it has been proved experimentally that the gold lost by volatilisation is inconsiderable compared with the absorption by the cupel, although it would appear that the loss by volatilisation, though small, is proportionally greater with an increase in the temperature of cupellation.

Much of the experimental work done in connection with the loss of gold in cupellation relates to gold alloyed with either silver or copper, and the results are not therefore altogether applicable to ternary alloys of gold, silver, copper, of which most of the industrial alloys consist.

A comprehensive series of experiments made by A. Riche, Directeur des Essais à la Monnaie de Paris, and recorded in his well-known book *L'Art de l'Essayeur* (Paris, 1888), appears to be the first record of experimental work relating to the assaying of industrial gold alloys. The results obtained by Riche have been confirmed more recently by the author.[1]

The loss of gold in cupellation is considerably influenced by the presence of copper, experience showing that an increase in the percentage of copper results in an increase in the loss of gold. According to Rose, the loss of gold is about 0·3 or 0·4 per 1000 when the copper in the gold does not exceed 10 per cent., and more if larger quantities of copper are present. The loss on the cupel is also dependent on the quantity of gold in the assay piece. In ordinary assays of gold alloys it is usual to take a constant weight of the alloy for all standards; hence, as pointed out by Beringer,[2] it will be obvious that as the weight of copper in the cupel charge increases, the weight of gold decreases. The silver, on the other hand, is always approximately two and a half times as much as the gold, whatever its quantity may be.

But the cupellation loss, as previously pointed out, is smaller with less gold and greater with more copper, and it so happens in assaying these alloys that under suitable conditions these two opposites may nearly neutralise one another.

In this connection some interesting results have been obtained by W. F. Lowe[3] of the Chester Assay Office.

He found that the gold recoverable from the cupels on which 20 grains of gold alloy had been assayed was almost identical, although the standards of the alloys varied from 916·6 to 375·0. The results are given in Table XLI.:—

Table XLI.—Gold absorbed by Cupel during Cupellation of Gold Assays. (W. F. Lowe.)

Standard.	Actual weight of gold in 20 grains of alloy cupelled.	Loss of gold per 1000 of actual gold present.	Weight of gold obtained from 1 cupel (= 4 assays of 5 grains each).	Absorption in part per 1000.
	grains.		grains.	
916·6	18·3	0·765	0·014	0·70
750·0	15·0	0·966	0·014 0·015	0·70 0·75
625·0	12·5	1·160	0·014 0·015	0·70 0 75
375·0	7·5	1·866	0·014 0·014	0·70 0·70

Mr Lowe remarks "that the above were all cupelled on different days, so that the temperature of the furnace would not account for this regularity; but as the amount of lead used is the same (about twenty times the weight of the sample taken) for all standards, this, most probably, is the cause of the regularity." The heavier losses shown in the third column, added by the author, are mainly due to the increase in the proportion of copper in the separate alloys.

By altering the weight of lead, or in any other manner modifying the working conditions, other figures would of course be obtained.

In practice it is more usual, as already explained, to vary the lead according to the fineness of the samples, and under these conditions the general experience appears to be that the loss due to absorption into the cupel is greater with alloys of low standard than with those of high standard.

[1] "The Assay of Industrial Gold Alloys," *Journ. Inst. of Metals*, vol. iii., 1910, pp. 98–137.
[2] *A Text-Book of Assaying*, 10th edit., p. 145.
[3] "Assaying and Hall-marking at the Chester Assay Office," *Journ. Soc. Chem. Ind.*, Sept. 1889.

In addition to the cupel losses there is the loss of gold by solution in acid during parting, but experience proves that this loss is comparatively small.

The total losses may or may not be counterbalanced by the silver retained by the gold after parting, which amounts to about 1 part per thousand under normal conditions of working. In practice it is found that with alloys rich in gold the silver retained by the cornet more than compensates for the various losses, and the surcharge is positive; but with alloys of low standard the losses are greater and the surcharge is negative, the cupellation loss being usually greater on account of the larger proportion of copper present, as previously pointed out.

Examples of the surcharge obtained in the assay of gold alloys of various standards by different assayers are given in the following Table XLII. The figures represent only the relative surcharges; the absolute amounts vary with the treatment, but when a very systematic and uniform method of working is adopted the variations are very slight.

Table XLII.—Surcharge per 1000 on Assays of Gold Alloys obtained by Different Assayers.

Standard.	Pelouze and Frémy.[1]	Rose.[2]	Lowe.[3]	Smith.[4]
916·6	...	+0·6	+0·26	+0·1 to +0·35
900·0	+0·25			
800·0	+0·50			
750·0	...	+0·5	0·0 to −0·13	0·0 to +0·4
700·0	0·00			
666·6				
625·0	...	+0·4	0·0 to −0·13	0·0 to +0·2
600·0	0·00			
546·0				
500·0	−0·50	0·0	...	−0·1 to −0·5
400·0	−0·50			
375·0	...	−0·6	−0·52 to −1·05	−0·2 to −0·6
333·3				
300·0	−0·50			
200·0	−0·50			
100·0	−0·50			

The necessity of using checks and of working all assays under exactly the same conditions in all cases where results of extreme accuracy are desired, is well known to all bullion assayers, and cannot be too strongly emphasised in dealing with the assay of gold alloys. The more closely a routine is adhered to, the more uniform will the surcharge remain from day to day.

To make the assays absolutely comparable, it is essential, especially when the alloys contain much copper, that the composition of the check should approximately correspond to the composition of the alloy being assayed.

When the approximate composition of an alloy is not known, a preliminary assay must be made to determine this, as described on page 335. It will be seen, however, from Table XXXVII., given on page 323, that there is considerable variation in the composition of the alloys of low standard; and when a large

[1] "Experiments on Synthetic Alloys of Gold and Copper," Traité de Chimie, 3rd ed., vol. iii. p. 1230.
[2] Compiled from results obtained at Royal Mint, Journ. Inst. of Metals, vol. iii., 1910, p. 131.
[3] Results obtained at Chester Assay Office, Journ. Inst. of Metals, vol. iii., 1910, p. 128.
[4] Compiled from a number of results obtained by the author.

number of different alloys of any particular standard have to be assayed it would be exceedingly laborious to have to make up checks to correspond in composition to each alloy. It is the usual practice, therefore, in such cases to make up the checks to conform to an alloy of average composition.

When assaying gold alloys of the legal standards, the author uses checks which correspond in composition to the following representative alloys:—

Table XLIII.—Composition of Representative Gold Alloys used for Checks.

	9 carat or standard of 375.	12 carat or standard of 500.	15 carat or standard of 625.	18 carat or standard of 750.	22 carat or standard of 916·6.
Gold	375·0	500·0	625·0	750·0	916·6
Silver . . .	156·5	166·0	94·0	125·0	41·4
Copper . . .	468·5	334·0	281·0	125·0	42·0
	1000·0	1000·0	1000·0	1000·0	1000·0
Ratio between copper and silver . }	Cu 3 Ag 1	Cu 3 Ag 1	Cu 3 Ag 1	Cu 1 Ag 1	Cu 1 Ag 1

In each case silver is added to make the total weight of silver equal to two and a half times that of the gold.

In practice it is usual (except in the case of alloys containing a large proportion of silver) to disregard any silver that may be present in the alloy being assayed, and to add the same total weight of silver to both checks and samples.

In cases where only copper is present as an alloying metal, and extreme precision is not required, cupellation is alone necessary to determine the gold, and the quantities of lead given in Table XXXVIII. are used for the purpose. The principal difficulty which the direct cupellation of such an alloy presents, consists in the fact that some gold is absorbed by the cupel when the temperature is high, and that it is impossible, as stated above, to remove the copper and lead when the temperature is low.

Assay for Gold and Silver (see also page 279).—Silver in gold alloys is usually determined by difference, the gold and silver being weighed together after cupellation, and the gold determined separately in the same or another portion of alloy.

When one assay piece only is used for the determination of both gold and silver, from 0·2 to 0·5 gramme of the alloy is cupelled with enough lead to remove all the copper, but *without* the addition of silver for parting purposes. In this respect it differs from the method just described, in which the copper is eliminated and the silver at the same time incorporated with the gold by a single cupellation with lead.

The gold-silver button resulting from the cupellation is carefully weighed and then subjected to inquartation and parting in the ordinary way. By subtracting the weight of the gold cornet from the weight of the gold-silver button, the proportion of silver is at once ascertained.

To obtain reliable results by this method considerable practice is required, and special attention must be given to the proportion of lead employed and to the temperature of cupellation. These must necessarily to a large extent be left to the judgment and experience of the assayer, based on a preliminary examination of the alloy.

The quantities of lead given in Table XL. are representative of those in general use for the cupellation designed for the removal of the copper, but some assayers prefer to use the quantities given for gold-copper alloys in Table XXXVIII., or even larger quantities. Experience has shown, however, that the copper can be more completely separated from gold alloys and less silver lost by cupelling at a slightly higher temperature with a small quantity of lead than by employing more lead and working at a lower temperature. As the loss in cupellation is largely dependent on the amount of gold and copper present, the temperature must be varied in accordance with the composition of the alloy in order that the loss may be minimised as much as possible.

In cases where the gold predominates, the finish of the cupellation should be effected at a higher temperature than in the case where the quantity of gold is small. For alloys of low standard the temperature at the beginning should be very little above that ordinarily employed for silver assays, and should be increased gradually until at the end it approaches more to that employed for the cupellation of gold. Assayers who use the larger quantities of lead invariably cupel at the lowest possible temperature, i.e. with formation of feathers. Gas muffles are undoubtedly the best form of furnace to use for these assays, as the temperature is far more readily controlled in these than in coke-fired furnaces.

The effect of varying quantities of lead in cupellation has been ascertained by Riche in the series of experiments already referred to. The following table gives the results of his experiments on 0·5-gramme synthetic alloys of gold-silver-copper cupelled in the quantities of lead recommended by D'Arcet for gold-copper and gold-silver alloys respectively (see Table XXXIX.). The results have been tabulated by the author for convenience of reference :—

Table XLIV.—Surcharge on Cupellation of Gold-Silver-Copper Alloys in varying Lead. (Riche.)

Composition in parts per 1000.			Amount of lead used for 1 part of alloy.	Cupellation surcharge. Parts per 1000.	Amount of lead used for 1 part of alloy.	Cupellation surcharge. Parts per 1000.
Gold.	Silver.	Copper.				
900	50	50	10	} +5·0 to +8·0	7	} +12·0 to +15·0
900	25	75	10		7	
800	150	50	16	No surcharge	10	
800	100	100	16	} Slight surcharge	10	} +8·0 to +10·0
800	50	150	16		10	
700	200	100	22	Very slight surcharge	12	
700	100	200	22		12	} +5·0 to 8·0
700	50	250	22		12	
600	300	100	24		14	No surcharge
600	200	200	24		14	
600	100	300	24	-2·0 to -5·0	14	} +4·0 to +9·0
600	50	350	24		14	
500	400	100	26		16–17	Assays approach exactness perhaps a little high
500	300	200	26		16–17	
500	200	300	26	-3·0 to -10·0	16–17	
500	100	400	26		16–17	
500	50	450	26		16–17	
400	600		34	...	16–17	Exact
300 and less	700		34	Exceeds -10·0	16–17	Exceeds -10·0

The author has repeated many of Riche's experiments, and confirmed his conclusions that the surcharge resulting from the cupellation may vary considerably according to the amount of lead employed and the temperature of cupellation : the latter is the more important factor. The "surcharge" in this case refers to cupellation losses or gains only, and does not include "parting."

The method previously mentioned of using less lead for the cupellation and of adding it in two separate quantities (viz. adding one-third after the other two-thirds have worked off) may also be adopted with advantage for direct cupellation without parting silver.

The resulting gold-silver buttons are always much brighter and more satisfactory, and, provided the quantity of lead and the cupellation temperature are carefully regulated, the loss of gold and silver on the cupel, although generally a little greater than when the lead is added in one charge, is not excessive.

It would appear from tests made by the author that when a suitable quantity of lead is employed the copper is more satisfactorily eliminated with two successive charges than with one charge.

With a little experience the best quantity of lead for two charges for any particular alloy is readily ascertained, and undue cupellation loss avoided. The temperature of cupellation should be kept low until the first charge has worked off, and then raised during the working off of the second charge.

To ascertain the percentage of silver in the alloy it is necessary to determine the gold, and, if this has not been done in a separate assay piece, the gold-silver button from the cupellation is carefully weighed, inquarted with silver, and parted.

The amount of silver to be added for inquartation can be ascertained from the preliminary assay, or by the touchstone, but an experienced assayer can frequently judge the quantity required with sufficient accuracy from the colour of the gold-silver button.

Buttons containing from 20 to 40 per cent. of silver have a greenish-yellow tint, but, when the silver amounts to 50 per cent., the yellow tint is scarcely perceptible. With more than 50 per cent. of silver the buttons are silver-white.

The inquartation is best effected by cupellation with the smallest convenient quantity of lead. The practice of alloying the silver by means of the blowpipe is to be deprecated.

The results obtained for gold by this method are approximately accurate when due care is exercised, but they are invariably lower than those obtained for the gold when determined in a separate assay piece by direct cupellation with parting silver, as previously described.

The difference is, no doubt, partly accounted for by the fact that, in the latter case, the presence of the parting silver facilitates the elimination of the copper, and also helps to minimise the loss of gold.

The actual differences between the two methods obtained with gold alloys of different standards is largely dependent on the amount of copper present. The author found the losses by the double cupellation method exceed the losses by direct cupellation by about – 2 parts per 1000 for alloys 375 fine, and about – 0·5 per 1000 for alloys 917 fine.

In the hands of an inexperienced assayer, the results obtained for the determination of the gold and silver by the double cupellation method may be far from accurate, and, since the losses and gains previously detailed are dependent on so many conditions, it is always necessary, in all cases where extreme accuracy is desired, to use checks. This method of double cupellation, or "parting assay," in which the gold and silver are determined in the same assay piece, is largely used by trade assayers for the valuation of lemel bars, etc., and, in the hands of experienced assayers who have ascertained the best working conditions for the particular class of alloys with which they have to deal, the results are sufficiently accurate for commercial purposes.

It is not the usual practice of trade assayers to use checks except in special cases, but it may be remarked that the increasing demand for greater accuracy in trade assays is causing checks to be more frequently used than was formerly the case. The results recorded above show to what extent the method is reliable when checks are not used.

In making assays by this method, it is well to remember that it is the weight of the gold which is the more important, because of its value, and the working conditions should be such that no undue loss of gold is occasioned.

As uniformity of temperature is very necessary with this method, a few assays only should be cupelled at the same time in order to avoid extreme variations of temperature, and the batch of cupels should be surrounded by empty cupels.

When the exact determination of the gold and silver in an alloy is required, the author prefers to determine the gold by direct cupellation with parting silver, and to determine the silver in a separate assay piece by direct cupellation with lead only, checks being used in both cases.

Another method sometimes recommended for the determination of silver in gold alloys, consists in cupelling the alloy with lead and the addition of parting silver, as already described, and then deducting the weight of the parting silver from the weight of the gold-silver button resulting from the cupellation.

In this case the silver added for parting must be accurately weighed, but, unless checks are worked off at the same time as the alloys, the results are of little value, except, perhaps, as a guide to the approximate composition of the alloy.

If checks are not used, the whole of the cupellation loss will be thrown on to the silver in the alloy. This will be evident from the following figures, which give the results of an assay on a 0·5-gramme sample of gold bar from clean lemel :—

Weight of gold and silver button from cupellation .	0·7660 gramme.
Less parting silver added .	0·4750
Gold and silver in alloy .	0·2910
Gold after parting .	0·1896
Difference = silver in alloy .	0·1014 = 202·8 per 1000

The total loss on the check during cupellation was ·0065 gramme, and, if this loss is added to the above result, it will make the silver in the alloy 0·1079 gramme or 215·8 part per 1000, instead of 0·1014 gramme or 202·8 parts per 1000.

Experiments conducted by the author show the average loss in the cupellation of alloys of different standards with parting silver direct to be as follows :—

Table XLV.—*Relative Losses on Gold Alloys cupelled direct, with the Addition of Parting Silver.*

Standard of alloy.	Loss in parts per 1000.
916·6	26·1
750·0	23·0
625·0	22·5
500·0	20·3
375·0	19·5

These figures represent the relative losses only; the actual losses will vary with the composition of the alloys, the proportion of lead, and temperature of cupellation.

Assay to determine Composition of Gold-Silver-Copper Alloy.—When the composition of an alloy is desired, the gold and silver are determined by the methods described, and the copper determined by difference. The following is an example. An ingot resulting from the melting of scrap gave—

	Parts per 1000.	Composition.	
Gold and silver .	683·8	Gold . . .	503·7
Gold	503·7	Silver . . .	180·1
		Copper (by difference) .	316·2
Difference = silver .	180·1		1000·0

$1000 - 683·8 = 316·2$ copper.

If zinc enters into the composition of an alloy, it will be detected on the cupel as a scoria during the first stages of the cupellation. When it is required to know the amount present, the alloy must be treated with aqua-regia and the zinc determined by any of the well-known methods.

Preliminary Assay.—When the composition of an alloy is quite unknown, a preliminary assay must be made to ascertain this approximately in order to determine the correct quantities of silver, copper, and lead to be added.

It very frequently happens, however, that the standard of the alloy is stated when the sample is sent for assay, and even when this is not the case a knowledge of the source of the metal will often be a sufficient guide for adjusting the lead, etc.

An experienced assayer will frequently be able to decide the approximate composition from the hardness and appearance of the metal and the effect of the action of nitric acid or other "test" acid. The colour alone is not a sufficient indication of the standard, as alloys of low standard are made to resemble alloys of higher standard by the careful regulation of the proportion of silver and of copper added to the gold. The "test" acid used by the author is that recommended by Wigley,[1] and consists of nitric acid, 8 parts; hydrochloric acid, 1 part; and water, 6 parts.

After considerable experience it is possible to determine, with a fair degree of accuracy, the standard of any given alloy, but it may be remarked that the action of the acid is not quite the same in every case with alloys of the same standard but with different ratios of copper and silver. For example, an alloy 375 standard containing a large percentage of copper is more readily acted upon than an alloy of the same standard in which the alloying metal is mainly silver. If these tests are not sufficient to indicate the quality of the gold, this may be determined by the touchstone or by cupellation. In the latter case about 0·25 gramme of the alloy is cupelled with three times its weight of silver and from 15 to 30 times its weight of lead, and the resulting button inquarted with silver and parted. Should the cupellation button be unsatisfactory, it must be recupelled with more lead.

It is the practice of some assayers to make the preliminary assay serve the purpose of an assay for the determination of the silver, and then to make a separate assay for gold based on the information obtained.

Reporting.—The results of assays of gold alloys are reported either decimally in the ordinary way or by the "carat" system, according to requirements. The former method is now being more generally adopted, but the older "carat" system is still used, as the generality of workers and dealers in gold and silver

[1] Wigley, *Art of Goldsmith and Jeweller.*

alloys are more conversant with this system of computation than with the decimal equivalent; it is, therefore, the usual practice of trade assayers in reporting the proportion of gold in jewellery alloys to refer always to the legal gold standard of 22 carat (916·6). These reports are readily understood when it is remembered that "standard gold" (916·6) contains in 24 parts, 22 of gold and 2 of other metal; these parts, as is well known, are called carats, the gold carat containing 4 grains and the grain being subdivided into 8 parts. In reporting how much gold in carats, grains, and eights is present in an alloy above or below the legal "standard," the conventional terms "betterness" and "worseness" are used, the metal being reported as Bʳ (abbreviation for better) or as Wᵒ (abbreviation for worse).

For example, a sample of 18-carat gold would be reported as 4 carats worse, the report reading thus :—

Gold Report.

						Carats.	Grains.
Wᵒ	.	.	.	.	.	4	...
Standard	.	.	.	22 carats.			

"Parting" assays, in which both the gold and silver are determined, are reported in ounces of fine gold and fine silver in one pound troy, the abbreviations F.G (fine gold) and F.S (fine silver) being very frequently employed thus :—

Parting Assay Report.

							Ozs.	Dwt.	Grs.
F.G	.	.	.	.	.	.	8	10	6
F.S	.	.	.	.	.	.	2	7	0
			In pound troy.						

The remaining 1 oz. 2 dwt. 18 grs. would be base metal, probably copper. The result is also sometimes reported in pennyweight per ounce.

The less cumbrous method of showing how many parts of gold or silver are contained in 1000 parts of the alloy is now universally adopted for all bullion assays, and will no doubt in time replace the above methods for jewellers' and goldsmiths' reports.

Remedy of Fineness.—It is not possible to make all standard alloys of exactly the legal standard, and the maximum difference or variation in fineness above or below the legal standard allowed by law is called the remedy.

In England the remedy allowed in the case of the gold coinage (916·6 fine) is 2 parts per mille [1] (i.e. 2 parts per 1000), and for silver coinage (925 fine) 4 per mille.

The remedy of fineness allowed in the countries which have adopted the standard 900 fine for both the gold and the silver coinage, is in most cases 1 part per mille, which is only one-half of that allowed on the alloys of higher fineness used in England.

II. Industrial Silver Alloys.

Composition.—The silver alloy known as "standard" or "sterling" silver is by far the most important of the industrial alloys of silver. It consists of 92·5 per cent. of silver and 7·5 per cent. of base metal, generally copper, and is the alloy fixed by Act of Parliament as the legal standard for the silver coinage and also for the manufacture of silver wares.

Another legal standard silver alloy for silver wares also exists, but is seldom used, as it is softer than sterling silver, less serviceable, and not so durable.

[1] *Thirty-eighth Mint Report*, 1907, p. 49.

This alloy of higher standard contains 95·84 per cent. of silver, and is known as the "Britannia" standard, owing to the fact that silver wares of this standard must be hall-marked with the "figure of a woman commonly called Britannia" instead of the lion, as in the case of sterling silver.[1]

The silver standards employed for coinage and for plate in other countries are also fixed by law, and are given in the tables on pages 339–341. It will be seen from the tables that the alloy 900 fine is more universally adopted, especially for coinage, than any other silver alloy.

As stated above, the silver is usually alloyed with copper, but "standard silver" alloys are also in use in which the silver is alloyed with zinc, cadmium, and tin: nickel in small quantity is also sometimes present. Silver alloys, 900 fine, and also of lower fineness down to 600 fine, are in use for the manufacture of jewellery and articles which are exempt from hall-marking.

Industrial silver alloys used as solders vary considerably in composition: they consist usually of alloys of silver, copper, and zinc, but tin and arsenic are also present in some solders. The following are given as examples of silver solders used by silversmiths[2]:—

	Silver.	Copper.	Zinc.	Tin.
	per cent.	per cent.	per cent.	per cent.
Hard solder	80	13·2	6·8	
Medium solder . . .	66·7	23·3	10·0	
Ordinary solder for plate . .	64·5	22·5	13·0	
Quick-running solder . . .	62·5	20·87	10·38	6·25
Common quick solder . . .	57·0	27·7	11·53	3·85

Sampling.—The remarks on the sampling of industrial gold alloys apply equally to industrial silver alloys. As previously stated, more or less liquation always takes place during the solidification of silver-copper alloys, so that it is not possible to obtain ingots absolutely uniform in composition. In preparing standard silver which is to be subsequently manufactured into plate, and hall-marked, it is the usual practice of manufacturers to add a small quantity of pure or virgin silver (usually about half a pennyweight to the troy pound) above that required by the legal standard, in order to counteract any irregularities in composition due to liquation, and also because commercial fine, or virgin silver is never perfectly pure. The variations in the silver content due to the non-homogeneous character of the alloy are to some extent modified by rolling the bars, but experience shows that the middle of a fillet or strip of standard silver (925 fine) often differs by at least 0·25 parts per 1000 from the edges. The actual differences are dependent on the width of the strip; but from the results of assays made by the author from time to time, it would appear that in the case of the fillets about 18 inches wide, most generally sold for silversmiths' work, the middle differs from the edges by, on an average, about 0·8 to 1·0 part per 1000.

In sampling silver wares by scraping, the first scrapings should be rejected, as a certain amount of enrichment may have taken place, due to the "pickling" in dilute sulphuric acid (or "boil") to which the articles are subjected in the process of manufacture.[3] In the case of finished wares, the surface may consist

[1] Britannia standard silver must be carefully distinguished from the alloy known as "Britannia metal"—an alloy of tin, antimony, and copper.
[2] E. A. Smith, "Silver Alloys of Industrial Importance," *Proc. Sheff. Soc. of Eng. and Metallurgists*, Nov. 1901.
[3] Consult M. Lire, *Ann. Chim.*, iv., tome ii. p. 131 (quoted in *Mint Report*, 1875).

of pure silver, as it is a common practice to "finish" silver wares by depositing a thin layer of silver by electro-deposition, as in electro-plating.

Preliminary Tests for Silver Alloys.—To determine whether an alloy, having the appearance of silver, consists of silver or of some white base-metal alloy, a dilute solution of silver nitrate (or sulphate) is invariably employed. The "test acid" is prepared by dissolving 2 grammes of silver nitrate in 30 c.c. of distilled water, or by dissolving 1·27 grammes of pure silver in a small quantity of dilute nitric acid, expelling the excess of acid by evaporation, and diluting to the required volume. A drop of this dilute silver nitrate has no effect on silver alloys above 925 fine, but with alloys below this fineness a brown stain is produced, which increases in intensity as the proportion of base metal in the alloy becomes greater. With base metals, the silver nitrate instantly produces a black stain, due to the precipitation of finely divided metallic silver by the base metal. The surface of the alloy should be cleaned by scraping before applying the test, so as to remove any silver that may have been electro-deposited upon it.

Methods of testing silver alloys by means of the touchstone and by oxidation are described on page 308.

The oxidation test is very useful in locating the "seams" or joints in silver wares, since the solder used is almost always of lower fineness than the article. For this purpose the article is gently warmed over a bunsen flame in the place where the seam is suspected.

Methods of Assay.—Silver alloys are universally assayed by the methods previously described in Chapter XIX.

The dry or cupellation method is still largely employed by trade assayers, but is gradually being replaced by the wet methods. For the cupellation assay 0·5 gramme of alloy is usually taken and cupelled with the quantities of lead given in Table XXXII., page 290. The loss of silver in cupellation varies with the amount of copper present in the alloy and with the temperature employed.

Table XLVI. gives the average loss of silver which occurs in the cupellation of alloys of silver and copper in different proportions. In all cases where great accuracy is desired, check assays must be used to determine the actual losses, and the assays corrected accordingly.

Table XLVI.—Average Cupellation Loss on Silver-Copper Alloys.

Fineness of alloy.	Silver obtained.	Loss of silver.
parts per 1000.	parts per 1000.	parts per 1000.
1000	998·97	1·03
975	973·24	1·76
950	947·50	2·50
925	921·75	3·25
900	896·00	4·00
875	870·93	4·07
850	845·85	4·15
825	820·78	4·22
800	795·70	4·30

Standard silver in which the silver is alloyed with zinc, cadmium, or tin may also be assayed by cupellation without increasing the quantity of lead; but to obtain satisfactory results a good supply of air must be allowed to pass through the muffle, and the temperature must not be allowed to get too low. With these precautions the whole of the scoria first formed is entirely removed before the completion of the cupellation and the buttons obtained are quite satisfactory.

The wet assay of industrial silver alloys is conducted as described for silver bullion.   When the composition of the alloy is entirely unknown, a preliminary cupellation assay is made to determine the approximate fineness.

Reporting.—The results of assays of silver alloys are usually reported in parts per 1000 to the nearest 0·5 per 1000, as in the case of silver bullion; but the old method of reporting trade assays of silver alloys as better or worse than the legal silver standard of 11 ounces 2 dwt. of fine silver per pound troy (*i.e.* 925 fine) is still in use.   Thus an alloy 926 fine, which is ¼ dwt. above standard, would be reported ¼ dwt. better; and an alloy 900 fine, which is 6 dwt. below standard, would be reported 6 dwt. worse.

Test for Cadmium, etc., in Silver Alloys.—The presence of cadmium in silver alloys may be tested by dissolving the alloy in dilute nitric acid, adding hydrochloric acid to precipitate the silver, and filtering off the silver chloride.   An excess of ammonium hydrate is then added to the filtrate: a deep blue coloration denotes the presence of *copper*.   Add to the blue solution an excess of potassium cyanide solution, until colourless, and pass a current of sulphuretted hydrogen. A bright yellow precipitate denotes the presence of *cadmium.*

To test for zinc, another portion of the hydrochloric acid solution from the filtration of the silver chloride precipitate is treated with sulphuretted hydrogen and the precipitated sulphides filtered off.   Ammonium hydrate is then added to the clear filtrate and a white precipitate denotes the presence of zinc.

Standard Alloys for Gold and Silver Coin.[1]—The following table gives the legal standards now in force for gold and silver coin in the principal countries of the world.   The standards given are in parts of precious metal per 1000, the remainder being copper.

	Gold coin.	Silver coin.
Abyssinia	...	835
Argentine	900	900
Austria-Hungary . . .	900	900 and 835
Belgium	900	900 and 835
Bolivia	...	900
Brazil	916·6	916·6
Bulgaria	900	900 and 835
Canada	...	925
Ceylon	...	800
Chile	916·6	500
China	...	900, 866, and 820
Colombia	900	900 and 835
Congo	900	900 and 835
Corea	900	800
Costa Rica	900	900
Crete	900	900 and 835
Curaçao	...	640
Cyprus	...	925
Denmark	900	800, 600, and 400
Dominica	...	900 and 835
Dutch East Indies . . .	...	720
Ecuador	900	900
Egypt	875	833·3
Finland	900	868 and 750
France	900	900 and 835
Germany	900	900

[1] T. K. Rose, *Precious Metals*, pp. 239, 240.

	Gold coin.	Silver coin.
Great Britain . . .	916·6	925
Greece . . .	900	900 and 835
Guatemala . . .	900	900 and 835
Hayti . . .	900	900 and 835
Holland . . .	900	945 and 640
Honduras (British) . .	...	925
Honduras . . .	...	900
Hong Kong . . .	...	800
India . . .	916·6	916·6
Italy . . .	900	900 and 835
Japan . . .	900	800
Mauritius . . .	...	800
Mexico . . .	...	902·7 and 800
Morocco . . .	...	900 and 835
Newfoundland . .	916·6	925
Nicaragua . . .	...	800
Norway . . .	900	800, 600, and 400
Panama . . .	900	900
Paraguay . . .	...	900
Persia . . .	900	900
Peru . . .	916·6	900
Portugal . . .	916·6	916·6
Roumania . . .	900	900 and 835
Russia . . .	900	900 and 500
Salvador . . .	900	900 and 835
Servia . . .	900	900 and 835
Siam . . .	...	900
South Africa . .	916·6	925
Spain . . .	900	900 and 835
Sweden . . .	900	800, 600, and 400
Straits Settlements . .	...	900 and 800
Switzerland . . .	900	900 and 835
Turkey . . .	916·6	830
United States . .	900	900
Uruguay . . .	...	900
Venezuela . . .	900	900 and 835

Standard Alloys for Gold Wares.—The following table gives the legal standards for gold plate in use in some of the principal countries of the world :—

Austria-Hungary . .	920, 840, 750, 580
Belgium . . .	800, 750.
Denmark . . .	585 and upwards
France . . .	920, 840, 750, 583
Great Britain . .	916·6, 750, 625, 500, 375
Germany . . .	583 and upwards
Italy . . .	900, 750, 500
Netherlands . .	916, 833, 750, 583
Portugal . .	916·6, 750
Russia (see note) .	979·1, 958·3, 854·1, 750, 583·3
Spain . . .	916·6, 750
Sweden and Norway .	975·7, 847·2, 763·9
Switzerland . .	750, 583
United States . .	750, 583·3 (not legalised, see page 343)

Note.—In Russia the legal standards of gold and silver plate are stated in the number of zolotnics in one pound of 96 zolotnics, the gold standards being 94, 92, 82, 72, and 56, and the silver standards 95, 91, 88, and 84. These figures are stamped on gold and silver wares marked at the assay offices in Russia.[1] In the accompanying tables the Russian standards are expressed in parts per 1000 to make them uniform with the others.

Standard Alloys for Silver Wares.—The following table gives the legal standards for silver plate in use in some of the principal countries of the world :—

Austria-Hungary	950, 900, 800 and 750
Belgium	900 and 800
Denmark	826
France	950 and 800
Germany	800 for plate. Any degree of fineness for jewellery, etc.
Great Britain	958·4, 925
Netherlands	934 and 833
Italy	950, 900, and 800
Norway and Sweden	828·1 and 812·5
Portugal	916·6 and 833. Various standards down to 800 for export
Russia	947·9, 916·6, 875
Spain	916·6 and 750
Switzerland	875 and 800
Turkey	900
United States	925 (not legalised)

Assay Offices for Hall-marking.[2]—British law requires that the quality of the metal of all gold and silver wares, with certain exceptions, including jewellery, shall be determined by assay at offices or "halls" in various parts of the kingdom duly authorised for that purpose; and that if the article be found equal to the required standard, it shall be stamped at those offices with a series of marks known as the "hall-mark."

The earliest of the hall-marking laws was enacted nearly six hundred years ago in the reign of Edward I.

The standards required by law for gold and silver wares in Great Britain are given in the tables above. The wares sent for hall-marking are carefully sampled by scraping or cutting so as to obtain a representative portion of the metal of which they are made, and the samples are subsequently assayed. If the ware, after testing, is found to comply with one of the various recognised standards, the marks prescribed by statute for that respective standard are then struck upon it. In the event of any ware failing to pass the official test, a fine is imposed, and the article is broken up and returned to the manufacturer. A description of the methods of assay for gold and silver alloys used for wares has already been given.

Every ware submitted for assay must bear the mark of the manufacturer before it can be accepted for hall-marking. This mark consists of the maker's initials or, in the case of a company, the initials of the name of the firm.

The maker's mark must be approved by the assay authorities and duly registered before being used.

---

[1] Consult *Monnaie Medailles et Bijoux*, A. Riche, Paris, 1889, p. 356.
[2] For information on hall-marks, consult C. J. Jackson, *English Goldsmiths and their Marks*, 1905.

The hall-marks struck upon the wares are usually placed quite close to the maker's mark and are designed to show :

1. The hall or assay office at which the ware was tested.
2. The standard or quality of the gold or silver as the case may be.
3. The year in which the ware was hall-marked.

Formerly another mark, known as the duty mark, and consisting of the sovereign's head, was struck upon the wares to denote the payment of duty, but this was discontinued in 1890.

Offices for hall-marking are at London (Goldsmiths' Hall), Birmingham, Chester, Sheffield, Dublin, Edinburgh, and Glasgow. Offices formerly existing at Exeter, York, Newcastle, Norwich, and other places have now been closed.

The marks to denote the place of assay are as follows :—

City Marks.

London	A leopard's head (the arms of the Goldsmiths' Company).
„	A lion's head erased (*i.e.* without the body), for silver wares of 958·4 standard.
Birmingham	An anchor.
Chester	A sword between three garbs.
Sheffield	A crown for silver wares.
„	A York rose for gold wares.
Dublin	A harp crowned.
Edinburgh	A castle.
Glasgow	A tree with a fish across the trunk, a bell hanging from one of the branches, and a bird on the top branch.

The Standard or quality is denoted by the following marks :—

Standard Marks for Silver.

Standards.	Marks.
925·0 (sterling or standard silver)	A lion passant.
958·4 (Britannia silver)	A figure of Britannia.

Standard Marks for Gold.

Standards.	Marks.
916·6 or 22 carat	A crown and figures 22.
750·0 or 18 „	A crown and figures 18.
625·0 or 15 „	Figures 15 and decimal figures ·625
500·0 or 12 „	„ 12 „ „ ·5
375·0 or 9 „	„ 9 „ „ ·375.

The Date Mark to denote the year of assay consists of a single letter of special design, and different for each assay office, which is used throughout the year, and is changed every year.

A complete hall-mark on any single piece of plate will therefore consist of four separate marks, viz. : (1) maker's mark ; (2) city mark ; (3) standard mark ; and (4) date letter.

Foreign silver or gold wares imported for sale must first be submitted for assay, and if of the required standard, they are stamped with distinctive marks, to denote that they are of foreign manufacture.

The city marks for foreign plate are as follows [1] :—

London	Sign of constellation Leo.
Birmingham	Equilateral triangle.
Chester	Acorn and two leaves.
Sheffield	Libra.
Dublin	Boujet.
Edinburgh	St Andrew's Cross.
Glasgow	Double block letter F inverted.

Standard Marks for Foreign Silver.

Standards.	Marks.
925 (sterling or standard silver) . .	Decimal figures ·925.
958·4 (Britannia standard) . . .	Decimal figures ·9584.

Standard Marks for Foreign Gold.

Standards.	Marks.
916·6 or 22 carat . .	Figures 22 and decimal figures ·916.
750·0 or 18 „ . .	„ 18 „ „ ·75.
625·0 or 15 „ . .	„ 15 „ „ ·625.
500·0 or 12 „ . .	„ 12 „ „ ·5.
375·0 or 9 „ . .	„ 9 „ „ ·375.

The date mark stamped on foreign gold and silver wares is the same as that used for wares manufactured in the United Kingdom.

Hall-marking offices are provided by most of the Governments of the civilised world, the most noteworthy exception being that of the United States. American silver wares of sterling fineness (925 fine) are usually stamped STERLING by the manufacturer under his own guarantee and name. The most prominent American goldsmiths stamp their name and the figures denoting the carat quality on gold wares made of 18- and 14-carat gold, which are the standard qualities most frequently used. The legal standards of gold plate in use in some of the principal countries of the world are given on page 338, and of silver plate on page 339.

---

[1] Hall-marking of foreign plate, Order in Council, 1906. See *London Gazette*, May 15, 1906.

CHAPTER XXII.

ASSAY OF AURIFEROUS AND ARGENTIFEROUS METALLURGICAL PRODUCTS.

In this chapter are included descriptions of the methods employed for the determination of gold and silver in some of the more important products and bye-products obtained in the metallurgical treatment of the precious metals. They are conveniently arranged in alphabetical order under three heads, viz. :— gold and silver in (1) base metals ; (2) metallurgical products other than metals ; (3) miscellaneous products.

I. Gold and Silver in Base Metals.

The chief base metals assayed for gold and silver are lead and copper, constituting base bullion, which has been fully dealt with in Chapter XX. Other commercial base metals, however, frequently contain more or less gold and silver, and the assayer is sometimes called upon to assay these.

Determination of Gold and Silver in Antimony.—Gold and silver have long been known to accompany antimony ores from Borneo, Australia, Portugal, and other places. Owing to the difficulty of separating the precious metals from antimony ores, the ingots of "star antimony" obtained by smelting the ores not infrequently contain appreciable amounts of gold and silver. Thus the following amounts of gold and silver were found in different samples of commercial metallic antimony assayed by the author [1] :—

Gold and Silver in Commercial Antimony.

Details of sample.	Ounces per ton (2240 lbs.).			
	Gold.		Silver.	
	ozs.	dwt.	ozs.	dwt.
1. Commercial "star antimony"	0	19·5	1	15
2. ,, ,, ,, . . .	1	12	0	5
3. ,, ,, ,, . . .	0	4·5	6	4
4. ,, ,, ,, . . .	0	13·5	0	8
5. Smelted from antimony ore from Servia . .	0	0·75	0	15
6. ,, ·, ,, Oporto, Portugal .	1	15	1	12

Metallic antimony may be assayed for gold and silver by the following crucible fusion method, which consists in oxidising part of the antimony with litharge, and using the lead thus reduced as a collecting agent, and oxidising any excess of antimony by means of nitre. The sample should be powdered in

---

[1] *Journ. Soc. of Chem. Ind.*, vol. xii., 1893, p. 316.

an iron mortar, to pass an 80-mesh sieve, and then well mixed. Satisfactory results are obtained with the following charge :—

Powdered antimony	25 grammes.
Litharge or red lead	50 ,,
Sodium carbonate	25 ,,
Nitre	10 ,,

The charge is fused for fifteen or twenty minutes at a dull red heat, and poured when tranquil fusion has taken place. The lead button, which should be malleable and weigh about 25 to 30 grammes, is cupelled direct, and the resultant gold-silver button parted in the ordinary way.

Determination of Gold and Silver in Bismuth.— A considerable portion of the bismuth of commerce is derived from the native metal, which frequently contains gold and silver. The precious metals are also found in most bismuth ores.

Commercial bismuth comes into the market in the form of circular "cakes," and may contain appreciable quantities of gold and silver. Samples of commercial bismuth from different sources, assayed by the author, gave the following precious metal contents [1] :—

Gold and Silver in Commercial Bismuth.

Details of sample.	Ounces per ton (2240 lbs.).			
	Gold.		Silver.	
	ozs.	dwt.	ozs.	dwt.
1. "Australian" bismuth	3	12	108	8
2. "American" bismuth	0	3	23	0
3. "German" bismuth	0	2	23	16
4. Commercial bismuth (source unknown) .	0	12	72	17

Samples of bismuth are prepared for assay by crushing to pass an 80-mesh sieve, and mixing. As bismuth is capable of being cupelled as readily as lead, the assay may be made by cupellation direct; but experience shows that the cupellation loss is greater than in the case of lead, and the buttons left on the cupel are liable to retain bismuth (see page 168). If cupellation is adopted, the sample need not be finely crushed ; a weighed quantity (say 25 to 30 grammes) is wrapped in a piece of sheet lead, and cupelled direct at a low temperature. To obtain silver buttons free from bismuth, it is advisable to add 1 or 2 grammes of lead when the cupellation of the bismuth is nearly complete. Checks should be used if great accuracy is desired.

It is more satisfactory to oxidise the bismuth by fusion with litharge or nitre, then flux it and collect the precious metals in a button of lead. For this purpose the powdered sample is fused with a charge similar to that given above for metallic antimony. A small quantity of bismuth may be retained by the lead, but the button from the fusion may be cupelled direct without difficulty.

Determination of Gold and Silver in Copper.—Metallic copper may be assayed for gold and silver by scorification, or by a combined wet and dry method, as previously described under "Base Copper Bullion," page 319.

Determination of Gold and Silver in Lead.—Argentiferous and auriferous lead are assayed by direct cupellation. The methods of sampling and assay-

---

[1] *Journ. Soc. of Chem. Ind.*, vol. xii., 1893, p. 816.

ing lead containing gold and silver are described under "Base Lead Bullion," page 315.

Determination of Gold and Silver in Tin.—As previously stated, gold and silver are sometimes found in tin ore, and pass into the bars of tin obtained by smelting the ore (see page 236). The bars may be sampled by drilling, and the drillings carefully melted under a layer of potassium cyanide. The resultant button is freed from slag by treatment with hot water, and then rolled into a thin strip.

A weighed quantity of the metal is then cut into small fragments and fused with the following charge :—

Metallic tin .	. . . . .	25 grammes.
Litharge or red lead	. . . .	50 ,,
Sodium carbonate .	. . . .	50 ,,
Nitre .	. . . . . .	10 ,,

The fusion of the charge is conducted at a dull red heat for fifteen to twenty minutes, and poured when tranquil. The lead button obtained should be sufficiently free from tin to permit of direct cupellation ; but if it is hard and white, owing to the presence of an appreciable amount of tin, it must be scorified with the addition of its own weight or more of lead prior to cupellation.

The assay may be conducted by scorification, but in this case a smaller quantity must be taken. From 2 to 5 grammes of tin are scorified with sixteen to twenty times its weight of lead and 5 grammes of sodium carbonate to form sodium stannate. The lead button is very liable to retain tin, and may require rescorifying with the addition of lead. As the amount of precious metals present is usually small, it is frequently necessary to scorify five or more separate lots of the sample and to rescorify the resultant lead buttons together so as to concentrate the gold and silver into one lead button, which is cupelled and parted in the ordinary way.

The determination of silver in argentiferous tin by treatment with hydrochloric acid or nitric acid and solution of the silver is sometimes recommended, but, from the author's experience, the method is not reliable.[1]

Determination of Silver in Zinc.—The following combined wet and dry method is recommended by K. Friedrich[2] for the determination of silver in commercial zinc. It is a combination of Pufahl's and Kerl's methods, the former consisting of solution in hydrochloric acid and fusion of the residue with potassium cyanide, followed by cupellation, and the latter consisting of fusion with excess of lead and borax and subsequent cupellation.

The method is as follows :—The sample is granulated, and a suitable quantity (100 to 1000 grammes, according to the expected silver content) weighed off into a beaker (or beakers in the case of large amounts). A small quantity of hydrochloric acid is added, and when nearly saturated it is poured off through a filter, and a second quantity poured on to the sample ; by adopting this plan solution is greatly hastened. As soon as, or just before, solution is complete, the whole residue is rinsed on to the filter and washed free from chlorides. The filter is placed in a crucible, dried and burnt ; to the residue from 7·5 to 15 grammes of assay lead and some borax are added, and the whole heated to fusion. The button of argentiferous lead is then cupelled, and the silver button weighed in the usual way. For accurate results a lead-silver button of about the same silver content should be cupelled at the same time as a check to determine the cupellation loss. The loss is allowed for in calculating the silver content of the zinc.

---

[1] Consult also Bannister, *Trans. Inst. Min. and Met.*, vol. xv., 1905-6, p. 518.
[2] *Zeit. Angew. Chem.*, 1904, v·l. xvii. 1636-1644.

As the result of experiments on alloys of known composition, Friedrich concluded that:

(1) The cupellation method gives accurate results even when zinc as well as lead is present in the alloy.

(2) If the excess of acid, after dissolving the zinc, be allowed to stand upon the residue, the silver gradually goes into solution: hence the direction to filter off before complete solution of the zinc.

(3) In the presence of hydrochloric acid or zinc chloride, very considerable quantities of silver chloride may be formed and volatilised during the burning of the filter: hence the need for thorough washing of the residue.

In the paper referred to, a table is given showing the amount of silver in grammes per metric ton in various samples of commercial zinc.

This method may be used for the determination of silver in the rich argentiferous zinc resulting from Parke's process for the desilverisation of lead; but in this case a smaller quantity, from 5 to 10 grammes, of material is taken for assay.

The amount of silver in the rich argentiferous zinc from Parke's process varies considerably; it may contain from 5 to 10 per cent. of silver or even more.

In some works this rich material is assayed by scorification, from 3 to 5 grammes being scorified with fifteen times its weight of lead and 1 or 2 grammes of borax. The temperature should be well maintained to produce a fluid slag.

II. Auriferous and Argentiferous Metallurgical Products other than Metals.

Determination of Gold and Silver in "Burnt Ore."—"Burnt ore" is the term given to roasted cupriferous iron pyrites. It usually contains small quantities of gold and silver: burnt ores from Spanish pyrites carry about 0·005 per cent. of silver and a very small quantity of gold. Burnt ores, which consist mainly of ferric oxide, are assayed by crucible fusion.

Beringer recommends the following charge which gives a fluid and satisfactory slag [1]:—

Ore	100 grammes.
Litharge	100 ,,
Charcoal	7 ,,
Borax	50 ,,
Fine sand	50 ,,

Mix; place in a large crucible; cover with salt; and melt down under cover. When fused, insert an iron rod into the molten charge for two minutes, then withdraw it, and pour the charge quickly into a large conical mould. The button of lead should weigh about 50 grammes. Cupel and weigh the silver. The litharge may be replaced by red lead, in which case another gramme of charcoal powder must be added.

The following combined wet and dry method may also be employed:—

"Take 100 grammes of the ore, place in a large beaker of 2½ litres capacity, and cover with 375 c.c. of hydrochloric acid. Boil for half an hour, until the oxides are dissolved and the residue looks like sand and pyrites; then add 20 c.c. of nitric acid, and boil till free from nitrous fumes. Dilute to 2 litres with water, and pass a current of sulphuretted hydrogen till the iron is reduced, the copper and silver precipitated, and the liquor smells of the gas. This takes about one hour and a half.

"Filter off the precipitate (rejecting the solution), and wash with warm water.

[1] *Text-Book of Assaying*, 11th edition, 1910, p. 116.

Dry and transfer to an evaporating dish, adding the ashes of the filter-paper. Heat gently with a bunsen burner until the sulphur burns, and then roast until no more sulphurous oxide comes off. When cold, add 30 c.c. of nitric acid, boil, and dilute to 100 c.c. Add 1 c.c. of very dilute hydrochloric acid (1 to 100), stir well, and allow to stand over night. Decant on a Swedish filter-paper, dry, and burn.

"Mix the ashes with 100 grammes of litharge and 1 gramme of charcoal, and fuse in a small crucible. Detach the button of lead and cupel. Weigh, and make the usual corrections for silver in the litharge used."

Determination of Gold and Silver in "Concentrates."—The methods of assaying pyritic and other concentrates resulting from the mechanical dressing of ores are the same as those given for ores containing sulphides (see Chapter XIII.). They may be assayed by crucible fusion direct or roasted before fusion.

Charges used for the direct fusion of concentrates obtained on the Rand are given on page 359.

Beringer recommends the following charge for the assay of pyritic concentrates after roasting:[1] take one assay ton (30 grammes) of the concentrates, roast, and fuse the oxidised product with :—

Litharge	45 grammes.
Charcoal	2 ,,
Sodium carbonate.	30 ,,
Borax	30 ,,

Fuse, detach the slag, cupel and part in the ordinary way.

Determination of Gold and Silver in Copper Matte.[2]—Copper regulus or matte consists of copper sulphide, with varying proportions of iron sulphide, obtained in smelting copper ores. The amount of gold and silver in copper mattes varies considerably : a matte of average grade will contain about 20 per cent. of copper and 5 ounces of gold and 40 ounces of silver per ton. High-grade mattes will contain from 50 to 60 per cent. of copper, and mattes high in gold and silver may contain 15 ounces of gold or more and as much as 200 ounces of silver per ton. The gold in copper mattes is generally determined by the scorification or "all fire" method, as the results are higher than those obtained by combined methods. Silver is sometimes determined by scorification, but more generally by a combined wet and dry method. Crucible fusion methods are also employed.

(a) *Scorification Method.*—Ten portions of the sample, each of 0·10 assay ton, are placed in 3-inch scorifiers with 50 grammes of test lead, one-half of which is mixed with the matte and the other half used as a cover. The charge is then covered with 2 grammes of a mixture of equal parts of borax, glass, and fine sand. The scorification is conducted at a moderate temperature, and usually occupies from twenty-five to thirty minutes. The resultant buttons, which weigh about 15 grammes, and are hard from the presence of copper, are cleaned from slag and rescorified with the addition of 25 grammes of test lead. The lead buttons obtained weigh from 10 to 12 grammes : they are cupelled separately, and the gold-silver buttons weighed separately, and then together. To determine the gold, the buttons are grouped into two lots of five each, and parted in dilute nitric acid in the ordinary way, as described on page 207.

As a correction for the cupellation loss is frequently allowed in practice, the cupels (after removing the unstained portion) are crushed to pass a 100-mesh sieve, and then re-fused in five lots of two each with the following charge[3] :—

[1] *Text-Book of Assaying*, p. 140.   [2] See Bibliography on page 320.
[3] A. H. Ledoux, *Am. Inst. Min. Eng.*, Oct. 1894.

Litharge	.	.	.	.	.	90 grammes.
Sodium carbonate	.	.	.	.	.	50 ,,
Borax	.	.	.	.	.	50 ,,
Argol	.	.	.	.	.	3 ,,

The lead buttons are cupelled and the gold-silver buttons weighed and parted. The weight of silver and gold thus obtained is added to the weights obtained in the original assay. Sometimes no correction is allowed. It is not usual to clean the slags. In calculating the results, allowance must of course be made for the silver in the test lead used.

(b) *Combined Wet and Dry Method.*—Weigh out, in duplicate, 1 assay ton of matte, place in large beakers of a litre capacity, provided with glass covers, and treat with 100 c.c. of distilled water and 50 c.c. concentrated nitric acid. Allow the violent action to subside, add 50 c.c. more of concentrated nitric acid; heat until red fumes cease to come off, dilute to 500 c.c., and add 10 c.c. of a concentrated solution of lead acetate, 5 c.c. of concentrated sulphuric acid, and 10 c.c. or more of salt solution, stir briskly, and let it stand over night to allow the lead sulphate to settle. The strength of salt solution frequently recommended contains 0·542 gramme of NaCl per 1000 c.c. (1 c.c. = 1 milligramme Ag), and an excess of 4 to 8 c.c. beyond that required to precipitate the silver assumed to be present should be added. The precipitate is filtered off and washed : the filtrate should be perfectly clear. The filter-paper and precipitate are dried, then wrapped in about 8 grammes of sheet lead and scorified with 40 grammes of test lead, with a cover of 1 gramme of borax glass. The button is cupelled and weighed, and then the gold separated by parting.

The cupels and slags are in many cases re-assayed and the resultant gold and silver added as a correction to the original assay. Some assayers use sodium bromide instead of sodium chloride, on account of the greater insolubility of silver bromide. The lead acetate is added to cause the precipitate of silver-chloride and the finely divided gold to settle quickly, and to enable them to be filtered effectively.

(c) *Crucible Fusion.*—Crucible methods are sometimes used for the determination of gold and silver in copper mattes, as they permit of a larger quantity of material being used.

The following method is used at the Standard Smelting Company, at Rapid City, South Dakota, for copper mattes containing up to 20 per cent. of copper and high in gold and silver [1] :—

The sample of matte is put through a 120-mesh sieve and well mixed. Then 0·25 assay ton of matte is fused with 3·5 assay tons of the following stock flux and a thin cover of borax glass :—

Stock flux.			Proportion of fluxes for one assay.		
Litharge	. .	70 parts	Matte .	. .	0·25 assay ton.
Silica	. . .	11 ,,	Litharge	. .	67·0 grammes.
Sodium carbonate	.	25 ,,	Silica .	. .	10·5 ,,
Nitre	. . .	5 ,,	Sodium carbonate		24·0 ,·
			Nitre .	. .	5·0 ,,

Fusion is effected quickly at a comparatively high temperature, giving a clean fluid slag and a bright lead button weighing approximately 20 grammes. The button is cupelled directly for gold and silver. The lead buttons are always more or less cupriferous, but no scorification is made before cupellation. The assay should be made in duplicate : for control assays four fusions, each of 0·25 assay ton of matte, are made. In this case the slag and cupel losses are

---

[1] Quoted by Fulton, *Manual of Fire Assaying*, 2nd edit., p. 140.

determined. The slag and cupel from one assay are crushed and fused together in the same crucible as was used for the original fusion. The slag and cupel from each assay is treated in this way, and the sum of the four corrections thus obtained added to the sum of the original buttons.

According to Fulton,[1] the average correction by this method, on copper matte of ordinary grade (*i.e.* about 20 per cent. copper, 5 ounces gold, and 40 ounces silver), is 2·5 per cent. for the gold and 5·5 per cent. for the silver.

The results obtained by this crucible method are usually slightly higher than those obtained by the scorification method.

Determination of Gold and Silver in Copper Anodes.—Copper containing gold and silver is cast into slabs to be used as anodes for the separation of the precious metals by electrolytic processes. The percentage of precious metals in copper anodes varies considerably. The methods of sampling and assaying copper anodes are described under "Base Copper Bullion," in Chapter XX.

Determination of Gold and Silver in Anode, Mud, Slimes, or Sludge.—Anode sludge, formed during the electrolytic refining of copper and lead, contains considerable quantities of gold and silver. After screening to remove nodules of metal, and filtering through muslin to remove excess of the electrolyte, the sludge is washed, dried, and heated for the extraction of the precious metals.[2]

The gold and silver content varies greatly according to the nature of the material treated. Slimes from the electrolytic refining of copper usually contain from 40 to 55 per cent. of silver, from 0·2 to 1·3 per cent. of gold, and from 10 to 20 per cent. of copper. Lead anode sludge frequently contains only a small amount of copper. Small quantities of platinum and metals of the platinum group are not infrequently present in electrolytic slimes and should be tested for, since they become concentrated, and can be detected in the sludge, when the quantity is too small to permit of detection in the copper anodes and in the ore from which the copper was obtained.

The following are examples of the precious metal content of anode muds:—

Gold and Silver Content of Anode Mud from Different Sources.[3]

	A.	B.	C.	D.
	per cent.	per cent.	per cent.	per cent.
Silver . . .	15·725	25·0	40·0	46·58
Gold . . .	1·450	0·05 to 1·0	2·0	0·15

(A) Raw "anode mud" from electrolytic copper refinery at Great Cobar Syndicate at Lithgow, New South Wales.[4]

(B) Raw "anode mud" from refining copper from ores of Butte Montana.

(C) Typical "anode mud."[5]

(D) Source not given.[6]

The gold and silver in anode mud is determined by the scorification or the crucible fusion method.

The moisture in the sample is determined by carefully drying 1 lb. (454 grammes) in a stove; not in an open pan. Assay results are returned

[1] *Loc. cit.*, p. 141.
[2] Consult *Met. and Chem. Eng.*, 1911, ix. pp. 417–420.
[3] Quoted by W. J. Sharwood, *Economic Geology*, vol. vi., 1911, p. 28.
[4] G. H. Blakemore, *Proc. Aust. Inst. M.E.*, 1910.
[5] L. Addicks, *Journ. Franklin Inst.*, 1905.
[6] A. Holland, * Amm. Chim. Anal. Appl.*, vol. iv,, 1899, pp. 123–125.

on the dry material. The sample is prepared for assay by grinding to pass a sieve of 90 mesh, and the metallics which are invariably left on the sieve are collected and assayed separately. The sample must be thoroughly mixed after grinding.

The assay of slimes by scorification is conducted as described above for copper matte. When great accuracy is desired ten scorifications should be made, and the slags and cupels should be assayed and corrections made for the gold and silver they contain.

The following mode of procedure is given by T. Ulke[1] as a commercial method of determining gold and silver in electrolytic slimes.

Two samples are weighed out and scorified in the usual manner. One of the lead buttons is cupelled, the resulting silver bead weighed, then wrapped up in lead-foil until it approximately equals the weight of the other lead button, and then both cupelled, and the resulting beads weighed. The loss in weight of the first bead gives the loss in cupellation, which must be added to the weight of the other bead as a correction. The slag from the scorification is re-scorified with borax and 16 grammes of test lead, and the resulting lead button cupelled. The weight of the bead thus obtained is added to the corrected weight of the second of the beads previously obtained, and the combined beads are parted, etc., in the usual way.

For a crucible fusion assay the charge given on page 352 for silver precipitate will be found to be satisfactory.

Holland[2] gives the following charge for the determination of the gold in electrolytic slime by crucible fusion :—

A weighed quantity, 12·5 grammes of the dried and pulverised slime, are well mixed with :—

Litharge	50 grammes.
Sodium carbonate	25 ,,
Borax glass	15 ,,
Nitre	10 ,,

The charge is placed in a crucible and carefully heated until quiet fusion results, then a mixture of 20 grammes of litharge and 0·4 gramme of charcoal added. When quiet fusion again takes place the contents of the crucible are poured, the slag detached from the lead button, which is then cupelled, and the gold-silver button parted in the usual manner.

For the determination of the silver, Holland recommends heating 5 grammes of the slime in a porcelain dish in a current of dry chlorine to convert the silver into chloride. The silver chloride is then dissolved out by potassium cyanide solution, and deposited by electrolysis during twenty-four hours with a current of 0·5 ampère.

Determination of Gold and Silver in Lead Dross.—Lead dross results from the melting of base lead bullion, and consists of oxides of lead and other base metals contaminated with metallic lead. The method of assay for lead dross is described on page 317.

Determination of Silver in Lead Fume.—Lead fume consists of the solid particles which escape from lead-smelting furnaces and are condensed. The fume is deposited in the state of fine powder or dust, varying in colour from grey to white. It contains considerable proportions of lead sulphide, oxide, sulphate, and carbonate, together with more or less silver, and also gold in minute proportions.

Lead fume requires care in handling, owing to its fine state of division and

[1] *Eng. and Min. Journ.*, vol. lxviii., 1899, pp. 727-729.
[2] *Amm. Chim. Anal. Appl.*, vol. iv , 1899, pp. 123-125.

tendency to form dust.   The fume is assayed by crucible fusion, the following charge being suitable :—

Lead fume	.	.	.	.	.	25 grammes.
Sodium carbonate	.	.	.	.	.	25 ,,
Argol	.	.	.	.	.	5 ,,
Borax	.	.	.	.	.	5 ,,

A piece of hoop-iron (or iron nails) is added to the charge to decompose the sulphur compounds.   The charge is fused at a red heat, the time usually occupied being about twenty minutes.   The resultant lead is freed from slag and cupelled.

Determination of Silver in Litharge and Red Lead.—As previously stated, the oxides of lead invariably contain silver, which must be determined when litharge or red lead is used in an assay charge.   The silver in litharge is determined by fusing 100 grammes with 1 gramme of charcoal and 20 grammes of borax.   In the case of red lead the same charge is used with 2 grammes of charcoal.   The lead is cupelled, and the button of silver weighed.   As the silver buttons are very small, the cupellation must be carefully watched towards the end of the operation.

Determination of Silver in Silver Precipitate (metallic silver).—This substance, which is also termed "cement" silver, consists of metallic silver which is precipitated from solution in the lixiviation processes of extracting silver.   The precipitate contains, in addition to metallic silver and gold, sulphates of lead and lime, oxides of zinc, copper, and iron, and more or less organic matter.   The amount of silver in these precipitates is very variable ; an average precipitate contains about 5 to 10 per cent. silver, 0·02 per cent. gold, 20 per cent. lead, and 2·5 per cent. of zinc.

Samples of silver precipitate are generally received in a dried condition, i.e. free from "water at 100° C."   Since, however, it rapidly absorbs water, it is advisable to determine the moisture before making the assay.   The sample should be assayed as soon as it is dried, and care should be taken in weighing it to prevent absorption of water.   The determination of the silver and gold is made by crucible fusion.   Scorification cannot be suitably adopted owing to the combined water contained in the precipitate.

Beringer[1] recommends the following crucible assay :—

Weigh up 5 grammes of the precipitate, mix with 100 grammes of litharge and 1 gramme of charcoal.   Melt in a crucible at a moderate heat, and pour. Detach the slag, replace it in the crucible, and when fused "clean" it by adding a mixture of 20 grammes of litharge and 1 gramme of charcoal.   When the fusion is again tranquil, pour, and cupel the two buttons of lead together. The cupellation loss should be determined by the use of checks or other method (see page 180), and added to the result of the assay.

It is also advisable to fuse the cupels and make allowance for the silver they contain.   The charges for the fusion of cupels have been given previously.   In calculating the percentage of silver, allowance must be made for the silver contained in the litharge used.

The gold is determined in the ordinary way by treating the cupellation button with dilute nitric acid.

Owing to the richness of these precipitates it is difficult to get close agreement between the assay results, and from five to ten fusions should be made and the average of the results taken.   The gold may also be satisfactorily determined by a combined method.   For this purpose take 25 grammes of

---

[1] *Text-Book of Assaying*, 11th edit., 1910, p. 116.

precipitate, treat with 100 c.c. dilute nitric acid (1 acid : 1 water), and, when all action has ceased, dilute and filter. Wash and dry the residue, burn the filter paper, then mix the residue and filter ash with 50 grammes of litharge, 1 gramme of charcoal, and 50 grammes of sodium carbonate. Transfer to crucible, fuse, and pour; detach the slag, cupel the lead button, and part in the ordinary way.

The insoluble residue usually contains sufficient silver for parting the button from cupellation. An exact determination of the silver in the precipitate cannot be made by treatment with nitric acid, since not all the silver usually passes into solution.

Determination of Silver in Silver Sulphide Precipitate.—This consists of silver sulphide obtained in the lixiviation process of silver extraction. The dried precipitate contains from 25 to 40 per cent. of silver, as silver sulphide mixed with impurities, principally base-metal sulphides, notably copper.

Considerable difficulty is experienced in sampling this very rich material. According to Furman,[1] the following method answers very well for sampling comparatively large lots. Spread the material evenly on a clean iron floor, and divide it into "squares" about 1 foot square. From each square dig out five samples, and put these together into one pile. Well mix the pile, spread out, and reduce as before. The final bulk sample, weighing about 5 lbs., is pulverised in a coffee mill and reduced still further. The portion taken for the assay sample is ground on a bucking plate until it passes a sieve of 80 to 100 mesh, and then very thoroughly mixed. The assay of silver sulphide precipitate can be made both by the scorification and crucible fusion methods.

Furman[2] recommends scorification. From 6 to 20 charges are run on each lot of sulphides. One-tenth assay ton of the dried sample of sulphides is scorified with 55 grammes of lead and a cover of 3 to 5 grammes of borax. One-half of the lead is put in the bottom of the scorifier and hollowed out; the sulphides are put into the hollow, and the remainder of the lead poured over them; the borax is then placed on top. The resultant lead buttons are cupelled separately at a low temperature.

The slags from all the scorifications are carefully collected, crushed to 20 mesh, and mixed with the following charge:—

Litharge	20 grammes.
Sodium carbonate	15 ,,
Argol	2 ,,
Cover of salt.	

Fuse, pour, and cupel the lead button.

The cupels are crushed (after removing unstained portions) to pass 60 mesh, then fused with the following or other suitable charge:—

Litharge	30 grammes.
Sodium carbonate	30 ,,
Borax glass	30 ,,
Argol	2 ,,

The average silver content of the slags and cupels is added as a correction to the average of the original assays.

In the crucible methods for the assay of silver sulphides, no metallic iron is employed to decompose the sulphides, but a sufficient amount of litharge or red lead is added to oxidise completely the sulphur in the charge, and leave an

---

[1] *Assaying*, 6th edit., 1908, p. 20.
[2] *Loc. cit.*, p. 262. Also F. P. Dewey, *Trans. Amm. Inst. Min. Eng.*, 1896, Colorado Meeting.

excess of lead oxide to supply the metallic lead to collect the gold and silver. The result of the crucible assay when carefully conducted is as accurate as that obtained by scorification. The following stock flux may be used :—

Stock flux.			Quantities for one assay using one-tenth assay ton of sulphide precipitate.		
Litharge	. .	600 grammes.	Litharge	. .	40 grammes.
Sodium carbonate .		350 ,,	Sodium carbonate .		25 ,,
Borax glass .	.	100 ,,	Borax glass .	.	7 ,,
Silica .	. .	100 ,,	Silica .	. .	7 ,,
Charcoal	. .	8 ,,	Charcoal	. .	0·75 ,,

With one-tenth assay ton of sulphides use 80 grammes of stock flux, and twice this amount for one-fifth assay ton.

The precipitate is mixed with two-thirds of the charge and the remainder of the charge used as a cover. Fuse the charge at a good red heat for twenty-five or thirty minutes, then pour, detach the slag from the lead button, and cupel. The cupel absorption is determined and added. It is also advisable to collect the slag and clean it : this may be done by adding the slag to the charge for the fusion of the cupels, thus cleaning the slag and cupels in one operation. Smelters do not, as a rule, allow the slag absorption, and if the fusion is properly conducted the values retained by the slag are very small with the above flux. The charges given for the assay of cyanide precipitate may also be used for silver sulphide precipitate (see Table XLVII., page 397). At least five fusions should be made, and the results averaged.

Determination of Gold and Silver in Tailings.—Tailings or tails consist of the finely pulverised ore which has been treated for the extraction of the values, but which still contains small quantities of gold and silver. The assay of tailings differs only from that of ores in the fact that a larger quantity must be taken, frequently from 3 A.T. to 5 A.T., for the assay, since they are always poor in gold, containing usually less than 1 pennyweight per ton.

It is not unusual to take the portion for assay without any further grinding, as tailings have already been crushed and passed through a sieve of at least 30 mesh. It is advisable, however, to crush all tailings samples to pass a 90- or 100-mesh sieve, as is done for the assay of the original ore or "heads." From 2·5 to 5 assay tons of material should be taken for the assay. The charges given for tails and residues on page 359 may be used.

Determination of Gold and Silver in Zinc-Gold Slimes.—The assay of these slimes is described on page 384.

III. Various Auriferous and Argentiferous Products.

This section includes miscellaneous bye-products obtained in the extraction of metals from their ores and also those resulting from the working of the precious metals.

Determination of Silver in Chloride Precipitate.—This consists of silver chloride precipitated from the pickling solutions, etc., used in silversmiths' work. The precipitate is usually very rich in silver, of which it may carry over 60 or 70 per cent. ; pure silver chloride contains 75·27 per cent. of silver. The precipitate is frequently moist when received, and must be carefully dried and the result of the assay reported on the dry material.

The assay is made by crucible fusion, as described above for silver sulphide precipitate, corrections being made for slag loss and cupel absorption.

Determination of Gold and Silver in Cyanide Plating Solutions.—The gold and silver in cyanide solutions used for electroplating are determined by evaporation in small lead basins or with litharge, and the precious metals subsequently separated by cupellation, as described for cyanide solutions on page 375.

For the determination of gold in gilding solutions 25 c.c. are usually taken. The gold content is in most cases from 10 to 15 dwt. per gallon.

For the determination of silver in plating solutions 50 c.c. are usually taken.

Determination of Gold and Silver in Lemel.—The term "lemel" is applied to the gold dust produced in the manufacture of goldsmiths' work by the processes of filing, cutting, turning, etc.

The lemel always contains impurities, such as organic matter, iron filings, pieces of iron wire, etc. It is first sifted to remove metal scrap, then burnt to remove organic matter, and the iron particles in the residue removed by means of a magnet.

The lemel is then melted with fluxes to obtain a "lemel" bar. The following charge may be used :—

Lemel	20 parts.
Sodium carbonate	3 ,,
Borax (dried)	1 ,,
Sodium chloride	2 ,,

The charge is melted at a bright red heat and when fused a few small pieces of nitre about the size of a pea are added at short intervals. When tranquil fusion results the contents are poured into an ingot mould and the metal cleaned from slag. The bar thus obtained should be re-melted with a little borax or charcoal as flux and then poured into an ingot mould and weighed.

Lemel bars vary considerably in composition, and their valuation is attended with difficulty, as they are rarely uniform in composition. The sampling and the assay of lemel bars is described in Chapter XXI.

Determination of Gold and Silver in Plumbago (Graphite) Crucibles.— Graphite crucibles, stirrers, etc., that have been used in melting bullion always contain considerable quantities of gold and silver. They are crushed and washed and sometimes amalgamated with mercury to separate shots of metal. The sampling of the crushed material (frequently termed "blacks") is difficult, and must be carried out with care. The portion selected for the assay sample should be crushed to pass a sieve of 100 mesh and the metallics carefully collected and assayed separately. The assay of this material is also troublesome owing to the large percentage of graphite it contains and the difficulty of effecting its oxidation readily. The graphite is usually oxidised by roasting or by the use of nitre. If the sample is roasted, this is best done in the muffle in a roasting dish sufficiently large to permit of the material being spread out into a thin layer. To facilitate the roasting, the sample must be very finely powdered and must be stirred constantly: even under these conditions the oxidation is very slow.

Samples of "blacks" may be assayed either by scorification or by crucible fusion, but in both methods it is usual to take only a small amount, generally one-tenth assay ton of the material, as the graphite may cause trouble if larger amounts are used.

Scorification is adopted at the Royal Mint, London.[1] The dried sample is first roasted and then scorified, the charge being :

Roasted material	5 grammes.
Lead	50 ,,
Borax	0·5 ,,

[1] T. K. Rose, *Metallurgy of Gold*, 5th edit., 1906, p. 467.

Half the lead and the whole of the borax is used as a cover. The slag from several scorifications is re-melted in a crucible and a mixture of litharge and charcoal thrown on to the surface of the fused material. In this way from 0·5 to 2·5 per cent. of the total values is recovered from the slag. The cupellation loss is usually from 0·5 to 1 per cent.

According to Loewy,[1] correct results may be obtained by scorifying without previous roasting, the charge being 5 grammes of material, 60 grammes of lead, and a little borax or glass powder as a cover.

Fulton recommends fusion in a scorifier with oxidising agents.[2]

From 0·05 to 0·10 assay ton is taken and mixed with a little more than one-half of its weight of nitre and 30 grammes of litharge, placed in a 2·5-inch scorifier, covered with 30 grammes of litharge and afterwards with a thin cover of borax glass, then placed in a muffle, and fused finally at a yellow heat. The lead buttons are cupelled, the gold-silver button weighed and parted as usual.

This material may also be assayed by the crucible method by using nitre and litharge to effect the oxidation of the graphite. The charge is as follows :—

Graphite material	.	.	.	.	0·1 assay ton.
Litharge	.	.	.	.	70 grammes.
Sand	.	.	.	.	5 ,,
Sodium carbonate	.	.	.	5
Nitre	.	.	.	.	.	5 to 11 ,, (according to the amount of graphite in the sample).

Cover of borax glass, about 1 gramme.

The following charges are given by M'A. Johnston as typical of those in use on the Rand, South Africa, for the direct fusion of samples of graphite crucibles [3] : -

	A.	B.
Ground product .	.	0·25 assay ton.	0·25 assay ton.
Sodium carbonate .	.	0·5 ,,	0·5 ,,
Litharge	.	4 0 ,,	2·0 ,,
Borax .	.	.	0·25 ,,	0·25 ,,
Sand	.	.	.	0·5 ,,	0·5 ,,
Nitre	.	.	.	nil.	about 0·3 ,,
Weight of lead button obtained	55 to 60 grammes.	Approx. 40 grammes.

The weight of lead obtained is dependent on the amount of graphite present in the material. Charge A is stated to give the higher and more correct result.

Determination of Gold and Silver in "Sweep."—Sweep consists of the sweepings of the floors and the refuse from the melting furnaces, etc., from refineries and from the workshops of goldsmiths and silversmiths. This material is very heterogeneous in character, and sometimes contains several per cent. of mechanically mixed carbon. The gold and silver are usually present in the metallic state, but the amount varies very considerably. The sampling of sweep presents considerable difficulty. The material is first "dressed" by burning to destroy organic matter. The mass is then ground sufficiently fine to pass a sieve of 20 mesh. The metallic portion left on the sieve is freed from iron by means of a magnet, and then melted with fluxes to obtain a bar, which is weighed, sampled, and assayed as described under "Lemel."

[1] Proc. Chem. and Met. Soc. of S.A., vol. xi., 1898, p. 205.
[2] Fire Assaying, 2nd edit., 1911, p. 145.
[3] Rand Metallurgical Practice, vol. i. p. 317.

The portion that has passed through the sieve is freed from iron with a magnet, then quartered down in the usual way. The final portion taken for "assay sample" is ground to pass a sieve of 100 mesh and thoroughly mixed. It is then assayed, and the yield of metal obtained from the metallic portion and the fine material respectively is calculated as previously described.

Samples rich in metal may be assayed by scorification. Take 5 grammes of the material and scorify with 10 to 15 times its weight of lead and 1 gramme of borax. Cupel, weigh, and part as usual. From five to ten scorifications should be made, and the average of the results taken.

Poor samples are best assayed by crucible fusion. No charge can be given to suit all samples of this material, but the following will be found to be generally satisfactory :—

Dressed sweep	25 to 50 grammes
Litharge	50 ,,
Charcoal	2 ,,
Sodium carbonate	30 ,,
Borax	20 ,,

If the sample contains much ferric oxide it may be necessary to increase the amount of charcoal in the charge. A piece of hoop-iron or iron nails should be added to remove any sulphur that may be present. The lead button is cupelled and the resulting gold-silver button parted. It is advisable to make three or four fusions according to the amount of precious metal present.

CHAPTER XXIII.

LABORATORY WORK IN A CYANIDE MILL.

THE daily work of an assayer in a cyanide mill varies considerably, especially in regard to the testing of the cyanide solutions, which has now become a part of the routine work. In some plants simple titrations to determine the strength of the cyanide solutions, with an occasional determination of the amount of gold in solution and assays of "headings" and "tailings," is all that is called for. On the other hand, at some modern mills a careful check is kept on all the ore and solution in every stage of the treatment, and the number of ore assays and other determinations on control samples, "headings," "tailings," slimes, solutions, bullions, and various products is necessarily very large.

The laboratory work in a cyanide mill may be classified under six heads viz.[1] : —

1. Mine samples.
2. Reduction works samples.
3. Solutions.
4. Bullions.
5. Bye-products.
6. Testing of ores for cyanide treatment.

1. Assay of Mine Samples.

Mine samples consist of stope and development samples from the mine, and in the majority of cases come to the assayers in a prepared state.

For the degree of accuracy required of them they should be passed through a sieve of at least 80 mesh, although with some ores a 60 mesh may be satisfactory. Before being assayed, the samples should be thoroughly mixed by rolling on a sheet of rubber or paper. On some mines it is customary for the sampler to pan a portion of every sample before handing it to the assayer. This, in practised hands, gives an idea of the value of the ore, and also enables the assayer to ascertain the pyritic content, etc., of the ore for fluxing purposes. The ores are assayed as described in Chapter XIV. Development or prospecting samples are assayed in duplicate. The gold content of mine samples is usually reported in pennyweight and decimals of a pennyweight per ton to the first place of decimal.

2. Assay of Reduction Works Samples.

The term "reduction works samples" is applied to samples of solids taken from a running plant, and may consist of :—

(a) Screen samples (or heads) and tailings (tails).
(b) Cyanide pulp samples.
(c) Spitz concentrates and residues.
(d) Sand charges and residues.
(e) Slime charges and residues.

---

[1] Whitby, *Journ. Chem. Met. and Min. Soc. of S. Africa*, 1906, vol. vi. p. 269.

Screen samples represent all the ores charged into the leaching tanks and all the tailings discharged from the plant. The samples are taken by means of a bucket or some mechanical sampling device, as described on page 100. The preparation of these samples for assay requires great care, as they are the most difficult of all the mill samples to obtain accurate results from. Each bucket must be kept separate and the contents dried and quartered down ; but, as a general rule, before quartering, the sample should be passed through a sieve of the same aperture as, or slightly larger than, that used in the battery, so that any coarse particles due to damaged screens may be detected and crushed to the same fineness as the remainder of the sample. When the bulk of the contents of each bucket has been sufficiently reduced by quartering, the final portion selected for assay is crushed to pass an 80- or 100-mesh sieve, and then well mixed and assayed in the same way as described for gold ores.

The most reliable results are obtained by keeping the assay portions from each bucket quite distinct and making separate assays, and then averaging all the results obtained.

In some mills, however, it is the practice to mix together the assay portions from several buckets and assay the sample thus obtained. This, of course, effects a saving of time, but in making a larger number of separate assays the errors which might arise from the presence of coarse gold are probably eliminated. Other reduction works samples are quartered down in the manner described above.

A greater degree of accuracy is required in the results of reduction works samples than with mine samples ; the assays are therefore made in triplicate or even in quadruplicate.

The following charges are given by Whitby [1] as typical of those used on the Rand, South Africa :—

General charge for screen and pulp samples :—

Ore	2·5 A.T. or	75 grammes.
Litharge	3·0 ,,	90 ,,
Sodium carbonate	2·5 ,,	75 ,,
Borax	0·5 ,,	15 ,,
Charcoal, a sufficiency.		

For *Spitzlutte concentrates and residues* :—

Ore	1·0 A.T. or	30 grammes.
Litharge	2·0 ,,	60 ,,
Sodium carbonate	1·0 to 1·25 ,,	30 to 35 ,,
Borax	0·25 ,,	8 ,,
No charcoal.		

For *Sand charges* (*i e.* originals) and *residues* for treatment by a concentrating plant, and also slime charges and residues for slime plant :—

Ore	2·5 A.T. or	75 grammes.
Litharge	2·0 ,,	60 ,,
Sodium carbonate	2·5 ,,	75 ,,
Borax	0·5 ,,	15 ,,
Charcoal, a sufficiency.		

When, however, no spitzlutte, fruevanner, or other concentrate is separately treated, the charge for sand originals and residues is necessarily altered, as they

[1] *Loc. cit.*

will then contain the bulk of the pyrites. In this case the following charge is used :—

For sands, originals, and residues containing pyrites.

Ore .	. . .	2·5 A.T. or	75 grammes.
Litharge .	. .	3·0 ,,	90 ,,
Sodium carbonate	. .	2·5 ,,	75 ,,
Borax .	. .	0·75 to 1·0 ,,	22·5 to 30 ,,
No charcoal.			

When much pyrites is present, iron nails should be added.

In assaying residues some assayers use as much as 5 A.T., or from 100 to 150 grammes.

The treatment of *slime samples* necessarily involves evaporation to dryness. With residues this means a loss of more or less gold, which is deposited on the side of the evaporating vessel, and not recoverable. The discrepancies in slime residue assays and the assays of slimes undergoing treatment are attributed by Whitby [1] to this cause. He states that the following method of preparing slimes residues for assay gives satisfactory results. To the slimes residues is added first a little cyanide and then small quantities of solutions of copper sulphate, dilute sulphuric acid, and sodium sulphite. After shaking vigorously, the sample is turned out into a shallow, enamelled drying pan and dried.

Residues (containing undissolved gold) resulting from the treatment of sand and slimes are transferred to three-gallon buckets filled with water and settled with lime water for six hours, the water is then syphoned off, and the residues dried and assayed.

The results of reduction works samples are reported in pennyweight and decimals of a pennyweight per ton to the second place of decimal ; consequently a fine assay balance must be used for the final weighings. This should be sensitive to $\frac{1}{200}$ milligramme with a 0·5 milligramme rider for beam work.

3. Assay of Cyanide Solutions.

Much of the information contained in this section has been derived from the authoritative work on cyanide solutions by Clennell.[2] Most of the methods used in the cyanide mills for the assay of the various cyanide solutions are technical methods in which extreme rapidity is aimed at rather than extreme accuracy. However, in the hands of competent assayers the methods give, as a general rule, good comparative results sufficiently accurate to be of assistance to the cyanide manager in the control of his plant.

There is little doubt that in many mills the assaying of solutions, with a view to their correct valuation, is not given the consideration that its importance deserves ; but there appears to be a growing tendency to give more attention to the testing of solutions than formerly.

One of the most important functions of the cyanide manager is to reduce the consumption of potassium cyanide to the minimum and raise the percentage of extraction to the maximum at least cost : consequently the testing of the solutions in the laboratory is absolutely necessary if economy in working is to be effected. The ordinary works tests usually consist in determining :—

    (*a*) The free cyanide present.

    (*b*) The total cyanide.

    (*c*) The total or protective alkali.

---

[1] *Loc. cit.*, p. 270.

[2] Consult Clennell, *Chemistry of Cyanide Solutions* ; M'A. Johnston, *Rand Metallurgical Practice* (chapters on Assaying and Testing) ; Argall, *Western Methods of Smelter Analysis*.

These tests and some few others will be used daily, while other tests are used only at long intervals. Complete analyses of working cyanide solutions are made in many mills at varying intervals.

The strength of the cyanide solutions used varies with different ores. "In the Transvaal, the strong solution usually contains from 0·25 to 0·35 per cent. of cyanide and the weak solutions from 0·05 to 0·15 per cent. A strength of 0·5 per cent. is, however, necessary in the treatment of some ores, and, on the other hand, it is sometimes better to use no solutions stronger than 0·1 per cent. or even more dilute."[1] In the case of silver ores it has been found that solutions containing from 0·25 per cent. to 0·75 per cent. of cyanide are necessary to attack the sulphides ; hence solutions of the following strengths are recommended for silver ores or for gold-silver ores : viz. 0·25 per cent., 0·30 per cent., 0·40 per cent., 0·50 per cent., 0·75 per cent.[2]

(a) **Determination of Free Cyanide.**—"The term 'free cyanide' is used to indicate the equivalent, in terms of potassium cyanide, of all the cyanogen which is present in the solution as simple cyanides of alkalies and alkaline earth metals, such as potassium, sodium, ammonium, calcium, and barium cyanides. It does not include cyanogen present in the form of double cyanides or hydrocyanic acid."[3]

The determination of the cyanogen is of importance, since this is the active solvent of the gold, and so long as the cyanogen present is readily soluble it is a matter of indifference whether the base be sodium or potassium (see remarks on commercial cyanide, page 380). The percentage of cyanides in solutions is invariably determined by Liebig's[4] well-known method, which is based upon the fact that when a solution of silver nitrate is added to an alkaline solution, containing cyanogen, no permanent precipitate of silver cyanide occurs until all the cyanogen has combined with the alkali and silver to form a double salt; thus in the presence of potassium the reactions are :—

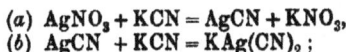

(a) $AgNO_3 + KCN = AgCN + KNO_3,$
(b) $AgCN + KCN = KAg(CN)_2$;

or, expressed in one equation :—

(c) $AgNO_3 + 2KCN = KAg(CN)_2 + KNO_3.$

As soon as the whole of the potassium cyanide has been converted into this double salt the addition of more silver solution results in its decomposition and the precipitation of silver cyanide ; thus :—

(d) $AgNO_3 + KAg(CN)_2 = 2AgCN + KNO_3.$

Hence the completion of the first stage can be taken as the end of the reaction, which will be indicated by the first appearance of a permanent precipitate (AgCN) causing a white turbidity or opalescence in the solution. If, therefore, the silver solution be of known strength, the quantity of cyanogen present is easily found. It will be noted from the above reactions that one molecule of silver nitrate is equivalent to two molecules of potassium cyanide. From reaction (c) it will be seen that 169·89 parts of silver nitrate are equivalent to 130·22 parts of potassium cyanide, or 1·3046 parts of silver nitrate equal to 1 part of potassium cyanide. The finishing-point may also be determined by the addition of potassium iodide as an indicator.

The first step in the reaction between silver nitrate and potassium cyanide

[1] T. K. Rose, *Metallurgy of Gold.*   [2] Argall, *Mill and Smelter Methods*, p. 96.
[3] Clennell, *Chemistry of Cyanide Solutions*, 2nd edit., 1910, p. 4.
[4] *Ann. der Chem. und Pharm.*, vol. lxxvii. p. 102.

is not interfered with by the presence of soluble iodides (and other haloids). If, therefore, potassium iodide is added to the solution, no silver iodide is formed until the cyanide has been entirely converted into the double salt. When that stage is complete, the further addition of silver nitrate results in the precipitation of silver iodide, which separates as a yellowish turbidity and is easily recognised. The end of the reaction is more clearly defined when potassium iodide is added, but both methods of determining the finishing-point are largely used in cyanide mills.

Standard Solutions. — Two standard silver-nitrate solutions of different strengths are usually employed in testing cyanide solutions: one for testing strong solutions, and the other for testing dilute solutions. The standard silver nitrate for strong solutions is prepared by dissolving 13·046 grammes of pure crystallised silver nitrate in distilled water, and diluting until the total volume of solution is 1000 c.c. Each c.c. of this solution is equivalent to 0·01 gramme of potassium cyanide. Hence, if we take 10 c.c. of the cyanide solution which is to be tested, every c.c. of the silver solution added will represent 0·1 per cent. of free cyanide, thus $\dfrac{0·01 \times 100}{10} = 0·1$ per cent.

Example.—Suppose 10 c.c. of the cyanide solution required the addition of 5 c.c. of standard silver nitrate to form a permanent precipitate : then, since 1 c.c. = 0·1 per cent., 5 c.c. will represent 0·5 per cent. free cyanide.

For testing dilute cyanide solutions it is more convenient to use a standard silver-nitrate solution of half this strength, *i.e.* containing 6·523 grammes of silver nitrate per litre. Each c.c. of this solution is equivalent to 0·005 gramme potassium cyanide (or 1 lb. KCN per ton) ; hence if we take 50 c.c. (a convenient quantity for dilute solutions) of the liquid to be tested, every c.c. of the standard silver nitrate added will represent 0·01 per cent. free cyanide—thus $\dfrac{0·005 \times 100}{50} = 0·01$ per cent.

Example.—If 50 c.c. of cyanide solution required the addition of 7·5 c.c. of standard silver solution, then, since 1 c.c. = 0·01 per cent., 7·5 c.c. will represent 0·075 per cent. of free cyanide. Sometimes a decinormal solution of silver nitrate is used, in which case 1 c.c. = 0·013022 gramme potassium cyanide ; hence, if we take 13 c.c. of the solution to be tested, 1 c.c. N/10 silver nitrate = 0·1 per cent. free cyanide.

The standard silver-nitrate solutions may also be conveniently prepared by dissolving the correct weight of pure or proof silver in dilute nitric acid, evaporating to dryness to expel the excess of acid, and then dissolving in water and diluting to the required volume.

Standardising of Solutions.—The silver-nitrate solution may be standardised by titration with sodium chloride. Weigh out 1 gramme of chemically pure dry sodium chloride, and dissolve in 100 c.c. of distilled water. Take 10 c.c. of this solution by means of a pipette and transfer to a small flask, dilute to 100 c.c., add a little sodium bicarbonate, and 2 c.c. of a solution of neutral potassium chromate as an indicator. Fill a burette with the standard solution of silver nitrate and run into the salt solution until a permanent faint blood-red colour is produced due to the formation of silver chromate. One c.c. of N/10 sodium chloride = 0·0058378 gramme of sodium chloride, equivalent to 0·016966 gramme of silver nitrate. From these values it is easy to calculate the amount of silver nitrate in the number of cubic centimetres of the solution required by a given quantity of sodium chloride, and hence the strength of the silver-nitrate solution ascertained.

Indicator Solution.—The strength of the potassium-iodide solution used as an indicator in testing cyanide solutions varies from a 1 per cent. to a 10 per cent. solution ; the latter strength is more usual. Three or four drops of this solution are added to the cyanide solution before titration.

Titration.—The volume of working cyanide solution taken for the determination of free cyanide varies according to the strength of the solution from 5 c.c. to 100 c.c. The solution is measured in a burette or pipette, then transferred to a small conical flask, the indicator solution added, and the silver-nitrate solution run in from a burette until it gives a precipitate of silver cyanide which does not redissolve on thoroughly agitating the solution in the flask. If the strength of the cyanide is approximately known, the standard solution may be added rapidly at first, and then drop by drop when finishing. The titration should be performed in a good light and the flask placed upon a dead-black surface, so that the first indications of a faint permanent precipitate may be more readily observed. If the cyanide solution used for the determination is measured in a pipette, it is advisable when measuring the desired quantity to place about 9 inches of rubber tubing over the end of the pipette as a safeguard against drawing cyanide solution into the mouth.

The practice varies considerably in different mills, both with regard to the strength of the silver-nitrate solution, and also the amount of the working cyanide solution taken for the test; but, as a general rule, silver solutions of the strength given above are used for the titrations and 50 c.c. or 100 c.c. of the weak cyanide solutions, and 5 c.c. or 10 c.c. of the strong solutions used for the tests. The weak solutions are generally titrated without dilution, but some chemists dilute all solutions containing, say, over 0·10 per cent. KCN, using 50 c.c. of solution diluted with 50 c.c. of water. Below this strength it is not advisable to dilute with water. Strong solutions (using 5 c.c. or 10 c.c.) are diluted to about 100 c.c. with water.

For very strong solutions containing, say, over 1·0 per cent. of potassium cyanide more dilution is necessary ; otherwise the silver cyanide may be precipitated in a granular form not readily soluble in excess of cyanide. In this case 10 c.c. of the solution are diluted with water to a total volume of 100 c.c. Then 10 c.c. of this diluted solution are used for the titration ; but in calculating the result, it must be borne in mind that the diluted solution contains only one-tenth of the original strong cyanide solution.

Titration of Turbid Cyanide Solutions.—There is no difficulty in observing the finishing-point with pure cyanide solutions, but when the solutions are at all turbid, they must be filtered : otherwise the end point cannot be observed with any approach to accuracy. "Solutions occurring in ore treatment sometimes contain very finely divided matter in suspension, which will pass through all ordinary filter papers, and cannot be removed even by repeated filtration. In such cases the liquid may generally be clarified, by the addition of lime,[1] which causes flocculation and settlement of the suspended matter, and allows the liquid to be filtered perfectly clear." This treatment is only permissible, however, with solutions free from zinc, and otherwise comparatively pure ; and when a perfectly correct titration of the free cyanide is required, it is necessary to add potassium iodide indicator after lime has been used.

Liebig's method works admirably with pure cyanide solutions, but gives uncertain and inaccurate results with ordinary working solutions, particularly in the presence of zinc. As, however, it is generally only necessary to obtain relative commercial results, and as a knowledge of the exact strength of the working solution in actual free KCN or its equivalent is not essential, this method is in general use.[2]

Titration of Cyanide Solutions containing Zinc.—Zinc occurs in cyanide

[1] Clennell, *loc. cit.*, p. 7.
[2] Argall, *Mill and Smelter Methods of Analysis*, p. 59.

solutions chiefly as double cyanide of zinc and potassium or other metal. When solutions containing much zinc are titrated as described above, the finishing point is almost always obscure and indefinite owing to the formation of a white flocculent precipitate, which occurs long before the whole of the potassium cyanide has been converted into the soluble double salt of potassium and silver $KAg(CN)_2$. The precipitate probably consists of simple (insoluble) zinc cyanide formed by decomposition of the soluble double cyanide thus [1] :—

$$K_2Zn(CN)_4 + AgNO_3 = KAg(CN)_2 + Zn(CN)_2 + KNO_3.$$

When it is desired to determine the "free cyanide" in solutions containing zinc, it is necessary to first determine the so-called "total cyanide" by the method given below, in which the cyanogen found represents not only the potassium cyanide, but also the double zinc compound. A determination of the zinc (page 377) is also made, and from the results obtained, the amount of "free cyanide" may be readily calculated, as 1 part of zinc corresponds to 4 parts of potassium cyanide. A similar allowance must be made if small quantities of copper are present.

The "free cyanide" in solutions containing zinc therefore equals: total cyanide less zinc × 4. The presence of cyanogen, in the form of ferrocyanides or thiocyanates, does not affect this result.

(b) **Determination of Total Cyanide.** —Strictly speaking, the term "total cyanide" should indicate the equivalent (as potassium cyanide) of all the cyanogen contained in the solution. In practice, however, it is generally taken to mean "the equivalent, in terms of potassium cyanide, of all the cyanogen existing in the form of simple cyanides, hydrocyanic acid, and certain readily decomposable double cyanides, notably that of zinc." Some other double cyanides, such as those of silver and copper, are generally excluded, together with ferro- and ferricyanides, thiocyanate, and similar bodies.[2] The total cyanide is generally determined in practice by titration with silver nitrate after the addition of an excess of caustic alkali. To obtain accurate results, it is necessary to use the potassium-iodide indicator, and it is convenient to add the caustic alkali to the indicator. An alkaline-iodide indicator is therefore prepared by dissolving 40 grammes of caustic soda and 10 grammes of potassium iodide in water and making up to 1000 c.c.

Fifty c.c. of the cyanide solution to be tested are taken and from 5 c.c. to 10 c.c. of the alkaline indicator added. The solution is then titrated with standard silver nitrate (6·523 grammes per litre) until a distinct yellow coloration is obtained, disregarding any white turbidity. The finishing point is generally quite sharp and definite, and if a white cloudiness occurs before the true finishing point, it may in most cases be removed by adding a little ammonium hydrate, which in moderate amounts does not affect the accuracy of the test. Each c.c. of silver nitrate added will represent 0·005 per cent. of potassium cyanide (equivalent to total cyanide).

On the Rand,[3] 5 c.c. of a 10 per cent. solution of caustic soda are added to 100 c.c. of the working solution with the addition of a few drops of a 10 per cent. solution of potassium-iodide indicator. In ordinary practice the 100 c.c. used for the determination of "free cyanide" are also employed for the determination of the total cyanide, after the addition of alkali and indicator.

(c) **Determination of Alkalies.**—Potassium cyanide and other simple cyanides of the same class are alkaline to ordinary indicators such as methyl orange and phenolphthalein, hence the whole of the alkali may be determined by titration

[1] T. K. Rose, *Metallurgy of Gold*, 5th edit., p. 367.
[2] Argall, *loc. cit.*   [3] *Rand Metallurgical Practice*, p. 324.

with standard acid, using methyl orange as an indicator. The double cyanide of zinc and potassium is likewise alkaline to methyl orange.

With phenolphthalein the end point is indefinite owing to the faint action of hydrocyanic acid on this indicator.

"For practical purposes it is most important to know the alkalinity exclusive of cyanide, as it is this alkali which is chiefly of use in preventing the unnecessary waste of cyanide by reactions due to the presence of base-metal compounds, termed 'cyanicides,' and to the carbonic acid of the air." This so-called protective alkali may be determined as described below.

The following explanation of the term "protective alkali," by Clennell,[1] will enable students to understand more clearly the method adopted for its determination.

"The term 'protective alkali' is based on the assumption that certain constituents of an ordinary working solution will be wholly or partially neutralised, on addition of a dilute mineral acid, or carbonic acid, before any decomposition of cyanide occurs. In the treatment of ores, the addition of lime or caustic soda is made with the object of protecting the cyanide from the decomposing effect of various substances contained in the ore, and of the carbonic acid in the air. Since a cyanide solution undergoes gradual decomposition with evolution of hydrocyanic acid, even when an excess of caustic alkali is present, the protection afforded is only partial and temporary. There is, however, a fairly definite point in the neutralisation of an ordinary solution (containing free cyanide, hydrates, carbonates, etc.) at which the decomposition, with formation of hydrocyanic acid, begins to take place with marked rapidity. In the absence of zinc this point corresponds with tolerable exactness to the alkalinity of the hydrates and carbonates present towards phenolphthalein, and if the amount of acid corresponding to this alkalinity be added, with agitation, to such a solution, the cyanide strength, as shown by *immediate* titration with silver nitrate, will remain apparently unchanged." In this case, therefore:

Protective alkali = alkalinity of hydrates + ½ alkalinity of carbonates.

In determining the alkalinity of working cyanide solutions, two methods are generally employed, one to determine the "total alkali," and the other to determine the "protective alkali." Both methods are based on the behaviour of different indicators towards the various substances such as free cyanide, hydrates, carbonates, etc.,[2] likely to be found in working solutions.

(a) *Total Alkali.*—This is usually defined as "the equivalent, in terms of caustic potash (KOH), of all the ingredients which are alkaline to *methyl orange*."

(b) *Protective Alkali.*—In practice this term usually means "the alkalinity which the solutions show to phenolphthalein, after sufficient silver nitrate has been added to convert all the cyanogen of the free cyanides into the double silver salt."[3]

The indicator solutions used in the tests are prepared as follows[4]:—

Methyl Orange.—"This substance dissolves in water, giving an orange-coloured solution, which is turned yellow by alkalies and a pink-red colour by acids. The solution for use as an indicator is prepared by dissolving 0·5 gramme of the solid in 500 c.c. of water; and 1 or 2 drops only should be employed, as its indications are less sensitive if much of it be used."

Phenolphthalein.—"A solution of this compound is prepared by dissolving 1 gramme of the solid in 100 c.c. of alcohol. The dilute neutral solution is

[1] *Chemistry of Cyanide Solutions*, 2nd edit., 1910, p. 61.
[2] For full details of the alkaline constituents in cyanide solutions and the behaviour of different indicators towards them, consult Clennell, *Chemistry of Cyanide Solutions*.
[3] Argall, *loc. cit.*, p. 59.   [4] Newth, *Chemical Analysis*, p. 317.

colourless, but on the addition of an alkali it becomes a deep red colour. This colour is immediately discharged when the liquid is acidified either with mineral or organic acids."

Titration of Total Alkalies.—A measured volume (say 50 c.c.) of the working cyanide solution is transferred to a flask and a few drops of a 0·1 per cent. solution of methyl orange added as an indicator. The liquid is then titrated with N/10 acid (any mineral acid being used, but nitric acid is preferable) until a faint permanent pink tint is produced. When the solution contains double cyanides of zinc, copper, silver, etc., a white precipitate forms at a certain stage on the addition of acid, consisting of the simple cyanides of these metals.

In such cases it is better to add an excess of the standard acid, or, in other words, continue adding acid until no further precipitation takes place, and then to determine the excess of acid in the filtrate, or a definite fraction of it, by titration with standard alkali.

In this case, however, the whole of the alkali metal of the double cyanides will be determined as forming part of the total alkali, as the double cyanides give precipitates which represent a consumption of acid proportional to the amount of such bodies as may be present, the probable reactions being : [1]

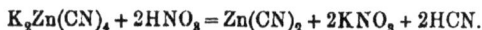

$$K_2Zn(CN)_4 + 2HNO_3 = Zn(CN)_2 + 2KNO_3 + 2HCN.$$

Hydrocyanic acid and carbonic acid do not affect methyl orange. Ferrocyanides, ferricyanides, and thiocyanates of potassium, sodium, etc., are neutral to all indicators. [2]

Titration of Protective Alkalies.—The protective alkalies may be determined (accurately in the absence of zinc) by the method devised by Clennell, which is based on the facts already pointed out, viz. :—

(a) That double cyanides of silver and potassium are neutral to phenolphthalein.

(b) That the amount of acid required to neutralise the hydrates and carbonates towards phenolphthalein is a measure of the protective alkali as defined.

(c) That the presence of cyanide or double cyanide of silver and potassium does not interfere with the titration of alkali, the double salt not being decomposed until the whole of the alkali is neutralised.

The method consists of titration with standard acid and phenolphthalein indicator, after addition of silver nitrate. The method is conducted as follows :—

Silver nitrate is added to a measured quantity of the working solution until a slight permanent turbidity is produced, then a drop of alcoholic 0·5 per cent. phenolphthalein solution added, and the pink solution thus obtained titrated (without filtering) with N/10 nitric acid (or any convenient standard acid) until the colour entirely disappears. The ordinary standard-silver solution may be used, so that the same measured portion of the liquid to be tested will serve for the determination of both " free cyanide " and " protective alkali," which is, of course, a great advantage in practice.

The amount of standard acid used measures the protective alkali.

All results are calculated as the equivalent of caustic potash (KOH). As previously stated, the result indicates generally :—

$$\text{Protective alkali} = \text{hydrate} + \tfrac{1}{2}\text{carbonate.}$$

That is, the equivalent of the hydrates in terms of N/10 acid, added to the equivalent of half the alkali metal in normal (mono) carbonates, in terms of N/10 acid.

The principal substances included in this test are hydrates, carbonates

[1] Clennell, *loc. cit.*, p. 68.   [2] Argall, *loc. cit.*, p. 60.

(converted in the titration to bicarbonates), ammonium hydrate, and a small portion of $K_2Zn(CN)_4$. (Clennell.)

The following are given by Argall as the reactions in a typical case :—

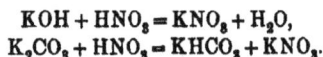

$$KOH + HNO_3 = KNO_3 + H_2O,$$
$$K_2CO_3 + HNO_3 = KHCO_3 + KNO_3.$$

Bicarbonates are neutral to phenolphthalein, hence are not determined. They have no protective influence in this case.

Influence of Zinc.—When solutions containing zinc are titrated with standard acid and phenolphthalein, only a small portion of the double cyanide of zinc and potassium, K_2Zn $(CN)_4$, is shown as alkali. Moreover, the addition of excess of silver nitrate after the first turbidity causes the $K_2Zn(CN)_4$ solution to become quite neutral to phenolphthalein at some point before the complete precipitation of the cyanide. Green[1] has pointed out that, when large quantities of zinc are present (in the absence of much hydrate or carbonate of the alkalies), the addition of silver nitrate causes the solution to become acid to phenolphthalein.

To obviate the difficulties due to the presence of zinc, Green has devised the following method by which the protective alkali is determined after the addition of silver nitrate and potassium ferrocyanide. In this method the total cyanide is first determined in the ordinary way by titration with silver nitrate, using the alkaline-iodide indicator.

Another portion, say 50 c.c. of the original solution, is now taken, an excess of ferrocyanide solution added, and then a little more silver solution than was used in the previous test, to ensure the complete conversion of all cyanide into silver salts. Phenolphthalein is then added, and the liquid titrated with N/10 standard acid, as described above.

The following reactions are supposed to take place (in the absence of alkalies and ferrocyanides) while the double cyanide of zinc and potassium is in excess (Bettell) :—

(a) $K_2Zn(CN)_4 + AgNO_3 = Zn(CN)_2 + KAg(CN)_2 + KNO_3$;

but in the presence of a sufficient excess of silver nitrate,

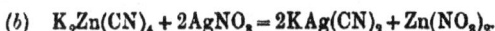

(b) $K_2Zn(CN)_4 + 2AgNO_3 = 2KAg(CN)_2 + Zn(NO_3)_2.$

When excess of ferrocyanide is added to a solution of the double cyanide of zinc and potassium, the following reaction takes place, according to Bettell : [2]—

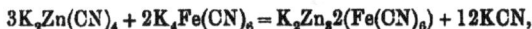

$$3K_2Zn(CN)_4 + 2K_4Fe(CN)_6 = K_2Zn_32(Fe(CN)_6) + 12KCN,$$

so that in the presence of an excess of silver nitrate the solution would be neutral by the conversion of the potassium cyanide into the double cyanide of potassium and silver, $KAg(CN)_2$.

Determination of Ferrocyanides.[3]—The efficiency of cyanide solutions for gold extraction depends not only on the amount of free cyanide present, but also on the amount of oxygen present. Hence any substance which is capable of absorbing oxygen will exercise an injurious effect, and it is sometimes desirable to be able to determine the amount of such deoxidising or reducing agent.

The chief reducing agents present in cyanide solutions are ferrocyanides and thiocyanates (sulphocyanides).

An approximate idea of the relative quantities of reducing agent in various solutions may be obtained by comparing the amount of standard permanganate

[1] L. M. Green, *Trans. Inst. Min. and Met.*, 1901, vol. x. pp. 29–37.
[2] *Journ. Chem. and Met. Soc. of S. Africa*, vol. i. p. 164.
[3] Clennell, *loc. cit.*, p. 70.

required to give a permanent tint to equal volumes of the different solutions, tested under the same conditions, and acidified in each case with sulphuric acid.

Pure potassium cyanide and other simple cyanides, when treated in this way, give hardly any reaction with permanganate, a single drop of N/10 potassium permanganate ($KMnO_4$) added to the acidulated cyanide solution being generally sufficient to establish a permanent pink tint, owing to the fact that the hydrocyanic acid liberated scarcely affects the permanganate at all. When, however, ferrocyanides, thiocyanates, sulphides, nitrites, formates, and some other substances are present, the pink colour at first disappears rapidly as each drop is added; sometimes, however, towards the finish, the colour fades only on standing for a few moments.

A very large number of methods have been devised for the determination of ferrocyanides in cyanide solutions.

Bettell [1] suggests the following method, in which the ferrocyanide is determined by adding a measured quantity of the solution to be tested to a known quantity of acid permanganate.   .

In the absence of organic matter, an acidified solution of a simple cyanide such as potassium cyanide or of a double cyanide (as $K_2Zn(CN)_4$), i.e. a solution of hydrocyanic acid, is not affected by dilute permanganate.

In the presence of zinc, the method is as follows :—

(a) A burette is filled with the cyanide solution to be tested, and the liquid run into 10 c.c. or 20 c.c. of N/100 permanganate, strongly acidified with sulphuric acid until the pink colour is just discharged.

The result = A (calculated for 50 c.c. of original solution).

(b) A solution of ferric sulphate or chloride is acidified with sulphuric acid and 50 c.c. of the cyanide solution poured in. After shaking for about half a minute, the Prussian blue precipitate which forms is separated from the liquid by filtration, and the precipitate and filter paper washed. The filtrate is next titrated with N/100 permanganate.

The result = B. Then : the equivalent of ferrocyanide = A -- B.

In this method the first titration estimates all the reducing agents, and the second all except ferrocyanides, hence the difference between the two results gives the ferrocyanide. The reaction of permanganate on ferrocyanide in the presence of sulphuric acid may be represented as follows :—

$$KMnO_4 + 5K_4Fe(CN)_6 + 4H_2SO_4 = 3K_2SO_4 + MnSO_4 + 4H_2O + 5K_3Fe(CN)_6.$$

N/100 permanganate contains 0·316 grammes of potassium permanganate per litre.

The ferrocyanogen is precipitated by an excess of ferric chloride (or sulphate), in the presence of an excess of hydrochloric acid (or sulphuric acid), as Prussian blue, thus :—

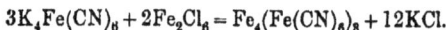

$$3K_4Fe(CN)_6 + 2Fe_2Cl_6 = Fe_4(Fe(CN)_6)_3 + 12KCl.$$

According to W. J. Sharwood,[2] the methods for the determination of ferrocyanides and thiocyanates based upon oxidation by permanganate are unreliable, and he advocates the following method, which consists in first decomposing the ferrocyanogen by evaporation with mineral acids, and then determining the iron contents in the residue. For this purpose 100 c.c. of the solution to be tested are evaporated twice with nitric acid, the residue dissolved in dilute sulphuric acid, and the iron present precipitated by the addition of excess of ammonium hydrate. The precipitate is then at once redissolved in hydrochloric acid, and the iron determined colorimetrically as thiocyanate, unless the quantity is sufficient

[1] *Proc. Chem. and Met. Soc. of S. Africa*, described by Clennell, *loc. cit.*, p. 75.
[2] *Eng. and Min. Journ.*, 1898, vol. lxvi. p. 216. Quoted by Clennell, *loc. cit.*, p. 85.

to allow of reduction by means of metallic zinc and titration with standard potassium permanganate, as in the ordinary method of iron assay.

The result is calculated as follows :—

Weight of iron × 7·562 = weight of crystals of potassium ferrocyanide
$(K_4Fe(CN)_6, 3H_2O)$.

Determination of Thiocyanates (Sulphocyanides). — Sharwood[1] gives the following colorimetric method for the determination of thiocyanates, which is based on the fact that a very small trace of thiocyanate is sufficient to produce in neutral or slightly acid solutions of ferric salts a very characteristic blood-red tinge (see Volhard method, page 303). A measured quantity, 10 c.c. or 20 c.c. of the solution to be tested, is acidulated with hydrochloric acid, a solution of ferric chloride added, and the blood-red colour produced compared with that produced by standard thiocyanate under the same conditions. The comparison is best made against a white background, so that the colour is seen by reflected light. When ferrocyanides are also present, in the presence of hydrochloric acid and ferric chloride, a precipitate of Prussian blue is formed which must be filtered off before the tint can be compared, and the precipitate well washed.

In using this method, Hurter[2] recommends the addition of zinc chloride to precipitate ferrocyanides as ferrocyanide of zinc before applying the coloration test.

The standard thiocyanate used in the above method may be adjusted to the required strength by titrating with potassium permanganate.

The reaction is :—

$$6KMnO_4 + 12H_2SO_4 + 5KCNS = 11KHSO_4 + 6MnSO_4 + 5HCN + 4H_2O.$$

From this equation it will be observed that 1 c.c. of $N/100KMnO_4 = 0·00016193$ gramme KCNS, hence it is preferable to use a N/10 solution of permanganate. In the case of ferrocyanides, 1 c.c. $N/100KMnO_4 = 0·0036831$ gramme $K_4Fe(CN_6)$.

Determination of Calcium (Lime). — Calcium is determined by dissolving the precipitated oxalate and then titrating with permanganate, or by igniting it and weighing as oxide. Clennell[3] recommends the following method, which he states is accurate enough for technical purposes. From 50 c.c. to 100 c.c. of the solution to be tested are acidified with 10 c.c. of hydrochloric acid, then boiled, and the precipitate of zinc ferrocyanide filtered off.

The filtrate is made alkaline with ammonium hydrate, again heated to boiling, and filtered if any turbidity appears. It is then mixed with a boiling solution of ammonium oxalate, allowed to stand until clear (say half an hour or so), filtered, and the precipitate of calcium oxalate washed with hot water, and dissolved and titrated with permanganate. For this purpose the precipitate, after washing free from ammonium oxalate, is returned to the flask in which it was originally precipitated, using a jet of hot water. The paper and funnel is moistened with 10 c.c. of 25 per cent. hydrochloric acid and the washings allowed to run into the flask. The solution is heated to boiling, diluted to 50 c.c. with distilled water, and 5 c.c. of concentrated sulphuric acid added. It is then heated to about 70° C. and titrated with N/20 permanganate (1·5803 grammes $KMnO_4$ per litre), of which 1 c.c. = 0·001 gramme calcium = 0·001 per cent. on 100 c.c. of solution.

The precipitate of calcium oxalate (CaC_2O_4) may also be dried and then converted by a gentle heat into calcium carbonate $(CaCO_3)$, in which form it may be weighed ; or it may be strongly heated until it is entirely changed into calcium oxide, and weighed as such.

---

[1] W. J. Sharwood, *Eng. and Min. Journ.*, 1898, vol. lxvi. p. 216.
[2] *Chem. News*, vol. xxxix. p. 25.
[3] *Loc. cit.*, Appendix, p. 195 ; also *Eng. and Min. Journ.*, 1905.

The following are the factors :—

$$(CaCO_3) \ 100 : (Ca) \ 40 = 1 : 0\cdot4000$$
$$(CaO) \ 56 : (Ca) \ 40 = 1 : 0\cdot71428.$$

Determination of Sulphates.—Sulphates are usually determined by the ordinary gravimetric method, by precipitating with barium chloride and weighing the resultant barium sulphate ($BaSO_4$).

Sharwood[1] recommends adding excess of hydrochloric acid to the cyanide solution to decompose cyanides, heating until all odour of hydrocyanic acid has disappeared, and filtering off any precipitate of zinc or copper ferrocyanides, Prussian blue, or silver chloride that may be formed. The clear filtrate is heated nearly to boiling, after a little hydrochloric acid has been added if necessary, and then a slight excess of barium chloride solution added. After standing several hours the precipitate of barium sulphate is washed carefully by decantation several times with hot water, and finally collected on a small filter, then washed with hot water until free from chlorides, dried, heated to dull redness, and weighed as barium sulphate $BaSO_4$. The dry precipitate should, if not too small, be detached as completely as possible from the paper, which should be burnt separately.

The factor is :—

$$(BaSO_4)233 : (SO_4)96 = 1 : 0\cdot4120.$$

Determination of Oxygen.—A correct determination of the amount of free oxygen or its equivalent would be of the greatest value in determining the efficiency of a cyanide solution, since in the cyanide process, as ordinarily carried out, the solution must contain oxygen in order to dissolve the gold. Unfortunately, no method has so far been suggested which does not involve more or less complicated and delicate manipulations. The only methods proposed for determining the amount of free oxygen in cyanide solutions are based upon well-known processes for the determination of dissolved oxygen in water.

The following method, devised by A. F. Crosse,[2] is a modification of Thresh's iodometric method for the determination of dissolved oxygen in water, and is the method in general use. Thresh's method[3] is based on the fact that a mere trace of nitrous acid will continue to liberate iodine in a solution of hydrogen iodide so long as air or oxygen has access to the solution. The reaction is represented by the following equation :—

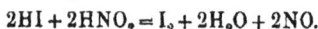

$$2HI + 2HNO_2 = I_2 + 2H_2O + 2NO.$$

The liberated NO then acts as an oxygen carrier, more HI being decomposed, thus—

$$2HI + O = H_2O + I_2.$$

When the reaction takes place in a closed vessel, it continues until all the free oxygen has been used up, and the amount of iodine set free corresponds to the amount of nitrous acid used, plus the amount of oxygen present. It is only necessary, therefore, to add a known quantity of sodium nitrite, together with a little acid and potassium iodide, to a definite volume of water, avoiding the presence of air, and to determine the amount of iodine liberated, in order to ascertain the amount of free oxygen in the water.

Ordinary working cyanide solutions contain substances which prevent the direct application of Thresh's method, but by preliminary treatment these substances can either be removed or neutralised, without affecting the oxygen in

---

[1] *Eng. and Min. Journ.*, 1898, vol. lxvi. p. 216.

[2] *Journ. Chem. and Met. Soc. of S. Africa*, 1898, vol. i. pp. 107–112 ; also pp. 125–127.

[3] J. C. Thresh, *The Examination of Water and Water Supplies*, 1904, p. 282 ; also Sutton's *Volumetric Analysis*, 8th edit., pp. 305–310.

the solution, and thus leave the solutions in a condition suitable for treatment by Thresh's method.

The following solutions are required for Crosse's modification of Thresh's method :—

(1) Combined nitrite and iodide solution, consisting of—
Sodium nitrite, 0·5 gramme.
Potassium iodide, 20 grammes.
Distilled water, 100 c.c.

(2) Dilute sulphuric acid—
Pure concentrated sulphuric acid, 1 part.
Distilled water, 3 parts.

(3) Clear fresh solution of starch.

(4) Sodium thiosulphate—
7·757 grammes per litre of water (1 c.c. corresponds to 0·25 milligramme of oxygen).

(5) Zinc sulphate—
200 grammes of zinc sulphate ($ZnSO_4,7H_2O$) dissolved and made up to 1 litre (1 c.c. of this solution corresponds to 0·2 gramme of $ZnSO_4,7H_2O$).

(6) Bromine water, consisting of—
Bromine, 1 part.
Water, 2 parts.

(7) Phenolphthalein as indicator (a solution of this compound is prepared by dissolving 1 gramme of the solid in 100 c.c. of alcohol).

The determination of the oxygen in cyanide solutions involves the following operations :—

I. Removal of all cyanides by means of zinc sulphate.

II. Testing the solution for iodine absorbents.

III. Determination of the oxygen in the clear solution decanted from the precipitate, by titration with sodium thiosulphate (i.e. Thresh's method).

IV. Correction for nitrites in the solution and reagents used.

V. Calculation of results.

In carrying out the method a large bottle (such as a " Winchester quart ") capable of holding about 2½ litres, and of known capacity, is filled with the solution to be tested, and well stoppered. The contents of this bottle are used for the actual analysis. A smaller, well-stoppered bottle of about 16 ounces capacity is also filled with the solution, which is used for the preliminary tests.

I. *Removal of all Cyanides.*—The solution to be tested is first prepared by the removal of all cyanides. This is effected by the addition of zinc sulphate, and the amount required is determined by a preliminary test, as follows :—

Take 100 c.c. from the small bottle, add a few drops of phenolphthalein as indicator, and titrate with the zinc solution (No. 5 above) until the alkaline action has disappeared, which is seen by the immediate discharge of the characteristic magenta tint. From the amount of zinc solution used, the quantity required for the large bottle, the contents of which are known, can then be calculated (1 c.c. = 0·2 gramme $ZnSO_4,7H_2O$).

The following reactions occur on the addition of zinc sulphate :—

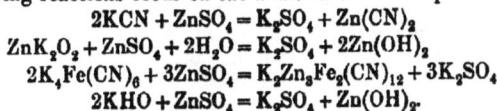

$$2KCN + ZnSO_4 = K_2SO_4 + Zn(CN)_2$$
$$ZnK_2O_2 + ZnSO_4 + 2H_2O = K_2SO_4 + 2Zn(OH)_2$$
$$2K_4Fe(CN)_6 + 3ZnSO_4 = K_2Zn_3Fe_2(CN)_{12} + 3K_2SO_4$$
$$2KHO + ZnSO_4 = K_2SO_4 + Zn(OH)_2.$$

Add 5 or 6 grammes of solid caustic potash to the large bottle and, when dissolved, add the required amount of crystallised zinc sulphate, taking care not to allow any air to enter the solution. Replace the stopper, shake the bottle well, and let it stand until the flocculent precipitate of zinc cyanide, etc., has settled and the supernatant liquid is quite clear: a little scum usually remains on the surface. If possible, let the solution stand overnight. When the precipitate has settled sufficiently the clear liquid is siphoned off without undue access of air. For this purpose the glass stopper is removed and quickly replaced by a rubber stopper, with two holes, and fitted like an ordinary wash-bottle with a glass-tube siphon passed through one hole, and a short bent tube through the other. To start the action of the siphon air is blown momentarily through the short tube. The end of the siphon immersed in the liquid should have a small bag of lint tied over it, to prevent the carrying over of any particles of precipitate.

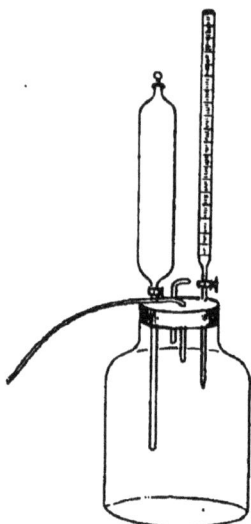

FIG. 162.

Two or three of the pipettes (290 to 300 c.c.) described below are now filled to the mark with the prepared solution, which is siphoned off without allowing access of air. The solution in these pipettes is used for the determination of the oxygen as described subsequently.

II. *Test for Iodine Absorbents.*—A preliminary test of the iodine-absorbing power of the solution (due to unprecipitated double cyanides) is necessary in order to ascertain the correct amount of bromine water to be added in making the test for oxygen. The preliminary test must be made with a quantity of solution equal to that used for the oxygen test; therefore the contents of one of the pipettes (say 290 c.c.), filled as described above, are run into a beaker, then 0·9 c.c. of dilute sulphuric acid (No. 2) and 1 c.c. of potassium iodide and starch added. Then add carefully from a burette dilute bromine water (No. 6) drop by drop until a blue colour is obtained, and note the number of drops required.

In this test potassium iodide is decomposed by the bromine water thus:—$KI + Br = KBr + I$, and the iodine thus liberated is absorbed by the iodine absorbents that may be present. When the absorption is complete the excess of iodine forms, in the presence of starch, the characteristic blue colour of iodide of starch. The number of drops of bromine added therefore represents indirectly the amount of iodine absorbed.

III. *Determination of the Oxygen.*—This operation must be conducted in an atmosphere free from oxygen, and a current of coal gas or carbon dioxide [1] (CO_2) is used to drive out all air in the apparatus used. This may be conveniently arranged by using a wide-mouthed glass bottle of about 500 c.c. capacity, closed by a rubber stopper pierced with four holes: two of which are fitted with glass tubes bent at right angles and used for the entrance and exit of the coal gas passed through the bottle during the experiment; one to receive the pipette containing the solution to be tested (with the addition of the necessary reagents); and the fourth to admit the burette containing the thiosulphate solution for titrating the liberated iodine. The apparatus is illustrated in fig. 162.

The pipette (or separator) is provided with a stopper at the top and a stopcock at the bottom, and a tube 3 or 4 inches long below the stopcock. The capacity of the pipette must be accurately determined. Usually the pipette holds when

[1] Suggested by Bettell, *Chem. and Met. Soc. of S. Africa*, vol. i., 1898, p. 127.

completely full 293 c.c , and has a mark on the neck at 290 c.c.   To displace the air in the bottle, one of the bent tubes is connected with the gas-supply and a current of gas is passed rapidly through the bottle until all the air has been expelled and the gas burns at the outlet tube with a full, luminous flame.   The gas is then turned low, but a constant stream is kept up until the conclusion of the experiment.

If carbon dioxide is used, instead of coal gas, it must be freed from oxygen by passing through a strong solution of potassium pyrogallate made by dissolving pyrogallic acid in caustic potash.

The additions of the dilute sulphuric acid, bromine water, and solution of nitrite and potassium iodide necessary for the reaction are made by means of small hand-pipettes, which are held vertically with the end below the surface of the cyanide solution contained in the large pipette.

The oxygen determination is conducted in the above apparatus as follows :—

The large pipette is filled with the prepared solution by means of the siphon, as previously described.   The same amount of dilute sulphuric acid (say 0·9 c.c. for 290 c.c. of solution) as used in the preliminary test, and the number of drops of bromine water shown by the preliminary test to be necessary, are now introduced, the stopper inserted, and the liquid mixed by turning the pipette over several times.   Next, 1 c.c. of the nitrite and iodide solution (No. 1) is introduced as described, and the stopper immediately replaced.   The stem of the pipette is now inserted through the stopper of the wide-mouthed bottle described above, which has previously been connected with the gas-supply, and also fitted with the burette containing the standard thiosulphate.   The apparatus is filled with gas and allowed to remain at rest for fifteen minutes, to enable the reactions to take place.   The stopper of the pipette is then removed, and the stopcock turned so that the mixture is allowed to run into the bottle.   The stopcock is turned off, and the free iodine (set free in proportion to the oxygen in the solution) is determined by running in the thiosulphate solution slowly until the colour of the iodine is nearly discharged.   About 1 c.c. of starch solution is then introduced through the pipette, thus causing the solution to assume a blue colour immediately.   The titration is now continued slowly until the blue colour disappears : at first the colour, after being completely discharged, returns on standing for a few minutes.   To ensure a slow delivery of thiosulphate solution, it is advisable to run it through a tube drawn out to a fine point, and attached to the stem of the pipette by means of a piece of rubber tubing.

The following reactions occur in the pipette [1] :—

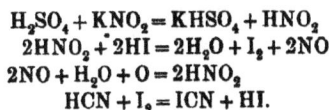

$$H_2SO_4 + KNO_2 = KHSO_4 + HNO_2$$
$$2HNO_2 + 2HI = 2H_2O + I_2 + 2NO$$
$$2NO + H_2O + O = 2HNO_2$$
$$HCN + I_2 = ICN + HI.$$

The reactions after the addition of solutions of thiosulphate and starch are :—

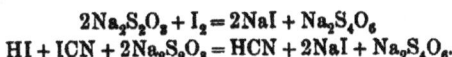

$$2Na_2S_2O_3 + I_2 = 2NaI + Na_2S_4O_6$$
$$HI + ICN + 2Na_2S_2O_3 = HCN + 2NaI + Na_2S_4O_6.$$

As previously stated, 1 c.c. of sodium thiosulphate solution of the strength given corresponds to 0·25 milligramme of oxygen.

In calculating the result, a correction has to be made for the amount of thiosulphate that represents the iodine set free by the reagents, and by any nitrites that may be present in the solution being tested.

[1] A. F. Crosse, *loc. cit.*

IV. *Correction for Nitrites in the Solution and Reagents used.*—The method of ascertaining the amount of the correction to be made is as follows :—

Take a very strong 350 c.c. flask, and pour into it the same quantity of solution as that used in the experiment (say 290 c.c.). Add a few drops of potassium hydroxide (KOH), and close the flask with a rubber stopper having one hole, through which is passed a glass tube with a glass stopcock. Boil the solution for a few minutes to expel the dissolved oxygen, and close the stopcock. Cool the flask, and when cold pour the liquid into the pipette, then add 0·9 c.c. of dilute sulphuric, the required amount of bromine water, and 1 c.c. of the potassium iodide and sodium nitrite solution as before. After mixing, insert the pipette in the stopper of the wide-mouthed bottle, allow to stand for fifteen minutes, and then titrate with thiosulphate in an atmosphere of coal gas, as previously described, adding about 1 c.c. of starch solution as an indicator at the finish. The quantity of thiosulphate required will give the correction for nitrites in the cyanide solution and for the reagents, as the same amount of acid and of iodine and nitrite solution is used in each case.

Qualitative Test for Nitrites.—The presence of nitrites may be tested by acidifying a little of the clear solution (*i.e.* after treatment with potassium hydrate and zinc sulphate) with dilute sulphuric acid and adding a few drops of potassium iodide and starch. If nitrites are present, iodine is liberated in accordance with the above equations, and gives the characteristic blue coloration due to the formation of iodide of starch.

V. *Calculation of Results.*—The amount of oxygen in the cyanide solution being tested is calculated from the following formula, and reported in milligrammes per litre :—

Let—

A = capacity of the pipette used, minus the quantity of reagents used, say 293 c.c. - 3 c.c. = 290 c.c.

B = correction for nitrites and oxygen in the reagents used.

C = quantity of thiosulphate used in the final determination.

x = milligrammes of oxygen per litre of the solution under examination.

Then—

$$\frac{(C - B) \times 0·25 \times 1000}{A} = x.$$

In actual analysis the following results were obtained by Argall [1] :—

C = 10·2 c.c.   B = 2·8 c.c.

Then—

$$x = \frac{(10·2 - 2·8) \times 0·25 \times 1000}{290} = 6·3 \text{ milligrammes of oxygen per litre.}$$

Precautions.[2]—In all stages of the analysis care should be taken to prevent the admission of air to the solution being tested. The bromine water in the lower part of the burette under the stopcock quickly deteriorates by loss of bromine, and should, therefore, be run off before beginning the titration.

Another and shorter method of determining the oxygen in cyanide solutions, proposed by Professor Prister,[3] consists in boiling off the air contained in the solution, collecting and measuring the gases in a Lunge nitrometer, and removing the oxygen by means of a strong solution of potassium pyrogallate. The residual gas (nitrogen, etc.) is again measured, and the difference represents the oxygen.

[1] *Loc. cit.*, p. 89.   [2] Argall, *loc. cit.*, p. 89.
[3] *Journ. Chem. and Met. Soc. of S. Africa*, 1904, p. 364.

Determination of Gold and Silver in Cyanide Solutions.—The amount of gold in cyanide solutions obtained in the cyanide process varies considerably. The solutions entering the precipitators may contain up to 10 dwt. of gold per ton on a rich mine, whilst the solution leaving the precipitation boxes may contain merely a trace of gold, less than 0·005 dwt.[1] For "rich" solutions 1 or 2 A.T. of the liquid may be taken, and for "poor" solutions 40 A.T. or more could be used. In general mine assay-office practice, however, it is advisable to adopt a fixed quantity, and it will generally be found that when using a balance turning to $\frac{1}{100}$ mg., 20 A.T. (584 c.c. for the ton of 2000 lbs.) of solution is a convenient amount.

Since the specific gravity of ordinary gold-bearing cyanide solutions is very little above that of water, exactly 10 A.T., 20 A.T., or even more, of solution may be measured out for assay. It may be well to point out that where liquids are being assayed, cubic centimetres are held to be equivalent to grammes, and the results may be stated as "so many parts by weight in so many parts by measure." By using measured quantities of solution corresponding to the assay-ton system, the weight of gold obtained from the assay gives at once, or with little calculation, the amount of gold per ton. The results of assays of cyanide solutions are reported in pennyweight and decimals of a penny-weight per ton of 2000 lbs.

Many methods have been proposed for the assay of gold in cyanide solutions, but for most accurate work the evaporation method is generally employed. This method,

FIG. 163.

however, entails much tedious work if some 15 or 20 determinations are required per day, and technical methods, in which rapidity is aimed at rather than extreme accuracy, have been devised and are in use in many mines where a large number of determinations have to be made.

The various methods may be summarised as follows :—

Evaporation Method (1).—A convenient volume of solution (preferably 10 or 20 A.T.) is measured out, poured into a glazed porcelain dish, about 10 grammes of litharge added, and the solution carefully evaporated to almost complete dryness on a hot plate. It is desirable to evaporate the solution somewhat before adding the litharge. The heat must be reduced towards the end of the operation, otherwise loss may result owing to "spurting," or from caking of the dried residue on the dish. Should caking take place, the hard scale may be removed by moistening with dilute nitric acid. Assays that spurt must be repeated. When dry, the residue is removed by means of a spatula, and transferred to a glazed paper. Any litharge adhering to the dish is removed by adding a little borax, and rubbing with a small piece of filter paper, thus thoroughly wiping out the dish. If the evaporation has not been carried too far, this can readily be done. The residue is then mixed with about 30 grammes of a flux consisting of [2]—litharge, 100 parts ; sand, 25 parts ; charcoal, 3 parts. Fuse the charge, cupel the resulting lead, and part as usual. Fusion is attained in about twenty minutes.

Evaporation Method (2).—Evaporation may also be performed in dishes or trays made of lead foil, about 4 inches × 2 inches, and 1 inch high, and weighing from 25 to 35 grammes (fig. 163). Such a dish is easily extemporised out of a piece of lead foil. For example, lead foil rolled to a thinness of No. 2 Birmingham metal gauge, and cut to 6 inches × 4 inches, will make a dish of the dimensions given above, and weighing about 25 grammes. From 1 to 2 A.T. (30 to 60 c.c.) of the solution are measured out, placed in the lead dish, and evaporated to

[1] M'A. Johnston, *Rand. Met. Pract.*, 1912, vol. i. p. 311.

[2] M'A. Johnston, *loc. cit.*, p. 311.

dryness, care being taken to prevent loss from spurting during the last few moments on the hot plate.

The dish with the dried residue is rolled up and then scorified and cupelled, the resulting bead being parted and weighed in the usual manner. This method has the disadvantage of permitting the use of comparatively small quantities of solution only. When the gold and silver are not present in sufficient quantity to give a bead of convenient size for weighing, with 1 to 2 A.T. of solution, a larger quantity must be used. If 20 or 30 A.T. are taken for assay, evaporation must be begun in glass or porcelain vessels, and finished in a lead tray after having been very carefully transferred when the solution has been sufficiently reduced in bulk.

Lead Acetate Method (Chiddey)[1].—In this method zinc shavings and a lead salt are used to produce a lead sponge on which the gold is precipitated and subsequently separated by cupellation. The quantities of the several reagents used, and the method of procedure adopted, vary somewhat with different assayers.

An adaptation of the method, which, according to M'A. Johnston,[2] gives more uniform results, comparable indeed with the evaporation assay, is to measure 584 c.c. (20 A.T.) of solution into a beaker of 500 to 700 c.c. capacity, boil, add 30 c.c. of saturated lead acetate solution, and about 3 grammes of zinc shavings, and again boil for half an hour. Add 25 c.c. of commercial hydrochloric acid, and keep at nearly boiling-point until the zinc is dissolved. The heating is continued until all effervescence has ceased, and the reduced lead has collected into a spongy mass. The spongy lead is washed by several decantations in water and then pressed into a ball, with the aid of a flattened glass rod, and then with the finger, care being taken to include any small loose pieces. The ball is dried slightly in a cupel tray over a hot plate, and then cupelled.

A. J. Clark [3] recommends the addition of 10 to 20 c.c. of lead acetate solution containing 10 to 20 per cent. of the salt, and then adds zinc dust (3 to 4 grammes) in the form of an emulsion or suspended in water. This latter is prepared by mixing 100 grammes of zinc dust with 300 c.c. of water in a bottle fitted with an ¼-inch glass tube passing through the cork. This mixture is shaken into a capsule of the proper size used as a measure. The lead sponge after treatment with 20 c.c. of strong hydrochloric acid, as described, is filtered on a "quick" paper, and washed twice. The sponge and filter paper are removed and as much water as possible squeezed out. It is then placed in a 2-inch scorifier with 10 to 15 grammes of test lead and 3 to 5 grammes of borax glass. Transfer at once to the muffle, burn the paper, scorify for only a few minutes, pour, cupel the lead button, part, and weigh. Unless silver is to be determined, silver foil should be added to the scorifier for inquartation, or a measured volume of dilute silver nitrate solution may be added to the cyanide solution from a burette.

Fulton states that "Clark's method gives somewhat better results on low-grade solutions than evaporation methods with litharge or litharge-bearing flux." It is also more rapid than evaporation.

Electrolytic Method (Crichton).[4]—In this method the gold is deposited on a lead cathode which is cupelled to obtain the gold. The beakers, each containing 10 or 20 A.T. of cyanide solution, are arranged in a row on a special stand. The anodes consist of $\frac{1}{16}$-inch carbon rods, held in position in the centre of each beaker by means of clamps fitted to a horizontal copper bar which runs parallel

[1] A. Chiddey, *Eng. and Min. Journ.*, lxxv., 1903, p. 473; see also Wilmouth, *Journ. Chem. Met. and Min. Soc. of S. Africa*, 1909, p. 140.

[2] *Rand Metallurgical Practice*, 1912, vol. i. p. 312.

[3] Allan J. Clark, Homestake Mining Co., described by Fulton, *Fire Assaying*, 2nd edit., 1911, p. 156.

[4] C. Crichton, *Journ. Chem. Met. and Min. Soc. of S. Africa*, 1911-12, pp. 90-92.

to and 6 inches above the line of beakers. Thus all the anodes can be lowered or lifted in one operation. The cathodes consist of cylinders of assay lead foil which fit just inside the beakers and have the lower edges serrated to allow better circulation. The cathodes are connected by means of insulated flexible wires with a copper rod which runs along the front side of the row of beakers. The terminals are made by almost severing narrow strips of the cathodes and folding the strips over the ends of the flexible wires.

The current for deposition of the gold is obtained from three 2-volt accumulator cells in series, and varies in amperage according to the strength of the cyanide solutions used. Thus when the strength is 0·3 per cent. KCN, 0·1 ampere is required; and when it is 0·02 per cent. KCN, the current is about 0·04 ampere. The time required for complete deposition of the gold is from three to six hours, the quantity of solution used being immaterial.

With 10 or 20 A.T. of solution complete deposition of the gold is effected in four hours.

When deposition is complete the carbons are raised clear of the beakers, the current switched off, the lead cathodes disconnected and removed to a hot plate to dry. When dry, these are folded into a small compass and cupelled with a little silver, and the buttons parted and weighed. To assist the deposition of the gold in weak solutions a little KCN should be added, and in impure solutions a little ammonium hydrate is advantageous. Care should be taken to see that the current is switched on during the time the anodes are in the solution. Assays made by this simple method gave results identical with those obtained by evaporation with litharge. The method has been in constant use for more than four years at the Kleinfontein Group Central Administration Assay Office (South Africa), and no inaccuracies in the results obtained have been detected.

Precipitation Methods.[1]—Various reagents have been suggested for precipitating the gold in cyanide solutions for assay purposes. De Wilde[2] adds a few drops of potassium ferrocyanide solution (5 per cent. strength), followed by sufficient acid cuprous chloride solution to bleach the first-formed precipitate of cupric ferrocyanide, so that the precipitate remains nearly white when stirred.

Whitby[3] recommends 25 c.c. of a 10 per cent. solution of copper sulphate, then 5 to 7 c.c. concentrated hydrochloric acid, and lastly, 10 to 20 c.c. of a 10 per cent. solution of sodium sulphite.

Seamon[4] employs aluminium foil, and aids the reactions by adding sulphuric acid and boiling.

The gold precipitate, however obtained, may be fused with litharge flux or scorified.

Reporting the Results.—In reporting assay values of cyanide and other solutions, the results should be given in parts by weight in a stated volume of the solution.[5] Usually the report is made in pennyweight per ton of solution. The use of the "fluid ton of 32 cubic feet" is recommended. It closely approximates to 2000 lbs., and is in common use.

Determination of Zinc.—"Zinc is generally supposed to occur in cyanide solutions as a double cyanide of the type of $K_2Zn(CN_4)$, though there is some evidence to show that this is partially dissociated in dilute solutions, which probably contain a portion of the zinc as the simple cyanide $Zn(CN)_2$. In certain cases potassium zincate, $Zn(OK)_2$, or some similar compound, may be present" (Clennell).

[1] See *Rand Metallurgical Practice*, vol. i., 1912, p. 312.
[2] White, *Journ. Chem. Met. and Min. Soc. S.A.*, Oct. 1909, p. 136.
[3] Whitby, *Journ. Chem. Met. and Min. Soc. S.A.*, vol. iii., 1902, p. 15.
[4] Seamon, *Western Chem. and Metall.*, 1909 (Aug.), p. 291.
[5] *Trans. Inst. Min. and Met.*, vol. xx., 1911, p. 530.

Two methods for the determination of zinc in cyanide solutions have been devised by A. F. Crosse, in both of which the zinc is precipitated as sulphide, and the zinc sulphide decomposed with an oxidising agent.

A solution of sodium sulphide (Na_2S) for the precipitation of the zinc is prepared by dividing a solution of sodium hydroxide (caustic soda) into two equal portions, saturating one portion with sulphuretted hydrogen and then mixing it with the other portion.

When zinc is precipitated from a cyanide solution by the direct addition of an alkaline sulphide, the precipitation is never quite complete, as zinc sulphide is slightly soluble in alkaline sulphides. In all cases the precipitation is best made in a hot solution.

Crosse's methods are conducted as follows : [1]—

(a) *Zinc Sulphide decomposed by Ferric Sulphate.*—This method is stated to be very exact and fairly quick for solutions containing not more than a trace of copper. The zinc sulphide is decomposed by ferric sulphate, a proportionate amount of which is reduced to ferrous sulphate, which is determined by titration with potassium permanganate as in the ordinary method of iron assay.

"Take 300 c.c. of solution, add about one gramme of KCN and the same quantity of pure caustic potash or soda, heat nearly to boiling-point, and then add a slight excess of sodium sulphide solution. The zinc will be quickly precipitated as sulphide, and should be collected on a filter paper and washed with hot water. Then place the filter paper in a wide-mouthed bottle of known capacity, between 250 and 300 c.c. This bottle must be provided with a well-fitting india-rubber bung through which is inserted a moderately wide tube, about 8 or 10 inches long. Then fill up the bottle with a weak solution of pure ferric sulphate, containing 5 per cent. to 7 per cent. of sulphuric acid, place the bottle or flask in a bowl of cold water, and raise the temperature to boiling-point. The reason for the glass tube will be apparent, as it allows for the expansion of the liquid.

"The zinc sulphide will have decomposed ($ZnS + Fe_2(SO_4)_3 = 2FeSO_4 + ZnSO_4 + S$) and reduced a proportionate amount of ferric sulphate to ferrous sulphate.

"When nearly cold, filter off the solution through a dry filter paper, take half the quantity contained in the bottle, and determine the amount of ferrous sulphate in the solution by titration with N/10 permanganate: 1 c.c. N/10 $KMnO_4 = 0\cdot003285$ gramme Zn." The reaction is :—

$$10FeSO_4 + 2KMnO_4 + 8H_2SO_4 = 5Fe_2(SO_4)_3 + 2MnSO_4 + K_2SO_4 + 8H_2O. $$

The results in the presence of ferrocyanides and thiocyanates were found by Crosse to be very satisfactory.

Owing to the slight solubility of zinc sulphide in weak cyanide solutions, an addition of 1 milligramme of zinc per 100 c.c. of solution taken should be made as a correction to the result obtained.

(b) *Zinc Sulphide decomposed by Iodine.*[2]—In this method an excess of standard iodine solution is added to decompose the zinc sulphide, and the excess of iodine determined by titration with a standard solution of sodium thiosulphate.

The solutions required are : (a) N/10 iodine, (b) N/10 thiosulphate, (c) 5 per cent. sodium sulphide. The method is as follows:—100 c.c. of the working solution is heated to about 70° C., and an excess of sodium sulphide added. The mixture is allowed to settle, then filtered, and the precipitate of zinc

[1] *Journ. Chem. and Met. Soc. of S. Africa*, vol. iii., 1902, p. 4. Quoted by Clennell, *loc. cit.*, p. 130.
[2] *Ibid.*, p. 165. Quoted from Mohr's *Volumetric Analysis*, p. 338.

sulphide well washed on the filter with hot water; the washing is continued until the wash-water shows no trace of sulphide; the filter paper is then transferred to a small 150 c.c. flask. This flask is fitted with a rubber cork having two holes; through one of which passes a drip funnel with tap, and through the other a short piece of glass tube, terminating in a small length of rubber tubing. A pinch-cock is attached to the latter. Sufficient N/10 iodine is then run into the flask to leave an excess of not more than 5 c.c. (30 to 35 c.c. N/10 iodine is usually sufficient). The cork is then replaced, and about 100 c.c. of very dilute hydrochloric acid is placed in the funnel. The tap is opened, and on pressing the pinch-cock air escapes and the dilute acid takes its place. The taps are then closed, and the whole apparatus shaken to thoroughly break up the filter paper. After a few minutes' standing the contents of the flask are titrated with N/10 sodium thiosulphate, and the excess of iodine determined.

If X = No. of c.c. N/10 iodine taken,

Y = No. of c.c. N/10 sodium thiosulphate required,

then $(X - Y) \times 0.003285 = $ grammes of zinc per cent.

The reactions are as follows:—

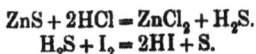

$$ZnS + 2HCl = ZnCl_2 + H_2S.$$
$$H_2S + I_2 = 2HI + S.$$

The method is rapid of execution and capable of great accuracy.

Determination of Copper.[1]—The presence of copper in the ore under treatment by the cyanide process, especially when the metal occurs as carbonate, has a very injurious effect, giving rise to high consumptions of cyanide. Hence in some cases it becomes a matter of importance to determine the quantity of copper in the solution.

The presence of copper, even in very small amount, may be detected by acidulating the cyanide solution with any mineral acid, and adding a few drops of dilute ferrocyanide solution, which gives the characteristic reddish-brown colour.

Ammonium hydrate, of course, gives no blue colour in solutions containing soluble cyanides, so that complete decomposition (e.g. by boiling with sulphuric and nitric acids) is necessary before this reagent can be applied for the detection of copper.

Copper may in most cases be rapidly determined by the colorimetric method with ammonium hydrate as follows:—

100 c.c. of the solution are strongly acidulated with HCl, heated to boiling, then 0.5 gramme potassium chlorate added, and the boiling continued until most of the chlorous gases are expelled. The liquid is then made alkaline with ammonium hydrate and filtered. The tint of the filtrate is compared with that of a similar volume of water, to which HCl and ammonium hydrate have been added in about the same quantities as used in the test, and to which standard-copper solution is added until the colours of the two liquids are alike.

For this purpose a standard solution of copper nitrate is prepared containing 0.1 per cent. of copper and about 1 per cent. of sulphuric acid. The ammoniacal copper solution from the cyanide liquid, diluted to a suitable volume (according to the amount of copper present), say 10 c.c. for every milligramme of copper, is placed in a clear glass cylinder, and a nearly equal volume of distilled water in a similar cylinder. To the latter is added the same amount of hydrochloric acid and ammonium hydrate as was used in the test, and then the standard-copper solution run in from a burette, with constant shaking, until the tint in the two cylinders is the same.

[1] Clennell, loc. cit., p. 133

Each c.c. of solution run in represents 0·001 gramme of copper. The best ·results are obtained when the amount of copper is between 5 and 15 milligrammes. With practice, the amount of copper present may be determined to within 0·0002 gramme.

The treatment of the solution with potassium chlorate described above is generally sufficient for the complete oxidation of all cyanogen compounds, but in a few cases it may be necessary to add a little bromine.

When much ferrocyanide is present, however, the decomposition should be effected by evaporation with nitro-hydrochloric acid, followed by sulphuric acid and boiling until white fumes of SO_3 are freely given off.

In cases where the quantity of copper present is sufficient, it may be determined accurately in the liquid by taking 100 c.c. or more of the solution, adding 5 c.c. of concentrated nitric acid and 5 c.c. of concentrated sulphuric acid, and boiling until white fumes are given off. This liquid is carefully diluted to about 100 c.c., then 10 c.c. of concentrated sulphuric acid are added, one or more sheets of aluminium foil introduced, and the solution boiled for ten minutes. The metallic copper thus precipitated is dissolved, and the copper determined by the ordinary analytical methods, such as titration with standard sodium thiosulphate.

Remarks on Commercial "Cyanide."—The "cyanide" used in cyanide mills and known in commerce as "potassium cyanide" actually consists of the mixed cyanides of sodium and potassium, containing varying amounts of impurities such as alkaline carbonates, sulphates, etc. It has been found that potassium cyanide carrying sodium cyanide is more soluble and more efficient than pure potassium cyanide, weight for weight, owing to the larger percentage of cyanogen it contains, as shown below. The combined cyanide can be made and sold more cheaply than the pure potassium cyanide, and is therefore invariably used. What is wanted in most cases is merely a soluble cyanide, and it is a matter of indifference whether the base be sodium or potassium.

It is customary to express the quality of a sample of commercial cyanide by stating that it contains so much per cent. of potassium cyanide; or, in other words, the quality is expressed by the cyanogen content, in terms of potassium cyanide. Since, however, sodium cyanide contains more cyanogen than is contained in an equal weight of potassium cyanide, this method of expressing the quality gives rise to expressions such as "130 per cent. cyanide." This is explained as follows :—

The molecular weight of sodium cyanide (NaCN) is 49, and of potassium cyanide (KCN) 65; and since 49 parts of sodium cyanide are equivalent to 65 parts of potassium cyanide, it is evident that a pure sample of sodium cyanide (= 100) would contain cyanide, equivalent to 132·7 per cent. of potassium cyanide.

The molecular weight of cyanogen (CN) is 26, therefore the percentage of cyanogen in sodium cyanide is :—

$$\frac{CN \times 100}{NaCN} = \frac{26 \times 100}{49} = 53\cdot1 \text{ per cent.,}$$

and the percentage of cyanogen in potassium cyanide is :—

$$\frac{CN \times 100}{KCN} = \frac{26 \times 100}{65} = 40\cdot0 \text{ per cent.}$$

Pure potassium cyanide containing 40 per cent. of cyanogen corresponds in the trade to "100 per cent. cyanide," and the percentage of cyanogen in commercial samples is calculated on this basis. Thus "98 per cent. cyanide," which is in common use, contains only 39·2 per cent. cyanogen, and "125 per cent. cyanide" contains 50 per cent. cyanogen.

Assay of Commercial Cyanide.— Since the "potassium" cyanide of commerce may contain impurities such as alkaline carbonates, and small quantities of alkaline chlorides, sulphates, and sulphides, it is usual to test it for its cyanide contents.

Determination of Free Cyanide.—Break off 20 grammes of the sample of cyanide to be tested, in clean fresh pieces, and weigh accurately to the nearest centigramme. The portion selected should be taken through the thickness of the cake so as to obtain a representative sample. Dissolve in water containing a little sodium hydrate, transfer the solution to a 2-litre flask, and dilute with distilled water to 2 litres. After thorough mixing, a measured portion, 50 or 100 c.c., is titrated, as previously described on page 360, with standard silver-nitrate solution and a few drops of potassium iodide as an indicator. It is well to bear in mind that the result does not give the actual potassium cyanide present, but the cyanogen, which may in reality be present as a sodium, calcium, ammonium, or other salt. This cyanogen is, however, reported in terms of its equivalent in potassium cyanide, as stated above. The percentage may be calculated by multiplying the number of c.c. of standard silver nitrate (10 c.c. equivalent to 0·1 gramme KCN) used by 40 (50 c.c. being one-fortieth of 2 litres) and dividing by the weight of commercial cyanide originally taken.

Detection of Impurities in Commercial Cyanide.—It is sometimes desirable to test for impurities in commercial cyanide, although it is seldom necessary to make exact determinations. The commonest impurity in commercial cyanide is carbonate of sodium or of potassium.

The various impurities may be detected in the following manner :—

Detection of Carbonates,—These may be tested for by dissolving, say, 2 grammes of the commercial cyanide in a little water and adding a solution of barium chloride. If carbonates are present a white precipitate of barium carbonate will be formed which, if filtered off, washed, and treated with hydrochloric acid, will dissolve with effervescence.

The amount of carbonic acid representing the carbonates present may be determined [1] by dissolving 10 or 20 grammes of the commercial cyanide in water, adding calcium nitrate, then shaking, and allowing the solution to stand. The precipitate of calcium carbonate is then filtered off, washed, ignited, and weighed as CaO. The weight of CaO multiplied by 0·786 gives the weight of CO_2 in the 20 grammes of cyanide taken.

Detection of Chlorides.—Chlorides may be detected by adding an excess of silver-nitrate solution, which gives a white curdy precipitate of silver chloride. Chlorides may be determined volumetrically as follows [2] :—

L. Siebold has shown "that chlorides, when present, may be conveniently determined in the same liquid in which the cyanide has been determined by neutralising the excess of free alkali (which should not be ammonia) by the cautious addition of dilute nitric acid, adding a few drops of a solution of neutral potassium chromate, and continuing the addition of the silver solution until the red tint due to the formation of silver chromate remains permanent. If cyanide only be present, the volume of silver solution now required will be exactly equal to that previously employed to obtain a permanent turbidity, whereas any excess over this amount represents the silver solution corresponding to the chlorides present."

[1] Seamon, *Chem. News*, 1910, p. 18, from *The Chemical Engineer*, vol. x., 1910.
[2] Quoted by Argall, *Mill and Smelter Methods*, p. 81.

Using a standard solution containing 13·039 grammes of silver nitrate per litre, the silver equivalents for complete precipitation will be as follows [1] :—

$$1 \text{ c.c. standard } AgNO_3 = \begin{cases} 0\cdot005 \text{ gramme potassium cyanide.} \\ 0\cdot002723 \text{ gramme chlorine.} \\ 0\cdot004489 \text{ gramme sodium chloride.} \\ 0\cdot005726 \text{ gramme potassium chloride.} \end{cases}$$

Detection of Sulphates.—Sulphates are detected by the formation of a white precipitate of barium sulphate obtained on adding a solution of barium chloride to the cyanide solution which has been previously acidulated with hydrochloric acid to effect the decomposition of the cyanides. The determination of sulphates is usually made by the ordinary gravimetric method with barium chloride. The cyanide solution may be completely decomposed by evaporating several times to dryness with nitric and hydrochloric acids, and finally with the latter alone. The residue is taken up with water and a little hydrochloric acid, the solution heated nearly to boiling, and barium chloride added in slight excess. Evaporation to dryness may be avoided by using the method recommended by Sharwood, described on page 370.

Detection of Sulphides. — Soluble sulphides produce a marked effect in checking the solution of the gold, consequently in South Africa the amount of sulphide which a commercial cyanide should contain is limited to the second place of decimals per cent.[2] A very small percentage of alkaline sulphides frequently occurs, however, in samples of commercial cyanide, and it is sometimes desirable to test for them, although it is seldom necessary to make exact determinations.

The presence of sulphides may be detected by adding a drop or two of a solution of lead tartrate to 4 or 5 c.c. of caustic soda solution and adding this alkaline lead salt to a clear solution of the commercial cyanide. If any sulphide is present, a brown coloration or a black precipitate will be formed according to the amount present.

"A solution of lead tartrate in excess of soda is found to be most suitable as a test for the presence of soluble sulphides, especially in the presence of cyanides, since ordinary lead salts yield a white precipitate of lead cyanide which may obscure the black or brown coloration" (Clennell).

The most delicate test for sulphides is to add a solution of nitro-prusside to the cyanide solution. The most minute trace of alkaline sulphide is shown by the appearance of a brilliant purple colour. Nitro-prussides are formed by adding nitric acid to a solution of potassium ferrocyanide, as described below.

Dr Loewy's method [3] for the determination of sulphides in cyanide solutions is to add sodium nitro-prusside, which gives the violet colour. He dissolves 2 grammes of the sample of cyanide to be tested in 100 c.c. of water, and compares it with a solution of cyanide of similar strength free from sulphides. He then adds 1 c.c. of a solution containing 5 per cent. of sodium nitro-prusside and a very little cyanide to each solution, and then adds a solution of sodium sulphide of known strength to the pure cyanide until the violet colour is of the same depth in each. The calculation of the result is then simple. The method is applicable to solutions in which the amount of sulphur lies between 0·0005 and 0·0015 per cent. The solutions are adjusted to bring the sulphur within these limits. The sodium nitro-prusside solution is prepared thus :—

Preparation of Sodium Nitro-prusside.[4]—Concentrated nitric acid is diluted

[1] Clennell, *loc. cit.*, p. 109.
[2] Salamon, *Journ. Soc. Chem. Ind.*, vol. xxix., 1910, p. 1426.
[3] Described by T. K. Rose, *Met. of Gold*, 5th edit., p. 371.
[4] Clennell, *Cyanide Solution*, p. 93.

with its own volume of water. This diluted acid is mixed with powdered potassium ferrocyanide in the proportion of :—

2 parts $K_4Fe(CN)_6 3H_2O$.
5 parts dilute HNO_3.

The salt dissolves to a coffee-coloured liquid evolving CO_2, N, CN, and HCN. It is warmed on a water bath until the liquid gives a dark green or slate-coloured precipitate instead of a blue precipitate with ferrous sulphate. It is then cooled, neutralised with sodium carbonate, and filtered.

It may be pointed out that the presence of sulphides is shown at once when a sample of commercial cyanide is being tested for its strength in cyanide, inasmuch as the first few drops of silver-nitrate solution produce at once a brown coloration due to the formation of silver sulphide. This coloration obscures and entirely vitiates the result of the titration. When sulphides are present they are removed by precipitation before titrating, by adding freshly precipitated lead carbonate and shaking the solution.

Detection of Silicates.—The presence of silicates can be detected and the amount determined by evaporating a solution of the commercial cyanide to dryness with hydrochloric acid. The residue is treated with water acidulated with hydrochloric acid and the insoluble residue of silica filtered off, washed, dried, ignited, and weighed.

4. Assay of Cyanide Bullions.

Bullion samples consist of (*a*) mill bars, (*b*) cyanide bars, (*c*) lead bars. Zinc-gold slimes are also included under this head.

(*a*) *Mill Bars.*—The term "mill bars" is given to the bars of gold resulting from the treatment of ores by amalgamation prior to cyanide treatment. The standard of mill bars is higher than that of bars obtained by the cyanide process. The bars are sampled and assayed as previously described in Chapter XVII. As a rule, 0·5 gramme is taken for assay and the assay made in duplicate.

(*b*) *Cyanide Bars.*—As previously pointed out, these bars vary greatly in composition and present considerable difficulty in sampling owing to the liquation which takes place on solidification. The methods of sampling and of assaying cyanide bars are fully discussed on page 254.

The assays are usually made on 0·5 gramme samples, and on account of the irregularity of composition it is advisable to make the assays in triplicate and to use checks.

Scorification, before cupellation, is sometimes recommended for cyanide bars of very low quality, say below 700 fine.

(*c*) *Lead Bars.*—These consist of bars of lead bullion obtained by melting the zinc-box precipitate (often called "gold slimes") with litharge and fluxes in a reverberatory furnace. The lead bars are usually soft, and contain up to about 8 per cent. of gold. The bars are subject to more or less liquation during solidification, and care is required in sampling, as discussed on page 247.

The methods of sampling in use on the Rand, South Africa, are as follows[1] :—

(1) Dip sample taken from the furnace before tapping.
(2) Dip sample taken from the kettle that receives the molten metal before casting into moulds (if such is used).
(3) Sampling the cast lead bars by drilling or cutting chips.

All three methods are said to give equally good results if the lead does not contain more than from 4 to 5 per cent. of gold and silver.

[1] *Rand Metallurgical Practice*, vol. i. pp. 280 and 319.

The usual method of assay on the Rand is to scorify from 0·5 to 1 A.T. with from 40 to 50 grammes of lead and a little borax and then to cupel and part in the ordinary way. If the approximate fineness of the lead bullion is known, the silver necessary for parting purposes can be added direct during the scorification or cupellation. The assays should be made in duplicate, and in cases where extreme accuracy is desired, checks should be used.

Details of the methods of sampling and assaying lead bullion are given in Chapter XX.

Assay of Zinc-Gold Slimes.—This is the precipitate formed by zinc acting on cyanide solutions of gold. The precipitate varies greatly in composition, and gold may be present in any proportion up to about 20 per cent.

The average composition in South Africa is approximately as follows [1] :—

Gold and silver   .   .   .   .   . 10 to 20 per cent.
Zinc   .   .   .   . 30 to 60   ,,
Base metals, silica, alumina |
Iron oxides, etc.   } .   .   . 20 to 40   ,,

The amount of impurity, especially of zinc, contained in the precipitate is largely dependent on the preliminary treatment to which it has been subjected. The sample is received in a dry condition, but before being assayed the moisture must be very carefully determined: it is preferable to make the same sample serve for both moisture and assay.

The precipitate may be assayed either by crucible fusion or by scorification ; both methods are in general use. The following is a suitable stock flux :—

Litharge   .   .   .   .   .   . 600 grammes.
Sodium carbonate   350   ,,
Borax glass 100   ,,
Sand 100   ,,
Charcoal   8   ,,

In some charges recommended for the assay of gold slime the sand is omitted, but in this case there is a greater tendency for the slag to retain gold and silver.

For one-tenth assay ton (3 grammes) of precipitate, take 80 grammes of the flux.

For one-fifth assay ton (6 grammes) of precipitate, take 120 grammes of the flux.

The weighed quantity of precipitate is mixed with two-thirds of the flux and the remainder used as a cover. The charge is fused in a rather hot fire for thirty minutes. The resultant lead button is freed from slag and then cupelled and parted as usual. The cupel and slag absorption is determined and added.

Since the composition of the precipitate is so variable, at least five or six assays should be made on each sample, and the average of the results taken.

A crucible-fusion charge used at the Homestake Mine for cyanide precipitate is given in Table XLVII., page 397.

If the scorification method is used, from 2 to 3 grammes of the precipitate are scorified with 40 grammes of lead and a cover of borax, and the resultant lead cupelled, etc.

Scorification is usually adopted for this class of material in the Rand, South Africa,[2] a preliminary assay being first made to determine the approximate percentage of fine gold. From 0·25 to 1·0 gramme is then scorified with

[1] Rose, *Metallurgy of Gold*, 5th edit., p. 298.
[2] *Rand Metallurgical Practice*, vol. i. p. 320.

fifteen times its weight of lead foil or shot, with the addition of the silver necessary for parting, as ascertained by the preliminary assay. At least five separate scorifications should be made.

Fulton and Crawford [1] have investigated the different methods for the determination of gold and silver in zinciferous precipitates.

Experimental assays were made of a well-mixed sample of gold and silver precipitate from the zinc boxes of the cyanide process, to obtain a rapid and accurate method of analysis. Scorification gave low results due to volatilisation, and so was discarded. The crucible method yielded variable results, and the combination wet and dry method gave results high in silver and low in gold, due probably to its solubility in nitrous acid as a result of the action of the nitric acid on the zinc present. To overcome this difficulty a combination method was tried, using sulphuric acid in place of nitric acid. One-tenth assay ton of precipitate was treated with 20 c.c. of concentrated sulphuric acid and 60 c.c. of water and boiled for one hour, after which the solution was cooled and diluted to 100 c.c. Normal sodium-chloride solution 75 c.c., and lead-acetate solution 20 c.c. were added, and the precipitate settled, filtered, washed, and dried. The paper was burnt off, and the residue scorified until half covered over in the scorifier, then poured and the button cupelled. The results were good and uniform, the slag corrections being small. The gold is lower than in the crucible assay, while the silver is higher, but the results are not so erratic, and it is not necessary to make more than two assays to get a reliable figure, as is the case with the crucible method. The original sample should be assayed at once or kept from the air, as the zinc oxidises quickly, and sufficiently to alter the percentage of gold and silver present. Care should be taken that the sulphuric acid is free from nitric acid.

5. Assay of Cyanide Bye-Products.

Under the head of bye-products are included :—

    (a) Black sands (mill concentrates).
    (b) Concentrates (Fruevanner concentrates).
    (c) Crushed plumbago crucibles and liners.
    (d) Slags and sweepings.
    (e) Litharge dross.

(a) Assay of "Black Sands" (Mill Concentrates).—This material consists largely of magnetite (Fe_3O_4), which is a heavy mineral and separates from the lighter gangue materials such as quartz. The sands should be passed through a sieve of 100 mesh before being assayed, the sieve being watched carefully for metallics.

The assay of black sands is made by crucible fusion in the same way as basic ores are assayed, as described in Chapter XIII.

Whitby gives the following charges for the previous assay of 1 A.T. of black sand obtained in gold mills on the Rand [2] :—

Ore	.	.	.	.	.	1·0 assay ton.	1·0 assay ton.
Litharge	.	.	.	.	.	3·0 ,,	
Red lead	.	.	.	.	.	...	2·5 ,,
Sodium carbonate	.	.	.	.	2·0 ,,	2·0 ,,
Borax	.	.	.	.	.	0·5 ,,	0·5 ,,
Weight of lead button obtained				50 to 58 grammes.	35 to 40 grammes.

[1] School of Mines Quart., 1901, vol. xxii. pp. 153-162 (abstract Journ. Soc. Chem. Ind. 1901, vol. xx. p. 749).
[2] Journ. Chem. Met. and Min. Soc. of South Africa, 1906, p. 271.

If one-half assay ton is taken for assay, half the above quantities of flux are used, in which case the lead button will weigh about 25 grammes.

(b) *Assay of Concentrates (Fruevanner).* — These consist of the heavy metallic minerals, mainly sulphides, that have been separated from the accompanying gangue or rock by concentration. Samples of concentrates should be crushed to pass 100-mesh sieve: metallics may be present. Concentrates are assayed by the methods described for pyritic material in Chapter XIV. In many gold mills, samples of concentrates are roasted before being assayed, but with proper precautions they may be satisfactorily assayed by direct fusion.

The following flux is recommended by Whitby for the direct fusion of concentrates obtained on the Rand [1] :—

Concentrates	.	.	.	.	.	0·5 assay ton.
Red lead	.	.	.	.	2·0	,,
Sodium carbonate	.	.	· .	1·5	,,
Borax	.	.	.	.	0·5	,,

The sulphur present will act as a reducing agent and give a button of lead weighing about 50 grammes.

(c) *Assay of Graphite Crucibles and Liners.*—Graphite (plumbago) crucibles are extensively used in cyanide mills for smelting the zinc-gold precipitate. They often contain a considerable amount of gold and silver, which filters into the pores of the crucible and also adheres to the surface. A large proportion of this may be separated as "metallics" by crushing the sample and passing it through a sieve of 100 or 150 mesh. The metallics are collected and assayed separately (see page 201). The fine material presents difficulty in assaying on account of the graphite and zinc it contains: it is well mixed and assayed as described on page 355. Typical charges used in South African cyanide mills for this material are given on page 357. The method of calculating the value of material containing metallics is given on page 216.

Detachable clay liners are sometimes used inside the graphite crucibles. These are crushed and sampled and assayed in the same way as graphite pots, but since they contain very little graphite, due allowance must be made for this fact in making up the charge for the assay, and the material fluxed as an ordinary siliceous ore.

(d) *Assay of Slags and Sweepings.*—The slags and sweepings which result from the smelting and casting of bullion always contain more or less precious metal. The methods of assay for these materials are described on page 356.

(e) *Assay of Litharge Dross.*—This consists of impure litharge obtained from the cupellation of the auriferous lead resulting from the smelting of zinc-gold slime. Litharge dross is assayed as described on page 317.

6. Testing of Ores for Cyanide Treatment.

The testing of small quantities of ore in the laboratory to determine the maximum extraction that can be looked for, and the consumption of cyanide, is an important part of the duties of the assayer in a cyanide works. "It is not unusual in large Customs works to make such tests on each lot of ore received." All materials, before being treated by the cyanide process, should be carefully examined and tested in the laboratory in order to ascertain their adaptability to cyanide treatment. A complete analysis of the ore should be made before treatment, and also from time to time, special care being taken to search for such elements as may be expected seriously to affect the treatment, for example, tellurium in combination with gold.

---

[1] Whitby, *loc. cit.*, p. 271.

Preliminary tests are made on the ore crushed to pass a 20- or 30-mesh sieve.

Determination of the Acidity of an Ore.—The acidity of an ore is due mainly to the presence of free sulphuric acid and soluble sulphates (chiefly iron) resulting from the decomposition of sulphides. These products are destructive to cyanide solutions, destroying more or less their alkalinity and in a proportionate degree reducing their efficiency. The soluble salts and acids that consume cyanide are termed *cyanicides*. To prevent the waste of cyanides due to the action of cyanicides it is necessary to neutralise them before treating the ore with cyanide solution.

In practice the acidity of ores is neutralised by the addition of either lime or caustic soda, and a preliminary test is made to determine how much lime or soda must be added for each ton of the ore in order to counteract the acidity. "Whether this acidity should be reported in terms of the lime or of the caustic soda required to neutralise it will depend on which of these reagents is to be used in the actual practice."

In most cyanide mills lime is used in preference to caustic soda, except in cases where tests have proved lime to be unsuitable. The amount of alkali or lime to be mixed with a charge of ore is determined by adding, little by little, an alkaline solution of known strength to a given weight of ore until the whole is neutral to litmus or other indicator.

The strength of the standard solution of caustic soda and the quantity of ore used for this determination vary considerably in different cyanide mills, but in many mills the standard solutions are so adjusted that each c.c. corresponds to 1 lb. of caustic soda or of lime per ton of ore, and the quantity of ore taken for the test is calculated accordingly.

The following may be taken as examples of the method of calculation :—

Example 1.—A normal solution of caustic soda contains 40 grammes NaHO per litre, but the solutions used in practice are usually weaker.

Suppose the strength of the solution used is one-fourth normal (N/4), it will contain 10 grammes of NaHO per litre, and 1 c.c. will contain 0·01 gramme of NaHO. Then, since 1 c.c. of the standard solution is equivalent to 0·01 gramme of caustic soda, if we take 2000 times this weight of ore for the test (*i.e.* 0·01 gramme × 2000 = 20·0 grammes), each c.c. of standard solution will be equivalent to 1 lb. of caustic soda per ton (2000 lbs.) of ore. It is frequently more satisfactory to take a larger quantity of ore for the determination, and if 200 grammes is taken then each c.c. of the standard solution will be equivalent to one-tenth (0·1) of a pound of caustic soda per ton of ore. If the acidity is to be reported in terms of lime, the result obtained for caustic soda (expressed in pounds per ton) is multiplied by 0·7, since 1 part of caustic soda is equivalent to 0·7 part of lime, thus (NaHO = 40, CaO = $\frac{56^1}{2}$ = 28, and $\frac{28\cdot0}{40\cdot0}$ = 0·7).

Therefore if the result of the determination of the acidity of an ore showed that it required the addition of 2·5 lbs. of caustic soda per ton, then the equivalent amount of lime required would be 2·5 × 0·7 = 1·75 lbs. per ton of ore.

Example 2.—As shown above, a normal solution of caustic soda contains 40 grammes of NaHO, which is equivalent to 28·0 grammes of lime. One c.c. of normal solution is therefore equivalent to 0·04 gramme NaHO or 0·028 gramme CaO, and 1 c.c. of a soda solution one-fifth normal (N/5) will be equivalent to $\frac{0\cdot04}{5}$ = 0·008 gramme NaHO or $\frac{0\cdot028}{5}$ = 0·0056 gramme CaO.

Then, since 1 c.c. of the N/5 standard solution is equivalent to 0·0056 gramme of lime, if we take 2000 times this weight of ore for the determination, viz.

—————

[1] Calcium being divalent, one-half of the molecular weight is taken.

0·0056 gramme × 2000 = 11·2 grammes, each c.c. of the standard solution will be equivalent to 1 lb. of lime per ton (2000 lbs.) of ore.

In practice ten times this weight of ore would be taken (*i.e.* 112·0 grammes), and 1 c.c. of the standard solution would in this case be equivalent to one-tenth of a lb. of lime per ton of ore.

If a normal solution of caustic soda is used for the titration and 56 grammes (the calculated quantity) of ore is taken, then 1 c.c. of the standard solution will be equivalent to 1 lb. of lime per ton of 2000 lbs.

The test for acidity may be made by placing 200 grammes of the pulverised ore into a tall, wide-mouthed stoppered bottle (of colourless glass), adding 1000 c.c. of water, and shaking well for five or ten minutes. A N/4 standard solution of caustic soda (10 grammes NaHO in 1000 c.c. distilled water) is then run in from a burette until the liquid in the bottle is neutral. A piece of litmus paper or a small quantity of litmus or phenolphthalein solution is used as an indicator.

As stated above, each c.c. of the standard soda solution will indicate that 0·1 lb. of caustic soda must be added to each ton of ore (2000 lbs.) to neutralise the cyanicides present, provided that 200 grammes of ore are used for the determination. For example: If 13·0 c.c. of standard solution of caustic soda were required, then 0·1 × 13 = 1·30 lbs. of caustic soda to be added to each ton of ore or 1·30 × 0·7 = 0·91 lb. of lime per ton of ore.

It must be remembered that the results obtained refer to pure caustic soda and pure lime, and since the soda and lime of commerce are not pure, a larger quantity will be required in practice than that indicated by the laboratory test.

It is also the practice in some mills in determining the acidity of an ore to add an excess of standard caustic soda solution, the additions of soda solution being continued until the indicator shows that the alkaline reaction is permanent. The amount of soda solution added is carefully noted and the excess of soda then determined by titrating back with standard acid until the solution is neutral. Thus:

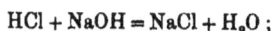

$$HCl + NaOH = NaCl + H_2O \; ;$$

or, if the return is to be made in terms of lime,

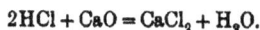

$$2HCl + CaO = CaCl_2 + H_2O.$$

Standard hydrochloric acid or sulphuric acid is usually employed.

If the ore is strictly neutral (*i.e.* free from cyanicides) the quantity of standard acid required will be the same as the quantity of standard soda used. If, as is frequently the case, the ore is acid, less standard acid will be used.

For example: If 14·0 c.c. of soda solution were used and only 10 c.c. of standard acid were required, then the ore will have done the work of the remaining 4 c.c. of acid. Therefore, to neutralise this acidity in the ore an addition of 0·28 lb. of lime per ton of ore will be required.

When the result of the test for acidity indicates that the ore requires the addition of more than about 3 lbs. of caustic soda per ton, it is usual to wash the ore first with water to remove the free sulphuric acid and soluble salts and then to test the acidity of the washed ore. An approximate determination of the acidity of an ore may be made [1] by taking 200 grammes of the crushed ore along with 200 c.c. of fresh water, and adding, little by little, small quantities of powdered lime until an alkaline reaction, using phenolphthalein as indicator, is obtained. The pulp is shaken after each addition of lime. From the amount of lime added is calculated the quantity necessary to be added to the ore bins. This test gives a figure which represents a quantity of lime slightly in excess of that actually required in practice, but it is advisable to err on the side of excess rather

---

[1] *Rand Metallurgical Practice*, vol. i. p. 346.

than deficiency in the addition of lime. A more accurate figure would, of course, be obtained by titrating a portion of the water in contact with the ore against standard alkali, as described above.

Acidity after Washing.—Take 200 grammes of ore and shake well with 1000 c.c. of water in a stoppered bottle as before, allow the ore to settle, and decant the clear liquid as far as possible. This wash-water is tested for cyanide consumption, etc., as directed below under "tests for extraction." The residue is then treated immediately, without drying, with an excess of a standard solution of caustic soda and well shaken for ten or fifteen minutes. About 500 c.c. of water are then added, the whole again shaken, and then the excess of soda determined by titrating back with a standard solution of acid, as described above. The acidity of the washed ore is due to insoluble basic salts, which are converted by the action of the caustic soda into ferric hydrate and soluble sulphates, thus :

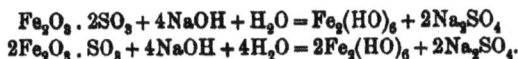

$$Fe_2O_3 . 2SO_3 + 4NaOH + H_2O = Fe_2(HO)_6 + 2Na_2SO_4$$
$$2Fe_2O_3 . SO_3 + 4NaOH + 4H_2O = 2Fe_2(HO)_6 + 2Na_2SO_4.$$

The water filtrate may be mixed with a measured quantity of cyanide solution and subsequently titrated with standard silver nitrate to determine how much cyanide is decomposed by it. In this way the value of the *water-wash* on the particular ore is determined.

Test for Cyanide Consumption.—To determine the amount of cyanide consumed, place 200 grammes of the ore (ground to 30 mesh) with 200 c.c. of a cyanide solution of known strength (say 0·1 or 0·2 per cent.) in a stoppered bottle and shake for fifteen minutes, then leave undisturbed for twelve hours, and finally shake again for five minutes. Most of the decomposition of cyanide (due to the presence of cyanicides) takes place after shaking for the first few minutes, but not the whole, hence it is necessary to let the solution stand as described.

If power is obtainable, agitation is carried on in suitable jars. Allow the ore to settle, filter off some of the liquor, and assay 20 c.c. for cyanide by titration with standard silver nitrate in the usual way.

The difference in the strength of the cyanide solution before and after extraction gives the percentage of cyanide consumed in treating 200 grammes of ore, and from this result the consumption of cyanide per ton of ore can be calculated. Thus, if the original solution contains 0·2 per cent. of cyanide, and the solution after treatment contains 0·17 per cent., showing a consumption of 0·2 − 0·17 = 0·03 per cent.; this will be equivalent to a consumption of $\dfrac{0·03 \times 2000}{100}$ = 0·6 lb. of cyanide per ton of ore (2000 lbs.) treated.

The filtrate may be tested for soluble sulphides, ferrocyanides, sulphocyanides (thiocyanates), copper, zinc, etc., and for its reducing power.

The tests recommended by Clennell[1] are as follows :—

(1) Add hydrochloric acid. A white turbidity may indicate the presence of soluble sulphides or of zinc.

(2) After adding a slight excess of hydrochloric acid, add a few drops of potassium ferrocyanide $K_4Fe(CN)_6$. A brown-coloured precipitate indicates copper, a white flocculent precipitate may indicate zinc.

(3) Add lead tartrate dissolved in excess of caustic soda : a brown or black coloration indicates soluble sulphides.

(4) Acidulate with sulphuric acid, and add dilute potassium permanganate drop by drop. If it is decolorised, it indicates that the solution has *reducing power*, the permanganate being affected by ferrocyanides and sulphocyanides,

[1] *Journ. Chem. Met. and Min. Soc. of South Africa*, 1898.

sulphides, organic matter, etc. If this reaction be strong, oxidation (*i.e.* aeration) will probably be necessary before or during the practical cyanide extraction.

Further Tests on Ores.—After making the above preliminary tests, experiments with 1 or 2 lbs. (0·5 to 1 kilogramme) of material are made to obtain information on the following points :—

(1) The most economical degree of fineness to which the ore may be crushed for subsequent cyanide treatment.

(2) The best strength of cyanide solution to use.

(3) The best proportion of this solution to the ore.

(4) The length of time necessary for satisfactory results.

(5) The necessity for using alkalies.

(6) The necessity for oxidation.

(7) The necessity for any preparatory treatment.

Extraction Tests.—These tests are preferably made in wide-mouthed, glass-stoppered bottles. To determine the best fineness of ore the tests are carried out by first crushing portions of the ore to pass a 10-, 20-, 30-, 40-, and 60-mesh sieve respectively, and subjecting each to treatment with cyanide solution a trifle above the probable working strength, under exactly similar conditions of quantity, time, etc. If the preliminary test for "acidity" has shown that the ore contains cyanicides, add to each test the amount of lime found to be necessary to neutralise acidity. If the ore contains much free acid, it should first be subjected to a preliminary wash with water, as described above. The wash-water is carefully decanted off, and then the cyanide treatment is begun.

The tests should be made in duplicate, one kilogramme (1000 grammes or 2 lbs.) of the pulverised ore and the required quantity of pulverised lime are well mixed, transferred to the stoppered bottle, 1 litre of cyanide solution (strength say 0·2 per cent. KCN) added, and the bottle placed on an agitator, frequently a revolving shaft, and agitated for six to twelve hours.

After agitation, allow the ore to settle, then decant the solution on to a filter. Test 10 c.c. of the filtrate for cyanide consumption by titration with standard silver nitrate. Now wash out the ore from the bottle on to the filter, wash twice with water, then dry and carefully sample and assay the residue (*i.e.* tailings) for gold in the usual manner. The ore before treatment (*i.e.* the heads) having been carefully sampled and assayed, the difference between this assay result and that obtained by assaying the residue gives the quantity of precious metals that have been removed by cyanide treatment. The results may be checked by recovering the gold from the filtrate by evaporation, as described on page 375.

To determine the Best Strength of Solution.—The ore of the fineness yielding the best result is then tried with cyanide solutions of different strengths.

Clennell recommends tests with solutions of 0·01, 0·05, 0·1, 0·2, and 0·5 per cent. potassium cyanide. The tests are carried out as described above. The results of these tests will indicate the best strength of cyanide solution to use.

To determine the Best Proportions of Solution to Ore.—The best strength of solution having been ascertained, tests are then made to ascertain the best proportion of this solution to the ore. For this purpose the ore of suitable mesh is treated with this solution in the proportions of 1, 1¼, 1½, 2, 3, and 4 assay tons of solution to 1 assay ton of ore.

The ore before and after treatment is assayed and the difference between the results noted. A comparison of the extraction results obtained for each test will show the best proportions of solution to ore to use.

Time of Contact with Solution.—The best proportion of solution to ore having been ascertained, the time of contact required to attain the best results on the ore is now determined.

A number of tests of ore of suitable mesh are treated with the proper proportion of solution of correct strength, in stoppered bottles as before. The solution is allowed to remain in contact with the ore for 1, 2, 3, 4, 7, and 14 days. A comparison of the assay results from each test will give the time required for extraction.

Percolation Tests.—The rate at which a cyanide solution will percolate through a mass of ore is largely dependent on the nature of the ore and the size of the particles. With coarse ore and clean sand the percolation is always quicker and more uniform than with finely pulverised material, clayey ore, or mixtures of sand and slime. "A physical examination of the ore will give a good idea of the screen aperture through which it must be passed in order to obtain a good extraction. For example, if the ore is a porous or cellular oxidised product, perhaps crushing through a 0·44-inch screen aperture will suffice; if of dense and solid structure it should be crushed to pass screen apertures varying from 0·024 to 0·018 inch. Flinty material, pyritic and telluric ores, silver ores, etc., may have to be reduced to impalpable powder to obtain the desired extraction." [1]

The method of ascertaining the fineness of the ore that gives the best extraction has been described above.

Percolation tests are made in duplicate in glass percolators (such as that described and illustrated, fig. 114, on page 241) holding about 4 lbs. of ore. A false bottom is made of sand as described, or it may be constructed of a perforated board with a piece of canvas stretched across it to act as a filter. The strength of the cyanide solution to be used for the percolation tests on a particular ore is previously determined by bottle tests.

The dry ore, after being thoroughly incorporated with the required amount of lime, is placed evenly and carefully on the false bottom of the percolator and gently pressed around the sides to prevent channelling.

The cyanide solution of correct strength is then carefully added on top until the ore is thoroughly submerged. The solution is allowed to stand on the ore for four hours, and then allowed to percolate through by opening the stop-cock. Allow the ore to drain dry and then let it stand to aerate it: this will probably occupy a day. Then treat again with cyanide solution and allow to drain dry as before; after the second day's treatment turn the ore out from one of the percolators when drained dry, and cut out a sample for assay, to determine the extraction, and return the remainder to the percolator. Keep up this treatment and sampling daily, until the final extraction is reached so far as one percolator is concerned.

Allow the duplicate to run on without being disturbed, but it must be drained dry and allowed to aerate daily.

The percolation tests (as just outlined) should follow the usual cyanide practice, as follows:—

(1) Treat with water, or alkaline wash, if required.
(2) Treat with weak cyanide solution.
(3) Treat with strong cyanide solution (see page 361).
(4) Treat again with weak cyanide solution.
(5) Treat with water to wash.   Finish.

If tests have shown that the preliminary wash before cyanide treatment is unnecessary, the strong solution is added at once, followed by weak solutions and finally water to wash. When the solutions are run in on top of the ore and allowed to percolate downwards there is a tendency, especially with fine ore, for packing to take place, and to obviate this the solution may be run in from below and allowed to rise through the ore. This must be performed slowly to prevent the formation of holes or channels in the charge.

[1] Argall, *Mill and Smelter Methods*, p. 93.

Every different ore presents its own problems which must be worked out for that particular ore, but laboratory tests, if carefully and intelligently made, form a most useful and reliable guide to actual practice.

Guided by the results of the experimental tests, a more practical trial, on about 1 ton of ore, is then made.

In reporting the results of cyanide tests on ores the following information should be given:—

Character of the ore, with analysis if necessary.
Size of mesh to which ore is crushed.
Weight of ore taken for the test.
Assay value before treatment.
    ,,   ,,   after (say 4) hours treatment.
    ,,   ,,   ,, (,, 8) ,,   ,,
    ,,   ,,   ,, (,, 12) ,,   ,,
    ,,   ,,   ,, (,, 16) ,,   ,,
Lime used per ton of ore.
Strength of potassium cyanide solution used.
Amount of solution used per ton of ore treated.
Potassium cyanide consumed per ton.
Total time of treatment, percolation, and washing.
Percentage of gold and silver extracted.

Laboratory Work at the Homestake Mine, South Dakota.

The following recent description by Messrs A. J. Clark and W. J. Sharwood,[1] of the laboratory work carried out at the Homestake Mining Company, Lead, South Dakota, may be given as an example of the tests made in actual practice in connection with the cyanide process.

SAMPLING AND ASSAYING.

Mill Feed is not directly sampled, but in metallurgical summaries is calculated from tailings assays, and bullion actually recovered.

Sand Heads.—The feed to the sand plants is sampled by means of a horizontal slotted pipe, pivoted at one end, which is swung at intervals by hand across the stream of pulp. An improved sampler has lately been installed, in which the slotted pipe is suspended from a small car running on rails, and is moved through the stream parallel to itself and at a uniform rate, by a sprocket chain driven by a winch. Three portions of 1·5 A.T. each are assayed from each vat charged, and the gold buttons are weighed progressively.

Sand Tails or Residues are sampled after the final draining by making a number of vertical holes with a pipe [2] having a slot with one lip projecting to form a cutting edge, and a conical solid point. A pole is kept in the pipe until it reaches the vat bottom, and the sample is taken by rotating the pipe. The sampler is then withdrawn and laid over a trough equal in length to the depth of the vat sampled, and divided by partitions into three equal sections, into which the residue is discharged, giving a "top," "middle," and "bottom" section. Each of these is mixed, and a portion is cut out and dried, half of which is taken to the assay office in a closed can, and half reserved in case of accident. Two portions of 2 A.T. are assayed from each sample, and the two buttons are weighed

[1] *Trans. Inst. Min. and Met.,* vol. xxii., 1912–13.
[2] This sampler is a modification of the one described by Merrill, *Trans. I.M.M.,* vol. vii. p. 224, and is in use at Marysville, Montana. A similar one used in Nevada and described in *Min. and Sci. Press,* Nov. 7, 1908, vol. xcvii. p. 636, was adapted from Homestake practice.

together. The assays of the three layers, and their average, are all reported, and the average is used with the corresponding "charge" assay in computing probable extraction.

Slime Heads or Charge Samples are drawn from the main feed-pipe, one being taken when each press commences filling, and a general sample being cut from the accumulation of each twenty-four hours. Four melts are made of 2 A.T. each. The result is checked by daily samples taken hourly with a cup from the "No. 1 sludge" and "No. 2 sludge" streams.

Slime Residues.—The slime plant is divided into three sections of nine or ten presses each. In each section there are placed three residue sample buckets, A, B, and C, corresponding to the "fast," "average," and "slow" leaching rates of the charges. During the sluicing of each charge a cup sample is caught at the outlet cocks and put into the sample bucket corresponding to its leaching rate. Three fusions of 2 A.T. each are made on each of the three samples caught daily.

Cutting down Samples.—All slime samples are de-watered on a horizontal laboratory filter-press, having a working surface 18 inches square. Compressed air is used, as no vacuum is available. The thin, tough cake obtained is rolled into a cylinder on the cloth and cut into sections with a knife or spatula. Sand samples, after draining if necessary, are cut down with a trowel while moist. Dried samples, either of sand, slime, or crushed ore, are cut down with the Jones riffle.

Pan and Tube-Mill Samples.—Assay samples of the feed to pans and tube-mill are taken with slotted-top hand samplers, and are used for assay and sizing tests. Similar samples of the tails or ground product are taken from the tipple discharge, which prevents contamination from amalgam plates. Three samples of 1·5 A.T. each are taken for assay.

Sampling and Assay of Solutions.—At the sand plants it is the practice to fill a vat with pregnant solution to a fixed point, and then precipitate it, so that the tonnage of each precipitation is known. The pregnant sample is taken by a drip system, a small iron pipe, tapped into the main and controlled by an iron screw, dripping into an enamelled iron bucket, from which a sample is dipped into a bottle after the vat is full. This is occasionally checked by taking a "dipper" sample from the full vat.

At the slime plant continuous pumping is practised, and a similar sample is taken for each eight-hour shift, the tonnage being computed from the speed of the electric pump.

The "barren" solution samples, corresponding in periods to the various "pregnants," are taken by a similar drip system from the launders leading from the precipitation presses. A special barren sample is usually taken to cover the first half-hour of a tank precipitation, and occasionally hand samples are taken half-hourly from suspected sections of presses, if precipitation is not entirely satisfactory. As a further check on the barren solutions, daily samples are taken from the storage vats in which they are collected.

Common quart-glass bottles are used to convey solution samples to the assay office. By holding these lightly on an emery wheel a label is ground on them, so that they can be legibly marked with a lead-pencil.

Solution samples are assayed by precipitation with zinc dust in presence of lead acetate, followed by heating and addition of hydrochloric acid. The lead sponge obtained is filtered off and scorified with the addition of metallic silver, and then cupelled and parted. This is a modification of the Chiddey method, and has been in use since 1904. In the case of rich solutions 250 c.c. are taken, and from "barrens" 500 c.c.; a special table is used to convert the milligrammes of gold obtained to the basis of dollars per ton.

Precipitates.—During a clean-up the wet precipitate from the press is well mixed in the shallow collecting tray, and then shovelled into boxes. Usually every fifth shovel is thrown upon an iron plate, where it is systematically quartered down at once to a sample of 2 lbs. to 4 lbs.

An alternative method is to take a tryer sample (four or five cores) from each of the filled boxes, which is further reduced by the tryer.

The boxes are weighed on a platform scale as soon as filled, and again when delivered for acid treatment.

The final sample is put in an iron jar with air-tight cover, and is taken to the assay office. Here it is weighed in a shallow pan, dried at 100° C., and moisture is calculated to 0·1 per cent. The dried precipitate is then roughly crushed and quartered down on a Jones riffle sampler to about 100 grammes or 150 grammes. This sample is ground fine in a Wedgwood mortar, without sifting, and samples of 0·1 A.T. (usually triplicates) are weighed out for assay as quickly as possible, weighing to the nearest milligramme or half-milligramme. Better results are obtained by grinding without sifting, as dusting, oxidation, and absorption of moisture are thus avoided, and there are no actual metallics as in zinc-box product.

For the assay of precipitates a fusion method is used, one-third of the flux being retained for use as a cover, with a further addition of a little borax glass. The slag and cupel of each sample are fused together with a special flux, and the resulting lead cupelled; the correction button thus obtained is weighed with the main buttton. These are inquarted and cupelled with proofs, etc., and parted in a standard platinum apparatus. When fusions are performed in a muffle this method gives results which agree closely among themselves, and also with parallel determinations made by the best "combination" acid-scorification methods.

Bullion Bars.—Samples are taken from two opposite corners of each bar, with a pneumatic drill in the case of mill bullion, and by chipping with a chisel in the case of bars from the cyanide plants. After sampling, the bars are weighed to the nearest hundredth of an ounce, stamped with a consecutive number and the weight, and, finally, sacked for delivery to the United States assay office. The cyanide bars average 980 fine. One assay (0·5 gramme) is run on each sample for gold; silver is determined by cupellation of only one sample from each bar. The fineness is reported to the nearest quarter-millième for gold and half-millième for silver, which conforms to the practice of the United States mint, to which the bullion is sold.

Amalgamation Tests.—When samples of ore show higher values than $3·00 per ton, they are frequently tested by amalgamation. 100 grammes of the finely-ground ore are shaken for two hours in a bottle with 20 grammes of pure mercury and 150 c.c. water.[1] The mercury is then separated and dissolved by nitric acid, and the residual gold is inquarted, parted, and weighed. Its value is calculated as a percentage of the original assay. Duplicates are found to agree well, and the average of a large number of such tests agrees closely with the percentage recovered in the stamp-mill.

Panning Tests.—A considerable number of mine samples are tested by panning, 2-lb. lots being crushed by a small stamp operated by compressed air. Formerly the mine work was almost entirely controlled in this way, but owing to the increased percentage of sulphide in certain parts of the mine, it is difficult to see the free gold in some samples, even when it is present in comparatively coarse particles.

---

[1] For description of amalgamation test and illustration of agitator used, see Fulton, *Manual of Fire Assaying*, 2nd edit., 1911, pp. 154–156.

CHEMICAL TESTS.

Available Cyanide.—Ten c.c. (or 20 c.c.) are measured with a pipette into a tube. About four drops of 5 per cent. potassium ferrocyanide solution are added, and the solution is titrated with standard silver nitrate, adjusted to read directly to the percentage of KCN (or of NaCN). A heavy glass tube (combustion tube) is used, 0·7 inch internal diameter, with wall about $\frac{1}{16}$ inch thick, and either 5 or 7 inches long; ordinary test-tubes being too fragile.

At the slime plant, with weak solutions containing but little zinc, a few drops of 2 per cent. potassium iodide solution are used as indicator.

Traces of Cyanide.—In sand leaching the effluent is at first turned from the sewer into the "low" precipitation sump as long as a trace of cyanide is recognisable. This time can be estimated within an hour or so, and, as it approaches, the effluent is occasionally tested with silver nitrate. For this purpose 100 c.c. are taken, 5 c.c. of 2 per cent. potassium iodide are added, and then silver nitrate solution run in; as long as the first drop or two of silver solution shows a turbidity the effluent is allowed to run to waste.

Alkalinity is determined by titrating a 20-c.c. sample with one fifth-normal sulphuric acid, using phenolphthalein as indicator. It is expressed as the number of c.c. of tenth-normal acid required to neutralise 100 c.c. of the solution tested; one unit corresponding to 0·0028 per cent. CaO. "Protective alkalinity," in terms of lb. CaO per ton, can be calculated from the total alkalinity thus found by correcting for the cyanide present. It is read from a curve or table based on the following formula.

Protective alkali (lb. CaO per ton) $= 0·056a - 8·6k = 0·056a - 11·4n$; where a is the total alkalinity as above determined, k the percentage of KCN, or n the percentage of NaCN.

Analyses of typical cyanide solution from the various plants of the Homestake Mine are given in Table XLVIA., page 396.

Other Determinations.—All carloads of cyanide and zinc dust are sampled by opening a certain number of the boxes or barrels. In the case of cyanide the percentage, calculated as KCN, is determined, and also the alkalinity towards methyl orange; and it is usually tested qualitatively for sulphide. Shipments of zinc dust, and samples of untried brands, are examined for percentage of lead, zinc oxide, and "precipitating efficiency" toward dilute silver potassium cyanide. They are also sized over 100- and 200-mesh sieves. Total zinc and insoluble are sometimes determined.[1]

Lime is occasionally examined for available CaO by the sugar test, but the kiln product has been found uniformly good since proper arrangements have been made for its prompt distribution.

Blast-furnace slag is assayed for gold, silver, and lead. Besides gold and silver, zinc and "insoluble" are usually determined in the cyanide precipitate. The reducing power of solutions is determined once or twice daily by titration with permanganate, in addition to the determinations of alkalinity, cyanide strength, and gold content. More elaborate analyses of solutions and of precipitate have not proved of any practical value.

Assay Offices.—The main assay office of the Homestake Company at Lead includes the laboratory, refinery, and electro-plating equipment, and serves the mine and cyanide plant No. 1, and here all assays of bullion and precipitate are made. A small assay office is operated by one man at cyanide plant No. 2, and another at the slime plant.

Sample Grinders.—At the slime plant no grinding is necessary; at the other

---

[1] W. J. Sharwood, *Journ. Chem. Met. and Min. Soc. of S.A.*, Feb. 1911, vol. xii. p. 332.

offices disc grinders are in use, and ore samples have a preliminary crushing with a small Gates crusher and bell grinder.

Table XLVIA.—Analyses of some Typical Cyanide Solutions at Homestake Mine.

Percentage = gramme in 100 c.c.

(1) Solution used about three years on sand coming mainly from oxidised ore : weak solution, Plant No. 2. (2) Same : strong solution, Plant No. 2. (3) Solution used about five years on tailings from mixed ore, mostly unoxidised ; weak solution, Plant No. 1. (4) Same : strong solution, No. 1. (5) Solution after using about nine years, sampled after a considerable period of deep level, highly sulphuretted ore. The high reducing power is noteworthy ; it at one time reached 550, or about 0·9 per cent., KCNS. Weak solution, Plant No. 1. (6) Weak solution, Slime Plant.

Sample No.		1	2	3	4	5	6
Usual plant determinations :—							
"Available KCN"	%	0·075	0·12	0·05	0·115	0·065	0·05
"Total KCN"	,,	0·135	0·165	0·09	0·15	0·095	0 06
Reducing power [1]	.	25·5	24·3	117·5	117·0	333·0	14·0
Alkalinity [2]	.	11·5	21·0	11·5	20·0	14·0	9·0
"Protective alkali" calc. as "Lb. CaO per ton"		0·0	0·144	0·214	0·131	0·225	0·074
Additional determinations :—							
Total CN (by HgO)	%	0·058	0·076	0·048	0·058	0·040	0·025
CNS (thiocyanate)	,,	0·024	0·024	0·102	0·105	0·319	0·012
SO₄ (sulphate)	,,	0·014	0·020	0·040	0 041	0·041	0·010
S (sulphide)	.	0·0	0·0	0·0	0·0	0·0	0·0
S in other forms (by difference)	,,	...	...	...	...	0·002	0·004
Ca	,,	0·066	0·062	0·051	0·049	0·072	0·028
Na [3]	,,	...	...	...	0·063	0·188	
K	,,	...	...	...	0·044	0·148	
Zn	,,	0·032	0·023	0·028	0·022	0·014	0·004
Cu	,,	0·006	0·006	0·003	0·005	0·004	Trace.
Fe (in ferrocyanide)	,,	0·00015	0·00015	0·0002	0·0002	0·0002	0·00005
Alkalinity :— [2]							
Direct by phenol-phthalein	.	11·5	21·0	11·5	20·0	14·0	9·0
Back titration ,,	.	25·0	31·5	24·0	33·5		
Direct by methyl orange	.	32·0	39·0	28·0	36·0		
Back titration ,, ,,	.	37·0	41·0	34·0	42·0		

Hg, Mg, Al, Si : traces found in all solutions examined for them ; Mn not found.

Assay Furnaces.—Gasoline is used as fuel in the assay furnaces, and ordinary fusions are made by direct heating in furnaces holding twenty 30-gramme or 35-gramme crucibles. The fusion of precipitate samples and the scorification of solution samples are carried out in the same muffles as are used for cupellation. Gas, where available, is used for drying samples, distillation of water, etc. Cupels are made of bone-ash, No. XXX., using Her cupel-making machines. Crucibles of either English or Colorado make are used.

Table XLVII. gives the assay fluxes and charges in general use, and Table XLVIII. shows the number of fire assays per month.

[1] No c.c. N/10 permanganate consumed by 100 c.c.
[2] No c.c. N/10 acid neutralised by 100 c.c.
[3] Mixed cyanide, rather high in K, was in use during the periods represented.

Table XLVII.—Standard Assay Fluxes : Homestake Assay Office.

Material.	Weight of charge. Assay tons.	Size crucible. "Grammes."[1]	Weight lead but ton. Grammes.	Litharge.	Soda ash.	Borax glass.	Silica.[3]	Flour.[3]	Nitre.[3]	Borax glass cover.	Special flux ingredients.
Cyanide 1 : sand charge (SiO₂ 55, S 3-5, Fe 10-17 per cent.).	1·5	30 or 35	35-45	45-50	35	10	8	...	2-4	5-8	7 to 8 grammes. ¼-inch iron brads on top of charge.[3]
Cyanide 1 : sand residue	2·0										
Slime : (less than 1 per cent. S).	2·0	30 or 35	35-40	60	48	12	4	2-4	...	5-8	
No. 2 sand charge or residue : partly oxidised (S about 2 per cent.)..	1·5 or 2·0	30 or 35	40	50	50	10	4	...	...	5-8	
Ores, variable in composition	1·0	20	25+	{ 40 50	32 40	10 12	4 5	1·2 1·5 }	...	5	(Iron brads or nitre if required.
Cyanide precipitate	0·1	20	25±	50	25	25	10	1·8	...	5	One-third of flux reserved as additional cover.
Precipitate correction : one precipitate slag and one cupel, after removing white portion of bone-ash.	...	20 or 30	30	24	18	18	18	3·0	...	5-8	Fluorspar 36 grammes.

Table XLVIII.—Monthly Average of Assays made at the several Homestake Assay Offices for One Year (July 1910 to June 1911).

Assay Office.	Number of assays per month.				
	Pot.	Solution.	Bullion.	Amalgamation.	Total.
Lead, Assay Office	1,970[3]	688	71	52	2,781
Cyanide, Plant No. 2, Assay Office .	506	342	...	...	848
Slime, ,, ,, ,, .	588	510	...	...	1,098
	3,064	1,540	71	52	4,727

The Homestake plant consists of 1000 stamps, crushing over 125,000 tons per month, of which 57 per cent. goes for treatment to the sand plant, and 41 per cent. to the slime plant, between 1 and 2 per cent. of the slime being wasted. The ore carries about 0·2 ounce of gold per ton.

Number of Assays made in a Cyanide Mill.—The actual number of assays made in a cyanide mill is dependent upon the size of the mill, etc., and necessarily varies considerably. The figures for the Homestake Mill, crushing

[1] The rating of Colorado crucibles by "grammes" capacity is supposed to indicate the weight of lead ore they will carry with the usual flux.
[2] Varied somewhat from time to time as the character of ore, etc., varies.
[3] About one-quarter are mine samples.

over 125,000 tons per month, are given above in Table XLVIII. As an additional example the following figures are given: they represent the number of determinations made per month on a producing mine in Rhodesia, South Africa, crushing 7000 tons per month.

Development samples . . . . 200 assays.	Amount taken 2 A.T.
Stope samples . . . . . . 400 ,,	,, 2 ,,
Screen pulp samples . . . . . 200 ,,	,, 2 ,,
Battery tailing samples . . . . 100 ,,	,, 2 ,,
Cyanide mill :—	
Sand charges . . . . . . 30 ,,	,, 4 ,,
Slime charges . . . . . . 30 ,,	,, 4 ,,
Sand residues . . . . . . 30 ,,	,, 4 ,,
Slime residues . . . . . . 30 ,,	,, 4 ,,
Sand solutions (entering precipitating boxes) 30 ,,	,, 10 ,,
Sand solutions (leaving precipitating boxes) 30 ,,	,, 10 ,,
Slime solutions (entering precipitating boxes) 30 ,,	,, 10 ,,
Slime solutions (leaving precipitating boxes) 30 ,,	,, 10 ,,
Samples of bullion . . . . . 8 ,,	,, 0·5 gramme.

Total assays per month   .   1148

With the exception of the bullion samples, only one assay of each sample is made as a general rule.

CHAPTER XXIV.

PLATINUM AND THE METALS OF THE PLATINUM GROUP.

Physical and Chemical Properties of Platinum.—Symbol, Pt; atomic weight, 195·0; specific gravity, 21·5; melting-point, 1745° C.

When molten, platinum absorbs oxygen, which is given off on cooling, sometimes with sufficient rapidity to cause the metal to spit like silver. Platinum volatilises readily in the electric furnace. At a red heat it may be welded with great ease. It can be melted by the oxyhydrogen flame.

Commercial Forms of Platinum.—Metallic platinum comes into the market in the form of small ingots, bars, sheets, and wire, also in the form of a black spongy mass known as spongy platinum or platinum sponge, obtained by heating ammonium platinic chloride.

Price of Platinum.—The price of platinum has been steadily increasing for some years, and in 1911 it reached the highest level yet recorded, viz. £9, 5s. per ounce (£9, 10s. to £10 in the United States). This is due to the fact that the supply has not kept pace with the growing demand, and the price of the crude platinum has also tended to increase by reason of the decline in the yield, which has now fallen to as low as 0·07 ounce of crude platinum per ton of gravel [1] (see Platinum Ores, page 400).

Platinum Alloys.—Platinum readily alloys with many metals. The most important alloys are those with iridium. The addition of 2 per cent. of iridium is found greatly to increase the hardness and raise the melting-point of platinum. An alloy containing 10 per cent. of iridium resists the corrosive action of chemical reagents to a greater extent than pure platinum. The alloys of platinum with gold are discussed on page 67, and with silver on page 72.

Platinum and lead form readily fusible brittle grey alloys, from which the platinum can be extracted by cupellation. In ordinary furnaces, however, the alloys freeze before the whole of the lead has been removed, and it is necessary to finish the operation with the aid of the oxyhydrogen flame. [2]

Solubility.—Platinum is not acted on either by pure hydrochloric, nitric, or sulphuric acid. It is dissolved by aqua-regia (and other mixtures which evolve chlorine) with the formation of platinic chloride, $PtCl_4$, but far less readily than gold, so that gold, after being fused so as to be firmly adherent to a platinum vessel, can be slowly dissolved by dilute aqua-regia at moderate temperatures without much injury to the vessel. Advantage is sometimes taken of this fact to remove gold that accidentally adheres to the platinum cups in the parting assay of gold when the "gold cornets" are annealed at too high a temperature.

When platinum is alloyed with silver, copper, lead, zinc, and some other metals, it is attacked and partly dissolved by nitric acid, with the formation of platinum nitrate. The amount dissolved varies with the nature of the alloyed metal and the amount of platinum in the alloy. The action of nitric and of

[1] *Chem. Trade Journ.*, 1912, l. p. 171.　　[2] T. K. Rose, *Precious Metals*, p. 268.

sulphuric acid on alloys of platinum and silver is of importance in the assay of platinum, and is discussed on pp. 412, 415.

Platinum is oxidised when fused with caustic alkalies, or with potassium nitrate, and is also attacked by fused alkaline cyanides forming platinocyanides. In the form of sponge, it is dissolved by boiling potassium cyanide with the evolution of hydrogen, and formation of a double cyanide.

Reactions of Platinum.[1]—*Sulphuretted Hydrogen*, H_2S, slowly produces a dark brown precipitate of platinic disulphide, PtS_2. The precipitate forms more quickly on heating. It is insoluble in nitric or hydrochloric acid, soluble in aqua-regia; difficultly soluble in normal ammonium sulphide, but more speedily in the yellow sulphide, with which it forms a sulpho salt $(NH_4)_2PtS_3$. Heated out of contact with air, it is decomposed into PtS and S.

Ammonium Sulphide, $(NH_4)_2S$, gives the same precipitate.

Ammonium Chloride, NH_4Cl, produces a light yellow crystalline precipitate of ammonium platinic chloride, $2NH_4Cl,PtCl_4$. From dilute solutions a precipitate is obtained only after evaporation to dryness on a water bath. The precipitate is somewhat soluble in water, insoluble in alcohol.

Potassium Chloride, KCl, also produces a yellow crystalline precipitate of potassium platinic chloride, $2KCl,PtCl_4$, analogous in its appearance and properties to the precipitate just described. Ammonium platinic chloride, however, leaves on ignition *only* spongy platinum which distinguishes it from potassium platinic chloride, which on ignition leaves spongy platinum and potassium chloride, Pt + 2KCl. The decomposition of the ammonium platinic chloride, by heating to dull redness to drive off the ammonium chloride and chlorine to obtain spongy platinum, is made use of for the determination of platinum and also on a large scale for the separation of platinum from its ores and residues.

Sodium Chloride, NaCl, forms with platinic chloride a double chloride, which is, however, soluble in water, and is obtained in needle-shaped crystals by evaporation.

As in the case of gold, the most characteristic reactions of platinum are based upon the fact that it is readily reduced from its compounds and precipitated in the metallic state, although not so readily as gold salts.

The following reactions may be compared with those for gold on page 69.

Ferrous Sulphate, $FeSO_4$, gives a brown precipitate of metallic platinum, but the action only takes place on prolonged boiling.

Oxalic Acid, $C_2H_2O_4$.—Platinic chloride is not reduced by oxalic acid.

Sulphurous Acid, H_2SO_3, does not precipitate platinum from solutions of platinic chloride.

Stannous Chloride, $SnCl_2$, produces a brown coloration, by the reduction of the platinic chloride to platinous chloride $PtCl_2$, but no precipitate is formed.

Potassium Nitrite, KNO_2, gives no precipitate at first, but on standing, yellow crystals of a double nitrate of potassium and platinum are deposited.

Metallic Zinc precipitates metallic platinum.

Separation of Platinum.—Whenever platinum and gold are contained in a solution, together with other metals that are precipitated with sulphuretted hydrogen, it is preferable to remove the gold by means of oxalic acid (which does not reduce platinic chloride) and then to evaporate the solution down with ammonium chloride to precipitate the platinum.

Ores of Platinum.

Platinum ore or native platinum occurs in flattened or angular grains in alluvial deposits and river sand, principally in the Ural Mountains, Brazil, New

---

[1] Valentin's *Practical Chemistry*, 10th edit., 1908, p. 221.

Granada, California, British Columbia, New South Wales, and in many other localities.

The richest and most extensive deposits occur in or near the Ural Mountains. Usually, but not invariably, the gravels are also auriferous, and other minerals occurring with the platinum are zircon, spinel, corundum, magnetite, and osmiridium.[1]

Platinum is also found in the form of sperrylite, arsenide of platinum, $PtAs_2$, associated with nickel sulphide at Sudbury in Ontario. It sometimes occurs in fahlers, zinc blende, lead ores, copper ores, etc.

The amount of platinum in alluvials is gradually declining : thus the amount of crude metal found in the Urals in 1879 was 5 ounces per ton, in 1909 the average was below 2 dwt. per ton, and in 1911 the yield had fallen to as low as 0·07 ounce per ton as previously stated.

The platinum grains occurring native are alloys containing the so-called platinum metals, iridium, osmium, palladium, rhodium, ruthenium, and also iron and copper. The chief impurities are iridium, which may form more than half the alloy, and iron which has been known to amount to 19 per cent. of the mass. Usually, however, the grains contain from 70 to 80 per cent. of platinum.[2]

The following analyses of native platinum concentrates from different sources are given by Kemp [3] :—

Table XLIX.—Analyses of Native Platinum from Different Sources.

| | Russia | | Colombia El Choco district | California | Oregon | New South Wales | Borneo | British Columbia | Brazil | Australia (probably New South Wales). |
	Goroblagodat district.	Nijni Tagilsk district.								
	per cent.	per cent.	per cent.	per cent.	per cent.	per cent.	per cent.	per cent.	per cent.	per cent.
Platinum	82·90	76·28	82·70	76·29	51·45	75·90	82·60	72·07	55·44	60·60
Iridium	0·33	2·44	1·44	1·70	0·40	1·30	0·66	1·14	27·79	1·65
Rhodium	1·54	2·37	2·08	1·35	0·65	1·30	...	2·57	6·86	1 68
Palladium	1·24	0·22	0·93	0·99	0·15	...	...	0·19	0·49	1·65
Osmium	0·54	0·20	0·23	0·33	...	...	...	...	...	0·40
Ruthenium	...	...	...	...	...	...	...	...	...	...
Osmiridium	1·17	0·73	2·62	9·04	37·30	9·30	3·80	10·51	...	25·50
Gold	...	...	0·74	0·71	0·85	...	0·20	...	...	1·80
Insoluble	0·17	1·56	...	...	...	...	...	...	...	...
Iron	10·72	13·75	7·60	5·93	4·80	10·15	10·67	8·59	4·14	4·43
Copper	0·56	1·31	0·51	1·91	2·15	0·41	0·13	3·39	3·30	1·10
Sand	...	...	1·54	1·76	3·00	1·22	...	1·69	...	1·20
Manganese	...	...	0·08	...	...	...	...	...	...	...
	99·17	98·17	100·47	100·01	100·25	99·58	98·06	100·15	98·02	100·01
Number of samples analysed to obtain average analyses.	3	17	5	4	1	1	1	1	1	2

---

[1] T. K. Rose, *Precious Metals*, p. 265. [2] T. K. Rose, *loc. cit.*, p. 206.
[3] Kemp, *U.S. Geol. Surv.*, Bull. 193, 1912, pp. 18–19. Consult also Thorpe's *Dictionary of Applied Chemistry*, vol. ii. p. 253.

The Metals of the Platinum Group.

The five metals iridium, osmium, palladium, ruthenium, rhodium, belong to the natural group of elements known as the *platinum metals*, because they all occur associated together in platinum ore. Of these, ruthenium and rhodium are the most rare.

Platinum ore, or native platinum, contains all these elements in the metallic state.

Amongst the grains of crude platinum there are also found grains which consist essentially of an alloy of platinum and iridium (containing from 30 to 50 per cent. of iridium) known as platiniridium ; and also particles of an alloy of osmium and iridium (called osmiridium), which contains from 30 to 40 per cent. of osmium as well as small quantities of rhodium and ruthenium.

The platinum metals are all white lustrous metals having high melting-points. They are unacted upon by air or oxygen at ordinary temperatures ; and, with the exception of osmium (which burns when strongly heated, forming the tetroxide), they are scarcely oxidised by oxygen at any temperature. The metals have high specific gravities, osmium being the heaviest known element.

With the exception of palladium, which dissolves in hot nitric acid, the platinum metals are not acted upon by any ordinary acids.

Aqua-regia is without action upon rhodium and iridium ; it acts slowly upon ruthenium, while it converts osmium into the tetroxide.

Industrial Uses of Platinum Group Metals.[1]—The following information will be of assistance in determining the source of the various materials containing platinum metals which may be submitted for assay.

Iridium.—Most of the iridium produced is probably employed for hardening platinum, the iridium content of the alloys ranging from 5 to 20 per cent. Such alloys are hard and difficult to work when containing from 10 to 20 per cent. of iridium, and are not attacked by aqua-regia when the proportion of iridium exceeds 30 per cent. Iridium alloys are used for scientific and technical purposes, such as thermo-couples, but the pure metal, owing to its brittleness, is very difficult to work. On account of its hardness, and consequent resistance to wear, it is used for protecting the points of gold pens. The metal has been increasing rapidly in value within recent years.

Osmium.—Except in the manufacture of certain kinds of incandescent lamps, osmium is but little employed. When alloyed with iridium, it is used to make the " diamond points " of gold pens and also for compass bearings.

Palladium.—The most extensive use of . palladium is probably in the manufacture of certain alloys employed in dental work. The metal is also used in watch-making, for the circles of astronomical instruments, and for soldering platinum metals. The demand is considerably greater than the supply.

Rhodium.—This metal is chiefly used for the thermo-elements employed in high temperature determinations.

Physical and Chemical Properties.[2]—The properties of the individual metals of the platinum group are summarised as follows :—

Iridium.—Symbol, Ir ; atomic weight, 193·0 ; specific gravity, 22·38 ; melting-point, 1950° C.

Iridium is chiefly obtained from crude platinum, in which it generally exists alloyed with osmium as iridosmine. Iridium may form more than half the

[1] *Journ. Ind. Eng. Chem.*, 1911, iii. 354-355.
[2] The author is indebted to Valentin's *Practical Chemistry*, 10th edit., 1908, for much of the information on the reactions of the platinum metals.

alloy; but, as stated above, the grains usually contain 70 to 80 per cent. of platinum. The alloy remains behind undissolved when the platinum ore is treated with aqua-regia in the form of white, metallic-looking, hard grains.

SOLUBILITY.—Iridium is insoluble in every acid, and therefore differs from platinum in not being dissolved by dilute aqua-regia. However, when in a very finely divided state, it is slowly attacked by concentrated aqua-regia. Fusion with acid potassium sulphate oxidises but does not dissolve it (distinction from ruthenium). It is likewise oxidised to the trioxide, Ir_2O_3, when heated with a fused mixture of sodium nitrate and hydroxide, or when fused with NaHO, with access of air; but the residue, consisting of a compound of Ir_2O_3 and sodium, is only slightly soluble in water.

On treating the residue with aqua-regia a dark-coloured solution is obtained of the double chloride, $2NaCl,IrCl_4$.

REACTIONS OF IRIDIUM.—(A solution of the double chloride, $2(NaCl),IrCl_4$, may be used for the following reactions.)

Suphuretted Hydrogen, H_2S, first decolorises the dark-coloured solution owing to the reduction of the $IrCl_4$ to Ir_2Cl_6, and simultaneously precipitates sulphur. The further passage of the gas throws down the trisulphide Ir_2S_3 as a dark brownish precipitate.

Ammonium Sulphide $(NH_4)_2S$ also gives the same precipitate, readily soluble in excess.

Caustic Alkali, NaHO or KHO, added in excess, colours the solution to a greenish tint, and throws down a little brownish-black double chloride of potassium and iridium. On heating the liquid with exposure to the air, it acquires at first a reddish tint, which changes afterwards to a deep azure-blue owing to the precipitation of iridic hydroxide, $Ir(HO)_4$, and when evaporated to dryness and taken up with water a colourless solution is obtained, and an indigo-blue deposit of iridic dioxide is left undissolved. This reaction serves to distinguish iridium from platinum.

Ammonium Chloride and Potassium Chloride precipitate the double chlorides $2NH_4Cl,IrCl_4$ and $2KCl,IrCl_4$ respectively. They are both dark-brownish red precipitates, insoluble in strong solutions of the precipitant.

Reducing Agents, such as ferrous sulphate, oxalic acid, stannous chloride, etc., reduce these double salts, especially when in hot solutions, giving similar compounds containing the lower chloride of iridium, and having the composition expressed by the formulæ $6NH_4Cl,Ir_2Cl_6$ and $6KCl,Ir_2Cl_6$ respectively. The solution is at the same time decolorised, and the double chloride gradually crystallises out on cooling.

When iridic hydroxide, $Ir(HO)_4$ (bulky, indigo-coloured), is suspended in a solution of potassium nitrite and the solution saturated with sulphurous acid and boiled, with renewal of the water, as long as SO_2 is given off, the whole of the iridium is converted into an insoluble brownish-green iridic sulphite, $Ir(SO_3)_2,4OH_3$. This reaction may be used for the separation of iridium from platinum.[1]

Metallic Zinc precipitates black metallic iridium.

Iridium Salts are reduced by alcohol in alkaline solutions to iridious compounds which are soluble in hydrochloric acid.

Osmium.—Symbol, Os; atomic weight, 191; specific gravity, 22·47; melting-point about 2500° C.

Intense white heat (the oxyhydrogen flame) volatilises the metal but does not melt it. When strongly heated in contact with air, the finely divided metal burns and is converted into osmic anhydride, OsO_4, commonly called *osmic acid*.

_____
[1] Valentin, *ibid.*, p. 897.

All compounds of osmium yield the metal when ignited in a current of hydrogen.

Osmium occurs chiefly as a natural alloy of osmium and iridium (osmiridium) in platinum ores, and remains behind undissolved when the ores are treated with aqua-regia, in the form of hard, white, metallic-looking grains.

The chief source of the world's supply of osmiridium is in the neighbourhood of the Savage River, on the west coast of Tasmania.[1] The return of the Mines Department states that 409 ounces were produced in the first quarter of 1912. The metal, which is used in the production of metallic filament electric lamps, sells at £7, 10s. per ounce as against 85s. per ounce for gold.

The following interesting information on iridium and osmiridium is given by F. W. Horton[2] in discussing the presence of iridium in American placer platinum.

Russian and Colombian crude platinum generally contain from 1 to 3 per cent. of iridium. The total Russian production of osmiridium for the last ten years is officially reported as only 308 ounces. The amount of iridium and its natural alloys imported into the United States amounts only to about 3 per cent. of the weight of platinum imported, notwithstanding the great demand for the former metal. Iridium is usually associated with osmium and platinum. On Trinity River, Trinity County, California, coarse nuggets are found of the following average composition: osmiridium 69·7 per cent., platinum 22·7 per cent., and 0·4 per cent. of gold. On treating this material with aqua-regia, it disintegrates and small flat scales of osmiridium remain undissolved. The nuggets seem, therefore, to consist of osmiridium cemented by platinum. A 17-ounce sample from China Flat, fifty miles farther down the Trinity River, gave, on analysis, the following figures: osmiridium, 74·883; platinum, 17·640; iron, 6·376; copper, 0·568; gold, 0·074 per cent.; nickel, traces; chromium and palladium, none. The osmiridium contained: iridium, 55·04; osmium, 44·28; iron, 0·59; silica, 0·09 per cent.; rhodium and ruthenium, none. Analyses show that crude platinum from this district contains in general 69 to 75 per cent. of osmiridium, of which over 50 per cent. is iridium; and, on the whole, the crude platinum contains about 40 per cent. of iridium as compared with 20 per cent. of platinum. This river basin has long been worked for gold, and has also produced much platinum and other allied metals, though the greater part of the output of the platinum metals has been, until recently, thrown away as worthless. For comparison, the following analyses of crude platinum from Oroville, California, are given: (1) osmiridium, 23·36; platinum, 65·66; iridium, 0·66; gold, 0·72 per cent.; (2) osmiridium, 32·01: platinum, 56·19; iridium, 1·02; gold, 4·16 per cent.

Osmiridium is attacked by mixing it with common salt, or potassium chloride, and exposing it in a glass or porcelain tube to a current of moist chlorine gas. Osmic acid is formed, which volatilises below 212° C., and can be condensed and fixed by passing the fume into a solution of an alkali. (Iridium remains behind in the tube as a double chloride, $2KCl,IrCl_4$. This salt is obtained in reddish black regular octahedra by recrystallisation from water.) The alkali solution is evaporated with excess of ammonium chloride, and leaves, on ignition of the dry residue and extraction with water, metallic osmium, as a black or grey powder, with metallic lustre.

SOLUBILITY.—Red-fuming nitric acid, or aqua-regia, acts upon osmium and oxidises it to osmic tetroxide (or osmic anhydride), OsO_4. When very intensely ignited, osmium is rendered insoluble in acids, and has to be fused with nitre and then distilled with nitric acid, when OsO_4 (osmium tetroxide) distils over.

[1] *Chem. and Drug.*, July 20, 1912.
[2] F. W. Horton, *Eng. and Min. Journ.*, 1912, xciv. 873-875.

This oxide is remarkable for its peculiar, exceedingly irritating, and offensive odour, resembling that of chlorine and bromine. It exerts a most injurious effect upon the eyes, and is extremely poisonous. This vapour is given off when any osmium compound is heated with nitric acid, and serves as a characteristic test for the metal.

The tetroxide is soluble in water, giving a neutral solution, and is precipitated from its solutions by all metals, even by mercury and silver, as a black precipitate. On heating a mixture of finely divided osmium, or of the sulphide, with potassium chloride in a stream of chlorine gas, a double chloride, $Os_2Cl_6KCl,3OH_2$, is obtained, which crystallises from water in dark red-brown regular octahedra. The salt is insoluble in alcohol.

The solution of this double chloride is more stable than that of the osmium chloride.

REACTIONS OF OSMIUM (solution of double chloride).

Sulphuretted Hydrogen, H_2S, or sulphides, give a brownish-black sulphide, OsS, which only separates when a strong acid is present. The precipitate is insoluble in ammonium sulphide.

Caustic Alkali, KHO or $NaHO$, as well as their carbonates, produce a brownish-red precipitate of hydrated osmic dioxide, $Os(HO)_4$. On fusing the double chloride, $Os_2Cl_6KCl,3OH_2$, with sodium carbonate, dark grey OsO_2 (osmium dioxide) is obtained.

The neutral solution obtained by dissolving the tetroxide in water has powerful oxidising properties; thus:

It oxidises ferrous sulphate, the osmium compound being reduced to the state of hydrated dioxide, $OsO_2,2H_2O$ (or $Os(HO)_4$), which is thrown down as a black precipitate.

Sulphurous acid, or a sulphite added to the solution, produces a series of colour changes from yellow to green, and lastly dark blue, the colour of the osmious sulphite, $OsSO_3$, which then separates out.

Stannous chloride produces a brown precipitate, soluble in hydrochloric acid to a brown fluid.

Palladium.—Symbol, Pd; atomic weight, 106·5; specific gravity, 11·4; melting-point, 1535° C.

Palladium is the most fusible of the so-called platinum metals. The metal oxidises when heated in air, the surface becoming coloured from films of oxide. It absorbs hydrogen to a large extent. A solution of iodine produces a black stain on palladium, but has no effect on platinum.

Palladium occurs in small quantity in crude platinum, and alloyed with gold and silver in a gold ore found in Brazil. It also occurs in small quantity in certain copper deposits. To extract the metal the gold dust is fused with the addition of silver, the resulting alloy granulated and heated with nitric acid, which dissolves the silver and palladium. The silver is removed as chloride by the addition of sodium chloride, and the palladium may then be precipitated as palladious cyanide by means of mercuric cyanide, and the palladious cyanide, $PdCN_2$, decomposed by ignition.

SOLUBILITY.—The best solvent for palladium is aqua-regia. It is sparingly soluble in pure nitric acid, but dissolves more readily in fuming nitric acid, forming palladious nitrate, $Pd(NO_3)_2$. If the solution be diluted with water, especially if the amount of free acid present is only small, a brown-coloured precipitate is produced consisting of a basic nitrate.

Palladium dissolves slightly in boiling concentrated sulphuric acid.

The metal is readily attacked by fusing with hydrogen potassium sulphate.

REACTIONS OF PALLADIUM.— *Water* precipitates a brown basic salt from solutions containing a slight excess of acid only.

Sulphuretted Hydrogen, H_2S, or ammonium sulphide, $(NH_4)_2S$, throws down from acid or neutral solutions black palladious sulphide, PdS, insoluble in alkali sulphide, but soluble in boiling hydrochloric acid, and readily soluble in aqua-regia.

Caustic Alkali, KHO or NaHO, precipitates a yellowish-brown basic salt, soluble in excess.

Soluble Carbonates precipitate brown palladious hydroxide, $Pd(HO)_2$, soluble in excess, reprecipitated on boiling.

Ammonium Hydrate, NH_4HO, or carbonate produce no precipitate from the nitrate solution, but decolorise the dark brown solution, forming double pallad-ammonium salts. A somewhat transient blue colour is formed by ammonium hydrate alone.

Potassium Iodide, KI (and other soluble iodides), gives in very dilute solutions a black precipitate of palladious iodide, PdI_2, somewhat soluble in excess of KI. This is a most characteristic reaction for palladium.

Mercuric Cyanide, $HgCN_2$, gives a yellowish-white gelatinous precipitate of $PdCN_2$, readily soluble in KCN and in ammonium hydrate. Slightly soluble in hydrochloric acid. It leaves on ignition a spongy mass of metallic palladium. This is also a characteristic reaction for palladium.

Ammonium Chloride, NH_4Cl, does not readily precipitate palladium salts.

Potassium Chloride, KCl, precipitates a brown-red double chloride, 2KCl, $PdCl_2$, insoluble in absolute alcohol; soluble in water to a dark red fluid.

Ammonium Thiocyanate, NH_4SCN, gives no precipitate even after the addition of SO_2. This serves to distinguish it from copper.

Stannous Chloride, $SnCl_2$, produces a brownish-black precipitate, soluble in hydrochloric acid to an intense green solution.

All palladium compounds are decomposed when ignited.

Ruthenium.—Symbol, Ru; atomic weight, 101·7; specific gravity, 12·26; melting-point, 1800° C.

Ruthenium is found in small quantity only in crude platinum, and remains in the insoluble residue resulting from the treatment of platinum ore with aqua-regia. It is stated by E. de Hautpick [1] that ruthenium is never found in nature in the same minerals with platinum and palladium, and osmium is found in them very seldom. However, osmio-ruthenium minerals are found under conditions similar to those of native palladium. They are sometimes even separated along with platino-palladium bodies. Ruthenium is a greyish-white metal, closely resembling iridium and very difficultly fusible. When heated in the air it becomes covered with bluish-black ruthenic oxide, Ru_2O_2, insoluble in acids.

SOLUBILITY.—When pure it is insoluble in acids, and is scarcely acted upon by aqua-regia. It is even unacted upon when fused with hydric potassic sulphate.

It is attacked either by fusion with KHO and nitre, or potassium chlorate, and is converted thereby into potassium ruthenate, K_2RuO_4, a dark green mass, soluble in water to an orange-coloured fluid, which stains the skin black, from the separation of black ruthenic oxide. From this, acids (nitric) throw down the black hydroxide.

Ruthenic hydroxide is soluble in HCl, giving a solution of ruthenic chloride, Ru_2Cl_6, an orange-yellow-coloured liquid, which on heating is resolved into HCl and a hydrated oxide.

Ruthenium is also rendered soluble by ignition with potassium chloride in a current of chlorine gas, being thus converted into potassium ruthenic chloride, 2KCl, RCl_4.

REACTIONS OF RUTHENIUM (solution of ruthenic chloride, Ru_2Cl_6).

Sulphuretted Hydrogen, H_2S, produces at first no precipitate, but by prolonged

[1] Article on "Osmiridium," by E. de Hautpick, *Min. Journ.*, August 24, vol. xcviii. p. 851, 1912.

passing of the gas the solution acquires an azure-blue tint, and deposits brown ruthenic sulphide, Ru_2S_3. This reaction is very delicate and most characteristic of ruthenium.

Ammonium Sulphide, $(NH_4)_2S$, produces a brownish-black precipitate difficultly soluble in excess.

Caustic Alkali, KHO, gives a black precipitate of ruthenic hydroxide, $Ru_2(HO)_6$, insoluble in alkalies, but soluble in acids.

Potassium Thiocyanate, KCNS, produces (in the absence of other platinum metals) after some time a red coloration, which gradually changes to purple-red, and, upon heating, to a fine violet tint. This is another very characteristic reaction of ruthenium.

Alkali Chlorides produce in concentrated solutions crystalline glossy violet precipitates of the double chlorides, difficultly soluble in water, insoluble in alcohol. They are decomposed on boiling with water, with separation of black ruthenious oxychloride.

Potassium Nitrite, KNO_2, forms a double salt, $3KNO_2.Ru(NO_2)_3$, readily soluble in an excess of the precipitant. On the addition of a few drops of colourless ammonium sulphide the solution assumes a dark red colour, changing to brown, without precipitation of sulphide.

Metallic Zinc reduces ruthenic chloride, Ru_2Cl_6, to ruthenious chloride, $RuCl_2$, producing a fine azure-blue coloration, which subsequently disappears, ruthenium being deposited in the metallic state as a black powder.

Rhodium.—Symbol, Rh; atomic weight, 103.0; specific gravity, 12.10; melting-point, $1660°$ C.

The metal is oxidised at a red heat. It is found, occasionally to a considerable extent, in the insoluble residue resulting from the treatment of platinum ore with aqua-regia.

SOLUBILITY.—When pure and in a compact state, it is unacted upon by the strongest acids, even of aqua-regia; but when alloyed with other metals, as with lead, copper, bismuth, and platinum in certain proportions, it is soluble in aqua-regia. When, however, alloyed with gold or silver it does not dissolve.

It is oxidised by fusion with dry potassium hydroxide and nitre.

Fusion with hydrogen potassium sulphate converts it into soluble potassium rhodic sulphate, $K_6Rh_2(SO_4)_6$. When the metal is mixed with sodium chloride and ignited in a current of chlorine, a double chloride of sodium and rhodium, $3NaClRhCl_3 + H_2O$, is formed, which, like the double sulphate, is easily soluble in water, forming a rosy-red solution.

REACTIONS OF RHODIUM (solution of potassio-rhodic sulphate).

Soluble Sulphides produce, from a hot solution, a brown precipitate of rhodic sulphide, Rh_2S_3, insoluble in ammonium sulphide, but soluble in boiling nitric acid.

Caustic Alkali, KHO or NaHO, gives (with potassio-rhodic chloride) a yellowish-brown precipitate of rhodic hydroxide, $Rh_2(HO)_6$, soluble in excess; in other rhodic salts this precipitate appears only on boiling.

KHO produces at first no precipitate from a solution of rhodic chloride, but on the addition of alcohol a brown precipitate of rhodic hydroxide is given. This is a distinguishing reaction of rhodium.

Ammonium Hydrate (NH_4OH) gives a yellow flocculent precipitate, only formed, however, after some time, soluble in HCl, forming a rhodamine salt.

Potassium Iodide, KI, produces a slight yellow precipitate.

Potassium Nitrite, KNO_2, gives (with rhodic chloride) an orange-yellow precipitate, which is slightly soluble in water, and only very slowly decomposed by strong hydrochloric acid. This is a characteristic reaction of rhodium.

Metallic Zinc precipitates black metallic rhodium.

Distinguishing Test.—Rhodium is distinguished from the other platinum metals by its insolubility in aqua-regia, its solubility when fused with hydrogen potassium sulphate, $KHSO_4$, and the behaviour of its chloride with caustic potash, KHO, and alcohol.

Bullion containing Platinum and Metals of the Platinum Group.

Platinum frequently occurs in the gold and silver bullion which has to be treated by the ordinary methods of refining, usually the sulphuric acid process. In an alloy of gold and silver, containing a small proportion of platinum, nearly all the silver is dissolved by the sulphuric acid, leaving the platinum associated with the gold. This partially refined gold holding the platinum is melted and assayed, to determine the amount of platinum and gold it contains.

Liquation in Bullion containing Platinum.—It has been found in practice that the results of assays of samples taken in the ordinary way by cutting a small proportion from one end of a bar or ingot of platinum gold alloy, do not indicate the actual percentage of gold and of platinum existing in the entire mass. It is therefore evident that the platinum has been redistributed by liquation during the cooling and solidification of the mass. The liquation of alloys of gold and platinum has been investigated by E. Matthey,[1] who gives the following results (Table L.) obtained from six platinum-gold ingots of different qualities, as they occurred in the course of refining commercially by sulphuric acid. Each of these bars, after melting and assaying, was separately treated with a view to extracting the total amount of gold contained, as a check on the assay report.

It will be at once seen that the higher percentage of gold indicated by the assay of a portion cut from one end of the ingot is *not* borne out by the actual amount of fine gold obtained by refining, which, of course, truly represents the proportion of gold existing in each bar.

Table L.—Results of Assays of Platinum-Gold Ingots. (*Matthey.*)

No. of bar.	Weight of bar in troy ozs.	Platinum by assay. Parts per 1000.	Gold by assay. Parts per 1000.	Gold per 1000 by fine gold actually obtained.	Difference between gold by assay and gold obtained. Parts per 1000.
42	728·5	111·0	825·0	812·0	13·0
67	355·0	120·0	660·0	630·0	30·0
109	589·5	120·0	800·0	780·0	20·0
126	435·0	045·0	850·0	845·0	5·0
149	480·5	086·0	842·0	830·0	12·0
188	473·0	110·0	830·0	821·0	9·0

These results prove that the percentage of gold in the outer portions of ingots of platinum-gold alloys does not represent the true percentage of gold in the alloy, and that liquation takes place to a considerable extent in such bars, the platinum becoming concentrated towards the centre of the mass. The bars from which the above results were obtained contained, in addition to platinum and gold, either silver or copper, and in some cases both silver and copper, but the same tendency to liquate is observed in bars consisting of platinum and gold only. Thus Matthey melted 900 parts of fine gold with 100 parts of pure platinum, and, after repeated meltings, cast the alloy into a

---

[1] E. Matthey, *Proc. of Royal Society*, 1890 ; also *Chemical News*, vol. lxi., 1890, p. 111.

special iron mould, so as to obtain the metal in the form of a disc. The result was, as in the previous cases, liquation of the platinum towards the centre of the mass, the gold in 1000 parts being 900 on the exterior against 845 at the centre, and the platinum in 1000 parts 98·0 on the exterior and 146·0 at the centre.

Osmiridium in Gold Bars.[1]—Gold sometimes contains osmium and iridium, which remain together during melting and refining as an alloy commonly known as osmiridium.

When the gold is melted the osmiridium settles to the bottom owing to the fact that a mixture of these metals does not seem to form a true alloy with gold, and being of high density and very infusible, the particles, unfused or partly fused, settle through the liquid gold. To separate the osmiridium the greater portion of the molten gold is carefully but rapidly poured into a mould; the remainder, which contains almost all the osmiridium, is allowed to cool in the crucible until it has solidified, and then assayed for osmiridium. An alternative plan is to allow the whole charge to solidify in the crucible, and then cut off the lowest portion, which is set aside.

As osmiridium settles better from an alloy chiefly consisting of silver than from pure gold, the rich "bottoms" are melted several times with silver, the lowest part being cut off each time. The gold is thus gradually replaced by silver, which eventually forms by far the greater part of the mass. The alloy is then granulated and the granules sampled for assay.

The osmiridium is extracted commercially by parting the granulated metal in sulphuric acid and treating the resulting powder of gold and osmiridium with aqua-regia, by which the gold is dissolved, and the osmiridium separated as a black powder.

Iridium is, however, not invariably separated from the gold bars in which it is contained, and traces can be observed in some of the refined commercial bars met with in London.[2]

[1] T. K. Rose, *Metallurgy of Gold*, 5th edit., p. 389.   [2] T. K. Rose, *ibid.*

CHAPTER XXV.

THE ASSAY OF PLATINUM.

Introduction.—The materials containing platinum and metals of the platinum group that are most frequently presented for assay, are alloys of platinum used in jewellery and dentistry, and scrap metal, sweeps, etc., resulting from the working of the alloys in the different manufacturing processes. Black sands and ores of gold and other metals contain platinum in small quantity, but true platinum ores are rarely received for assay. Electrolytic slimes are not infrequently assayed for platinum (see page 350).

An accurate determination of platinum is difficult and tedious and requires considerable experience : it is complicated by the presence of gold and metals of the platinum group.

The methods of assay in use have been worked out by Chaudet,[1] Riche and Forest.[2] They are similar in principle to those employed for the assay of gold in ores and bullion and involve concentration of the platinum in lead, which is cupelled and the resultant button of precious metals parted to separate the gold; but when platinum is present the parting operation has to be modified to effect its separation.

Platinum alone is not soluble in nitric acid, but when alloyed with other metals, such as silver, which dissolve in this acid, it is rendered soluble ; so that in parting gold in nitric acid in the ordinary way, platinum, if present, will be partly dissolved and will pass into solution with the silver. If, however, the parting is effected in sulphuric acid, the silver only is dissolved, the platinum remaining in the residual gold.

Advantage is taken of these facts in the assay of platinum, the alloy of silver-gold-platinum being first parted in sulphuric acid to remove the silver and the residue of gold and platinum weighed. The residue is then re-alloyed with more silver and parted in nitric acid, whereby the platinum is dissolved with the silver, and the gold left as an insoluble residue. In order to separate the platinum satisfactorily, it is necessary that the several metals shall be present in certain well-established ratios and the parting operation repeated several times, as described later.

General Remarks on the Assay of Platinum.

The following remarks on the behaviour of platinum and of metals of the platinum group in the several assay operations will help to explain the principles upon which the assay of platinum is based. They are conveniently grouped under three heads, viz. (1) fusion, (2) cupellation, (3) parting.

(1) **Fusion.**—Ores and materials containing platinum are fused with lead or lead compounds to collect the platinum and other precious metals.

---

[1] Chaudet, *L'Art de l'Essayeur*, Paris, 1835.
[2] Riche (A.) and Forest, *L'Art de l'Essayeur*, Paris, 1892.

Platinum and most metals of the platinum group readily alloy with lead, so that during the ordinary crucible fusion or scorification they are collected and concentrated in the resultant lead button. Osmium and ruthenium, if present in any considerable amount, may be only partially collected, as they do not alloy readily with lead and also tend to become oxidised and pass into the slag.

(2) Cupellation of Platinum.—The presence of platinum and of the platinum group of metals raises the melting-point of the gold or silver button left on the cupel, so that it is necessary to conduct the cupellation at a higher temperature in order to prevent freezing and to remove the lead as far as possible. The cupellation requires a higher temperature in proportion as the percentage of platinum is greater.

According to results obtained by Professor Stansfield in the laboratory at the Royal Mint, it is impossible to free buttons containing over 50 per cent. of platinum from lead at the ordinary temperatures attainable in a muffle. In the cupellation of pure platinum, buttons weighing from 0·002 to 0·01 gramme retain about 10 per cent. of their weight of lead, and those weighing from 0·04 to 2 grammes retain about 33 per cent.[1] From experiments by Sharwood[2] on the cupellation of platinum-silver alloys, it is evident that the lead retained after cupellation decreases with an increase in the ratio of silver to platinum, as shown by the results given in Table LI. The alloys were cupelled at the temperatures ordinarily employed for gold cupellation.

Table LI.—Lead retained in the Cupellation of Platinum-Silver Alloys and of Platinum-Silver-Gold Alloys. (Sharwood.)

No.	Composition of alloy.			Lead retained.	Character of button left on the cupel.
	Platinum taken.	Silver taken.	Gold taken.		
	mgs.	mgs.	mgs.	mgs.	
1	100	...	...	37·5	Hard silvery.
2	100	25	...	31·0	,,
3	100	50	...	26·2	Dull grey.
4	100	100	...	25·0	,,
5	100	101	48	24·0	,,
6	100	205	48	22·0	Smooth, brittle.
7	100	206	6	10·0	,, ,,
8	100	310	...	10·0	Slightly crystallised.
9	100	427	...	5·0	Smooth and silvery.
10	100	470	19·4	2·0	,, ,,

The lead is almost entirely eliminated with a ratio of 10 parts of silver to 1 part of platinum.

The retention of lead appears to depend mainly upon the fusibility of the button, oxidation ceasing for practical purposes when the button solidifies.

The appearance of the buttons from cupellation is considerably altered by the presence of platinum and the platinum metals. With platinum alone, or with little silver, the cupelled button is of a dull grey; slight increases in the silver give a crystalline surface: with two or more parts of silver the button closely resembles a normal silver button, but is flatter, and has a somewhat steely appearance. The button is usually brittle if the silver is less than three times the weight of the platinum.[3] The buttons containing less than 16 parts of silver do not flash on solidification.[4]

[1] Quoted by Rose, *Metallurgy of Gold*, 5th edit., p. 493.
[2] *Journ. Soc. Chem. Ind.*, vol. xxiii., 1904, p. 412.   [3] Sharwood, *loc. cit.*
[4] Schiffner, *Mineral Industry*, vol. viii., 1899, p. 397.

The effect of the presence of metals of the platinum group on the appearance of the cupellation button is similar to that of platinum, but not identical. Iridium, if present, being very dense, always sinks to the lower surface of the button while in a fluid state.

(3) **Parting of Platinum.**—Much experimental work has been done in connection with the action of acids on platinum alloys, but the results are very conflicting, and this important part of platinum assaying still requires investigation.

Some few facts, however, appear to be definitely established. The action of sulphuric acid and nitric acid is as follows :—

(a) *Action of Sulphuric Acid on Platinum Alloys.*—When alloys of platinum and silver are attacked by boiling concentrated sulphuric acid, the silver dissolves as silver sulphate and the platinum remains undissolved as a black residue.

To ensure the complete removal of the silver, however, the proportion of the two metals must be at least 10 parts of silver to 1 part of platinum. According to results obtained by Thompson and Miller,[1] silver remains with the platinum residue when the alloy contains 20 per cent. or more of platinum, the amount of silver remaining undissolved increasing with the increase in the percentage of platinum present. The following results (Table LII.) were obtained with platinum-silver alloys of different composition, 300 milligrammes of the alloy being taken in each case and heated for fifteen minutes in 10 c.c. of concentrated sulphuric acid. The solution was then decanted off and the residue re-treated for fifteen minutes with 5 c.c. more of concentrated sulphuric acid. The residues were finally tested for silver and the amount determined. In some cases traces of platinum were found in the acid filtrate, showing that platinum is slightly soluble in sulphuric acid.

Table LII.—Action of Concentrated Sulphuric Acid on Platinum-Silver Alloys. (Thompson and Miller.)

Approximate composition of alloys used.		Actual composition of alloys used.		Silver retained in platinum residue.
Platinum.	Silver.	Platinum.	Silver.	
per cent.	per cent.	per cent.	per cent.	per cent.
10·0	90·0	10·39	89·61	Trace
20·0	80·0	20·59	79·41	0·59
30·0	70·0	31·46	68·54	0·98
40·0	60·0	37·89	62·11	2·24
50·0	50·0	57·05	42·95	2·70

The state in which the platinum is left after parting is dependent on the proportion of silver in the alloy. With more than five parts of silver the platinum is left in a fine state of division (but not as colloid), and loss may take place in decanting unless great care is exercised.

Any lead that may be present in the cornets, as the result of cupellation, will be converted into lead sulphate, which forms a white deposit on the residual platinum. It must be removed by treatment with ammonium acetate before weighing.

When ternary alloys of platinum, gold, and silver are attacked by boiling in concentrated sulphuric acid, the silver dissolves and leaves the gold in the

[1] Thompson and Miller, "Platinum-Silver Alloys," *Journ. Amer. Chem. Soc.*, 1906, vol. xxviii. pp. 1115-1132.

platinum residue; but experience shows that the silver is only satisfactorily removed when the platinum, gold, and silver are present in definite ratios. Very little work has been done, however, to ascertain the best ratios to be employed.

A number of experiments were made by H. Carmichael [1] with a view to separating the silver from various ternary alloys of gold, platinum, and silver in which the metals were present in different ratios. The results were as follows, (Table LIII.):—

Table LIII.—Separation of Silver from Alloys of Platinum, Gold, and Silver by means of Sulphuric Acid Parting.

No.	Weight of metal used, milligrammes.			Ratio of metals.			Weight of residue. Platinum + gold.	Weight of silver retained in residue.
	Platinum.	Gold.	Silver.	Platinum.	Gold to platinum.	Silver to platinum.		
							mgs.	mgs.
I.	100	100	500	1	1	5	204·7 / 204·7	4·7 / 4·7
II.	100	100	500	1	1	5	204·0 / 200·6	4·7 / 0·6
III.	5	100	300	1	20	60	105·5 / 105·4	0·5 / 0·4
IV.	5	100	300	1	20	60	105·3 / 105·2	0·3 / 0·2
V.	5	50	300	1	10	60	55·3 / 55·2	0·3 / 0·2
VI.	5	25	300	1	5	60	30·3 / 30·3	0·3 / 0·3

Note.—All the alloys were cupelled and then treated as follows:—

No. I. alloy. Parted in strong sulphuric acid.

No. II. alloy. Parted in strong sulphuric acid and the residue treated with strong nitric acid.

No. III. alloy. Parted in dilute sulphuric acid.

No. IV. alloy. Parted in dilute sulphuric acid.

No. V. alloy. Parted in dilute sulphuric acid, washed and treated with strong nitric acid.

No. VI. alloy. Parted in dilute sulphuric acid, washed and treated with strong nitric acid.

The residual platinum was treated with nitric acid in the cases stated with the object of dissolving out, if possible, the small amount of silver still remaining in the cornet.

It may be remarked that in the above table the ratio of silver to platinum alone is given, but in adjusting the proportion of silver for parting ternary alloys it is usual to consider the ratio of the silver to one part of platinum and gold together.

The above results show that the proportion of silver may be decreased if the proportion of gold to platinum is increased.

In practice it is usual to employ about 3 parts of silver to 1 part of platinum and gold together in order to obtain a satisfactory cornet, as the cornet is very liable to break up if this proportion is much exceeded. The silver may therefore be regarded as a constant quantity, and the ratio of platinum to gold is regulated

[1] *Journ. Soc. Chem. Ind.*, vol. xxii., 1903, p. 1325.

accordingly, the ratio generally employed being 1 part of platinum to 10 parts of gold.

As previously stated, 1 part of platinum requires about 10 parts of silver for sulphuric acid parting, and gold requires about 2½ parts of silver. But since the proportion of silver is fixed at three times the weight of the platinum and gold together, it becomes necessary to dilute the platinum with gold to such an extent that the ratio of silver to platinum is such, that practically no silver remains in the cornet. Experience shows that this is accomplished by diluting the platinum with ten times its weight of gold as stated.

This proportion is arrived at thus :—

Parts of silver.

1 part of platinum requires 10 parts of silver   .   .   . 10
1 part of gold requires 2½ parts of silver, therefore 10 parts
require 10 × 2½ = 25 parts   25

Total   .   . 35

Thus 11 parts of alloy (1 part platinum + 10 parts gold) would require 35 parts of silver, which is approximately three times the proportion of platinum and gold together, i.e. 11 × 3 = 33, the actual quantity that would be used.

Satisfactory results may be obtained with a ratio of about 8 of gold to 1 of platinum, but if the ratio is less than this the cornets retain more silver. Some assayers prefer to work with a larger ratio of silver and smaller ratio of gold to platinum, such as those given in Table LIII.

The figures given in the table show that equally good results may be obtained by varying the ratios. Thus the surcharge in Experiments V. and VI. is the same, viz. 0·3 milligramme, for the same weight of platinum (5 milligrammes); but in Experiment V. the platinum was alloyed with ten times its weight of gold, and the silver added was equal to about five times the gold + platinum, while in Experiment VI. the platinum was alloyed with only five times its weight of gold, and with silver equal to ten times the gold + platinum.

It is doubtful whether any ratios of metals can be selected that will give cornets which are entirely free from silver.

The parting should be effected in dilute sulphuric acid, since it is proved by the researches of Conroy[1] and others that hot concentrated sulphuric acid, even when pure, exerts a marked solvent action on platinum, especially at a temperature of about 250° C. to 280° C. The temperature of boiling pure sulphuric acid is 338° C.

According to Steinmann,[2] boiling concentrated sulphuric acid dissolves considerable quantities of platinum from platinum-silver alloys, the loss being usually from 20 to 30 parts per 1000 parts of platinum, but may be as much as 50 parts. By using slightly diluted acid the silver can be extracted without loss of platinum. Steinmann recommends rolling the alloy to a thickness of 0·2 millimetre, and heating for fifteen minutes with dilute sulphuric acid (100 vols. of concentrated acid to 22 vols. of water) at a temperature not exceeding 240° C. The treatment with acid is repeated twice, and the residual platinum is washed, dried, and weighed.

---

[1] "The Action of Sulphuric Acid on Platinum," J. T. Conroy, Journ. Soc. Chem. Ind., vol. xxii., 1903, p. 465; "The Dissolution of Platinum by Sulphuric Acid," M. Delepine, Bull. Soc. Chem., series 3, vol. xxxv., No. 1, 1906 (abstract Chem. News, 1906, vol. cxciii. pp. 108-109); and also "Action of Boiling Sulphuric Acid on Platinum," Le R. W. M'Cay, Eighth Int. Cong. Appl. Chem., 1912, Sect. I., Orig. Comm., i. pp. 351-359.

[2] A. Steinmann, "Platinum Assay," Schweiz. Woch. Chem. Pharm., 1911, vol. xlix. pp. 441-444, 453-457 (abstract Journ. Soc. Chem. Ind., 1911, vol. xxx. p. 1216).

Some assayers use slightly diluted acid consisting of 90 vols. of concentrated acid to 10 vols. of water, and boil for fifteen minutes or until all visible action has ceased. The residual platinum is then treated once or twice with strong sulphuric acid. The use of dilute acid lessens the tendency of the cornet to break up, especially if from 4 to 5 parts of silver are used for parting

The sulphuric acid used for parting must be free from nitric acid and hydrochloric acid.

The action of concentrated sulphuric acid on metals of the platinum group is as follows :—

Palladium is attacked during the parting and goes into solution with the silver, giving an orange-coloured liquid. The solution of the palladium is generally considered to be complete, but this is doubted by some assayers.

The other metals of the platinum group are not dissolved by sulphuric acid and will therefore remain in the platinum residue.

(b) *Action of Nitric Acid on Platinum Alloys.*—As previously stated, platinum, when alloyed with silver, is soluble to some extent in nitric acid, so that it can under certain conditions be separated from gold by parting in nitric acid.

Table LIV.—*Solubility of Platinum-Silver Alloys in Nitric Acid of* 1·10 *sp. gr. and of* 1·40 *sp. gr.* (*Thompson and Miller.*)

Composition of alloy.		Parted in nitric acid, 1·10 sp. gr.		Parted in nitric acid, 1·40 sp. gr.	
Platinum.	Silver.	Platiniferous residue.[2]	Platinum dissolved.[1]	Platiniferous residue.[2]	Platinum dissolved.[1]
per cent.	per cent.	per cent.	per cent.	per cent.	per cent.
0·5	95·5	0·42	0·08	0·22	0·28
1·0	99·0	0·85	0·15	0·42	0·58
2·0	98·0	1·74	0·26	1·09	0·91
3·0	97·0	2·19	0·81	1·81	1·19
4·0	96·0	2·98	1·02	2·42	1·58
5·0	95·0	3·56	1·44	2·62	2·38
10·0	90·0	...	...	4·53	5·47
13·0	87·0	3·33[3]	9·67	5·79	7·21
14·0	86·0	4·26	9·74	4·97	9·03
15·0	85·0	4·32	10·68	7·93	7·07
16·0	84·0	4·55[3]	11·45	11·54	4·46
18·0	82·0	4·54[3]	13·46	11·65	6·35
20·0	80·0	...	...	13·94	6·06
25·0	75·0	16·62	8·38	20·66	4·34
30·0	70·0	...	...	29·29	0·71
31·5	68·5	38·58	See Table LV.	...	...

The platinum goes into solution in the nitric acid in colloidal form, giving to the solution a brown to blackish colour according to the amount of platinum present. The formation of colloidal platinum is very marked, especially with alloys containing about 20 per cent. of platinum,[4] the solution in this case being opaque and almost black. With different percentages of platinum the intensity of the colour decreases, while with 10 per cent. the solution is almost colourless. On allowing the platinum solution to stand for two or three days, the colloidal platinum separates as a very fine black powder, and the solution becomes almost colourless.

[1] Difference between original percentage of platinum and percentage of residue.
[2] This includes silver retained (see also Table LII. for silver in residues).
[3] Average of two results.   [4] Thompson and Miller, *loc. cit.*

The results of experiments made by different workers to determine the extent of the solubility of platinum in nitric acid are conflicting. The most complete and most recent investigation on the action of nitric acid on platinum-silver alloys has been conducted by Thompson and Miller,[1] who worked on alloys containing from 0·5 to 57 per cent. of platinum. The tests were made with nitric acid of two different strengths, viz. 1·1 sp. gr. and 1·4 sp. gr., but it was found that in no case would the alloy dissolve completely in the acid; some platinum was always dissolved with the silver, and some silver always remained with the platinum residue.

The proportion of platinum passing into solution varies with the increase in the ratio of silver to platinum and also on the strength of the acid, as shown by the results given in Table LIV., page 415.

The platinum residue invariably retains silver ; the amount of which increases with the increase in the percentage of platinum in the alloys treated. Thus Thompson and Miller obtained the following results with alloys containing more than 10 per cent. of platinum (Table LV.) :—

Table LV.—Showing Amount of Silver retained in Platinum Residue from Treatment of Platinum-Silver Alloys with Nitric Acid of 1·10 sp. gr. (Thompson and Miller.)

Composition of alloy.		Total residue.	Silver in residue.	Platinum in residue.	Platinum dissolved.
Platinum.	Silver.				
per cent.	per cent.	per cent.	per cent.	per cent.	per cent.
10·39	99·61	3·86	0·27	3·59	6·80
20·59	79·41	8·58	1·81	6·77	13·82
31·46	68·54	36·59	12·09	24·50	6·96
37·89	62·11	49·18	13·64	35·49	2·40
57·05	42·95	65·16	12·19	52·97	4·08

From a study of the cooling curves and microstructure of the silver-platinum series of alloys, Thompson and Miller conclude that the alloys consist of a number of platinum-silver compounds of different solubilities, the relative proportions of these compounds in the alloys being dependent on the rate of cooling. These conclusions help to explain many of the irregular results obtained in parting with nitric acid. Thus an alloy containing 31·46 per cent. of platinum when *rapidly cooled* and treated with nitric acid 1·10 sp. gr. gave 36·59 per cent. of residue (including silver retained), while the same alloy *very slowly cooled* gave 33·58 per cent. of residue (including silver retained), a difference of 2 per cent.

The solubility of platinum in nitric acid is influenced by the presence of gold in the platinum-silver alloys, the actual amount of platinum dissolved being dependent on the ratio that exists between the three metals present.

As gold very frequently accompanies platinum in the materials submitted for assay, the action of nitric acid on ternary alloys of gold-platinum-silver is of considerable importance to assayers, but at present very little data is available to show what ratio of metals will give the best results. The figures obtained by the few investigators who have worked on these ternary alloys present such wide differences, that it is impossible to decide what individual metal is responsible for any given set of results.

[1] Thompson and Miller, "Platinum-Silver Alloys," *Journ. Amer. Chem. Soc.*, 1906, vol. xxviii. pp. 1115-1132.

From a few tests made by Sharwood[1] it would appear that an increase of gold, with the platinum and silver remaining constant, seems to decrease the dissolving of platinum, but more extended work is necessary on these lines.

As an example of the action of nitric acid on ternary alloys of gold-platinum-silver, the following figures (Table LVI.) may be given. They are the results of experiments by H. Carmichael[2] to determine the amount of platinum dissolved by nitric acid from gold-platinum-silver alloys of different composition.

The alloys were cupelled, rolled, and parted first in dilute nitric acid 1·16 sp. gr. and then in stronger nitric acid 1·26 sp. gr. The resulting cornets were washed, dried, and weighed.

Table LVI.—Parting of Platinum-Gold-Silver Alloys in Nitric Acid.
(Carmichael.)

Weight of metals used, milligrammes.			Ratio of metals.			Weight of cornet (gold + platinum).	Weight of platinum retained by cornet.[3]
Platinum.	Gold.	Silver.	Platinum.	Gold.	Silver.		
						mgs.	mgs.
20	100	300	1	5	15	102·7	2·7
15	100	400	1	6·6	26·6	{ 101·2 / 100·2	1·2 / 0·2
10	100	300	1	10	30	{ 100·8 / 100·4	0·8 / 0·4
10	100	500	1	10	50	{ 100·2 / 100·2	0·2 / 0·2
10	200	600	1	20	60	100·0	0
14	200	800	1	14·3	57·1	{ 200·8 / 200·8	0·3 / 0·8
14	300	900	1	21·4	64·3	300·0	0
7	100	400	1	14·3	57·1	100·2	0·2
5	100	500	1	20	100	100·0	0

In these experiments it was found that 400 milligrammes of added silver parted as successfully as 500 milligrammes, and at the same time gave a more compact cornet not so liable to break up.

The result of the last experiment appears to indicate that it is possible to use ratios of the metals such that the alloy may be successfully parted by one treatment in nitric acid, but since this conclusion is contrary to experience it needs confirmation. Thompson and Miller concluded, as the result of the experiments already referred to, that the separation of platinum from gold, in one operation, by means of alloying with silver and parting with nitric acid, is impossible.

In practice, as a general rule, no attempt is made to remove all the platinum from gold by one parting, but authorities differ as to the best ratios of metals to be employed to effect the solution of the platinum.

From the result of the last experiment given in the above Table LVI., it seems that the platinum is most satisfactorily removed when there is 7 per cent. of platinum to gold, or a ratio of 1 to 14, and the silver amounts to about four times the platinum and gold together. Although more platinum may be removed in one operation by using a large proportion of gold, this will result in producing very large cornets if much platinum is present, and if the total weight of the metals is reduced so as to obtain a cornet of convenient size, the original weight of sample that will have to be taken for assay may be too small to be truly

---

[1] *Loc. cit.*   [2] *Loc. cit.*
[3] This may also include a little silver retained.

representative. Also the addition of more than 3 parts of silver renders the cornets very liable to break up, although 4 parts or even 5 may be satisfactorily employed if the strips of alloy are not rolled too thin. In this case, however, the cornets are very porous and tender, and there is a tendency for more silver to be retained.

In considering the ratios of metals to be used, it is well to draw attention to the results of some experiments by Carmichael, from which it appears that the action of mass plays a part in the separation of the platinum. Thus, an alloy of 7 milligrammes of platinum, 100 milligrammes of gold, and 400 milligrammes of silver parted successfully, but when double the quantity of each metal was taken, the cornet retained 0·3 milligramme of platinum ; by increasing the gold, however, to 300 milligrammes the alloy parted successfully, practically all the platinum being removed. This would show that as the weight of platinum to be parted increases, it is desirable to increase the ratio of gold to platinum. This fact has been recognised in practice, as there is more or less agreement amongst assayers that the total weight of platinum present in an assay should not exceed 200 milligrammes, and that the quantity of material taken for assay should be such that about this weight or less of platinum is present. Some assayers use a smaller ratio of gold to platinum, say 5 to 1, for the parting when a small quantity of platinum only is present, and increase the ratio to 10 to 1 or even more when much platinum is present.

The ratios of gold, platinum, and silver most generally employed for nitric-acid parting are the same as those given for sulphuric-acid parting, namely, 1 part of platinum to 8 or 10 parts of gold, and silver equal to 2½ or 3 times the weight of the platinum + gold. These ratios ensure a satisfactory cornet, and, as shown in the above table, permit of a considerable proportion of the platinum being dissolved in the first parting. To separate the remaining platinum it is necessary, after weighing the residue of platinum and gold, to alloy it with more silver by cupellation and then to part again in nitric acid. The inquartation and parting is repeated until all the platinum has been removed.

The extent to which the platinum has been separated in each parting is judged by the colour of the residue or cornet obtained. When much platinum is present the cornet is black, but as the platinum is gradually removed in successive partings the colour of the cornet changes, passing through a series of steel-grey tints until finally it assumes the characteristic yellow colour of pure gold.

The action of nitric acid on metals of the platinum group that may be present is as follows :—

Palladium is dissolved in parting if the weight of silver is at least three times that of the palladium, yielding an orange-coloured solution ; but double parting appears to be necessary to ensure complete solution.[1]

Iridium, if present, always sinks to the bottom of the cupelled button, as it is very dense and is not usually fused at the temperature of the muffle, but occurs in the state of fine black crystalline particles. Hence when the button is rolled into a cornet with the lower face outwards, as directed on page 269, iridium shows itself as black sooty spots or streaks which are seen by a lens to fill up depressions in the surface of the gold. Iridium, being unacted upon by nitric acid, remains with the gold, consequently cornets containing iridium require special treatment, as described subsequently. The same remarks apply when osmiridium is present, as this alloy is not dissolved in nitric acid, and therefore remains with the gold on parting.

Osmium, if present alone, is dissolved.

---

[1] Rose, *Metallurgy of Gold*, 5th edit., p. 494.

Rhodium is slightly dissolved during parting in nitric acid, but most of it remains with the gold.

Ruthenium is not dissolved.

Apparatus for Sulphuric-Acid Parting.—The parting of cornets containing platinum by boiling in concentrated sulphuric acid is a delicate operation, as the boiling-point of sulphuric acid (338° C.) is so high that glass vessels are liable to be cracked by cold draughts. Moreover, the acid boils explosively with heavy bumping, which may occasion loss by ejection and even lead to the breaking of the flask. Boiling sulphuric acid is also terribly corrosive and inflicts dangerous wounds. On this account it is frequently advised that the platinum frame described and illustrated on page 263 should be used with a platinum boiler for sulphuric-acid parting, but experience proves that the platinum is attacked during the parting, and in some cases in sufficient amount to give a brown tint to the liquid. The solution of the platinum is probably accelerated by galvanic action between the gold cornets and the platinum cups.

Under these conditions it is advisable to use a parting frame and cups made of fused silica, or to use glass.

If the ordinary glass-bulb parting flask (figs. 134A and B, page 262) is used, it must be protected from the air and heated at the sides more than at the bottom. Sometimes

Fig. 164.

small clay peas or pieces of carbon are added to assist the acid to boil quietly.

Quennessen,[1] of the Paris Mint, has introduced a new form of burner for heating the flasks, which, it is stated, obviates the drawbacks mentioned above and permits of sulphuric acid being boiled readily without bumping. The burner is shown in section in the accompanying illustration (fig. 164). The gas comes out through the holes arranged round the interior of a crown, so that the heating is annular and the boiling is effected, not at the bottom as in the burner illustrated in fig. 135, page 262, but round the sides of the flask. Thus ejections during parting are prevented. To enable the heat to be applied at various heights the bottom of the flask is held by a copper support in the form of a cup, which is brazed to the upper part of a screw so that it can be raised or lowered as desired.

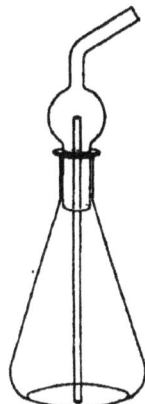

Bumping during the boiling of sulphuric acid may be satisfactorily prevented by boiling in small conical flasks fitted with the device described on page 270. A flask devised by the author for sulphuric-acid parting is shown in fig. 165. The flask is provided with a glass hood in the form of a bulb with a short tube for the outlet of acid fumes. The bulb fits loosely into the neck of the flask and is readily removed by means of a pair of wooden tongs. A capillary tube consisting of a piece of glass tubing about ½ inch bore and sealed at about 1 inch from one end is placed in the flask, so that it rests loosely on the bottom as shown. With this simple device the acid boils quietly, and bumping is entirely prevented. The author has also used a similar capillary tube in the ordinary glass-bulb parting flasks with very satisfactory results (fig. 166). In this case the neck of the parting flask should be cut down to 4 or 5 inches and the capillary

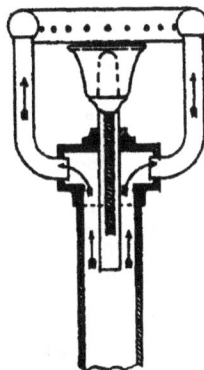

Fig. 165.

[1] "New Gas Burner for Heating Parting Flasks," L. Quennessen, *Chem. News*, 1903, vol. lxxxviii. p. 66 (illustrated).

tube, which merely rests in the flask, should be sufficiently long to project about ¾ inch above the mouth of the flask.

When glass-parting apparatus is used, the flask, after boiling, is allowed to cool almost completely, as otherwise it would crack during decantation of the acid liquid owing to sudden cooling.

In working with sulphuric acid, caution is also necessary in decanting the hot acid and in washing the cornets. Sulphuric acid has a powerful affinity for water, and on mixing with water it causes hissing from the energy of the combination, and great heat is evolved. In diluting the acid it should always be poured into cold water, and not water added to the acid. When either the acid or water has been previously heated, the action on mixing is very violent, causing spurting, and if special precautions are not taken a serious accident may result. Therefore, after boiling cornets in sulphuric acid the liquid is decanted as completely as possible into a dry vessel, or very slowly poured into a considerable bulk of cold water, so as to avoid danger of breakage from the heat generated. In washing the cornets warm water must be used, as cold water is very liable to break the flask owing to the high temperature to which it has been raised by the boiling acid. The warm wash-water should not be poured into the strong acid first decanted, for the reasons stated, but poured into a separate vessel.

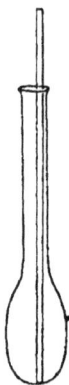

Assay of Platinum in Ores.—It may be remarked at the outset that the determination of platinum in ores, alloys, etc., is dependent on so many conditions that it is necessary in many cases to run a preliminary assay to ascertain approximately the amount of platinum present, and also whether, and to what extent, it is accompanied by gold and silver.

Fig. 166.

As previously stated, the platiniferous sands and gravels which constitute the true platinum ores are seldom received for assay, but the rarity of the metal, and its high price due to its useful application to chemical and other purposes, has caused search to be made for it in all classes of ores and metallurgical products. The assayer is therefore frequently asked to test for platinum in ores of gold, silver, and other metals. In many cases the quantity of platinum (and some of the metals of the platinum group) is so small that it may readily escape detection in the ordinary assay, as carried out for gold and silver ores, as it would pass into solution in the nitric acid on parting. Parting in sulphuric acid is therefore necessary in order to determine whether platinum and any of the platinum metals are present.

Methods of Assay.—To determine the platinum in ores, crucible fusions are made on 50 grammes of ore as described for gold and silver ores, with sufficient litharge in the charge to give a lead button of at least 30 grammes. If the ores are very low grade, four or five fusions are made, and the resultant lead buttons scorified into one button weighing about 20 grammes.

The lead button is then cupelled at a moderate temperature until about two-thirds of the lead has been oxidised, when the temperature is raised to that ordinarily employed for gold, the heat being well maintained towards the end of the cupellation in order to keep the metal molten and make it acquire a round form. The resultant button is carefully examined, and if any appreciable amount of platinum is present the surface of the button is crystalline or rough. The appearance of buttons containing different proportions of platinum has been described on page 411. The description is equally applicable whether the platinum is alloyed with silver or gold or with both. By a careful inspection of the button an experienced assayer will frequently be able to decide the approximate amount of platinum and of silver and gold present.

The button is cleaned and weighed, and the weight obtained gives the total quantity of precious metals in the ore. The subsequent treatment of the button is dependent on its composition.

Platinum and Silver.—If the button contains platinum and silver only, it is wrapped in 2 or 3 grammes of sheet lead, with the addition of silver, if necessary, to bring up the ratio to 10 parts of silver to 1 part of platinum, and then cupelled at a high temperature as before.

The resultant button is brushed and weighed, then flattened, annealed, and, if large and not too brittle, it is rolled into a cornet. It is then parted by boiling for fifteen minutes in about 15 c.c. of slightly diluted sulphuric acid (9 acid, 1 water). The acid is then decanted and the residue re-treated with 5 c.c. of concentrated sulphuric acid and boiled for ten minutes. The acid is again decanted, and the residue washed with hot water in order to dissolve the silver sulphate. If the residue is finely divided, it should be collected on a small ashless filter and thoroughly washed with hot water. The filter paper is dried and carefully transferred to a small annealing cup, and heated in the muffle to burn the filter paper and anneal the residue. After annealing, the platinum is transferred to the pan of a delicate assay balance and weighed. From this weight the platinum content of the ore is calculated in ounces per ton in the usual manner. The weight of the platinum subtracted from the weight of the precious metal button first obtained gives the silver. Owing, however, to the high temperature of cupellation there is a considerable loss of silver, and if accurate results for silver are required the assay must be repeated and a check assay run beside it. In this case the silver requisite to bring the ratio up to 10 to 1 is added at once to the lead button and one cupellation only made. The loss of silver in the check assay will, of course, give the correction to be added to the result obtained for silver in the ore.

The residue from the parting is usually assumed to be platinum, but it may contain small quantities of metals of the platinum group, notably iridium and osmium. The detection of these is described on page 424.

Platinum, Gold, and Silver.—When gold is also present in the button of precious metal resulting from the cupellation of the lead from the crucible fusion, it is necessary to part first in sulphuric acid to determine the silver, and then in nitric acid to determine the platinum and gold. If very little platinum is present, the cornet from the sulphuric-acid parting will be of the characteristic gold colour, but with an increase in platinum the colour will be grey or black. The parting in nitric acid is repeated, if necessary, until the weight of the gold cornet becomes constant, which may not occur until the fourth or fifth parting if much platinum is present.

As previously stated, the ratio of the gold to platinum is of importance for successful parting, and it is often necessary to add a certain accurately weighed quantity of pure gold when inquarting with silver, so that the metals may be present in correct ratios, i.e. 8 to 10 of gold to 1 of platinum. If the proportion of gold, etc., in the precious-metal button from the crucible fusion cannot be judged from its appearance, it becomes necessary to make a prior test to determine this as described on page 424. When the approximate composition of the button has been ascertained, it is weighed and then alloyed by cupellation with the requisite quantity of silver, and of gold if necessary. The alloy obtained is parted in sulphuric acid and the residual metal weighed. This is realloyed with silver by a second cupellation and parted in nitric acid, the residual metal being again weighed. The partings are performed in the manner described for alloys on page 412.

The amount of silver, platinum, and gold in the ore is calculated as follows :—

The weight of the precious-metal button first obtained gives the combined weight of platinum, silver, and gold. The residue from the sulphuric-acid parting consists of gold and platinum, and the weight of this residue (less any gold added) subtracted from the weight of the original precious-metal button gives the silver in the ore.

The residue from the nitric-acid parting consists of gold only (see, however, page 418), and the weight of this (less any gold added) subtracted from the weight of the platinum-gold residue gives the platinum, which is dissolved and is thus determined by difference.

Dewey's Method.[1]—With regard to the above methods, Dewey has pointed out that the results obtained are often indecisive and sometimes erroneous when the ores contain only small amounts of platinum. If any considerable amount of platinum be present, there will be a decided difference between the weighings of the two residues, but with very small amounts of platinum the difference is very slight and is no real evidence of the presence of platinum. In fact, the weight of the second residue may equal or possibly exceed the weight of the first, even when traces of platinum are present. Dewey therefore omits the parting in sulphuric acid and recommends the following method, which is based on the fact that when the gold buttons obtained in the regular course of assaying are parted in nitric acid in the usual way any platinum that is present in small amounts only will readily go into solution in the acid.

To this solution is added a limited quantity of very dilute solution of hydrogen sulphide (1 part of the strong solution diluted with from 10 to 20 parts of water), sufficient to precipitate the platinum, and three to five times as much silver, and after standing for three to four hours, or preferably overnight, the precipitate is collected, dried, the paper burnt off, and the residue wrapped in a small piece of thin lead-foil and cupelled. The resulting bead is parted in strong sulphuric acid, leaving the platinum usually in the form of sponge, which is washed, annealed, and weighed.

Assay of Platiniferous Copper Ores at Wyoming.—The following method is described by Dart[2] for the assay of the platiniferous copper ores of the Rambler Mine, Wyoming. The high-grade ores contain about 25 per cent. of copper and low-grade ores about 2 per cent.

The first tests showed that various copper minerals carried from 0·1 to 0·7 ounce of platinum per ton, while a quantity of covellite ore carried from 0·4 to 1·4 ounce per ton, this platinum occurring as sperrylite ($PtAs_2$). Later investigations gave evidence of the presence, in considerable quantities, of other members of the platinum group of metals.[3]

The assay of these ores is conducted as follows :—

Eight portions of the ore, 1 assay ton each, are fused with suitable fluxes. If much copper be present, the resulting buttons are scorified until soft, and then cupelled. The weight of the beads obtained gives the total quantity of precious metals in the ore. The beads are combined to two lots of 4 assay tons each, and recupelled with ten times their weight of silver. The resulting beads are parted with 12 per cent. nitric acid, and finally with concentrated nitric acid. Silver, palladium, and platinum are dissolved ; gold remains, and is weighed. The bulk of the nitric-acid solution should be kept as small as possible. The silver is precipitated as chloride, by the addition of hydrochloric acid, and removed by filtration. The filtrate is evaporated to dryness, and a few drops of hydrochloric acid and 10 c.c. of water are added. If silver chloride separates out, it is filtered

[1] *Trans. Amer. Inst. Min. Eng.*, 1912, p. 439.
[2] *Met. and Chem. Eng.*, 1912, vol. x. pp. 219-220.
[3] *Ibid.*, 1911, vol. ix. pp. 75-78. (Analyses of various platinum metal concentrates are given.)

off and added to the main portion. The filtrate is rendered ammoniacal, acidified with formic acid, and boiled for thirty minutes. Platinum and palladium are reduced, and can be filtered off, washed, ignited, and weighed. The precipitate is dissolved in aqua-regia, evaporated to dryness, taken up again with hydrochloric acid and 50 c.c. of hot water, saturated with ammonium chloride crystals, and, after addition of 10 c.c. of alcohol, is allowed to stand twenty-four hours. The ammonium platinic chloride is filtered off, washed with ammonium chloride solution, then ignited slowly and weighed.

The Assay of Platinum in Sands.—As the amount of platinum in platiniferous sands and gravels is usually only a few grains per cubic yard, it is necessary to concentrate the metal by panning or washing a weighed quantity in the same way as described for gold-bearing sand, etc., on page 243.

The "concentrates" thus obtained are scorified with 20 to 25 times their weight of granulated lead, until the final lead button weighs about 20 grammes. This is cupelled, and the resultant precious-metal button treated as described for ores on page 420, but in this case the gold residue from the nitric-acid parting may contain iridium, osmiridium, and rhodium, and must be attacked by dilute aqua-regia to separate these metals, as described below.

The platinum, etc., in the lead button may also be determined by dissolving in dilute nitric acid direct, as described below for the treatment of rich ores.

When the platiniferous sand is rich and gives a comparatively large quantity of concentrates, some difficulty is experienced in obtaining a representative sample for assay purposes owing to the large proportion of metallics present and to the fact that the "metallic" grains may be composed of different metals. As previously stated, rich platiniferous sands or concentrates are rare and are seldom presented for assay.

To obtain a representative sample with such material, it is usual to melt the grains of platinum, etc., with lead so as to obtain a more or less homogeneous alloy, and to take a suitable quantity of this lead alloy for the assay.

For this purpose it is best to take the whole of the concentrates, or, if the quantity is large, from 25 to 50 grammes, and, after carefully weighing, to fuse with six times its weight of metallic lead and suitable fluxes to form a fluid slag with the gangue. The charge when melted is poured and the brittle lead-platinum alloy is carefully freed from slag.

Osmiridium, when present, is not taken up readily by the lead, but tends to sink to the bottom and is separated. In order to ensure a uniform alloy the metal is remelted under charcoal, well stirred when thoroughly molten, and then granulated as fine as possible by pouring into a large volume of cold water from a considerable height. The resultant granules are carefully collected, dried, any large pieces broken up, and then weighed. The sample of platiniferous lead thus obtained is well mixed and is ready for assay. If the approximate amount of platinum in the lead is not known, 1 or 2 grammes must be cupelled to determine this.

An amount containing approximately 200 milligrammes of platinum is then accurately weighed out, transferred to a scorifier, and scorified with 50 grammes of lead until a button weighing about 10 grammes is obtained. This is rolled out into a thin strip, placed in a large beaker with 200 c.c. of dilute nitric acid, 1·08 sp. gr. (81 vols. water and 19 vols. nitric acid), and heated until all action ceases.[1] The acid is decanted and the amorphous black residue carefully washed with hot water. A fresh quantity of acid is then added and heated to boiling to dissolve out any lead that may still be present. The acid is again decanted and the residue washed thoroughly with hot water, then washed into

[1] Consult E. H. Miller, *School of Mines Quarterly*, 1896, vol. xvii. p. 26; also G. Matthey, *Proc. Roy. Soc.*, 1879, p. 463.

an annealing cup, and dried, annealed, and weighed. The residue consists of platinum, gold, iridium, iridosmium, and most of the rhodium; also any ruthenium and unalloyed osmium that may have escaped oxidation during the scorification. The solution contains the silver and palladium and a small proportion of the rhodium. The residue is now transferred to a small beaker and gently warmed with dilute aqua-regia (1 aqua-regia to 5 water) for ten minutes, whereby the complete solution of the gold and platinum is effected. The solution is decanted and reserved and the residue washed, dried, annealed, and weighed. This second residue consists of iridium, iridosmium, and rhodium; small quantities of ruthenium may also be present. The solution containing the gold and platinum is evaporated just to dryness, taken up with a drop or two of hydrochloric acid and a little distilled water. The gold is then precipitated by adding a strong solution of oxalic acid and warming for about half an hour, so as to obtain the gold in a coherent form. The precipitate of metallic gold is washed, dried, and weighed, or it may be filtered, washed, and dried, and then transferred together with the filter paper to a plate of sheet lead, silver added equal to 2½ times the gold present, and the whole wrapped up and cupelled. The resultant button is flattened, parted in nitric acid in the usual way, and the gold annealed and weighed. The weight of the gold, subtracted from the difference in weight between the first and second residues, gives the amount of platinum present. If the second residue is boiled with strong aqua-regia, the iridium is dissolved with some osmium and ruthenium if these metals are present, leaving a third residue consisting of iridosmium and rhodium if present. The residue is washed, dried, annealed, and weighed. The weight deducted from the weight of the second residue gives the iridium. The method thus outlined gives fairly good results for platinum, gold, silver, iridium, and iridosmium (plus rhodium if present), which are the chief constituents of the metallic grains from platiniferous sands. As previously stated, rhodium and ruthenium are amongst the rarer metals of the platinum group, and are rarely present.

Considerable chemical knowledge is necessary to effect the separation of the metals of the platinum group. The complete analysis of crude platinum and platiniferous grains is made by the ordinary chemical methods given by Fresenius, Crookes, and others.[1] Descriptions of some methods of separation will be found on page 431.

Assay of Platinum in Alloys and Bullion.—The alloys most frequently submitted for assay consist of small bars obtained by the treatment of "lemel" and scrap, etc., from the workshops of jewellers and dentists, etc.

The alloys may contain platinum, gold, silver, and copper, or other base metal. In some cases iridium may be present.

It is necessary to determine first of all the approximate composition of the alloy, and then to make an exact assay.

Preliminary Assay.—Take 0·25 gramme of the alloy and cupel with 1 gramme of lead at a high temperature to remove base metals, and if the button is flat it is again cupelled with more lead. When a rounded button is obtained it is weighed and the weight accepted as that of the gold, platinum, and silver together. The difference between this weight and that of the metal taken for assay gives the base metal present. Allowance must, however, be made for considerable loss of silver as a result of the high temperature employed.

The button is then wrapped in sheet lead and cupelled with three times its weight of gold (accurately weighed), and silver equal to three times the weight of the button plus the gold added. The alloy thus obtained is flattened,

[1] Consult Fresenius, *Quantitative Chemical Analysis*; Crookes, *Select Method of Chemical Analysis*; P. E. Browning, *The Rarer Elements*.

annealed, and parted by boiling in concentrated sulphuric acid for fifteen minutes, which dissolves the silver and leaves the gold and platinum as a residue. The residue is washed, dried, annealed, and weighed. The difference between this weight and that of the original cupellation button is silver. The residue is inquarted by cupellation with four times its weight of silver, and the resulting button rolled and parted in dilute nitric acid as in the ordinary gold-bullion assay.

If the colour of the cornet inclines to steel grey instead of the characteristic colour of pure gold, it is desirable to inquart, re-cupel, and part a second time, so as to obtain a closer approximation of the amount of platinum present. In most cases one parting in nitric acid will give a result sufficiently near for a preliminary assay.

The cornet represents the original gold contents (after allowing for the amount added), and the platinum is taken by difference.

Thus, to take an example :—

Weight of alloy taken for assay 0·25 gramme.

Weight after cupellation (silver + platinum + gold) = 0·215 gramme. The difference between this and the weight of alloy taken gives the *base metal* 0·250 − 0·215 = 0·035.

Weight of residue after parting in sulphuric acid (platinum + gold) = 0·833, of which 0·645 was added as pure gold ; therefore 0·833 − 0·645 = 0·188, and this, deducted from the cupellation button, gives the *silver* 0·215 − 0·188 = 0·027.

Weight of residue after parting in nitric acid (gold only) = 0·770, of which 0·645 was added as pure gold ; therefore 0·770 − 0·645 = 0·125 equals the *gold* in the alloy, and this weight, subtracted from that of the above platinum + gold residue, gives the *platinum*, thus 0·188 − 0·125 = 0·063.

The approximate composition of this alloy is thus determined to be as follows :—

Gold	0·125 × 4 = 50·0	per cent.
Platinum	0·063 × 4 = 25·2	,,
Silver	0·027 × 4 = 10 8	,,
Base metal (probably copper) .	0·035 × 4 = 14·0	,,
	100·0	

Final Assay.—The amount of alloy taken for assay is usually 0·5 gramme, but as it is advisable not to have more than about 200 milligrammes (0·2 gramme) of platinum present, the weight taken should be proportional to the platinum present.

The alloy is cupelled with lead to remove base metals, the amount of lead required for cupellation being about double the amount which would be required in a gold-bullion assay, as given in Table XXX., page 267. The cupellation must be finished at a very high temperature, and if the button is not satisfactory it must be recupelled with 1 or 2 grammes of lead.

The weight of the cupelled button is of course compared with a check assay made up in accordance with the approximate composition of the alloy, previously determined, and this is done step by step through all the operations. If there is a surcharge in the check of more than 1 or 2 parts per 1000 in the first cupellation, showing retention of base metal (copper or lead), the buttons are recupelled with fresh lead as stated.[1] The cupellation button is cleaned and weighed, and the difference between this weight, after correction, and the weight of the alloy taken gives the base metal present.

[1] Rose, *Precious Metals*, p. 274.

The button is then recupelled with the addition of gold if necessary, and with silver equal to three times the weight of platinum and gold in the alloy plus any gold that is added. Thus, using the example given above. Suppose the corrected weight of the gold-platinum-silver cupellation button from 0·25 gramme of alloy was 0·220, then 0·25 – 0·220 = 0·030 base metal.

The amount of gold and silver required to be added for parting purposes would be determined thus :—

    Platinum present   = 0·063 gramme approximately.
    Gold already present = 0·125   „   „
    Silver   „   „ = 0·027   „   „

One part of platinum requires 10 parts of gold, therefore 0·063 requires 0·063 × 10 = 0·63 gramme gold, and this weight minus the gold already present gives 0·63 – 0·125 = 0·505 gramme, the weight of gold to be added. This must be accurately weighed.

One part of platinum + gold requires 2½ parts of silver, therefore 0·063 + 0·125 + 0·505 (added gold) = 0 693 parts require 0·693 × 2½ = 1·732 grammes silver, and this weight minus the silver already present gives 1·732 – 0·027 = 1·705 grammes, the weight of silver to be added.

The amount of sheet lead required for this second cupellation is about 5 grammes.

The cupellation is conducted as described above and the button cleaned, hammered, and rolled into a cornet, as described in the assay of gold bullion. It is very important to thoroughly anneal the fillet or strip before rolling into the form of a cornet. The cornet is boiled in slightly diluted sulphuric acid (9 vols. of concentrated acid to 1 vol. of water) for fifteen minutes, after which the liquid is decanted as completely as possible and the cornet is washed twice with warm distilled water, care being taken not to add the wash-waters to the strong acid first decanted for the reasons stated on page 420. After washing, the cornet is again boiled for ten minutes in concentrated sulphuric acid and then washed, transferred to an annealing cup, dried, annealed, and weighed.

If, after washing, a white deposit of lead sulphate is seen upon the cornet, it must be treated with a warm, strong solution of ammonium acetate and then again washed with distilled water. Lead may be present in consequence of the employment of too low a temperature for cupellation.

Some assayers boil the cornets a second time in concentrated sulphuric acid, and this is very desirable when the cornets are large.

The weight of the cornet gives the amount of platinum and gold in the alloy plus the added gold, in this case, say, 0·696.

Then 0·696 minus 0·505 added gold equals 0·191 gold + platinum in the alloy, and the difference between this weight and the weight of the gold-platinum-silver cupellation button gives the silver in the alloy thus :—

$$0·220 – 0·191 = 0·029 \text{ silver.}$$

For the determination of the gold, the cornet is recupelled with 2½ times its weight of silver and 3 grammes of sheet lead.

The resulting button is hammered, rolled, well annealed, and then parted in nitric acid of two strengths as described in the assay of gold bullion on page 268, and the gold cornet weighed. If, as generally happens, there is a surcharge in the check assays of more than 1 or 2 parts per 1000 showing the retention of platinum, the inquartation and parting is repeated. When the cornet retains an appreciable amount of platinum it will be steel grey in colour, and the inquartation and parting must be repeated until the cornet obtained is of the characteristic pure gold yellow. The final weight is then, say, 0·6306, and after deducting 0·505, the result 0·1256 is obtained for the amount of gold present in 0·25 of the original alloy. The platinum is taken by difference : thus the gold plus platinum from the sulphuric-acid parting was 0·191, and subtracting the weight of gold from the nitric-acid parting 0·1256, we get 0·0654 as the amount of platinum present in 0·25 of the original alloy.

It is assumed that the necessary corrections as shown by the check have been made in all the operations.

When iridium is present in the alloy it remains with the gold, and the cornet requires special treatment, as described on page 424.

The final result of the assay will be reported in parts per 1000 as follows :—

Gold	502·4	per 1000
Platinum	261·6	,,
Silver	116·0	,,
Copper (by difference) . . .	120·0	,,
Total . .	1000·0	

According to Trenkner,[1] the error in assaying alloys of gold and platinum, which is considered by the Paris Mint to amount to 1 per cent., may be greatly lessened by the following method, which gives the results within 0·5 per 1000, or closer in experienced hands. The alloy is subjected to a preliminary assay to determine the approximate composition. Then 0·5 gramme of bullion if finer than 250 per 1000, or 1 gramme if baser, is mixed with pure silver in the proportion of about 10 of silver to 1 of gold and platinum. From 3 to 15 grammes of lead are added, and the whole cupelled. Allowing for added silver, the weight of the button gives silver, gold, and platinum together. The silver loss during cupellation is best determined by control assays. The addition of so much silver results in the entire removal of lead from the obtained button. No appreciable amount of gold or platinum is lost. The button is parted, without flattening, in 25 c.c. of concentrated sulphuric acid, nearly boiling, for thirty minutes. The residue is coherent, and is washed, and dissolved in aqua-regia, and the solution evaporated to small bulk, diluted, and the silver chloride filtered off. To the filtrate is added 15 c.c. of hydrochloric acid, and 1 gramme of hydrazine hydrochloride. After one hour, with occasional stirring, the gold is filtered off, ignited, and weighed. As it often contains some platinum, it is cupelled with 5 to 8 parts of silver, and the button parted in nitric acid. Silver and platinum dissolve. The pure gold is weighed. Platinum is precipitated from solution by an excess of ammonium or potassium hydrate, ignited, and weighed.

Hollard and Bertiaux[2] find that by the usual methods for the analysis of platinum-gold-silver alloys the results are too low for the gold and too high for the platinum and silver, but are very accurate for the gold and platinum together. The following method is recommended by them for the determination of the silver :—The alloy, cut into small pieces or rolled out to thin foil, is dissolved in aqua-regia (1 vol. of nitric acid + 5 vols. of hydrochloric acid), the solution is evaporated to the consistence of a syrup, and then evaporated to dryness three times with the addition of nitric acid, the residue boiled with a few c.c. of water, 2 c.c. of nitric acid, and 2 drops of hydrochloric acid, the liquid diluted to 100 c.c., the silver chloride filtered off, washed well, dissolved in 30 c.c. of a 20 per cent. solution of potassium cyanide, the solution diluted

[1] *Métallurgie*, 1912, vol. ix. pp. 103-105.
[2] *Ann. Chim. Anal. Appl.*, 1904, vol. ix. pp. 287-292 (abstract *Journ. Soc. Chem. Ind.*, 1904, vol. xxiii. p. 952).

to 150 c.c., and electrolysed (see page 309). The separated silver is dissolved in nitric acid, and determined by titration with thiocyanate solution, as described on page 303.

Assay of Alloys containing Iridium.—As previously stated, iridium, if present, always sinks to the bottom of the button of precious metals obtained by cupellation, and when the button is rolled into a cornet with the lower face outwards the iridium may be seen with a lens, in the form of black spots or streaks. When the presence of iridium is suspected, it is advisable to clean the lower surface of the cupelled button by immersion in hydrochloric acid to remove all bone-ash adhering to it and leave the surface as clean as possible. The iridium may be accompanied by rhodium and osmium. When gold containing one or more of these metals is parted in the ordinary way with nitric acid any osmium present is dissolved, but only a small quantity of rhodium goes into solution with the silver, and iridium is unacted upon and remains with the gold.

The residue, consisting of gold, the iridium, and most of the rhodium, is dried and annealed as usual and weighed. It is then attacked with *dilute* aqua-regia (1 aqua-regia to 5 water), when the gold is dissolved together with only traces of the other metals. The residue of platinum metals is washed, dried, annealed, and weighed. The difference between this weight and the previous weight of the cornet gives the gold.

To confirm, the gold solution may be evaporated and the gold precipitated with oxalic acid or other precipitant. The precipitate obtained is washed, dried, and weighed. The iridium being in a finely divided state may be removed from the residue by digesting with strong aqua-regia, leaving iridosmium and rhodium as a final residue.

Assay of Silver-Platinum Alloys.—One of the most important silver-platinum alloys is that known as "dental alloy," which is used for dental purposes. Two qualities of this alloy are in general use in England, the "first" quality consisting usually of 2 parts of silver and 1 part of platinum, and the "second" quality of 3 parts of silver and 1 part of platinum, but the composition varies slightly with different makers. The following are analyses of the two qualities :—

	Best quality.	Second quality.
Silver . . .	67·35 per cent.	74·97 per cent.
Platinum . . .	30·57 ,,	25·00 ,,
Copper . . .	2·05 ,,	...
Gold . . .	·03 ,,	...
	100·00	99·97

Some qualities contain about 33⅓ per cent. of platinum. Dental alloy is used also by jewellers.

For the determination of the platinum, 0·5 gramme of the alloy is cupelled at a high temperature with from six to ten times its weight of lead and the resulting silver-platinum button weighed. The difference between this weight and the weight of alloy taken is base metal. The button is then alloyed by cupellation with enough silver to make the total silver equal to ten times the weight of the platinum. The alloy thus obtained is flattened, rolled into a cornet, and parted in slightly diluted sulphuric acid in the usual way.

The platinum residue is washed, dried, annealed, and weighed. The weight of the platinum deducted from the weight of the first cupellation button gives the silver.

Checks should be used if accurate silver results are desired.

Silver-platinum alloys containing less than 20 per cent. of platinum may be parted direct in sulphuric acid, but Thompson and Miller[1] have shown that the results are incorrect for alloys containing 20 per cent. or more of platinum, unless correction is made for the undissolved silver remaining with the platinum (see Table LV., page 416).

Determination of Silver in Platinum Alloys.—The silver in alloys containing platinum is usually determined by cupellation, as described in the assay of platinum alloys. In some cases it may be desirable to determine silver in the wet way. When the composition of the alloy being assayed is such that all the silver passes into solution on parting in sulphuric acid, the silver contained in the strong sulphuric acid solution may be determined by diluting to about 900 c.c., partly neutralising with ammonium hydrate, and then precipitating the silver as chloride by the addition of a strong solution of common salt or a little hydrochloric acid. The precipitate is filtered off, washed, dried, and weighed as chloride.

According to Richards,[2] the silver can be satisfactorily determined in the sulphuric acid solution by titrating with a standard solution of thiocyanate (Volhard's method, page 303), but it is stated by Thompson and Miller[3] that this method is not successful in the presence of the large amounts of sulphuric acid from the solution of the alloy. Better results can be obtained, according to Thompson and Miller, by precipitating the silver from the acid solution by means of aluminium foil, dissolving in nitric acid, and titrating, but the method is not entirely satisfactory.

Richards[4] recommends the following method for determining the silver. The silver sulphate solution is diluted, rendered slightly alkaline with ammonium hydrate, and then warmed with the addition of a little glucose. The precipitate of metallic silver is filtered, washed, dried, and weighed.

When silver is retained in the platinum residue from the sulphuric-acid parting it is necessary to dissolve the residue in aqua-regia and then evaporate down several times with nitric acid, followed by taking the solution gently down almost to dryness to remove most of the free acid, and to ensure that the free acid that remains is nitric and not hydrochloric.

The solution is then diluted, and the silver precipitated by the addition of common salt in the usual way. The precipitate is contaminated with platinum salts, and is therefore filtered and washed free from chlorides. It is then dissolved on the filter with dilute ammonium hydrate and the silver reprecipitated in the filtrate by adding nitric acid and a few drops of hydrochloric acid. It is then filtered and weighed in the usual way and the weight added to the silver obtained from the sulphuric acid solution.

When the separation of the silver in platinum alloys is effected by parting in nitric acid, the silver solution always contains more or less platinum in colloidal state (see page 415), which interferes somewhat with the determination of the silver by ordinary methods. If filtration is attempted, only a small part of the colloidal platinum is retained, and on washing with water the precipitate passes through the paper readily and acts in the manner common to all colloidal precipitates.

To remove the platinum, Thompson and Miller[5] recommend the addition of sodium (or ammonium) nitrate free from chloride and heating. After standing, the solution is filtered and the precipitate washed with 1 per cent. nitric acid (i.e. 1 c.c. of 1·42 acid in 100 of water), which is found to free the precipitate

[1] Loc. cit.   [2] Analyst, 1902, vol. xxvii. pp. 265-268.
[3] Thompson and Miller, Journ. Amer. Chem. Soc., vol. xxviii., 1906, p. 1125.
[4] Loc. cit.   [5] Loc. cit.

from silver without the formation of any colloidal platinum. The silver in the filtrate can then be determined as chloride in the usual way.

Attempts have been made to directly determine the silver in the solution from nitric acid parting, in the presence of the colloidal platinum, by titration with potassium thiocyanate. Some tests made by Sharwood[1] have shown, however, that when colloidal platinum is present, titration indicates more silver than is actualy present. In the experiments the excess of standard solution consumed corresponded very closely to 4 molecules of potassium thiocyanate for each atom of platinum in solution, but the end of the reaction is rendered very indefinite by the presence of platinum.

It may be remarked that if the colloidal precipitate is dried, ignited, and weighed for the determination of the platinum, slight explosions will be noticed during the ignition owing to the precipitate reacting with the filter paper.

Another wet method of determining silver in platinum alloys has been given on page 427.

Assay of Platiniferous Products.—The following are the chief platiniferous materials included under the head of products.

Determination of Platinum in "Sweeps."—This consists of the sweepings, refuse, etc., resulting from the working of platinum. The material is burnt, then crushed, and sampled as described for gold and silver sweep on page 356. Should a small quantity of "metallics" separate out during sampling, the particles should be wrapped in lead, cupelled, and assayed. The content of platinum is then allowed for in reporting the total value of the sweep. The method of calculating the value of samples containing metallics is given on page 216.

The finely crushed material passed through a 90-mesh sieve is assayed by crucible fusion in the same manner as described for gold and silver sweep on page 356, and the platinum determined by the method given on page 420 for platiniferous ores.

If platinum only is present, it may be left on the cupel as a flat irregular mass instead of in the form of a globule, in which case the cupel should be replaced in the muffle and the platinum collected by recupelling with a little more lead and the addition of silver equal to about five times the weight of platinum judged to be present. The appearance of the button thus obtained will afford a better indication of the actual amount of platinum present, and more silver can then be added to make the total silver equal to ten times the weight of platinum. The alloy thus obtained is parted in sulphuric acid as usual. The platinum content is reported in ounces and decimals of an ounce per ton of sweep.

With small lots of sweep, of about 1 cwt. or less, it is usual to report the total weight of precious metal, in ounces and decimals of an ounce, in the entire lot or parcel.

Determination of Platinum in Dental Sweep.—The sweep from the workshops of mechanical dentists always contains platinum, gold, and silver, and is contaminated with lead, zinc, and mercury. It varies very considerably in its content of precious metals which are present in the metallic condition. The material is alternately crushed and sifted to remove the metallic portion, until finally it passes a 90-mesh sieve. The metallic portion termed "siftings" is melted with a little soda and borax and cast into a bar. If zinc and lead are present together, the metal separates into two layers, the zinc rising to the top on solidification, in which case the bar is remelted and the metal granulated by pouring into water. The granules are collected, dried, and weighed. Large granules are hammered out and cut up into small pieces and the whole thoroughly mixed. It is then sampled and assayed and the yield of precious metals calculated.

[1] *Journ. Soc. Chem. Ind.*, 1904, vol. xxiii. p. 414.

The amount taken for assay is dependent on the platinum content, which must be approximately ascertained by cupelling 0·5 gramme with 2 grammes of lead and weighing the button obtained. If much zinc is present, the portion taken for assay should be scorified with about ten times its weight of lead. The button of lead obtained is then cupelled and the resulting precious-metal button assayed for gold, silver, and platinum as previously described.

The fine sifted portion of the sample is thoroughly mixed and assayed as described for gold and silver sweep. The yield of metal obtained from both "metallics" and fine material is calculated as shown on page 216.

Dental sweeps are generally received in comparatively small lots, and the total precious metal content of the parcel is reported as stated above. If the lot is large, say over 1 cwt., the value is reported in ounces ·and decimals of an ounce per ton in the usual way. When the bar resulting from the melting of the metallic portion is large, its precious-metal content is frequently reported independently in ounces and decimals per pound.

Determination of Platinum in Electrolytic Slime or Mud (see page 350).— This material is assayed by scorification, as described for the determination of gold on page 350. The button of precious metals obtained from cupellation is then assayed for platinum as described for platiniferous ores on page 420. As the amount of platinum present is usually very small, it is frequently necessary to make five or more separate scorifications and to rescorify the resultant lead buttons together until one button of about 20 grammes is obtained, which is cupelled and treated in the usual way. The precious-metal contents are reported in ounces and decimals of an ounce per *ton* of slime.

Remarks on the Determination of Metals of the Platinum Group.[1]

The separation of the metals of the platinum group is very difficult, and requires an expert knowledge of the chemistry of the individual metals comprising the group. The reactions of the platinum metals are given in Chapter XXIV., page 402.

The action of acids on these metals is of importance in considering their separation.

As previously stated, ordinary acids are without action on these metals in the pure state, with the exception of palladium, which readily dissolves in hot nitric acid. When, however, the platinum metals are alloyed with excess of silver, and the alloys are treated with acid, they are left in a very finely divided state and are more or less attacked by nitric acid and by sulphuric acid, as stated on page 412.

The chief solvent for platinum and the platinum metals is aqua-regia. Platinum is readily attacked by dilute aqua-regia with the formation of the tetrachloride. Concentrated aqua-regia converts osmium into the tetroxide, slowly acts upon ruthenium, but is without action upon iridium if the metal has been strongly heated, but slowly attacks it if it is very finely divided. Rhodium in the pure state and the alloy iridosmium are not attacked by aqua-regia. Rhodium and iridium, when alloyed with much platinum, are partly, and in some cases completely, dissolved by aqua-regia.

In separating the platinum metals in the ordinary course of analysis it is well to remember that palladium, rhodium, osmium, and ruthenium differ from gold, platinum, and iridium, by the insolubility of their sulphides in a solution of sodium or ammonium sulphide. Iridium may be distinguished from platinum by the deep blue colour given when caustic alkali is added to a solution of iridic

[1] Consult also Fresenius, *Quantitative Chemical Analysis*; Crookes, *Select Methods of Chemical Analysis*; P. E. Browning, *The Rarer Elements.*

chloride (or the double sodium salt). Palladium is distinguished by the insolubility of its iodide, and osmium by the volatility of its oxide on boiling with nitric acid.

The following are among some of the more recent methods recommended for the separation of the metals of the platinum group.

Determination of Platinum and Gold.—In the analysis of a solution containing gold and platinum and metals of the platinum group, it is preferable to first remove the gold in the metallic state by oxalic acid. The precipitated gold is filtered off, washed, dried, ignited, and weighed. The solution is then evaporated down with ammonium chloride to precipitate the platinum as ammonium platinum chloride. The precipitate is filtered off, washed with alcohol, dried, and transferred with the filter paper to a weighed crucible. It is then heated, gently at first, to avoid volatilisation of the platinum chloride, then more strongly until completely decomposed. The spongy metallic platinum thus obtained is cooled and weighed. With large quantities of platinum it is advisable to perform the heating in an atmosphere of hydrogen.

Vanino and Leeman [1] recommend the following method of separating gold and platinum. The solution containing the metals is treated with hydrogen peroxide after the addition of caustic potash or soda. The gold is precipitated in a few minutes, even in the cold, as a black deposit which under the action of heat agglomerates and becomes of a reddish-brown colour.

Platinum and iridium if present remain in solution.

Volumetric Determination of Platinum.—The following volumetric method for the determination of platinum is suggested by H. Peterson. [2]

The metallic chloride or alkaline platinic chloride is poured into a sufficiently concentrated and cold solution of potassium iodide, and the iodine, which is set free, is determined by titration with standard sodium thiosulphate with starch as an indicator.

The reactions are :—

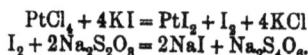

$$PtCl_4 + 4KI = PtI_2 + I_2 + 4KCl$$
$$I_2 + 2Na_2S_2O_3 = 2NaI + Na_2S_4O_6.$$

From these equations it will be seen that two molecules of sodium thiosulphate (316) correspond to one atom of platinum (194).

Gold may also be determined in the same manner (see page 285).

The accuracy of the method has been proved by titrating solutions containing known quantities of platinum and gold.

Determination of Platinum and Iridium. [3]—For the determination of platinum in alloys containing iridium, the substance is dissolved in a mixture of 1 vol. of nitric acid of specific gravity 1·32 and 2 vols. of hydrochloric acid of specific gravity 1·18. After carefully expelling the excess of nitric acid, the solution is heated to 120° C., water is added, and the metals precipitated by magnesium. The deposited metals are dried, ignited, and heated to dull redness in a current of hydrogen. After cooling, the excess of magnesium is removed by extraction with sulphuric acid (1 in 10), and the residue treated with aqua regia diluted with three times its volume of water. The solution contains platinum but no iridium; the platinum is precipitated as the double ammonium chloride, and the precipitate ignited, preferably in the presence of a reducing agent, such as oxalic acid or dextrose, or, still better, the charred filter paper. In carrying out this method, Quennessen found that the iridium precipitated by magnesium is soluble in dilute sulphuric and acetic acids, and is probably in the form of an oxide. The precipitate, while still moist, or after drying at 100° C., gives a

[1] *Chem. News*, 1900, vol. lxxxii. p. 70 (*Berichte*, vol. xxxii., 1900, p. 1698).
[2] *Zeits. anorg. Chem.*, 1899, vol. xix. p. 59.
[3] L. Quennessen, *Chem. News*, 1905, vol. xcii. pp. 29–30.

yellow solution gradually changing to violet with dilute sulphuric acid, and a green solution with acetic acid. After being heated in air at 440° C. it dissolves in sulphuric acid, forming a violet-coloured solution, and after heating at 800° C. it gives a blue solution.

Determination of Osmium and Palladium.[1]—To separate osmium from platinum, iridium, ruthenium, and rhodium, the solution is first treated with zinc or magnesium, and then hydrogen peroxide is added to the black precipitate thus formed. Osmium dissolves, and can be subsequently obtained as the pure tetroxide, whereas the other metals are insoluble.

Palladium may be separated from the other metals of the platinum group by adding freshly precipitated silver iodide to a solution of their chlorides. This, unlike potassium iodide, reacts only with palladium chloride, yielding the black insoluble iodide, and leaving the other metals in solution. The palladium may be extracted from the precipitate by the aid of potassium iodide or thiocyanate solutions, or by means of aqua-regia.

Determination of Osmium.[2]—The general gravimetric method of estimating osmium is by reduction in a current of hydrogen, with a hot copper spiral to prevent the carrying over of osmium tetroxide. This method is not applicable to some of the osmium haloids, and it is not very convenient for solutions. The method of Paal and Amberger (Ber., 1907, xl. 1378) is then applied. This depends on reduction of the osmium compound by means of alcohol, formaldehyde, or hydrazine, to insoluble osmium dioxide. The precipitate thus obtained is always partly colloidal and passes readily through the filter. This difficulty can be avoided by producing in the filtrate a barium sulphate precipitate, which carries down the colloidal suspended oxide. It is found, however, that if the original solution is made exactly neutral after precipitation, and warmed for some time on the water-bath before filtering, the osmium dioxide is completely precipitated, and can easily be filtered off and weighed, after careful drying in a current of carbon dioxide. The volumetric method of Klobbie (Chem. Centr., 1898, xi. 65) is applied to solutions of osmium tetroxide which is distilled from a mixture of the original substance with sulphuric and chromic acids in a current of oxygen. The osmium tetroxide is absorbed in alkali, carefully neutralised in the cold, treated with excess of potassium iodide, and titrated with standard sodium thiosulphate, using starch as an indicator. The colour of the osmium solution being very dark, the end point is not very sharp. One molecule of osmium tetroxide liberates four molecules of iodine. Another volumetric method consists in oxidising the original substance to osmium tetroxide by means of potassium permanganate in sulphuric acid or 40 per cent. hydrofluoric acid solution. The permanganate, method, however is chiefly useful as affording evidence of the state of oxidation of a previously determined percentage of osmium.

Determination of Palladium.—Palladium is usually determined by precipitation as palladious cyanide. The solution of palladious chloride is neutralised almost completely with sodium carbonate, and a solution of mercuric cyanide added. The solution is heated gently for some time, until the odour of prussic acid has gone off, when the yellowish-white precipitate of palladious cyanide separates out. With dilute solutions the precipitate forms only after long standing. Wash first by decantation, then on the filter, dry thoroughly, heat first cautiously, then more strongly to decompose the precipitate, and weigh the metallic palladium.

[1] N. A. Orlow, Chem. Zeit., 1906, vol. xxx. pp. 714-715 (abstract Journ. Soc. Chem. Ind., 1906, vol. xxv. p. 779).
[2] O. Ruff and F. Bornemann, Zeit. anorg. Chem., 1910, vol. lxv. pp. 429-456 (abstract Journ. Soc. Chem. Ind., 1910, vol. xxix. p. 800).

434 THE SAMPLING AND ASSAY OF THE PRECIOUS METALS.

The precipitate of palladious cyanide is distinguished from other metallic cyanides by the reaction common to all palladium salts, namely, that when heated they decompose, leaving a spongy residue of metallic palladium.

By the use of mercuric cyanide, palladium may be separated from the precious metals and also from most of the common metals, except lead and copper.

Bibliography.

1879. N. W. PERRY, "Assay of Platinum Alloys containing Base Metal, Silver, Platinum, Gold, and Osmium," *Eng. and Min. Journ.*, 1879, vol. xxviii. Gives a description of an assay method in which it is stated that after removal of base metal by cupellation the platinum can be dissolved completely in nitric acid along with the silver, provided the platinum is alloyed with at least twelve times its weight of silver. (The inaccuracy of this method is pointed out by Thompson and Miller, *loc. cit.*)

1899. H. PETERSON, "Volumetric Determination of Gold and Platinum," *Zeits. anorg. Chem.*, 1899, vol. xix. p. 59. Potassium iodide solution is added to the solution of gold chloride or platinic chloride and the liberated iodine titrated by standard sodium thiosulphate with starch as indicator.

1899. E. PRIWOZNIK, "Parting of Gold Platinum Alloy," *Oesterr. Zeits. Berg.-Hutt.*, 1899, vol. xlvii. pp. 356–358 (abstract, *Journ. Chem. Soc.*, 1899, vols. lxxvii–lxxviii. p. 111). Method used at Imperial Austrian Assay Office. Filings of the alloy are treated first with nitric acid to separate silver and the residue then treated with dilute aqua-regia, which dissolves the gold but does not perceptibly attack the platinum. If the alloy is composed of gold, silver, and platinum, each in large proportions, it is first fused with three times its weight of zinc and treated with sulphuric acid; the residue is dissolved in aqua-regia. The platinum is precipitated with ammonium chloride and the gold with ferrous sulphate.

1901. LEIDIE AND QUENNESSEN, "Determination of Platinum and Iridium in Platinum Ores," *Journ. Pharm. Chim.*, 1901, vol. xiv. pp. 351, 355 (abstract *Journ. Soc. Chem. Ind.*, 1901, vol. xx. pp. 1242–1243). The ore is treated with hot aqua-regia and the metals subsequently separated by analytical methods.

1901. OEHMICHEN, "Platinum-Gold-Silver Assay," *Berg. und Huttenm. Zeit.*, 1901, vol. lx. p. 137 (abstract, *Journ. Soc. Chem. Ind.*, 1901, vol. xx. p. 507). The alloy is parted in sulphuric acid and residue of platinum-gold is boiled twice for five minutes with nitric acid (1·3 sp. gr.), washed, and dried. The gold is separated by repeated quartation and parting in nitric acid if the amount of platinum is small. If platinum is in excess, gold is added and the alloy dissolved in aqua-regia and the gold precipitated by ferrous sulphate.

1902. P. A. E. RICHARDS, "Determination of Platinum, Gold, and Silver in Dental Alloys," *Analyst*, 1902, vol. xxvii. pp. 265–268. Describes methods for alloys containing (a) platinum and silver only; (b) gold, platinum, and silver; (c) gold, platinum, silver, and tin. The alloys are treated with concentrated sulphuric acid and the residue dissolved in nitro-hydrochloric to separate the platinum, which is then precipitated as ammonium platinum chloride. The gold is precipitated with ferrous sulphate. The silver is estimated by titration with thiocyanate.

1903. H. CARMICHAEL, "Notes on the Separation of Gold, Silver, and Platinum," *Journ. Soc. Chem. Ind.*, 1903, vol. xxii. pp. 1324–1325. Gives the results of parting gold-silver-platinum alloys of different composition in nitric acid and in sulphuric acid.

1904. W. J. SHARWOOD, "The Cupellation of Platinum Alloys containing Silver, or Gold and Silver," *Journ. Soc. Chem. Ind.*, 1904, vol. xxiii. p. 412. A short series of experiments to ascertain the extent to which lead is retained in the cupellation of platinum alloys, and the extent to which the platinum passes into solution during parting.

1904. HOLLARD and BERTIAUX, "Analysis of Platinum-Gold-Silver Alloys," *Ann. Chim. Anal. Appl.*, 1904, vol. ix. pp. 287–292 (abstract *Journ. Soc. Chem. Ind.*, 1904, vol. xxiii. p. 952). Describes separation of the metals by wet methods after solution in aqua-regia.

1906. J. F. THOMPSON and E. H. MILLER, "Platinum-Silver Alloys" (Separation of Platinum from Gold, Iridium, etc.), *Journ. Amer. Chem. Soc.*, 1906, vol. xxviii. pp. 1115-1132. Research on the solubility of alloys of the platinum-silver series in sulphuric and in nitric acids. (References are given in this paper to earlier experiments by other workers on the solubility of platinum-silver alloys).

1911. A. STEINMANN, "Platinum Assay," *Schweiz. Woch. Chem. Pharm.*, 1911, vol. xlix. pp. 441-444, 453-457. Draws attention to the errors in the platinum assay due to solution of the metal when parting in concentrated sulphuric acid. By using slightly diluted acid parting can be effected without loss of platinum.

1912. TRENKNER, "Determination of Gold, Silver, and Platinum," *Métallurgie*, 1912, vol. ix. pp. 103-105. Describes method used at the Paris Mint for the assay of platiniferous gold bullion which reduces the error of assay to 0·5 per 1000.

1912. F. P. DEWEY, "The Direct Determination of Small Amounts of Platinum in Ores and Bullion," *Trans. Amer. Inst. Min. Eng.*, 1912. Gives details of method of separating the small quantities of platinum that pass into the nitric-acid solution obtained in the regular course of assaying on parting gold from silver.

1912. A. C. DART, "Assay of Ores containing Platinum Metals," *Met. and Chem. Eng.*, 1912, vol. x. pp. 219-220. Describes method of assay for platiniferous copper ores in use at Wyoming.

1912. H. ARNOLD, "A Method for the Analysis of Platinum Alloys," *Zeit. and Chem.*, 1912, vol. li. pp. 550-554. Describes the method for the analysis of alloys containing platinum, copper, silver, nickel, and iron, which depends on the fact that chlorides of copper, nickel, and iron are readily soluble in alcohol.

1912. E. V. KOUKLINE, "Analysis of Platinum Ore," *Rev. Mét.*, 1912, vol. ix. pp. 815-824, abstract *Journ. Soc. Chem. Ind.*, 1912, vol. xxxi. p. 1036. Criticises the methods previously proposed for the analysis of platinum ores and alloys, and describes a new method.

1912. F. W. HORTON, "Iridium in American Placer Platinum," *Eng. and Min. Journ.*, 1912, vol. xciv. pp. 873-875. Gives the analyses of crude platinum, osmiridium, etc., from different sources.

1912. KEMP, "Analyses of Crude Platinum from American and Foreign Sources," *U.S. Geol. Surv. Bulletin*, cxciii., 1912, pp. 18-19.

APPENDIX.

TABLE OF ATOMIC WEIGHTS [1] AND OTHER CONSTANTS.

Metals.	Symbol.	Atomic weight. 0—16.	Specific gravity.	Melting points.
				°C.
Aluminium	Al	27·1	2·56	657
Antimony	Sb	120·2	6·71	632
Arsenic	As	75·00	5·67	450
Barium	Ba	137·4	3·75	850
Bismuth	Bi	208·50	9·80	266
Cadmium	Cd	112·4	8·60	322
Cæsium	Cs	133	1·88	26
Calcium	Ca	40·1	1·57	780
Chromium	Cr	52·1	6·80	1482
Cobalt	Co	59·00	8·50	1464
Copper	Cu	63·6	8·82	1084
Glucinum (Beryllium) . .	Gl	9·1	1·93	Below 960
Gold	Au	197·20	19·32	1064
Indium	In	114	7·42	155
Iridium	Ir	193·00	22·42	1950
Iron	Fe	55·9	7·86	1505
Lanthanum	La	138·9	6·20	810
Lead	Pb	206·9	11·37	327
Lithium	Li	7·03	0·59	186
Magnesium	Mg	24·36	1·74	633
Manganese	Mn	55·00	8·00	1207
Mercury	Hg	200·0	13·59	39
Molybdenum . .	Mo	96·00	8·60	2500
Nickel	Ni	58·7	8·80	1427
Osmium	Os	191	22·48	2500
Palladium	Pd	106·5	11·50	1585
Platinum	Pt	194·8	21·50	1745
Potassium	K	39·15	0·87	62
Radium	Rd	225	...	...
Rhodium	Rh	103·00	12·10	1660
Rubidium	Rb	85·4	1·52	38
Ruthenium	Ru	101·7	12·26	1800
Silver	Ag	107·93	10·53	962
Sodium	Na	23·05	0·97	95
Strontium	Sr	87·6	2·54	800
Tantalum	Ta	183·00	12·8	2910
Tellurium	Te	127·6	6·25	440
Thallium	Tl	204·1	11·85	303
Thorium	Th	232·5	11·10	1450
Tin	Sn	119·0	7·29	232
Titanium	Ti	48·1	3·54	2000 ?
Tungsten	W	184·00	19·10	3100
Uranium	U	238·5	18·70	...
Vanadium	V	51·2	5·50	1680
Zinc	Zn	65·4	7·15	419
Zirconium	Zr	90·6	4·15	1500

[1] *International Atomic Weights,* 1913.

RELATION OF IMPERIAL TO METRIC STANDARDS.

Standards of Mass.

1 grain	=	0·0648 grammes.	
1 ounce troy	=	31·1035 ,,	{ 480 grains. / 1·0971 ounces avoir.
1 ounce avoir.	=	28·350 ,,	{ 437·5 grains. / 0·9114 ounce troy.
1 pennyweight	=	1·5552 ,,	(24 grains).
1 pound troy	=	373·25 ,,	{ 5760 grains. / 12 ounces troy. / 0·822857 lb. avoir.
1 pound avoir.	=	453·593 ,,	{ 7000 grains. / 16 ounces avoir. / 1·215277 lbs. troy.

1 hundredweight (cwt.) = 50·8 kilogrammes (= 112 pounds).
1 British ton (2240 lbs.) = 1016 kilogrammes.
1 American ton (2000 lbs.) = 906 kilogrammes.

Standards of Capacity.

1 fluid ounce	= 0·0284123 litre	= 28·417 cubic centimetres.
1 pint	= 0·5682454 ,,	= 568·336 ,, ,,
1 gallon	= 4·545963 litres.	
1 U.S.A gallon	= 0·83254 imperial gallon.	

RELATION OF METRIC TO IMPERIAL STANDARDS.

Standards of Mass.

1 milligramme	=	0·015 grain.	
1 gramme	=	15·432 grains.	{ 0·032151 ounce troy. / 0·0352736 ounce avoir. / 0·0022046 pound.

1 kilogramme = 2·204622 pounds.
100 kilos (1 quintal) = 1·968 cwt.
1000 kilos (1 tonne) = 0·9842 British ton.

Standards of Capacity.

1 cubic centimetre (water at 4° C.) = 1 gramme.
1 litre (1000 c.c.) = 1·7598 pints.

USEFUL DATA.

Table to Convert Metric Weights into Avoirdupois and Troy Weights.

As 1 gramme is equal to 15·432 grains, or ·03527 avoirdupois ounce, or ·03215 troy ounce; to convert:

Grammes	into grains	multiply by	15·432
Centigrammes	,, grains	,,	0·15432
Milligrammes	,, grains	,,	0·01543
Kilogrammes	,, avoirdupois ounces	,,	35·2739
Grammes	,, avoirdupois ounces	,,	·03527
Kilogrammes	,, avoirdupois pounds	,,	2·2046
Kilogrammes	,, troy ounces	,,	32·1507
Grammes	,, troy ounces	,,	·03215

Table to Convert Avoirdupois and Troy Weights into Metric Weights.

As 1 grain is equal to 0·0648 gramme, and 1 avoirdupois ounce is equal to 28·3495 grammes, and 1 troy ounce is equal to 31·1035 grammes ; to convert :

Grains	.	.	.	into grammes	.	.	.	multiply by	0·0648
Grains	.	.	.	,, centigrammes	.	.	,,	6·4799
Grains	.	.	.	,, milligrammes	.	.	,,	64·799
Avoirdupois ounces	.	,, kilogrammes	.	.	,,	0·2835
Avoirdupois ounces	.	,, grammes	.	.	,,	28·3495
Avoirdupois lbs.	.	,, kilogrammes	.	.	,.	0·4536
Troy ounces	.	.	,, kilogrammes	.	.	,,	0·0311
Troy ounces	.	.	,, grammes	.	.	,,	31·1035

Table to Convert English Linear Measure into Metric Linear Measure.

As one inch is equal to 0·0254 metres ; to convert :

Inches	.	.	.	into metres	.	.	.	multiply by	0·0254
Inches	.	.	.	,, centimetres	.	.	,,	2·5339
Inches	.	.	.	,, millimetres	.	.	,,	25·3997

Table to Convert Metric Linear Measure into English Linear Measure.

As one metre is equal to 39·370 inches ; to convert :

Metres	.	.	.	into inches	.	.	.	multiply by	39·370
Centimetres	.	.	,, inches	.	.	,,	0·3937
Millimetres	.	.	,, inches	.	.	,,	0·03937

Area of circle	=square of diameter multiplied by 0·7854.
Area of sphere	=square of diameter multiplied by 3·14159.
Solid contents of sphere	=one-sixth of the cube of the diameter multiplied by 3·14159.
Area of curved surface of cylinder	=diameter × 3·14159 × length.
Solid contents of cylinder	=square of diameter × length × 0·7854.

Comparison of Thermometer Scales.

Centigrade.	Fahrenheit.	Réaumur.
Water freezes at 0°.	Water freezes at 32°.	Water freezes at 0°.
Water boils at 100°.	Water boils at 212°.	Water boils at 80°.

To convert Centigrade degrees into degrees of Fahrenheit, multiply by 9, divide by 5, and add 32.

To convert Fahrenheit degrees into degrees of Centigrade, subtract 32, multiply by 5, and divide by 9.

To convert Réaumur degrees into degrees of Fahrenheit, multiply by 9, divide by 4, and add 32.

To convert Fahrenheit degrees into degrees of Réaumur, subtract 32, multiply by 4, and divide by 9.

To convert Réaumur degrees into degrees of Centigrade, multiply by 5, and divide by 4.

To convert Centigrade degrees into degrees of Réaumur, multiply by 4 and divide by 5.

HYDROMETERS.

A quick and convenient method of determining the density or specific gravity of a liquid is by the use of the hydrometer. The method is based upon the fact that a floating body is buoyed up more by a heavy liquid than by a light one. The hydrometer consists of a hollow glass bulb or float, carrying above it a thin tube, with a scale marked on it, or on a paper enclosed in it. A smaller bulb is also blown beneath, into which mercury or lead shot is put, so as to adjust the weight and at the same time cause the instrument to float in a vertical position. The liquid to be examined is poured into a tall glass jar, and the hydrometer immersed so that it floats freely ; the specific gravity may then be read off from the stem, which is so graduated that the number level with the surface of the liquid shows the specific gravity of the liquid. It is manifest that the denser the liquid, the higher the instrument will float in it, the weight of the liquid displaced being always equal to that of the hydrometer.

The temperature of the liquid must be adjusted to the temperature required by the hydrometer, which is usually 15·5° C. The graduation of hydrometers is not made to any uniform system. Those marked in degrees Baumé or Twaddell, or according to specific gravity, are most commonly used. Hydrometers are only accurate within comparatively small ranges of density, hence they are usually supplied in sets as given below. Two different sets are used, one for liquids lighter than water, the other for liquids such as acids, that are heavier.

Baumé's Hydrometer.—The degrees on Baumé's hydrometer agree among themselves in being at equal distances along the stem, but they are proportional neither to the specific gravity nor to the percentage of salt in the solution. They may be converted into an ordinary statement of specific gravity by the use of the following formulæ :—

(1) For liquids heavier than water,

$$\text{Specific gravity} = \frac{144\cdot3}{144\cdot3 - \text{degrees Baumé}}.$$

For example, 30° Baumé corresponds to a specific gravity of 1·262, thus

$$\frac{144\cdot3}{144\cdot3 - 30°} = \frac{144\cdot3}{114\cdot3} = 1\cdot262.$$

(2) For liquids lighter than water,

$$\text{Specific gravity} = \frac{146}{136 + \text{degrees Baumé}}.$$

For general purposes a hydrometer graduated according to specific gravity is the most convenient. In these instruments the distances between the divisions become less as the densities increase. With the aid of the above formulæ a Baumé hydrometer may readily be converted into one showing the actual specific gravity.

To reduce specific gravity to degrees Baumé the formula is :—

$$\frac{144\cdot3}{\text{Specific gravity}} - 136 = \text{degrees Baumé}.$$

Twaddell's Hydrometer.—The graduations of this instrument correspond to equal differences of density. Each degree is 0·005 in excess of unity (water = 1000).
Thus,

1° Twaddell = 1·005 specific gravity
2°   ,,   = 1·010   ,,
7°   ,,   = 1·035   ,,

The relation between the specific gravity of a liquid and the reading of the instrument may be expressed as follows :—

To find the specific gravity, multiply the reading in degrees Twaddell by 0·005, and add 1. For example 31° Twaddell corresponds to a specific gravity of 1·155, thus

$$31 \times 0\cdot005 = 0\cdot155 + 1\cdot000 = 1\cdot155.$$

The degrees Twaddell may also be converted into specific gravity by multiplying the reading by 5, adding 1000 (water = 1000) and cutting off three decimal places.
For example,

25° Twaddell = 1·125 specific gravity.

To reduce specific gravity (water 1000) to Twaddell, deduct 1000 and divide the remainder by 5.
The following sets of hydrometers are those most generally supplied :—

Twaddell.		Baumé, for heavy liquids.	Baumé, for light liquids.
No. 1	0 to 24 = 1·00 to 1·12 sp. gr.	0 = 1·000 sp. gr.	60 = ·706 sp. gr.
2	24 to 48 = 1·12 to 1·24 ,,	10 = 1·075 ,,	50 = ·761 ,,
3	48 to 74 = 1·24 to 1·37 ,,	20 = 1·161 ,,	40 = ·817 ,,
4	74 to 102 = 1·37 to 1·51 ,,	30 = 1·263 ,,	36 = ·837 ,,
5	102 to 138 = 1·51 to 1·69 ,,	40 = 1·385 ,,	30 = ·871 ,,
6	138 to 170 = 1·69 to 1·85 ,,	50 = 1·532 ,,	26 = ·892 ,,
		60 = 1·714 ,,	20 = ·928 ,,
		70 = 1·946 ,,	10 = 1·000 ,,

NITRIC ACID.

Table showing the Percentage, by Weight, of Real Acid (HNO_3) *in Aqueous Solutions of Nitric Acid of different Specific Gravities.   Temperature,* 15° C. (60° F.).

Specific gravity.	HNO_3 per cent.	Specific gravity.	HNO_3 per cent.	Specific gravity.	HNO_3 per cent.
1·530	100·0	1·405	66·0	1·205	33·0
1·527	99·0	1·400	65·0	1·198	32·0
1·524	98·0	1·395	64·0	1·192	31·0
1·520	97·0	1·390	63·0	1·185	30·0
1·516	96·0	1·386	62·0	1·179	29·0
1·513	95·0	1·380	61·0	1·172	28·0
1·509	94·0	1·374	60·0	1·166	27·0
1·506	93·0	1·368	59·0	1·159	26·0
1·503	92·0	1·363	58·0	1·152	25·0
1·499	91·0	1·358	57·0	1·145	24·0
1·495	90·0	1·353	56·0	1·138	23·0
1·492	89·0	1·346	55·0	1·132	22·0
1·488	88·0	1·341	54·0	1·126	21·0
1·485	87·0	1·335	53·0	1·120	20·0
1·482	86·0	1·329	52·0	1·114	19·0
1·478	85·0	1·323	51·0	1·108	18·0
1·474	84·0	1·317	50·0	1·102	17·0
1·470	83·0	1·311	49·0	1·096	16·0
1·467	82·0	1·304	48·0	1·089	15·0
1·463	81·0	1·298	47·0	1·083	14·0
1·460	80·0	1·291	46·0	1·077	13·0
1·456	79·0	1·284	45·0	1·071	12·0
1·452	78·0	1·277	44·0	1·065	11·0
1·449	77·0	1·270	43·0	1·060	10·0
1·445	76·0	1·264	42·0	1·053	9·0
1·442	75·0	1·257	41·0	1·047	8·0
1·438	74·0	1·251	40·0	1·041	7·0
1·435	73·0	1·244	39·0	1·034	6·0
1·431	72·0	1·238	38·0	1·028	5·0
1·427	71·0	1·232	37·0	1·022	4·0
1·423	70·0	1·225	36·0	1·016	3·0
1·418	69·0	1·218	35·0	1·010	2·0
1·414	68·0	1·212	34·0	1·004	1·0
1·410	67·0				

SULPHURIC ACID.

Table showing the Percentage, by Weight, of Real Acid (H_2SO_4) in Aqueous Solutions of Sulphuric Acid of varying Specific Gravity. Temperature 15° C. (60° F.).

Specific gravity.	H_2SO_4 per cent.	Specific gravity.	H_2SO_4 per cent.	Specific gravity.	H_2SO_4 per cent.
1·838	100·0	1·568	66·0	1·247	33·0
1·840	99·0	1·557	65·0	1·239	32·0
1·841	98·0	1·545	64·0	1·231	31·0
1·841	97·0	1·534	63·0	1·223	30·0
1·840	96·0	1·523	62·0	1·215	29·0
1·838	95·0	1·512	61·0	1·206	28·0
1·836	94·0	1·501	60·0	1·198	27·0
1·834	93·0	1·490	59·0	1·190	26·0
1·831	92·0	1·480	58·0	1·182	25·0
1·827	91·0	1·469	57·0	1·174	24·0
1·822	90·0	1·458	56·0	1·167	23·0
1·816	89·0	1·448	55·0	1·159	22·0
1·809	88·0	1·438	54·0	1·151	21·0
1·802	87·0	1·428	53·0	1·144	20·0
1·794	86·0	1·418	52·0	1·136	19·0
1·786	85·0	1·408	51·0	1·129	18·0
1·777	84·0	1·398	50·0	1·121	17·0
1·767	83·0	1·388	49·0	1·113	16·0
1·756	82·0	1·379	48·0	1·106	15·0
1·745	81·0	1·370	47·0	1·098	14·0
1·734	80·0	1·361	46·0	1·091	13·0
1·722	79·0	1·351	45·0	1·083	12·0
1·710	78·0	1·342	44·0	1·075	11·0
1·698	77·0	1·333	43·0	1·068	10·0
1·686	76·0	1·324	42·0	1·061	9·0
1·675	75·0	1·315	41·0	1·053	8·0
1·663	74·0	1·306	40·0	1·046	7·0
1·651	73·0	1·297	39·0	1·039	6·0
1·639	72·0	1·289	38·0	1·032	5·0
1·627	71·0	1·281	37·0	1·025	4·0
1·615	70·0	1·272	36·0	1·019	3·0
1·604	69·0	1·264	35·0	1·013	2·0
1·592	68·0	1·256	34·0	1·006	1·0
1·580	67·0				

HYDROCHLORIC ACID.

Table showing the percentage, by Weight, of Real Acid (HCl) *in Aqueous Solutions of Hydrochloric Acid of different Specific Gravities. Temperature,* 15° C. (60° F.).

Specific gravity.	HCl per cent.	Specific gravity.	HCl per cent.	Specific gravity.	HCl per cent.
1·2000	40·78	1·1410	28·54	1·0798	16·31
1·1982	40·87	1·1389	28·13	1·0778	15·90
1·1964	39·96	1·1369	27·72	1·0758	15·49
1·1946	39·55	1·1349	27·32	1·0738	15·08
1·1928	39·14	1·1328	26·91	1·0718	14·68
1·1910	38·74	1·1308	26·50	1·0697	14·27
1·1893	38·33	1·1287	26·10	1·0677	13·86
1·1875	37·92	1·1267	25·69	1·0657	13·45
1·1857	37·51	1·1247	25·28	1·0637	13·05
1·1846	37·11	1·1226	24·87	1·0617	12·64
1·1822	36·70	1·1206	24·46	1·0597	12·23
1·1802	36·29	1·1185	24·06	1·0577	11·82
1·1782	35·88	1·1164	23·65	1·0557	11·41
1·1762	35·47	1·1143	23·24	1·0537	11·01
1·1741	35·07	1·1123	22·83	1·0517	10·60
1·1721	34·66	1·1102	22·43	1·0497	10·19
1·1701	34·25	1·1082	22·02	1·0477	9·78
1·1681	33·84	1·1061	21·61	1·0457	9·38
1·1661	33·43	1·1041	21·20	1·0437	8·97
1·1641	33·03	1·1020	20·79	1·0417	8·56
1·1620	32·62	1·1000	20·39	1·0397	8·15
1·1599	32·21	1·0980	19·98	1·0377	7·75
1·1578	31·80	1·0960	19·57	1·0357	7·34
1·1557	31·40	1·0939	19·16	1·0337	6·93
1·1536	30·99	1·0919	18·76	1·0318	6·52
1·1515	30·58	1·0899	18·35	1·0298	6·11
1·1494	30·17	1·0879	17·94	1·0279	5·51
1·1473	29·76	1·0859	17·53	1·0259	5·80
1·1452	29·36	1·0838	17·12	1·0239	4·89
1·1431	28·95	1·0818	16·72	1·0200	4·01

NORMAL SOLUTIONS.

In volumetric analysis standard solutions are designated *normal solutions* (distinguished by the letter N) when they are of such strength that 1 litre contains 1 gramme of replaceable hydrogen or a weight of the reagent in grammes, equal to the *chemical equivalent* of that reagent. Therefore, a normal solution of sulphuric acid, which is a dibasic acid, will contain :—

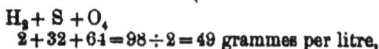

$$H_2 + S + O_4$$
$$2 + 32 + 64 = 98 \div 2 = 49 \text{ grammes per litre,}$$

or 0·049 gramme per cubic centimetre.

In the case of alkalies, etc., the normal solution must contain the chemical equivalent (weighed in grammes) of the metal or group of elements that replace the hydrogen. Therefore a normal solution of sodium hydroxide (caustic soda) will contain :—

$$Na + H + O$$
$$23 + 1 + 16 = 40 \text{ grammes per litre,}$$

or 0·04 gramme per cubic centimetre since the valency (or combining power) of sodium and hydrogen are the same.

A normal solution of barium hydroxide (barium hydrate) would contain :—

$$Ba + H_2 + O_2$$
$$137 + 2 + 32 = 171 \div 2 = 85\cdot5 \text{ grammes per litre,}$$

or 0·085 gramme per cubic centimetre since the valency of barium is twice that of hydrogen.

From the above remarks it will be obvious that equal quantities of any normal acid solution, and of a normal alkali solution, will neutralise each other. Also the quantity of any solution of a single acid or an alkali that is neutralised by the addition of 1 c.c. of a normal solution will contain 0·001 gramme of the hydrogen equivalent of its active constituent.

It is frequently convenient to use standard solutions the strength of which bears a definite ratio to that of the normal solution. Thus solutions of *one-half, one-tenth*, and *one-hundredth* of the strength of the normal solution are called respectively *semi-normal, deci-normal*, and *centi-normal*, and are distinguished by the signatures $\frac{N}{2}$, $\frac{N}{10}$, and $\frac{N}{100}$.

Solutions that are stronger than the normal solution are designated by prefixing the numeral denoting the number of times it is stronger than normal ; thus a solution twice the normal strength is designated 2N, and a solution five times normal strength 5N.

NORMAL ACIDS.

Sulphuric Acid. — H_2SO_4 equivalent = 49. Normal sulphuric acid contains 49 grammes of H_2SO_4 per litre. Ordinary concentrated acid, sp. gr. 1·84, has a strength of only 36 times normal, and may be designated 36N. Therefore one volume of strong acid diluted with 30 volumes of water = normal strength.

Nitric Acid. — HNO_3 equivalent = 63. Concentrated acid, sp. gr. 1·42 = 16N. One volume of strong acid diluted with 11 volumes of water = normal strength.

Hydrochloric Acid. — HCl equivalent = 36·5. Concentrated acid, sp. gr. 1·16 = 10N. One volume of strong acid diluted with 5 volumes of water = normal strength.

REAGENTS OF NORMAL STRENGTH.

The following normal reagents are prepared by dissolving the equivalent weight in grammes of the various salts in water, and diluting to 1 litre (1000 c.c.).

Ammonium oxalate $(NH_4)_2C_2O_4 2H_2O$, equivalent weight		. .	80·0
Ammonium thiocyanate NH_4CNS	,, ,,	. .	76·0
Potassium cyanide, KCN	,, ,,	. .	65·0
Potassium hydroxide, KOH	,, ,,	. .	56·0
Potassium iodide, KI	,, ,,	. .	166·0
Sodium carbonate, $Na_2CO_3 10H_2O$	,, ,,	. .	143·0
Sodium chloride, NaCl	,, ,,	. .	58·5
Sodium hydroxide, NaOH	,, ,,	. .	40·0
Silver nitrate, $AgNO_3$	,, ,,	. .	170·0
Sodium thiosulphate, $Na_2S_2O_5 5H_2O$	,, ,,	. .	124·0

NITRO-HYDROCHLORIC ACID, OR AQUA-REGIA.

This mixture termed aqua-regia (because it dissolves the noble metals) usually consists of 1 part by measure of nitric acid and 3 parts of hydrochloric acid. Its solvent action depends upon the fact that it contains free chlorine, liberated by the oxidising action of nitric acid on the hydrogen of the hydrochloric acid, thus :—

$$3HCl + HNO_3 = Cl_2 + NOCl + 2H_2O.$$

The mixture is best prepared when required, and is chiefly used for the solution of gold and platinum, and certain sulphides, which do not dissolve in either nitric or hydrochloric acid separately. When solutions in aqua-regia are evaporated, chlorides are left.

METHOD OF TESTING WEIGHTS.[1]

"As a rule, the weights of the best makers possess greater accuracy than is required of them in ordinary quantitative processes; nevertheless their examination should never be omitted. The readiest method of detecting errors in the values of the denominations is to place one of the gramme weights on the pan of a delicate balance, adjusted to perfect equilibrium, and equipoise it with pieces of brass or small shot, and finally tin foil. The weight is then removed and replaced by the second gramme weight, and the balance caused to oscillate. If the excursions on either side of the zero are of equal amplitude, the weights are equivalent; if not, the deviation must be noted. It should not exceed 1 division of the scale from the zero-point. The third gramme weight is next tried in the same way, and it is then replaced by platinum weights of the smaller denominations to make up 1 gramme, and the balance again caused to oscillate, every deviation from the equilibrium being carefully noted. In the same way the 2-gramme piece is compared with two of the single grammes, the 5-gramme piece with the 2+1+1+1 gramme pieces, and each of the 10-gramme pieces with the 5+2+1+1+1 gramme pieces. The larger pieces also should not show greater variations than 1 division from the zero, since the value of 1-scale division with a heavy load on the pans is almost invariably greater than with a diminished load."

---

[1] T. E. Thorpe, *Quantitative Chemical Analysis.*

INDEX.